GROUNDWATER HYDROLOGY

Engineering, Planning, and Management

GROUNDWATER HYDROLOGY

Engineering, Planning, and Management

Mohammad Karamouz • Azadeh Ahmadi • Masih Akhbari

CRC Press
Taylor & Francis Group
Boca Raton London New York

CRC Press is an imprint of the
Taylor & Francis Group, an **informa** business

CRC Press
Taylor & Francis Group
6000 Broken Sound Parkway NW, Suite 300
Boca Raton, FL 33487-2742

First issued in paperback 2019

© 2011 by Taylor and Francis Group, LLC
CRC Press is an imprint of Taylor & Francis Group, an Informa business

No claim to original U.S. Government works

ISBN-13: 978-1-4398-3756-6 (hbk)
ISBN-13: 978-0-367-38298-8 (pbk)

Visit the Taylor & Francis Web site at
http://www.taylorandfrancis.com

and the CRC Press Web site at
http://www.crcpress.com

*To millions of people around the world with
no access to safe drinking water*

M. Karamouz
A. Ahmadi
M. Akhbari

Contents

Preface

The demand for freshwater is increasing as the world's population continues to grow and expects higher standards of living. Water conservation, better systems' operation, higher end use, and water allocation efficiencies have not been able to offset the growing demand. Many societies are struggling to bring supply and demand to a sustainable level. Although water is abundant on earth, freshwater accounts only for about 2.5% of global water reserves. Out of this amount, approximately 30% is stored as groundwater and the same amount is on the surface as rivers and lakes; the remaining reserves are held in glaciers, ice caps, soil moisture, and atmospheric water vapor. Groundwater is a source of vital natural flow. In arid and semiarid areas, groundwater may represent 80% or more of the total water resources. The public has a perception of groundwater as a reliable, clean, and virtually unlimited source of water supply. Even though there could be exceptions, it is a dependable source almost everywhere in the world.

The term "groundwater" is usually used for subsurface water that is located below the water table in saturated soils and geological formations. Groundwater is an important element of the environment; it is a part of the hydrologic cycle, and an understanding of its role in this cycle is necessary if integrated analyses ought to be promoted. Aquifers and watershed resources are two inseparable media for regional assessment of water and contamination movement, as well as soil–water interaction.

Water enters the formation of the earth from the ground surface by natural or artificial recharge. Within the subsurface, water moves slowly downward through unsaturated zones under the force of gravity and in saturated zones in the direction determined by the hydraulic gradient. It is discharged naturally to the streams and lakes or as a spring or by transpiration from plants. It can also be discharged by pumping from wells. Storage of groundwater could be significantly beneficial compared to surface storage, which may require massive infrastructure and is subject to considerable evaporation.

In this book, a compilation of the state-of-the-art subjects and techniques in the education and practice of groundwater are presented. The materials are described in a systematic and integrated fashion useful for undergraduate and graduate students and practitioners.

Chapter 1 provides an introduction to different aspects of groundwater science, engineering, planning, and management. It discusses the concepts of integration, sustainability, and public participation with an emphasis on laws and regulations governing groundwater operation and management.

Chapter 2 presents an overview of general concepts on the occurrence of groundwater as a part of the hydrologic cycle. The basic definitions of subsurface water and aquifers are discussed in this chapter. The occurrence of groundwater in karst areas and karst aquifers is also explained.

Chapter 3 discusses groundwater hydrology and the basic laws of groundwater movement. It describes some characteristics of aquifers and soil formations that govern the movement of groundwater such as hydraulic conductivity, transmissivity, homogeneity, and isotropy of soil layers. It then discusses flownets simply as procedures to solve flow equations. Statistics and probabilistic methods in hydrogeology are also explained. Finally, the analysis of time series and the methods of generating synthetic time series and forecasting data are introduced.

Chapter 4 presents the technical aspects of developing and solving governing groundwater flow equations. It provides a mathematical description of groundwater flow based on the continuity equation and the equation of motion. The utilization of groundwater by pumping from wells leads to a decline in the water level. The chapter takes a detailed look at the theoretical analyses of the aquifer system's response based on the physics of flow toward a well in confined and unconfined aquifers for steady and unsteady states. It also describes the pumping test to estimate the aquifers' parameters as well as the methods of well design and well construction.

Chapter 5 describes environmental water quality issues related to groundwater systems. It describes the factors that can affect groundwater solubility as well as the evaluation of the nonequilibrium status of groundwater; this is followed by a short discussion on different types of contaminants in groundwater. Pollution of groundwater is caused by many activities, including leaching from municipal and chemical landfills and abandoned dump sites, accidental spills of chemicals or waste materials, improper underground injection of liquid wastes, and placement of septic tanks. Finally, the chapter covers the principles of reactive and nonreactive transport of contaminants in both saturated and unsaturated zones.

Chapter 6 provides details of the specific information about groundwater flow modeling. In the application of numerical methods, partial differential equations of groundwater are transformed into a set of ordinary differential or algebraic equations. State variables at discrete nodal points are determined by solving the equations. Three major classes of numerical methods have been presented for solving groundwater problems, including finite difference method (FDM), finite element method (FEM), and finite volume method (FVM).

Chapter 7 introduces conceptual models to simulate groundwater systems. The management problem can be viewed and determined in an appropriate type, location, and setting for the control of desirable system outputs. This chapter provides a comprehensive description of planning tools and important issues related to groundwater management in order to optimize groundwater resources and allocation schemes. Optimization methods are powerful tools that could maximize water utilization, minimize the costs of groundwater operation, and minimize the adverse impacts of overexploitation of groundwater. This chapter presents a set of analytical tools, based on coupling groundwater simulation and optimization models, to assist the analyst in solving complex management problems. It also provides examples and case studies to illustrate the application of the topics presented.

Chapter 8 presents basic information about the conjunctive use of surface and groundwater. It describes the interactions between groundwater and different types of surface water with an emphasis on planning issues. It also discusses certain advantages of conjunctive use planning especially in semiarid regions and presents case studies.

Chapter 9 discusses aquifer restoration in the context of different groundwater pollution control issues and the utilization of groundwater treatment. First, it describes the processes affecting the amount of contaminant, as well as natural and mechanical means of reducing contaminants' concentration. It then explains the importance of groundwater monitoring and the corresponding processes.

Chapter 10 discusses different groundwater disasters and the vulnerability of aquifers to natural hazards and human impacts, such as drought, flood, widespread contamination, and land subsidence. It presents risk as an overlay of hazard and system's vulnerability as well as methods for risk analysis. It also discusses the planning processes for groundwater disaster management, including the development of contingency plans.

Chapter 11 presents the impact of climate change on the hydrological cycle. It evaluates the direct impacts of climate change on groundwater as affected by changes in precipitation, temperature, evaporation, and soil moisture. It also presents different downscaling models to adjust the resolution of data for climate change impact studies and assesses indirect impacts such as saltwater intrusion, sea level rise, and changes in land use. Finally, methods to cope with climate change with some applications to adapt to the impacts of climate change on groundwater resources are discussed.

This book can serve water communities around the world and add significant value to engineering and application of systems analysis techniques to groundwater engineering, planning, and management. It can be used as a textbook by students in civil engineering, geology, soil science, urban and regional planning, geography, and environmental science, and in courses dealing with the hydrologic cycle. It also introduces basic tools and techniques for engineers and policy- and decision-makers who plan future groundwater development activities toward regional sustainability.

It is our hope that this book can add significant value to the application of system analysis techniques for groundwater engineering and management around the world.

M. Karamouz
A. Ahmadi
M. Akhbari

Acknowledgments

Many individuals have contributed to the preparation of this book. The initial impetus for the book was provided by Professor Milton E. Harr through his class notes when he taught at Purdue University. Professor Jacques W. Delleur has been a role model for Mohammad Karamouz as his PhD co-adviser and, as an inspiring professor, published two outstanding books entitled *Time Series Analysis* and *Handbook of Groundwater Engineering*. The authors acknowledge the significant contribution of Navideh Noori in preparation of Chapter 11.

Many graduate students at Polytechnic Institute of New York, who attended the first author's groundwater course, were a driving force for completion of the class notes that formed the basis for this book. Their valuable input is hereby acknowledged.

The authors would also like to thank Dr. A. Moridi; the PhD students S. Nazif, H. Goharnejad, Z. Zahmatkesh, and M. Taherioun; and the MS students S. Semsar, B. Ahmadi, A. Abolpour, M. Fallahi, B. Rohanizade, S. Ahmadinia, F. Emami, S. Safaie, and M. Sherafat. Special thanks to Mehrdad Karamouz, Mona Lotfi, Mina Akhbari, and Omid Ebrahimi for preparing the cover page, figures, and artwork for this book.

Authors

Mohammad Karamouz is a research professor at Polytechnic Institute of New York University and a professor at the University of Tehran. He is an internationally known water resources engineer and consultant. He is the former dean of engineering at Pratt Institute in Brooklyn, New York. He is also a licensed professional engineer in the state of New York, a fellow of the American Society of Civil Engineers (ASCE), and a diplomat of the American Academy of Water Resources Engineers.

Dr. Karamouz received his BS in civil engineering from Shiraz University, his MS in water resources and environmental engineering from George Washington University, and his PhD in hydraulic and systems engineering from Purdue University. He served as a member of the task committee on urban water cycle in UNESCO-IHP VI and was a member of the planning committee for the development of a 5 year plan (2008–2013) for UNESCO's International Hydrology Program (IHP VII).

Among many professional positions and achievements, he also serves on a number of task committees for ASCE. In his academic career spanning 27 years, he has held positions as a tenured professor at Pratt Institute (Schools of Engineering and Architecture in Brooklyn) and at Polytechnic University (Tehran, Iran). He was a visiting professor in the Department of Hydrology and Water Resources at the University of Arizona, Tucson, 2000–2003. He was also the founder and former president of Arch Construction and Consulting Co. Inc. in New York City.

His teaching and research interests include integrated water resources planning and management, groundwater hydrology and pollution, drought analysis and management, water quality modeling and water pollution, decision support systems, climate forecasting, and conjunctive use of surface and groundwater. He has more than 300 research and scientific publications, books, and book chapters to his credit, including three books entitled *Urban Water Engineering and Management* published by CRC press in 2010, *Water Resources System Analysis* published by Lewis Publishers in 2003, and a coauthored book entitled *Urban Water Cycle Processes and Interactions* published by Taylor & Francis Group in 2006. Professor Karamouz serves internationally as a consultant to both private and governmental agencies such as UNESCO and the World Bank.

Azadeh Ahmadi is an assistant professor in the Department of Civil Engineering at Isfahan University of Technology (IUT). She received her BS in civil engineering and her MS in water resources engineering from IUT, and her PhD in water resources management from the University of Tehran. She was a research associate in the Department of Civil and Environmental Engineering at the University of Bristol through a UK fellowship. Ahmadi is an international expert and consultant in the application of systems engineering in the planning of aquifer–river–reservoir systems. Her teaching and research interests include groundwater modeling, water engineering and management, systems analysis, conjunctive use of surface and groundwater, conflict resolution, climate change impacts studies, analysis of risk and uncertainty, and development of decision support systems.

Masih Akhbari is a PhD candidate in water resources management at Colorado State University. He received his master's degree in environmental engineering from Amir Kabir University of Technology (Tehran Polytechnic) in 2005. He is a member of the American Society of Civil Engineers (ASCE), the Iranian Water Resources Association, and the Iranian Society of Environmentalists. Since the completion of his master's studies, Masih has been involved with two research groups and has contributed to several research projects related to the environment and water resources. He was a research associate at the University of Tehran working on two projects, master planning and environmental monitoring, sponsored by the World Bank and Iran Department of Environment (DOE). He has worked as an international consultant in a variety of water and environmental engineering projects for two years. His teaching and research interests include integrated water resources management, groundwater hydrology, quality issues and restoration techniques for contaminated aquifers, planning and management aspects of groundwater, conflict resolution in water resources systems.

1 Introduction

1.1 INTRODUCTION

Increasing demand for water, higher standards of living, depletion of resources of acceptable quality, and excessive water pollution due to urban, agricultural, and industrial expansions have caused intensive environmental, social, and political predicaments. In the last century, the population of the world has tripled, nonrenewable energy consumption has increased by a factor of 30, and the industrial production has multiplied by 50 times. Continuously, we are seeking a balance between our physical being, ability to manage our water resources, and the limitations imposed by the environment. Although progress has improved the quality of life, it has caused significant environmental destruction in such a magnitude that could not be predicted. Now "the environment" is fighting back. The impacts of climate change on natural water resources have been devastating in many regimes around the world. More frequent and intensive floods and droughts have changed the ability of the manmade system to operate and provide services to the public. Also, it has changed the way we have planned and managed our surface and groundwater resources.

A question that should be answered is whether in the next decades, development could be done in a way that is economically and ecologically sustainable. We cannot answer this question unless we have a vision of the future and our planning schemes are environmentally responsible and sensitive toward the major elements of our physical environment, namely, air, water, and soil. Water among these elements is of special importance. Excessive use of surface and groundwater and misusing and polluting these vital resources by residential, agricultural, and industrial wastewater have threatened our well-being. Groundwater is more vulnerable to pollution because it is invisible to monitor and regulatory issues related to groundwater misuse, and act of contaminating groundwater are not easy to enforce.

Planning for sustainable development of water resources means water conservation, waste and leakage prevention, improved efficiency of water systems, improved water quality, water withdrawal and usage within the limits of the system, water pollution within the carrying capacity of the streams, and water discharge from groundwater within the safe yield of the system. It is not easy to determine and monitor the safe yield of the aquifers. Aquifer safe yields can not be enforced for many technical, operational, and political resources. So the aquifers are subject to conditions of over expectation in many parts of the world especially in arid and semiarid regions.

Water is a sustainable resource and the need for integrated water resources management (IWRM) is in every state and national agenda. Water more than ever, as one element in the formation and conservation of the environment, is receiving global

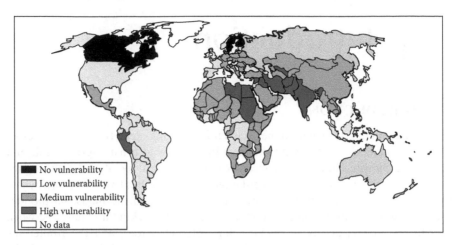

FIGURE 1.1 Nations' vulnerability to water scarcity. (From WRI (World Resources Institute), *World Resources, 1998–1999. A Guide to the Global Environment*, Oxford University Press, New York, 384 p., 1998.)

attention. Water security and water disaster preparedness are the subjects of many global concerns. Figure 1.1 shows different nations' vulnerability to water scarcity using a composite index based on available water resources and current use, reliability of water supply, and the national incomes (World Resources, 1998).

Groundwater is widely used for irrigation in countries with arid and semiarid climate. The total irrigated land by groundwater in the United States is 45%. In Asia and Africa, more than 60% of land mass is irrigated by groundwater. Libya's irrigated farming is primarily from low-quality groundwater resources, several kilometers deep.

1.1.1 WATER AVAILABILITY

In studying groundwater and its practical use for different purposes, it is important to realize it is not an important resource but one of the main components of the environment that is less affected by short-term climate and hydrologic variability. According to a global model Water, GAP-2 (Doll, 2002), 36% of the river runoff is formed by groundwater so separating groundwater shares from surface water resource is not easy to assess. The primary input for estimating the volume of water naturally available to a given nation is an information database (AQUASTAT) that has been developed and maintained by Food and Agriculture Organization of the United Nations (FAO) in 2007. It is based on a water balance approach for each country and the quantity of available water resources (FAO, 2003). This database has become a common reference tool used to estimate each nation's renewable water resources. The FAO has compiled an index called Total Actual Renewable Water Resources (TARWR). This index is a measure of the water resources availability for development from all sources within a region. It is a total-volume figure expressed in km³/year; divided by the total population. The index estimates the total available water resources per person in each country, considering a number of individual resources as follows:

- Adding all internally generated surface and groundwater from precipitation falling within a country's boundaries
- Adding flow entering from other countries, both surface and groundwater
- Subtracting any potential shared resource from surface and groundwater system interactions
- Subtracting the water that leaves the country through surface and groundwater resources, required by an existing treaty and/or a set by international guideline

TARWR gives the maximum amount of water that could actually be available for a country on a per capita basis. From 1989, this method has been used to make an assessment of water scarcity and water stress. It is important to note that the FAO considers what is shared in surface and groundwater resources. However, as discussed by UNESCO (2006), these volumes do not consider the socioeconomic criteria that are applied to societies, nations, or regions that are developing those resources. Water pricing and net costs can vary when developing surface and groundwater sources. Therefore, the reported renewable volume of water will be less available for a variety of economic and technical reasons. Simply, a lot of water is not used because of the circumstances that control the supply and the demand. The following factors should be considered when using the TARWR index (UNESCO, 2006):

- Roughly 27% of the surface water runoff is floods, and this share of water is not considered as a usable resource. Even though it contributes to the recharge of mainly overexploited aquifers, they are mostly discharges away from a short- and midterm access to the water users. But despite of that, floods are counted in the nations' TARWR as part of the available, renewable annual water resource.
- Seasonal variation in precipitation, runoff, and recharge, is not well reflected in annual figures. It is important in the assessment of supply and demand spatial and temporal distribution for basin-scale policy of decision making and strategy setting.
- Many sizable countries have different climatic zones as well as scattered population and the TARWR does not reflect the ranges of these factors that can vary in different regions in that country.
- TARWR has no data to identify the volume of water that sustains ecosystems, namely, the volume that provides water for forests, direct rain fed agriculture, grazing, grass areas, and environmental demand.

It is also documented that not all of the renewable freshwater resources can be used and controlled by the population of a country. It is estimated that only about one-third of the renewable freshwater resources can be potentially controlled even with the most technical, structural, nonstructural, social, environmental, and economic means. The global potentially useable water resources (PUWR) of the renewable freshwater resources are estimated to be around 9,000–14,000 km³ (Secker, 1993; UN, 1999). A part of the primary water supply (PWS) is evaporated, other part returns to rivers, streams, and aquifers and in many instances withdrawn again for different uses. This is known as the recycled portion of PWS. The PWS and the recycled water supply, adds up to 3300 km³ that constitute the water used in different sectors such as agriculture, industry, public, and water supply.

1.1.2 GROUNDWATER AVAILABILITY

Groundwater is by far the most abundant source of freshwater on continents outside Polar Regions, followed by ice caps, lakes, wetlands, reservoirs, and rivers. About 1.5 billion people depended on groundwater for their drinking water supply (WRI, 1998). It is estimated that about 20% of global water withdrawals comes from groundwater (WMO, 1997).

According to the United Nations Environment Programme (UNEP), annual global freshwater withdrawal has grown from 3790 km³ in 1995 to about 4430 km³ in 2000. If the annual global water withdrawal is expected to increase with a rate of 10%–20% every 10 years, reaching approximately 5240 km³ in 2025. The share of groundwater is expected to increase at a slower rate due to already over drafted aquifers in many points of the world.

There are many advantages in storage of groundwater compared to the surface storage:

1. Minimum evaporation losses—It is limited to
   a. Groundwater close to the surface by capillary fringes
   b. "Phreatophytes"—plants feed on capillary fringe
2. Quality may benefit from filtering action (however, may be too high in dissolved solids). There is general improvement of water quality because of the porous media filtration of airborne and surface runoff contaminants and pathogens.
3. Outflow is gradual (good regulation in underground reservoir).
4. No other missive structure such as dams is needed (however, may involve high pumping costs).
5. Land use above the groundwater resources can be continued without change (there is no submergence of houses, abstraction to infrastructure, and property development and agricultural development.

Groundwater quality is under continuous threats. Unless protected, groundwater quality deteriorates from saline intrusion, pollution from agricultural and urban activities, and uncontrolled wastewater, solid and hazardous waste disposal. The following activities should be undertaken:

- Improve the understanding of groundwater contribution to the hydrologic cycle and evaluate the changes to groundwater storage and water table fluctuations.
- Raise the awareness of decision makers, water users, and public on the importance of groundwater in order to encourage protection and sustainable use.

- Assess the impacts of economic development on groundwater resources and support international collaboration for nations, regional needs.
- Quantify climate change impact on groundwater resources including sea levels rise, and intrusion of saltwater.

1.2 GROUNDWATER SYSTEMS

The groundwater system is a part of the hydrologic cycle and should be studied on a watershed scale. Combination of many elements and factors affecting groundwater, so related and connected, call for a holistic view to groundwater challenges within a hydrologic cycle.

Groundwater systems are traditionally developed as sources of water for different uses of domestic, industrial, or agricultural purposes. The generally good quality of the water and its accessibility in many regions of the world, where surface water is nonexistent or extremely costly to develop, have been important factors in stimulating the development of this relatively low-cost, reliable water resource. An example of groundwater system representation is shown in Figure 1.2. See Wills and Yeh (1987) for more details.

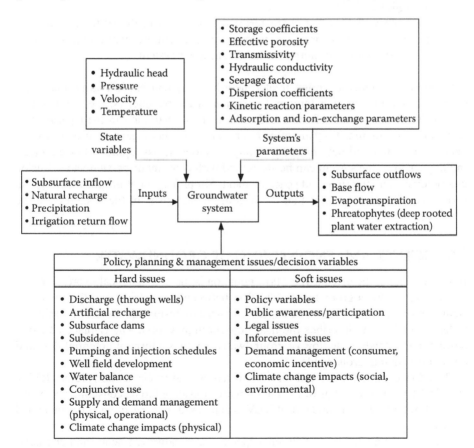

FIGURE 1.2 Groundwater system.

A groundwater system is defined as a system that can be isolated from upstream and downstream aquifers for the computation of its water balance and pollution characteristics:

1. The set of inputs to the system that are partially or completely controlled. Including precipitation, subsurface inflows, natural and artificial recharge, and irrigation return flows, interaction with streams.
2. The outputs, including subsurface outflows to adjacent aquifers, discharges to surface waters, naturally occurring springs, and evapotranspiration losses.
3. The parameters of the groundwater system that defines the flow, quality, and thermal properties of the aquifer system, such as the hydraulic conductivity, transmissivity, storativity, storage coefficient, and dispersion/advection parameters.
4. The control or decision variables within the context of policy, planning, and management. These decisions detail the pumping, injection, and artificial recharge variables of the ground and surface water system. The decision variables are bounded by policy, legal, physical, and operational constraints.
5. The state variables that characterize the condition of the system, such as the hydraulic head, pressure, velocity, or temperature distribution or the concentrations of pollutants in the groundwater system.

Groundwater systems are often treated as a lump system but in reality, they have distributed parameters. For temporal variation of the parameters, the lump system can be justified but most of the aquifers have spatial variation in their parameters that cannot be ignored. They are often referred to as nonhomogeneous properties that need to be tested and determined in many locations across aquifers. For engineering applications, the system can be simplified with constant dispersion and diffusion characteristics in any time and space element. Simplification of groundwater system should be done to the extent that the overall characteristic of the system could be preserved.

1.3 SCIENCE AND ENGINEERING OF GROUNDWATER

In hydrogeology, the groundwater flow equation is the mathematical expression that describes the flow of groundwater. The flow between two states of equilibrium (transient flow) in groundwater can be expressed by the diffusion equation. This is similar to heat transfer that describes the heat flow through a solid. The Laplace equation describes the steady-state flow of groundwater which is a form of potential flow and has analogs in numerous fields.

The flow equation is derived for a small representative elemental volume (REV). Within REV, the properties of the medium are considered to be constant. The continuity of water flowing in and out of REV is expressed in terms of an equation called Darcy's law.

Figures 1.3 and 1.4 show different aspects of groundwater science and engineering. A DSR (Driving force, States, and Responses) framework is used in these figures

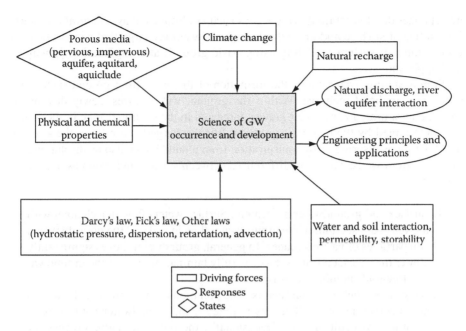

FIGURE 1.3 Science framework of groundwater (GW) occurrence (hydrogeology).

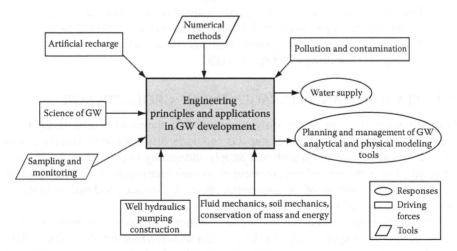

FIGURE 1.4 Basic engineering principles and applications in groundwater (hydrological and hydraulic design).

to define the hydrogeology (science of groundwater) and hydrological and hydraulic design in groundwater.

Figure 1.3 shows certain components of groundwater science in understanding its occurrence and development. Some principles including Darcy's law for hydraulic head/pressure simulation, and Fick's law for water quality simulation, the physical/chemical properties of the aquifer and porous media are transformed to the equations for clarifying the groundwater occurrence. The state of the system are changes due to certain driving forces such as climate change, natural disasters, interaction of surface and groundwater, groundwater recharge that need to be predicted through system modeling using basic groundwater engineering principles and applications.

Simply stating, water enters the formation of the earth from ground surface by natural or artificial recharge. Within the ground, water moves slowly downward through unsaturated zone under gravity force and in the saturated zone in the direction determined by the hydraulic gradient. It is discharged naturally to the streams and lakes or as a spring or by transpiration from plants. It can also be discharged by pumping from wells. The main differences between surface and groundwater are as follows:

1. Surface and groundwater movement. Surface waters flow in a channel with a given cross section, while groundwater flow in soil and the pores between soil particles of various shapes. In general, groundwater moves significantly slower than surface waters. Flow is fully laminar in most of the groundwater (Reynolds number, $Re \approx 1$).
2. Slow movement of contaminants. The pollutions can be prevented, but it is continuously polluted. If it gets polluted then, groundwater natural treatment is a difficult process. Occasionally, the only alternative process for treatment of groundwater is to discharge, treat, and recharge it back.
3. Filtration process considering groundwater flows through the soil layers, the biological contaminants will be treated through these natural layers.
4. Sampling and monitoring of groundwater is difficult which must be done by drilling and installing observation wells.

1.4   PLANNING AND MANAGEMENT OF GROUNDWATER

Groundwater management is broadly concerned with the evaluation of the environmental, hydrologic, and economic impacts and trade-offs associated with the development and allocation of groundwater supply and quality to competing water uses' demands. The planning and management of groundwater problems is done based on a system's representation of the underlying physical, chemical, and hydraulic transport processes occurring within the groundwater basin.

The inputs, boundary conditions, initial conditions, and, possibly, the parameters of the groundwater system may also be considered as random variables. These state variables of the system are also random. For example, if the annual groundwater recharge and/or demands for groundwater are random events, then water levels fluctuations in wells will follow a stochastic process.

In the analysis of groundwater system, it is also useful to consider three classes of management problems: the detection problem, the parameter estimation or the inverse problem, and the prediction or simulation problem. In the detection problem, the set of inputs to the system is unknown. The identification of the recharge or leakage in semi-confined or unconfined aquifers, from the response properties of the aquifer system, is an example of the instrumentation/detection problems. Parameter estimation is the key to groundwater simulation and modeling.

Laboratory or field data could be combined with theoretical assessment (Willis and Yeh, 1987).

In prediction problem, usually water table/piezometric surface fluctuation is predicted. Several monitoring wells and piezometers should be installed to calibrate the prediction models. At least three monitoring wells are needed to find the groundwater direction and to perform any meaningful field monitoring.

The water resources including groundwater and surface water should be managed considering the issue of sustainability. It can be addressed if we focus on the availability of resources for key services and their economical values. The consequence of losing access to such services that should be borne by the stakeholders should be equitably assessed. For example, if the water table of an aquifer drops significantly, it imposes a high equity impact on a region that could be measured only if we look at broad spectrum of social, economical, and environmental impacts over a long-term horizon.

Many farmers in the developing world can only afford low yield and shallow wells. As water level drops, these farmers cannot afford the costs of well deepening and new pumps. As a result, the food production and economic development can be affected and the benefit of access to groundwater lost. The drop in water table could increase the probability of water pollution migration and adverse impacts.

Figure 1.5 shows certain components of planning and management of groundwater systems. The groundwater system can be classified by considering the driving forces, responses and tools. The driving forces include socioeconomic issues, environmental issues, laws, standards, regulation, conflicts, and climate change. Management tools should be developed and implemented in order to develop the policies related

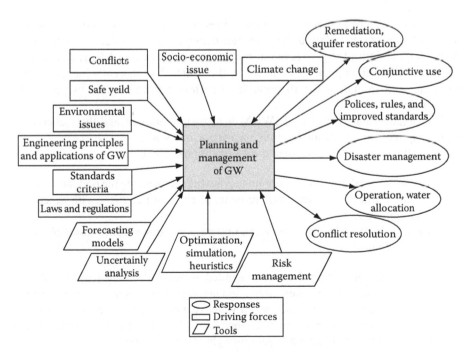

FIGURE 1.5   Basic principles of planning and management in groundwater, GW.

to conflict resolution, water allocation, disaster management, conjunctive use, aquifer remediation and restoration, and modifying and improving the standards as the system responses.

Optimization and simulation, forecasting, uncertainty analysis, risk assessment models are useful tools for resource allocation, monitoring, assessment, and dealing with probability and randomness in groundwater systems.

1.4.1 Integrated Water Resources Management

According to Rogers and Hall (2003), in the IWRM schemes, politics, and the traditional fragmented and sectored approach to water is made between resource management and the water service delivery functions. It should be noted that IWRM is a political process, because it deals with reallocating water, the allocation of financial resources, and the implementation of environmental goals. For an effective water governance structure, much more work has to be done.

This applies to both the management of water resources providing water services. There is a general agreement that IWRM is the only passage to sustainable management use of water. Even though, there is extended debate on the process and there is no generally accepted framework.

For example, IWRM is put into practice in Orange County Water District (OCWD) in California. It includes the use of groundwater recharge, alternative sources, conservation management, construction of levees in the river channel to increase infiltration and injection of treated recycled water to form a seawater intrusion barrier.

In IWRM, there are two incompatibilities: the needs of ecosystems and the needs of growing population. The shared dependence on water for both needs makes it quite natural to give ecosystems full attention within IWRM. At the same time, the Millennium goals of 2000 set by the United Nation's calls for safe drinking water and sanitation for all population suffering from poverty by year 2015 (Falkenmark, 2003).

The fundamental objective of IWRM is to sustain humans and the planet's life support systems and it should be done on a watershed scale.

The following goals are suggested by Falkenmark (2003).

* Satisfy societal water consumption needs while minimizing the pollution load
* Meet environmental flow requirements of river and aquifers with acceptable water quality
* Secure "hydro-solidarity" between upstream and downstream stakeholders and ecological needs

IWRM is sometimes referred to as integrated water cycle management (IWCM) such as in Australia, where water supply and management in general are the two top priorities for all levels of government, all water-related agencies, water utilities, and everybody else. This is understandable since Australia is the driest continent, and in the middle of experiencing a 1000 year drought (Kresic, 2008).

1.4.2 CONFLICT ISSUES IN GROUNDWATER

There are many conflicts over groundwater allocation and over exploitation of groundwater on a regional scale. There are also conflicts over shared aquifers between politically divided regions. Transboundary river basins are in most cases well defined and there are agreements/treaties for sharing their surface water resources. But transboundary aquifers are not well defined and managed by decision makers and authorities in both sides of the borders. International agreement about shared aquifers has limited practical implications and does not address the groundwater and its spatial distribution as well as how the limited resources should be allocated and discharged.

The International Association of Hydrogeologists (IAH) established a commission to investigate the issue of transboundary groundwater in 1999. It is a program supported by UNESCO, FAO, and UNECE (United Nations Economic Commission for Europe). One of the main objectives of the program is to support cooperation among countries in order to develop technical knowledge and conceptual understanding to alleviate potential conflicts. It is intended to train, educate, inform, and provide input to decision and policy makers, Puri (2001).

1.4.3 ECONOMICS OF WATER

Water as an economic good was first initiated at the Earth Summit of 1992 in Rio de Janeiro, Brazil. It was discussed during the Dublin conference on Water and the Environment (ICWE, 1992), and became one of the four Dublin Principles as an economic value in its competing uses and should be recognized as an economic good. Within this principle, it is vital to recognize first, the basic right of all human beings to have access to clean water and sanitation at an affordable price. Past failure to recognize the economic value of water has led to wasteful and environmentally damaging uses of this vital resource. Managing water as an economic good is an important way of achieving efficient and equitable use by encouraging conservation and protection of water resources.

Van der Zaag and Savenije (2006) explained the overall concerns over the true meaning of "water as an economic good" based on two schools of thought. The first school is the pure market driven price. Its economic value would arise spontaneously from the actions of willing buyers and willing sellers. This would ensure that water is allocated to uses based on its highest value. The second school interprets that the process of allocation of scarce resources in an integrated fashion which may not involve financial transactions (McNeill, 1998).

A unique property of water is that it belongs to a system and it always affects the users and any change in upstream water will affect the entire system downstream. Groundwater and surface water interchange their resources on a continuous basis so groundwater could be in surface water at some other point and vice versa.

Temporal and spatial variability of water resources is also constantly changing due to short- and long-term climate and land-use changes. Water could have negative economic value in case of floods. All of these make it difficult to establish the value of external impacts on water use.

The price of water could be defined as the price water users are paying per unit volume of water delivered in a unit of time (e.g., cubic meters/month). In most cases, neither water user customers nor self-providers pay the full price (value) of water, which should be equal to the real value of water and should include

- Capital cost of water withdrawal and distribution system including well installation and well field development, pumps, reservoirs, and water transfer facilities
- Cost of operation and maintenance of the system (O&M cost) including water treatment, water infrastructure maintenance, staff, and administrative costs
- Investments for augmenting the existing system
- Source protection cost reflecting its intrinsic value (water quality, reliability) including wellhead protection
- Environmental cost
- Sustainability cost

Kresic (2008) described the full price and value of water as shown in Figure 1.6. It is very difficult to assess the environmental and sustainable costs. They are quite related and a good assessment of tangible and nontangible environmental costs in shorter terms will help us to assess a longer time cost that should be paid to assure sustainability between supply and demand.

Groundwater use typically should be based on the current and future functions of groundwater as well as its expected values/costs. Groundwater use can generally be divided into drinking water, ecological, agricultural, industrial/commercial, and recreational. Drinking water use includes both public supply and individual (household or domestic) water systems. Ecological use commonly refers to groundwater functions, such as providing base flow to surface water to support wildlife

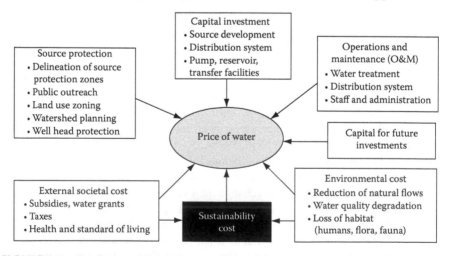

FIGURE 1.6 Components of full water price equal to full water value. (From Kresic, N., *Groundwater Resources: Sustainability, Management, and Restoration*, McGraw-Hill Professional, 852 p., 2008.)

and water species. Agricultural use generally refers to crop irrigation and livestock water demand. Industrial/commercial use refers water need in industrial process, especially for cooling purposes in manufacturing or commercial uses. Recreational use generally refers to as it impacts surface water; however, groundwater in karst aquifers may be used for recreational, such as cave observation purposes. These are considered "beneficial uses" of groundwater. Furthermore, within a range of reasonably expected uses and functions, the most beneficial groundwater use refers to the use or function that warrants the least stringent groundwater cleanup or remediation.

Groundwater value is typically considered: for its current uses and future as well as its intrinsic value. Current use value depends on need. Groundwater is more valuable where it is the sole source of water, or less costly than treating and utilizing surface water. Current use value can also consider the "costs" associated with impacts from contaminated groundwater on surrounding area (e.g., drinking supply wells and adjacent surface water). Future values refer to the value people place on groundwater to be used in the future. Society places an intrinsic value on groundwater; just knowing clean groundwater exists and will be available, irrespective of current or expected uses and it is a vital recourse to the preservation of the ecosystem. While the real value of groundwater is difficult to quantify, it will certainly increase with the increase in the expense of treatment of surface water cost and with continuing development and increasing demand.

1.4.4 GROUNDWATER SUSTAINABILITY

The term "sustainable development" was popularized by the World Commission on Environment and Development in its 1987 report entitled "Our Common Future". The objective of the Commission was to address the environmental and developmental problems in an integrated fashion.

It has had three general objectives:

- Examine the critical environmental and developmental issues and how to deal with them
- Formulate new forms of international cooperation that will influence policies in the direction of many changes needed
- Raise the level of understanding and commitment to action of all adversaries including institutions and governments

Commissioners from 21 countries analyzed the report, with the final report accepted by the United Nations General Assembly in 1987 (UNESCO, 2002). In various publications, the findings of the commission and the final resolution of the UN General Assembly were summarized as a single sentence describing sustainable development as "meeting the needs of the present without compromising the ability of future generations to meet their own needs." This sentence was criticized for failing to address the natural environment.

Educating the public considered as default of "stakeholders" about various choices is the crucial path toward sustainable development. Groundwater is a perfect example of many misunderstandings, by both the public and policy makers about the meaning of sustainability. Because there is a strange connotation about groundwater: as soon as

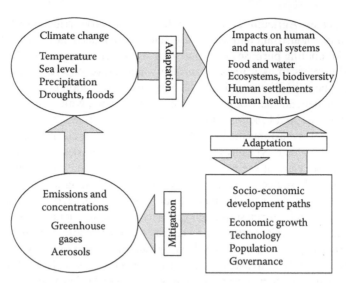

FIGURE 1.7 Sustainable development, adaptation, and migration interactions. (From IPCC, *Fourth Assessment Report, Intergovernmental Panel on Climate Change (IPCC)*, 2007, http://www.ipcc.ch).

it surfaces and it can be seen it is no longer groundwater, it is hard to debate publically when certain groundwater policies are announced and stated by certain agencies.

The development of groundwater often provides an affordable and rapid way to alleviate poverty and ensure food security in the underdeveloped and developing countries. Further, more by conjunctive use of groundwater and surface water, thoroughly IWRM strategies can serve to foster efficient use and sustainability and enhance the longevity of water supply by improving the allocation and end user efficiencies.

Instances of poorly managed groundwater development and the inadvertent impact of land-use practices have produced adverse effects on water quality, impairment of aquatic ecosystems, lowered groundwater levels and, consequently, land subsidence and the drying of wetlands. As it is less costly and more effective to protect groundwater resources from degradation than to restore them, improved water management will diminish such problems and save money.

To make the notion of groundwater as a sustainable resource requires the responsible use, management and governance of groundwater. In particular, actions need to be taken by water users, who sustain their well-being through groundwater abstraction; decision makers, both elected and none elected; civil society groups and associations; and scientists who must advocate for the use of sound science and engineering in support of better management.

International panel on climate change (IPCC, 2007) shows a broader perspective of sustainable development as showed in Figure 1.7. They have combined the impacts of socioeconomic development paths with technology, population, and governance through adaptation by human and natural systems.

The objective is to sustain food, water, ecosystems, biodiversity, and human health and prevent migration. At the same time, human and natural system should adapt to climate change impacts such as more frequent droughts and floods. The emission of greenhouse gases is the direct by-product of socioeconomic development that should be mitigated. This general framework could be adopted to foster groundwater sustainability on a more regional and global scale.

1.4.5 Supply and Demand Side Management

In the last three decades, the "ecological" or "holistic" paradigm comes to the forefront of dealing and solving man-made problems. This new way of thinking began out of new discoveries and theories in science, such as Einstein's Theory of Relativity, Heisenberg's Uncertainty Principle, and Chaos Theory. Contrary to Newton's mechanistic universe, these new theories have taken away the foundation of the opposing paradigm, the Cartesian (Newtonian) model Karamouz et al. (2010). The world has tried to "control the environment" for many centuries through the mechanistic way of thinking. The new way of thinking calls for "live with the environment," which is the essence of horizontal and creative thinking in this paradigm shift. It also demands a change on the way we perceive and understand our connecting world.

Water resources management is not an exception in this dramatic shift of paradigm. Throughout the history of water resources management, many methodologies have reflected some aspects of ecological thinking and have been shaping a water vision of the new paradigm.

For creating an opportunity for action, a paradigm shift in water management, away from the entrenched supply-side focus toward a comprehensive demand-side management, a soft path planning approach should be established. For instance, the recent NRTEE report, "Environmental Quality in Canadian Cities: The Federal Role," could easily be extended to explicitly include urban groundwater management. A schematic as shown in Figure 1.8 was proposed by Ashton and Haasbroek

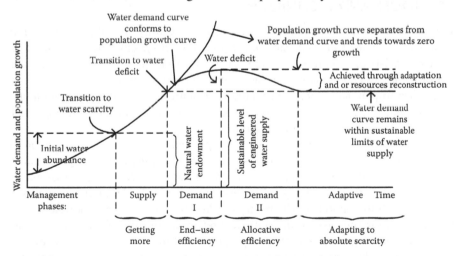

FIGURE 1.8 Schematic representation growth of water demand and supply. (From Ashton, P.J. and Haasbroek, B., Water demand management and social adaptive capacity: A South African case study, in *Hydropolitics in the Developing World: A Southern African Perspective*, eds. Turton, A.R. and Henwood, R., African Water Issues Research Unit, Pretoria, South Africa, 21 p., 2001.)

(2001) for representation of the growth in water demand and management. Two periods of demand I and demand II are defined to bring the normal growth of demand, as a result of population growth, to the level of sustainable limits of water supply.

Developing water-saving technologies should be considered in developing water demand management strategies. These methods cover a wide range of solutions, such as dual-flush toilets, flow restrictors on showers and automatic flush controllers for public urinals, automatic timers on fixed garden sprinklers, moisture sensors in public gardens, and improved leakage control in domestic and municipal distribution systems. All these measures are practical, but regulations and incentives are needed for their implementation. The water losses rate in many water distribution systems may vary from 10% to 60% of the supplied water and higher values are reported for the developing countries. To achieve the desired level of efficiency in water use, more attention should be given to decrease the amount of water losses and nonrevenue water which generally includes

- Leakage pipes, valves, tanks, meters and other components of water supply and distribution systems
- Water used in the treatment process (back-wash, cooling, pumping, etc.) or for flushing pipes and reservoirs (see Karamouz et al., 2010 for more details)

The cost of water and supplying services and technologies have a major influence on the level of water demand especially in developing countries. In rural areas, the distance from households to the standpipes, or the number of persons served by a single tap or well is the main factors that control the water demand.

1.5 TOOLS AND TECHNIQUES

Simulation and optimization models of groundwater quality or quality can be formulated with mathematical expression of the groundwater system. The model-building process consists of several interrelated stages as outline in Figure 1.9. Groundwater hydrology is broad, complex, and due to invisible nature of its media, many tools are needed to plan and manage groundwater resources, protection, mitigation, and development. Figure 1.9 shows a framework for combination of tools and methods needed for groundwater resources planning.

Models are based on the available hydrologic and hydrogeologic information. In the first phase of the modeling process, well logs, pumping test data, groundwater contour maps, water levels, precipitation, stream flow, and recharge information are compiled and statistically analyzed to determine the probable aquifer parameters, recharge conditions, initial and boundary conditions, and current groundwater pumping schedule. On the basis of the statistical reliability of this information, a choice has to be made regarding the most appropriate and realistic type of mathematical model. The model choice problem, the second phase of the process, is characterized by a hierarchy of mathematical models (Willis and Yeh, 1987). For example, in the case of limited or imprecise information regarding the underlying physical processes occurring in the aquifer system, a statistical or time series approach may be warranted. The analysis of trends, periodicities, and the correlation structure of the

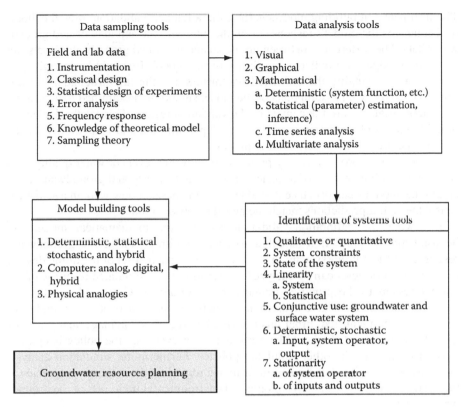

FIGURE 1.9 Framework of tools needed for groundwater resources planning. Kisiel, C.C. and Duckstein, L., Operation research study of water resources—Part II: Case study of the Tucson basin (Arizona), *Water Resources Bulletin*, 6(6), 857, 1970.

historical records may further define the mechanisms responsible for groundwater movement and water quality variations.

From a planning perspective, however, mathematical models of the groundwater flow or mass transport will consist of

1. Mathematical expresses in term of algebraic, integral, or differential equations characterize the flow or transport processes in groundwater. These equations relate the variables such as hydraulic head or mass concentration to the policy or decision variables such as pumping or injection rates that provide control over the variables. The process is also bounded by the parameters that define the flow, properties, and quality of the system such as storage coefficients, hydraulic conductivity, and dispersion coefficients.
2. Initial conditions that show the state of the system prior to development or operation of the aquifer system.
3. Inputs to the system such as recharge, precipitation, subsurface inflows (recharge), and outputs such as discharges to surface waters or evapotranspiration.

The next phase in building models is the calibration and validation. Model calibration compares the model's response or predictions with the actual collected groundwater data. The differences between predicted and historical groundwater levels can be used to judge the predictive capability of the model. It should be brought to a minimum by changing the estimated parameters and the initial/boundary conditions. The underlying assumptions of the model should be reexamined to determine their appropriateness in the context of the validation results. This feedback element of calibration and validation process is developed through and is facilitated by the iteration nature of the model-building process.

The validation process usually predicts groundwater levels or water quality concentrations with a new set of historical or generated surface and groundwater data. Again, the model's performance is evaluated on the basis of how well, in a statistical sense, the model predictions match the new data observation.

Following the calibration–validation process, the mathematical model can be used for the prediction (simulation) and/or optimization of the groundwater resources of the aquifer system. In a simulation approach, a set of planning, design, or operational policies can be analyzed by examining the simulated response of the aquifer system to the proposed management alternatives. Costs and benefits can be determined and sensitive hydrologic or environmental areas of the groundwater systems can be identified. From this information, the "best" management practices (BMP) can be found. It should be emphasize, however, that the policy is optimal only in relation to the other simulated policies. Furthermore, simulation analysis provides only localized or limited information regarding the response properties of the system, and the possible hydrologic and environmental trade-offs of the groundwater system.

1.6 PEOPLE'S PERCEPTION: PUBLIC AWARENESS

"There could be no Great Society if the water, air and soil are dirty" as stated by Goldman (1967). He reiterated that "there is reason to believe that it is only the very wealthy countries, able to afford the luxury of clean water and air, which can make a fuss about it." It can be argued that until recently having clean water was a "luxury" as often cited by the delegates of developing countries at the United Nations.

In most of the developed countries, legislation for pollution control was introduced in 1960s and 1970s. In the United States, EPA (Environmental Protection Agency) was created in 1970 to administer environmental programs. It was an encouraging effort, but many details were left undefended. The United Nations focused Conference on the Human Environment in 1972 in Stockholm was a starting global effort. Many other United Nations conferences followed dealing with population, food, women's rights, desertification, human settlements continued to emphasize on environmental issues.

Eckholm (1982) describes the enormous task faced by the developing countries to provide reasonable clean water, disposal facilities, and the enforcement sanitary standards. He estimated that half of the people in the developing countries do not have reasonable access to safe water supplies and only one-fourth to adequate waste disposal facilities.

It will take extraordinary statesmanship and wisdom at national and international level, to balance the economic development with the amount of sustainable water resources and the carry capacity of rivers and aquifers to withstand any quality degradation.

High priority for environmental improvement will be difficult to impose when the social costs of welfare and unemployment have caused huge financial deficits for governments of the developed world and have brought many developing and under-developed countries to the brink of financial disaster.

Groundwater plays an important role in the life of many especially in the developing countries. The quantity of water is the major concern. The economic development has imposed considerable pressure on the water resources especially groundwater. People's perception generally is that groundwater is an unlimited, reliable, and safe source of water to drink. The impact of climate change has affected the groundwater recharge and the frequency of drought. That has changed the safe yield of aquifers and has caused considerable drawdown in many aquifers around the world.

In developed countries, groundwater quality issues are the governing factor and the major public concerns. There have been many UN and other international agencies as well as NGOs (nongovernmental organizations) initiatives to promote public awareness on the quality and quantity as well as health issues among infants and children exposed to groundwater contamination.

1.7   GROUNDWATER PROTECTION: CONCERNS AND ACTS

A groundwater and food security investigation by the international food and agriculture organization, FAO (2003) confirms that it is not possible to assess the extent to which global food production could be at risk from over abstraction and exploitation of groundwater resources. Indeed, the search for reliable groundwater level and abstraction data to determine depletion rates is very difficult to obtain with consistency, reliability, and adequate coverage.

The impacts of over exploitation and water-level declines have been reported widely. It should be noted that over abstraction of groundwater can lead to a wide range of social and economic problems, and environmental predicaments including

- Declines in base flows of the rivers and wetlands with significant damage to ecosystems and downstream users
- Increased pumping costs and energy consumption
- Land subsidence and damage to surface infrastructure
- Changes in patterns of groundwater flow exchange with the adjacent aquifer systems
- Less water for drinking, irrigation, and other uses, particularly for the low-income people
- Increases in agriculture vulnerability (food security) and vulnerability of other water user sectors to climate change and/or natural climatic fluctuations.

Warnings of a groundwater disaster (declining groundwater tables and/or polluted aquifers) have led to calls for urgent social, governmental, and management responses. There is a need to evaluate all of the indications, documented records and facts and figures to call for occurrence of a disaster and to identify the types of management responses and adaptive measures that could actually work.

Hötzl (1996) discussed the need to distinguish between resource and source protection. Even though both concepts are related, it is not practical to utilize a source without protecting the resource. For example, in European countries, groundwater is considered a vital resource that must be protected. Activities endangering groundwater quality are forbidden by different members of EU (European Union) including Germany through their regulation of WHG (1996). The European Water Framework Directive (The European Parliament and the Council of the European Union, 2000) calls for treating and protecting water like no other commercial product.

In the United States, many communities use groundwater as the sole source of water supply despite of high relatively annual precipitation rate. For example, in Long Island, just outside of New York City, most of the water supply is provided by groundwater resource. There have been many social, state, and federal concerns about the groundwater protection almost everywhere in the United States that have resulted in many regulatory acts as explained in the next section.

In many parts of arid and semiarid areas in Asia and Africa, groundwater is vital to the economic stability and social preservation and development plan of different regions and nation. All of these concerns have brought many actions for the use, protection, and development of groundwater resource. But, the increasing demand for water in all of these societies (developing or less developing countries) resulted in over exploitation of water resource as discussed earlier. It is very difficult if not impossible to rely on social concern to protect groundwater. There is a need for well-defined transparent laws and regulations that are enforceable, tailors to local and regional level of resources, treats, and vulnerabilities of groundwater.

1.7.1 CLEAN WATER ACT

Increasing public awareness and concern for control of surface and round water pollution led to enactment of the Federal Water Pollution Control Act Amendments of 1972. It was amended in 1977, and this law, became commonly known as the Clean Water Act, CWA, because basis for many other statues and regulations that followed especially related to groundwater that was not explicitly spelled out in CWA. The act established the basic structure for regulating discharges of pollutants into the surface waters and groundwater of the United States. It gave the United States Environmental Protection Agency (USEPA) the authority to implement pollution control programs such as setting wastewater standards for industrial use and discharge to groundwater.

The CWA set water quality standards (WQS) direct toward all contaminants primarily in surface waters but with applicability to groundwater. The CWA made it unlawful for any person to discharge any pollutant from a point source into navigable waters, and other natural water resource including groundwater unless a permit is obtained under its provisions. It also funded the construction of water waste treatment plants and recognized the critical problems posed by nonpoint source pollution.

The key elements of the CWA are establishment of WQS, their monitoring and, if WQS are not met, developing strategies for meeting them. Drinking water regulations set enforceable maximum contaminant levels for particular contaminants in drinking water or required ways to treat water to remove contaminants. Maximum contaminant levels are legally enforceable, which means that both USEPA and different states can take enforcement actions against water supply systems not meeting the standards. USEPA and states may issue administrative orders, take legal actions, or fine water utilities and In contrast, the National Secondary Drinking Water Standards are (NSDWS) recommended and they are not enforceable. However, states may choose to adopt them as enforceable standards or may relax the NSDWS.

1.7.2 GROUNDWATER PROTECTION

Protection of groundwater has limited enforceable regional and national acts and regulations. On a local (state or city) scale, the groundwater management acts could protect the right of local groundwater users as long as it is not in conflict with other local and national laws and regulations. The highest priority is to protect the groundwater for drinking water supply (USEPA, 2005). In the state of Kansas, groundwater act was ratified, K.S.A. 82a-1020 (2001). The water districts have been established to conserve and protect the groundwater resources and prevent for to maintain its economic value deterioration. There are many opportunities for the water districts as a result of the authorities that are given to them.

In the United States, groundwater as a resource is addressed in the federal level as well as through safe drinking water act (SDWA). As a part of the USEPA's cleanup programs in 2002, the Ground Water Task Force (GWTF) was established to improve the quality of USEPA cleanup program. This program deals with brown fields, federal facilities and leaking underground storage tanks and RCRA Corrective Action and Superfund. It also provides a discussion on the importance of groundwater resources and their vulnerability (GWTF, 2007).

The USEPA defines a sole or principal source aquifer as one which supplies at least 50% of the drinking water consumed in the area overlying the aquifer. Such area cannot have an alternative drinking water source(s) which could physically, legally, and economically supply all those who depend upon the aquifer for drinking water.

The Sole Source Aquifer (SSA) Program within SDWA is aimed at promoting the importance of these aquifers. Although USEPA has statutory authority to initiate SSA designations, it has a longstanding policy of only responding to petitions. Any individual, corporation, company, association, partnership, state, municipality, or federal agency may apply for SSA designation. A petitioner is responsible for providing USEPA with hydrogeologic and drinking water usage data, and other technical and administrative information required for assessing designated criteria.

If an SSA designation is approved, proposed federal financially assisted projects which have the potential to contaminate the aquifer are subject to USEPA review. Proposed projects that are funded entirely by state, local, or private concerns are not subject to USEPA review. Examples of federally funded projects which have been reviewed by USEPA under the SSA protection program include, public water supply wells and transmission lines; wastewater treatment facilities, and construction projects that involve disposal of storm water (Kresic, 2008).

In European Union, the European Groundwater Directive of 2006 extends the concept of overall resource protection including groundwater in details (The European Parliament and the Council of the European Union, 2006).

1.7.3 USEPA GROUNDWATER RULE

The EPA is promulgating a National Primary Drinking Water Regulation, the Groundwater Rule, to provide for increased protection against microbial pathogens in public water systems that use groundwater sources (USEPA, 2006). This regulation is in accordance with the SDWA as amended. DWR requiring disinfection as a treatment technique for all public water systems, including surface water systems and, as necessary, groundwater systems.

The groundwater rule establishes a risk-targeted approach to target groundwater systems that are susceptible to fecal contamination. The occurrence of fecal indicators in a drinking water supply is an indication of the potential presence of microbial pathogens that may pose a threat to public health. This rule requires groundwater systems that are at risk of fecal contamination to take corrective action to reduce cases of illnesses and deaths due to exposure to microbial pathogens.

In the 1999 Ground Water Report to Congress, the USEPA emphasized the need for more effective coordination of groundwater protection programs at the federal, state, and local levels. While the Clean Water Action Plan and states' watershed protection programs address groundwater, true coordination of groundwater management efforts has not been achieved in most states (USEPA, 1999). However, the agency also indicated that at the time 47 states have approved wellhead protection programs which mandate delineation of wellhead protection areas (WPAs) for public water supplies. WPA is a designated surface and subsurface area surrounding a well or well field for a public water supply and through which contaminants or pollutants are likely to pass and eventually reach the aquifer that supplies the well or well field. The purpose of designating the area is to provide protection from the potential contamination of the water supply. These areas are designated in accordance with laws, regulations, and plans that protect public drinking water supplies.

1.8   OVERALL ORGANIZATION OF THIS BOOK

In the following chapters, attempt has been made to develop a system view of groundwater fundamentals and model-making techniques through the application of science, engineering, planning, and management principles. In Chapters 2 through 4, the classical issues in groundwater hydrology and hydraulics are discussed followed by quality aspects in Chapter 5. In these chapters, the process of analyzing data, identification, and parameter estimation are presented. In Chapters 6 through 8, tools and model-building techniques and conjunctive use of surface and groundwater are described followed by aquifer restoration and remediation and monitoring techniques as well as analysis of risk in Chapter 9. Chapter 10 is an introduction to groundwater risk and disaster management. Finally, in Chapter 11, the impact of climate change on groundwater and the needed tools and methods for analysis of future data realization, downscaling of large-scale low-resolution data to local watershed and aquifer scale for impact studies are discussed.

PROBLEMS

1.1 Make an assessment of groundwater resources around the world and compared them with the U.S. share of total groundwater.
Hint: use UNESCO, Cambridge/Oxford L.C. World Resources, World Bank, EU Switch program, FAO, USGs, sites and publications.
1.2 In Problem 1.1, make an assessment of groundwater quality figures around the world.
1.3 Identify the federal and national agencies involved in management and protection of groundwater, At least one in each part of the globe.
1.4 Compare the driving forces dealing with groundwater science and engineering with those affecting groundwater planning and management.
1.5 Identify and compare the "response" issues in Problem 1.4, concentrating on improvement of standards and quality of life as well as economic, environmental and legal issues.
1.6 Explain IWRM for aquifers in the context of public awareness/participation and consideration of social impacts.
1.7 What is the estimated water cost for residential customers in east (New York City) and west (city of Los angles of the United States)? Compare it with the price of water in 5 other major cities around the world. Is the price of groundwater in selected areas different from surface water resources?
1.8 Briefly discuss a case of transboundary conflict over shared aquifer.
Hint: Example of Kura aquifer in Caucasus region (Puri, 2001) can be used.
1.9 Identify and briefly explain factors affecting groundwater sustainability.
1.10 In demand side management, explain the two periods of demand I and demand II in Figure 1.8.
1.11 Identify three local (city or county) agencies in the United States or in other countries with the charter of protecting and enforcement issues related to quality and quantity (discharge) of groundwater.

REFERENCES

Ashton, P.J. and Haasbroek, B. 2001. Water demand management and social adaptive capacity: A South African case study, in *Hydropolitics in the Developing World: A Southern African Perspective*, eds. A.R. Turton and R. Henwood, African Water Issues Research Unit (AWIRU) and International Water Management Institute (IWMI), Pretoria, 21 p.

Doll, P. 2002. Impact of climate change and variability on irrigation requirements: A global perspective. *Climatic Change*, 54, 269–293.

Eckholm, E. 1982. *Down to Earth: Environment and Human Needs*, Pluto Press, London, 173 p.

Falkenmark, M. 2003. *Water Management and Ecosystems: Living with Change, Global Water Partnership Technical Committee*. Global Water Partnership, Stockholm, Sweden.

Falkenmark, M. 2003. Water management and ecosystems: Living with change. TEC Background Papers No. 9, Global Water Partnership Technical Committee (TEC), Global Water Partnership, Stockholm, Sweden, 50 p.

FAO (Food and Agriculture Organization of the United Nations). 2003. *Review of World Water Resources by Country, Water Reports 23*, FAO, Rome, 110 p. www.fao.org/ag/agl/aglw/docs/wr23e.pdf

FAO-AQUASTAT, 2007. *AQUASTAT Main Country Database.* www.fao.org/ag/agl/aglw/aquastat/main

Goldman, M.I. 1967. *Controlling Pollution, the Economics of a Cleaner America.* Prentice Hall, Englewood Cliffs, NJ.

GWTF (GroundWater Task Force). 2007. Recommendations from the EPA GroundWater Task Force; Attachment B: Ground water use, value, and vulnerability as factors in setting cleanup goals. EPA 500-R-07–001, Office of Solid Waste and Emergency Response, pp. B1–B14.

Hötzl, H. 1996. Grundwasserschutz in Karstgebieten, *Grundwater*, 1(1), 5–11.

ICWE. 1992. *Dublin Statement and Report of the Conference*, January 26–31, 1992, Dublin, Ireland.

IPCC (Intergovernmental Panel on Climate Change). 2007. Summary for policymakers of the Synthesis Report of the IPCC Fourth Assessment Report; 23 p. Available at: http://www.ipcc.ch/.

Karamouz, M., Moridi, M., Nazif, N. 2010. *Urban Water Engineering and Management*, CRC Press, Boca Raton, FL, 602 p.

Kisiel, C.C. and Duckstein, L. 1970. Operation research study of water resources—Part II: Case study of the Tucson Basin (Arizona). *Water Resources Bulletin*, 6(6), 857–867.

Kresic, N. 2008. *Groundwater Resources: Sustainability, Management, and Restoration.* McGraw-Hill Professional, New York, 852 p.

K.S.A. 82a-1020, 2001. *Discussion and Recommendations for Long-Term Management of the Ogallala Aquifer in Kansas.* Ogallala Aquifer Management Advisory Committee, 23 p.

McNeill, D. 1998. Water as an economic good. *Natural Resources Forum*, 22(4), 253–261.

Puri, S. 2001. Internationally shared (transboundary) aquifer resources, management, their significance and sustainable management. A framework document International Hydrological program, IHP-VI, IHP *Non Serial Publications in Hydrology*, UNESCO, Paris, 71 p.

Rogers, P. and Hall, A.W. 2003. Effective water governance. TEC Background Papers No. 7, Global Water Partnership Technical Committee (TEC), Global Water Partnership, Stockholm, Sweden, 44 p.

Secker, D. 1993. *Designing Water Resources Strategies for the Twenty-First Century.* Discussion Paper 16, Water Resources and Irrigation Division, Winrock International, Arlington, VA.

The European Parliament and the Council of the European Union. 2000. Directive 2000/6/EC of the European Parliament and the Council of 23 October 2000 establishing a framework for Community action in the field of water policy, *Official Journal of the European Union*, December 22, 2000, L 327/1.

The European Parliament and the Council of the European Union. 2006. Directive 2006/118/EC of the protection of groundwater against pollution and deterioration, *Official Journal of the European Union*, December 27, 2006, L 372/19-31.

UN (United Nations). 1999. *World Population Prospect*, 1998 Revision. UN Department of Policy Coordination and Sustainable Development, New York.

UNESCO. 2002. Manual of the General Conference, including texts and amendments adopted by the General Conference at Its 31st Session, Paris.

UNESCO (United Nations Educational Scientific and Cultural Organization). 2006. *Water, a Shared Responsibility.* The United Nations World Water Development Report 2. UNESCO. World Water Assessment Program (WAPP), Paris, and Berghahn Books, New York, 584 p.

USEPA. 1999. *Compendium of ERT Groundwater Sampling Procedures.* EPA 540/P-91/007. Environmental Protection Agency, Washington, DC.

USEPA. 2005. *Guidance for Developing Ecological Soil Screening Levels (Eco-SSL): Exposure Factors and Bioaccumulation Models for Derivation of Wildlife Eco-SSL.* Environmental Protection Agency, Washington, DC.

USEPA. 2006. *Final Groundwater Rule, Fact Sheet.* EPA 815-F-06-003. United States Environmental Agency, Office of Water, Washington, DC, 2 p.

Van der Zaag P.V. and Savenije, H.H.G. 2006. Water as an economic good: The value of pricing and the failure of the markets. *Value of Water Research Report Series* No. 19. UNESCO-IHE Institute for Water Education, Delft, the Netherlands, 28 p.

WHG. 1996. *Gesetz zur Ordnung des Wasserhaushalts,* BGBI, Bonn, Germany.

Willis, R. and Yeh, W.W.-G. 1987. *Groundwater Systems Planning and Management.* Prentice Hall Inc., Englewood Cliffs, NJ.

WMO (World Meteorological Organization). 1997. *Comprehensive Assessment of the Freshwater Resources of the World.* WMO, Geneva, p. 9.

WRI (World Resources Institute). 1998. *World Resources, 1998–1999. A Guide to the Global Environment.* Oxford University Press, New York, 384 p.

2 Groundwater Properties

2.1 INTRODUCTION

Water in nature is in solid, liquid, and vapor form. It also exists in different places like atmosphere (atmospheric water), surface of the ground (surface water), and subsurface of the ground (groundwater and unsaturated zone water). The source of surface and subsurface water is snow and rain precipitation. A portion of the precipitation flows on the land (surface runoff), a portion of that goes back to the atmosphere because of evapotranspiration from surface water and the surface of plants, and the remaining portion infiltrates into the land (subsurface flow). The movement process of water in nature is called the *hydrologic cycle*. A schematic of the hydrologic cycle is shown in Figure 2.1.

Groundwater constitutes only about 0.62% of the entire amount of water in the globe (Table 2.1). Although it is a small portion of the world's total water resources, it is considered as an important resource to supply water demands due to the population increase and in consequence increasing needs for water supply.

In this chapter, the fundamental aspects of subsurface water are explained. It should be noted that all of the subsurface water is not considered as groundwater. Groundwater is a part of the subsurface water that totally saturates the soil pores and flows with a pressure more than atmospheric pressure. The difference between pressure in groundwater and atmospheric pressure is called *gage pressure*. At the groundwater table, the gage pressure is zero. Depending on the location, the depth of the groundwater varies. The zone between the land surface and the surface of groundwater is called *unsaturated* or *vadose water*. The water is under the influence of capillary and adhesion forces (surface tension) between soil and water molecules. Figure 2.2 shows the classification of the subsurface water which are explained with some details in this chapter. As it is shown, hydraulic head at point A in the saturated zone is equal to the pressure head ψ and elevation head z.

2.2 VERTICAL DISTRIBUTION OF SUBSURFACE

Subsurface water flows below the ground surface. The water behavior in these areas and the characteristics of these zones are as follows:

1. *Soil-water zone*: The main characteristics of this zone are
   a. Water content is usually less than saturation.
   b. It might be saturated during rainfall or irrigation.
   c. Thickness depends upon the type of soil and vegetation.

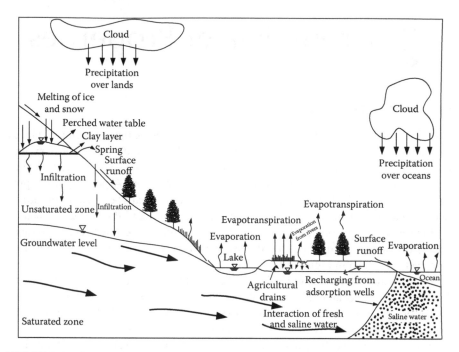

FIGURE 2.1   Schematic representation of the hydrologic cycle.

TABLE 2.1
Water Resources in the World

Water Resources	Volume (1,000 km³)	Percentage
Atmospheric water	13	0.001
Surface water		
Oceans saline water	1,320,000	97.2
Seas saline water	104	0.008
Lakes fresh water	125	0.009
Rivers fresh water	1.25	0.0001
Biosphere fresh water	29,000	2.15
Glacial water	50	0.004
Groundwater		
Water in Vadose zone	67	0.005
Groundwater in the depth less than 0.8 km	4,200	0.31
Groundwater in the depth more than 0.8 km	4,200	0.31
Total	1,360,000	100

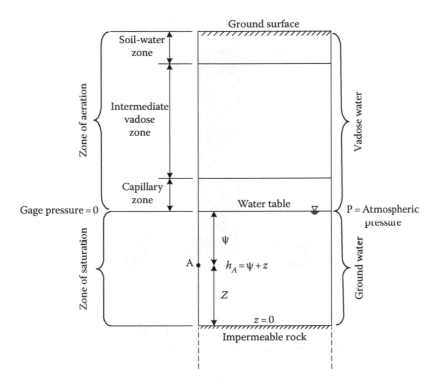

FIGURE 2.2   Classification of subsurface waters in a hypothetical section.

d. It consists of hydroscopic water, which sticks to soil and is unavailable to plants; capillary water, which is held by surface tension and is available to plants; gravitational water (excess and drain through soil).

e. Field capacity is the water held after gravity water drains away. In other words, field capacity is the water content of the soil when a thoroughly wetted soil has drained for approximately 2 days.

f. Water in root zone (after gravity water drains) varies between field capacity and "wilting point." Wilting point is the near dry condition, which, if prolonged, causes plants to wilt beyond recovery. It is also defined as the water content of soil when plants, growing in the soil, wilt and do not recover.

g. The amount of water held in the soil between field capacity and wilting point is considered to be the water available for plant extraction.

2. *Intermediate zone*: Water must move through this zone to reach groundwater. In this zone, excess water moves downward due to gravity force. But part of intermediate zone has pellicular water, which is nonmoving water and consists of hydroscopic and capillary water.

3. *Capillary zone*: The capillary zone (or capillary fringe) extends from the water table up to the limit of capillary rise of water. The water table is defined as the depth at which the water pressure equals atmospheric pressure. If a pore space could be idealized to represent a capillary tube, the

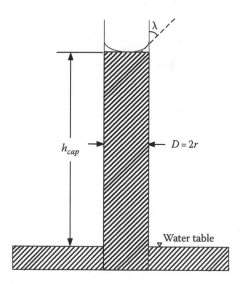

FIGURE 2.3   A schematic of capillary rise.

capillary rise h_{cap} (Figure 2.3) can be derived from an equilibrium between surface tension of water and the weight of water raised:

$$h_{cap} = \frac{2\tau}{r\gamma} \cos \lambda \tag{2.1}$$

where
 τ is the surface tension (kg/s²)
 γ is the specific weight of water (N/m³)
 r is the tube radius (m)
 λ is the angle of contact between the meniscus and the wall of the tube

For pure water in clean glass, $\lambda = 0$, and $\tau = 0.074$ kg/s² at 20°C, so that the capillary rise approximates

$$h_{cap} = \frac{0.15}{r} \tag{2.2}$$

h_{cap} is less than 2.5 cm for gravel, about 200 cm for silt and several meters for clay.
4. *Saturated zone*: It is the source of water supply. In the zone of saturation, groundwater fills all of the pores; therefore, the (effective) porosity shows a direct measure of the water contained per unit volume. Available water depends upon various factors such as porosity, specific yield, specific retention.
 All of these layers are shown in Figure 2.2.

2.3   AQUIFERS, AQUITARDS, AND AQUICLUDES

Dingman (1994) defined an *aquifer* as "a geologic unit that can store enough water and transmit it at a rate fast enough to be hydrologically significant." If significant quantities of water cannot be transmitted under ordinary hydraulic gradients in a saturated geologic formation, this formation is called an aquiclude. In general, it can be claimed that an aquifer is permeable enough to be economically exploited, while it is not economic to withdraw water from aquicludes.

Less-permeable geologic formations are termed as *aquitards*. Significant quantities of water might be transmitted through these media; however, their permeability is not sufficient to exploit water through wells within them. Very few geological formations have the characteristics of an aquiclude and most of the geologic strata are considered as either aquifers or aquitards.

Unconsolidated sand and gravel, permeable sedimentary rocks such as sandstone and limestone, and heavily fractured volcanic and crystalline rocks are most com mon aquifers. The most common aquitards are clay, shale, and dense crystalline rocks (Todd, 1980).

However, it is hard to specify whether a formation with given constituents is aquifer or aquitard. In other word, there is vagueness in the definition of aquifers and aquitards. For instance, in an interlayered sand–silt formation, the silt layer may be considered an aquitard; while if the formation is a silt–clay system, the silt may be considered aquifer. In general, aquifer has different meanings to different people and it might even have various meanings to the same person at different times.

2.4   TYPES OF AQUIFERS

Most aquifers are underground storage reservoirs made up of water-bearing permeable rock or unconsolidated materials. Water penetrates downward through the pores by gravitational forces until it reaches the saturated area. The water level, which is termed water table, is not always at the same depth below the land surface. Depending on the presence or absence of a water table, aquifers are classified as unconfined or confined while a leaky aquifer is considered as a semi-confined aquifer, which has the characteristics of both types of the aquifers.

2.4.1   Unconfined Aquifer

In an unconfined aquifer the top of the aquifer is defined by the water table which varies in fluctuation form and in slope. The water table fluctuation depends on changes in volume of water storage prepared by areas of recharge and discharge, pump from wells, and permeability. In Figure 2.4, the upper aquifer represents an unconfined aquifer. Using the water elevations in wells, the contour maps and profiles of the water table are prepared. Typically, the shallowest aquifer at a given location does not have a confining layer between it and the surface, and it is considered as an unconfined aquifer.

The perched water body is the groundwater accumulating above an impermeable stratum such as a clay layer. This term is a special case of an unconfined aquifer,

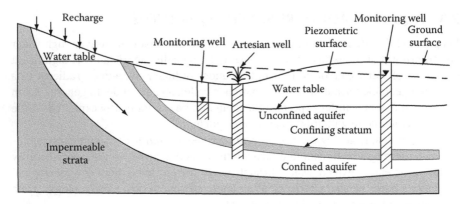

FIGURE 2.4   Schematic cross section illustrating unconfined and confined aquifers.

which refers to a small local area with an elevation higher than a regionally extensive aquifer. Clay lenses in sedimentary deposits often have shallow perched water bodies overlying them.

2.4.2   CONFINED AQUIFERS

Confined aquifers are restricted on the top by impermeable material. Water in a confined aquifer is normally under pressure greater than atmospheric pressure. This pressure in a well can cause the water level to rise above the ground surface, which is designated a flowing artesian well as shown in Figure 2.4. An area where water flows into the Earth to resupply a water body or an aquifer is known as a recharge area. Water fluctuations in wells discharging water from confined aquifers lead to variations in pressure rather than in storage volumes. Therefore, small change in storage volume will occur in confined aquifers and carry water from recharge areas to locations of natural or artificial discharge.

Figure 2.4 shows the potentiometric surfaces, which are known as isopotential level and piezometric surface. This surface represents the static head of groundwater and is defined by the level to which water will rise. In general, the piezometric surface at a point in a well, penetrating a confined aquifer, is shown by the water level at that point in the well. If the potential level of piezometric surface is higher than the land surface, it results in overflow of the well. When the piezometric surface is lower than the bottom of the upper confining bed, a confined aquifer becomes an unconfined aquifer.

2.4.3   AQUITARD (LEAKY) AQUIFER

Aquitards are zones within the earth that limit the groundwater flow from one aquifer to another. They are beds of low permeability along an aquifer. In general, the existence of leaky or *semi-confined* aquifers is more expected than confined or unconfined aquifers. This is because it is hard to find an aquifer that is completely confined or unconfined. Such characteristics are common in alluvial valleys, plains, or former lake basins.

2.5   GROUNDWATER BALANCE

For a particular area, the groundwater balance study is accomplished for the following purposes:

- To check if all flow components in the system are quantitatively accounted for and what components have the greatest bearing
- To calculate one unknown component of the groundwater balance equation, if the values of all other components are known with an acceptable accuracy
- To model the hydrological processes in the study area

To be able to estimate groundwater balance of a region, all individual inflows to or outflows from a groundwater system as well as changes in groundwater storage over a given time period must be quantified. The general form of water balance over a period of time is

Input to the system – outflow from the system = Change in storage of the system

$$(2.3)$$

To compute the groundwater balance of a system, the significant components are identified first. After that, the quantifying individual components are evaluated. Now these quantified components can be presented in the form of water balance equation.

Considering the various inflow and outflow components in a given study area, the groundwater balance equation can be written as

$$\Delta S = R_r + R_s + R_i + R_t + S_i + I_g - E_t + T_p - B_f - O_g \qquad (2.4)$$

where
ΔS is the change in groundwater storage
R_r is the recharge from rainfall
R_s is the recharge from canal seepage
R_i is the recharge from field irrigation
R_t is the recharge from tanks
S_i is the influent seepage from rivers
I_g is the inflow from other basins
E_t is the evapotranspiration from groundwater
T_p is the draft from groundwater
B_f is the baseflow, the part of the groundwater inflows to rivers
O_g is the outflow to other basins

The following data are required to accomplish a groundwater balance study in an area over a given time period:

- Rainfall data
- Land use data and cropping patterns

- River stage and discharge data
- River cross sections
- Canals discharge and distributaries data and the possible seepage from canals
- Tanks depth, capacity, area, and seepage data
- Water table data
- Groundwater draft (the number of each type of wells operating in the area, their corresponding running hours each month and their discharge)
- Aquifer parameters

Example 2.1

Assume that there is 900 mm/year rainfall in an area. The irrigation rate and evapotranspiration from irrigation are 180 and 420 mm/year, respectively. Also, 330 mm/year of the rainfall flows over land. The rate of outflow to the other basins is 200 and 150 mm/year flows to rivers.

Calculate the influent from seepage.

Assuming groundwater inflow and outflow remain unchanged, calculate the baseflow in the region if 80 mm/year water is pumped from the region.

Solution

a. When it rains, a portion of the precipitation, P, is evaporated, E_t, a portion flows over the land surface, O_f, and the rest of that recharges the groundwater, R_r. Therefore,

$$P + R_i = E_t + O_f + R_r \Rightarrow 900 + 180 = 420 + 330 + R_r$$

$$R_r = 330 \, \text{mm/year}$$

$$I_g + R_r = O_g + R_f \Rightarrow I_g = 200 + 150 - 330 = 20 \, \text{mm/year}$$

b. In this case, the inflow to groundwater, recharge from rainfall, and recharge from irrigation are the *inflows* and baseflow and the discharge from pumping are the *outflows* from the system:

$$I_g + R_i + R_r = O_g + B_f + Q_p \Rightarrow 20 + 180 + 330 = 200 + B_f + 80$$

$$\Rightarrow B_f = 150 \, \text{mm/year}$$

2.5.1 WATER BALANCE IN CONFINED AQUIFERS

In the water balance for the confined aquifer, the recharge and evaporation can be neglected in a short period (i.e., a day). Therefore, the daily water balance can be written as

$$W_{sc,i} = W_{sc,i-1} + W_{per} - W_{pc} \tag{2.4a}$$

where

$W_{sc,i}$ is the amount of water stored in the confined aquifer on day i (mm)

$W_{sc,i-1}$ is the amount of water stored in the confined aquifer on day $i-1$ (mm)

W_{per} is the amount of water percolating from the unconfined aquifer into the confined aquifer on day i (mm)

W_{pc} is the amount of water removed from the confined aquifer by pumping on day i (mm)

2.5.2 Water Balance in Unconfined Aquifers

Due to short-term input of recharge in unconfined aquifer and the interaction between surface and groundwater flow (base flow), the water balance for unconfined aquifers can be written as

$$W_{su,i} = W_{su,i-1} + R_r - B_f - W_{sd} - W_{per} - W_{pu} \tag{2.4b}$$

where

$W_{su,i}$ is the amount of water stored in the unconfined aquifer on day i (mm)

$W_{su,i-1}$ is the amount of water stored in the unconfined aquifer on day $i-1$ (mm)

R_r is the amount of recharge entering the aquifer on day i (mm)

B_f is the base flow to the main channel on day i (mm)

W_{sd} is the amount of water moving into the soil zone in response to water deficiencies on day i (mm)

W_{per} is the amount of water percolating from the unconfined aquifer into the confined aquifer on day i (mm)

W_{pu} is the amount of water removed from the unconfined aquifer by pumping on day i (mm)

2.5.3 Water Balance in Unsaturated Zone

The incoming terms of the unsaturated zone are infiltration, capillary rise from the saturated zone, and lateral inflow. The outgoing terms consist of evapotranspiration, baseflow, and percolation to the saturated zone. The water balance setup for a selected time interval is

$$\Delta S_{unsat} = W_{inf} + W_{cap} + I_g - W_{per} - E_t - B_f \tag{2.4c}$$

where

ΔS_{unsat} is the storage change in unsaturated zone for the reflected time interval (mm)

W_{inf} is the amount of recharge entering the zone by infiltration (mm)

W_{cap} is the amount of recharge entering the zone by capillary rise (mm)

The other terms were described before.

2.6   COMPRESSIBILITY AND EFFECTIVE STRESS

The change in volume or strain, induced in a material under an applied stress, can be described by compressibility. It is the inverse of the modulus of elasticity, which is the ratio of the change in stress, $d\sigma$, to the resulting change in strain, $d\varepsilon$. Therefore, compressibility can be defined as $d\varepsilon/d\sigma$. In aquifer systems, compressibility is considered for both the water and the porous media, which are explained in the following.

2.6.1   COMPRESSIBILITY OF WATER

If water is subjected to an increase in pressure, dp, a decrease in the volume of water, V_w, occurs. The compressibility of water, β, can then be expressed as

$$\beta = \frac{-dV_w/V_w}{dp} \tag{2.5}$$

The negative sign guarantees obtaining a positive number for the compressibility of water. This equation expresses a linear elastic relationship between the volumetric strain, dV_w/V_w, and the stress caused by the change in fluid pressure. In groundwater hydrology, the ratio shown in Equation 2.5 does not change over the range of fluid pressures. On the other hand, the impact of temperature on β is so small that can be ignored. Hence, β is considered as a constant in groundwater hydrology. Equation 2.5 can also be rewritten as

$$\beta = \frac{d\rho/\rho}{dp} \tag{2.6}$$

where ρ is the water density. The equation of state for water can then be obtained by integrating Equation 2.6, which gives

$$\rho = \rho_0 \exp\left[\beta(p - p_0)\right] \tag{2.7}$$

where ρ_0 is the water density at the datum pressure, p_0. If p_0 is considered equal to the atmospheric pressure, Equation 2.7 can be written as

$$\rho = \rho_0 e^{\beta p} \tag{2.8}$$

For a given fluid, if $\beta = 0$ and $\rho = \rho_0 =$ constant, the fluid is considered incompressible.

2.6.2   EFFECTIVE STRESS

If a unit mass of saturated sand is subjected to a stress, three mechanisms may cause a reduction in the volume of this saturated sand. These mechanisms are

1. Compression of the water in the pores
2. Compression of the sand grains
3. Rearrangement of the sand grains and formation of a more closely packed configuration

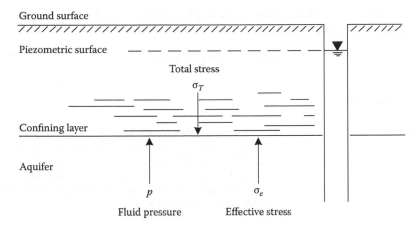

FIGURE 2.5 Total stress, effective stress, and fluid pressure on an arbitrary plane through a saturated porous medium.

The water compressibility controls the first mechanism and the second mechanism is almost negligible. Therefore, the third mechanism needs to be defined to obtain a compressibility term that reflects it. In this regard, the principle of effective stress, proposed by Terzaghi (1925), is required to be delineated first.

In Figure 2.5, consider the stress equilibrium on the confining layer. σ_T is the total stress, due to the weight of overlying rock and water, being imposed downward to the confining layer. This stress is held up by the soil grains and water pressure in the porous medium. Effective stress, σ_e, refers to that portion of the total stress that is held by the granular skeleton, i.e., the grains of the porous medium. The relation between the total and effective stresses is

$$\sigma_T = \sigma_e + p \tag{2.9}$$

or

$$d\sigma_T = d\sigma_e + dp \tag{2.10}$$

According to the fact that the weight of rock and water overlying each point in an aquifer system often remains constant through time, there is usually no change in total stress, $d\sigma_T = 0$. Therefore,

$$d\sigma_e = -dp \tag{2.11}$$

Equation 2.11 implies that an increase in the water pressure causes an equal amount of decrease in the effective stress and vice versa. Therefore, if the total stress remains constant, the effective stress and, in consequence, the resulting volumetric deformations are controlled by the water pressure. If z is the elevation head and h is the hydraulic head, $p = \rho g \psi$ and the pressure head, ψ, is $\psi = h - z$. Hence, the hydraulic head controls the changes in the effective stress at a point of interest:

$$d\sigma_e = -\rho g d\psi = -\rho g dh \tag{2.12}$$

Example 2.2

Determine the total stress acting on the top of the aquifer underlying a clayey aquitard, which is 30 m thick, and calculate the effective stress in the aquifer. Assume the bulk density of the aquitard is 2500 kg/m³ and there is an average of 40 m pressure head in it. If pumping causes a reduction of 1.5 m the hydraulic head in the aquifer, what will be the resulting changes in the water pressure and effective stress. Consider the density of water equal to 1000 kg/m³.

Solution

$$\sigma_T = \rho_b g b = 2,500 \times 9.806 \times 30 = 735,450 \, \text{N/m}^2$$

$$\sigma_e = \sigma_T - p \quad \text{and} \quad p = \rho_w g h.$$

Therefore,

$$\sigma_e = 735,450 - (1,000 \times 9.806 \times 40) = 343,210 \, \text{N/m}^2$$

The corresponding change in water pressure and effective stress due to the reduction of the hydraulic head in the aquifer by 1.5 m will be

$$\Delta p = \rho_w g \Delta h = 1,000 \times 9.806 \times (-1.5) = -14,709 \, \text{N/m}^2$$

$$\Delta\sigma_T = \Delta\sigma_e + \Delta p \Rightarrow 0 = \Delta\sigma_e + (-14,709) \Rightarrow \Delta\sigma_e = 14,709 \, \text{N/m}^2$$

2.6.3 COMPRESSIBILITY OF A POROUS MEDIUM

The compressibility of a porous medium can be expressed as

$$\alpha = \frac{-dV_T/V_T}{d\sigma_e} \tag{2.13}$$

where
V_T is the total volume of a soil mass ($V_T = V_S + V_v$)
V_S is the volume of the solids
V_v is the volume of voids

An increase in effective stress, $d\sigma_e$, causes a reduction in the total volume of the soil mass. In general, $dV_T = dV_S + dV_v$; however, the change in the volume of the grains is usually negligible, i.e., $dV_S = 0$, and therefore, $dV_T = dV_v$.

For the determination of soil compressibility, consider a sample of saturated soil as depicted in Figure 2.6a. Imagine a total stress $\sigma_T = L/A$ is applied to the sample

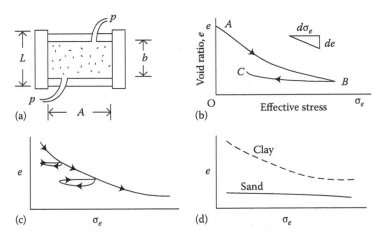

FIGURE 2.6 (a) Laboratory loading cell for the determination of soil compressibility. (b) Change of effective stress relative to void ratio. (c) Impact of repeated loading and unloading on effective stress. (d) Schematic curves of void ratio versus effective stress for clay and sand. (From Freeze, R. and Cherry, J., *Groundwater*, Prentice-Hall, Englewood Cliffs, NJ, 1979.)

through the pistons; while the sample is laterally confined by the cell walls. Water can escape through vents in the pistons to an external pool, which has a constant given pressure. While L is increased step by step, the reduction of the volume of the soil sample is measured. In the initial steps, the increased total stress, $d\sigma_T$, is held by the water under increased pressures. However, since water is drained from the sample, the stress will be transferred from the water to the soil grains. This transient process is termed consolidation. As consolidation is achieved, $dp = 0$ within the sample. In this condition, according to Equation 2.10, $d\sigma_e = d\sigma_T = dL/A$. Assuming that $dV_T = dV_v$, if the soil sample has an initial void ratio e_0 ($e = V_v/V_s$) and height b, Equation 2.13 can be rewritten as

$$\alpha = \frac{-db/b}{d\sigma_e} = \frac{-de/(1+e_0)}{d\sigma_e} \tag{2.14}$$

To determine the compressibility, α, a strain–stress plot is created in the form of e versus σ_e and the slope of this plot specifies α. In Figure 2.6b, the curve AB represents loading (increasing σ_e) and the curve BC corresponds to unloading (decreasing σ_e). The strain–stress relation is not linear or elastic. As shown in Figure 2.6c, for repeated loadings and unloadings, many fine-grained soils have hysteretic properties. Unlike the fluid compressibility β, the soil compressibility, α, is not a constant and is a function of the applied stress. Moreover, the previous loading history influences the soil compressibility as well.

A schematic comparison of the strain–stress curves for clay and sand has been expressed in Figure 2.6d. Since sand has a smaller α, the corresponding curve has a lesser slope. Furthermore, its linearity means that its value remains constant over a wide range of σ_e. Table 2.2 presents compressibility values for various types of geologic materials (Domenico and Mifflin, 1965; Johnson et al., 1968). In this table, values are expressed in SI units of m²/N or Pa⁻¹. As presented in this table, there is the same order of magnitude for the compressibility of water and the compressibility of the less-compressible geologic materials.

It can be concluded that the compressibility of some soils in expansion is much less than in compression (Figure 2.6b and c). The ratio of the soil compressibility in loading and unloading conditions for clays is usually in the order of 10:1; while it approaches 1:1 for uniform sands. If the compressibility value is significantly less in expansion than compression for a soil, volumetric deformations corresponding to increasing effective stress will be irreversible. In other words, this soil will not be, recovered when effective stresses consequently decrease. For example, if large compactions occur in a clay aquitard, it will not be recovered; but the small deformations in the sand aquifers are considerably recovered.

TABLE 2.2
Values of Compressibility for Different Soil Types

Soil Type	Compressibility, α (m²/N or Pa⁻¹)
Clay	10^{-6}–10^{-8}
Sand	10^{-7}–10^{-9}
Gravel	10^{-8}–10^{-10}
Jointed rock	10^{-8}–10^{-10}
Sound rock	10^{-9}–10^{-11}

2.6.4 EFFECTIVE STRESS IN THE UNSATURATED ZONE

From the Equation 2.12, it can be inferred that there is a linear relationship between the effective stress, σ_e, and the pressure head, ψ. However, this relation can be applied only in the saturated zone and not in the unsaturated zone (Narasimhan, 1975). Bishop and Blight (1963) suggest that in the unsaturated zone Equation 2.12 should be modified as

$$d\sigma_e = -\rho g \chi d\psi \qquad (2.15)$$

where the parameter χ depends on the degree of saturation, the soil structure, and the wetting-drying history of the soil. If $\psi > 0$, $\chi = 1$ and if $\psi < 0$, $\chi \leq 1$. For $\psi \ll 0$, $\chi = 0$. However, Narasimhan (1975) claimed that for $\psi < 0$, $\chi = 0$, $d\sigma_e = d\sigma_T$ and changes in pressure head in the unsaturated zone will not cause changes in effective stress. Equation 2.13 is yet useful to define the compressibility of a porous medium in the unsaturated zone.

2.7 AQUIFER COMPRESSIBILITY

Equation 2.14 and Figures 2.5 and 2.6 define compressibility in one dimension, which means it has been assumed that the vertical direction is the only direction that the soils and rocks are stressed. Equation 2.14 can be applied to an actual aquifer problem, if b is considered as the aquifer thickness rather than a sample height. In this condition, the parameter α expresses the vertical compressibility. α can be measured in a laboratory device (like the one in Figure 2.6a), while the soil cores are arranged to be vertical and loading is applied at right angles to any horizontal bedding. However, α usually varies from point to point in a horizontal direction

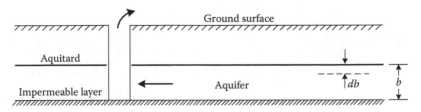

FIGURE 2.7   Land subsidence caused by groundwater pumping.

within an aquifer, since it is unlikely to find a completely homogeneous aquifer. The concepts of homogeneity and heterogeneity will be discussed in Chapter 3. However, since the changes of horizontal stresses are very small and negligible, the aquifer compressibility is usually considered as an isotropic parameter, which means that it does not change in all directions. It should be noted that the only direction that large changes in effective stress are expected is the vertical direction.

Consider the aquifer shown in Figure 2.7 has the thickness b. If the hydraulic head in the aquifer is reduces by an amount $(-dh)$ while the weight of overlying material remains constant, the increase in effective stress $(d\sigma_e)$ can be obtained by Equation 2.12 as $\rho g dh$. The aquifer compaction can then be calculated as

$$db = -\alpha b d\sigma_e = \alpha b \rho g dh \qquad (2.16)$$

In this equation, the minus sign shows that the decrease in head results in a reduction in the thickness b.

If groundwater is pumped out from an aquifer, the hydraulic head will be decreased. In other words, pumping stimulates horizontal hydraulic gradients toward the well in the aquifer. Therefore, the hydraulic head is reduced at the points closed to the well and in consequence, there will be an increase in effective stresses at those points which will result in the compaction of the aquifer. In contrast, there will be an increase in the hydraulic head, a decrease in effective stress, and in consequence aquifer expansion due to an aquifer recharge (pumping water into an aquifer). If the compaction of an aquifer–aquitard system is expanded to the ground surface, it will lead to land subsidence. Therefore, in the operation of well systems, the amount of discharge from an aquifer needs to be carefully determined to avoid undesired consequences.

2.8   AQUIFER CHARACTERISTICS

Some important characteristics of aquifers are porosity and void ratio, specific retention, specific yield, storage coefficient, specific storage coefficient, and safe yield. Specific yield is usually used in unconfined aquifer and specific storage is used in confined aquifers. These parameters are explained as follows:

2.8.1   Porosity and Void Ratio

The total volume of a porous medium, V_T, involves the volume of the solid portion, V_S, and the volume of the voids, V_v. The ratio of volume of voids over the total

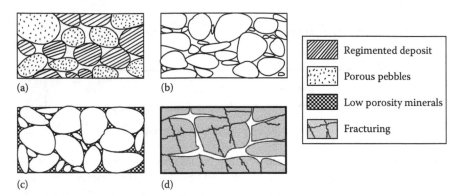

FIGURE 2.8 (a) Well-graded sedimentary deposit, with porous pebbles. (b) Poorly graded sedimentary deposit. (c) Well-graded sedimentary deposit with mineral filling. (d) Rock rendered porous by solution and fracturing.

volume gives the porosity, n, of the medium ($n = V_v/V_T$), which parameter is in the form of a decimal fraction or percentage.

The porosity of the medium in different rock and soil textures are shown in Figure 2.8. Well-graded sedimentary deposit with porous pebbles that has high porosity is depicted in Figure 2.8a. Poorly graded sedimentary deposit is shown in Figure 2.8b. Figure 2.8c shows well-graded sedimentary deposit containing mineral filling that has a very low porosity. Figure 2.8d shows rock rendered porous by solution and fracturing with high porosity.

Consider a cylinder full of gravel and its cross section at the elevation z. The porosity at elevation z, $n(z)$, in this cylinder can be calculated as

$$n(z) = \frac{A_P}{A} \tag{2.17}$$

where
A_P is the cross section of the porous medium (the area of the pores)
A is the area of the cylinder

Therefore, the porosity in the whole cylinder can be obtained as

$$n = \frac{1}{h}\int_0^h n(z)dz = \frac{1}{Ah}\int_0^h An(z)dz = \frac{1}{V_T}\int_0^h A_P(z)dz = \frac{V_V}{V_T} \tag{2.18}$$

and then,

$$n = 100\frac{V_V}{V_T} = 100\left(\frac{V_T - V_s}{V_T}\right) = 100\left(1 - \frac{V_s}{V_T}\right) \tag{2.19}$$

TABLE 2.3
Porosity of Different Geologic Materials (%)

Unconsolidated Deposits	Porosity	Effective Porosity
Gravel	25–40	13–44
Sand	25–40	10–46
Silt	35–50	10–39
Clay	40–70	10–18
Fractured basalt	5–50	—
Karst limestone	5–50	—
Sandstone	5–30	2–41
Limestone, dolomite	0–20	10–24
Shale	0–10	—
Fractured crystalline rock	0–10	—
Dense crystalline rock	0–5	—

$$n = 100\left[1 - \frac{\rho_b}{\rho_d}\right] \qquad (2.20)$$

where
V_V is the volume of pores
V_T is the volume of cylinder
V_s is the volume of solids
ρ_b is the bulk density (about 1400–1900 kg/m³)
ρ_d is the particle density (about 2650 kg/m³)

Table 2.3 presents the porosity ranges for various geologic materials (Davis, 1969). According to this table, gravels, sands, and silts, which are formed by angular and rounded particles, have lower porosities than soils made up of platy clay minerals.

Example 2.3

Given a porosity of 0.22 and a dry bulk density of 1.95 g/cm³, what is the particle density?

Solution

$$\rho_d = \frac{\rho_b}{(1-n)} = \frac{1.95}{1-0.22} = 2.50 \text{ g/cm}^3$$

Example 2.4

Consider the wet bulk density, ρ_{bw}, which includes the mass of water, or $\rho_{bw} = m/v = m_s + m_w/v$, where m is the total mass, including the mass of water, m_w, and the mass of solid, m_s, and V is the total, or bulk, volume. Given a wet bulk density, ρ_{bw}, of 2.25 g/cm³ and a dry bulk density, ρ_d, of 1.95 g/cm³, and assuming the material as completely saturated with water, what is the porosity?

Solution

$$\rho_{bw} = \frac{m_s + m_w}{V} = \frac{m_s}{V} + \frac{m_w}{V} = \rho_d + \frac{m_w}{V}$$

$$m_w = \rho_w V_w$$

$$\text{Assuming } V_v = V_w \Rightarrow \frac{m_w}{V} = \frac{\rho_w V_v}{V} = \rho_w n$$

$$\text{Therefore,} \quad \rho_{bw} = \rho_d + n\rho_w \Rightarrow$$

$$n = \frac{\rho_{bw} - \rho_d}{\rho_w} = \frac{2.25 - 1.95}{0.998} = 0.3 \; (\rho_w = 0.998 \text{ g/cm}^3 \text{ at } 20°C)$$

In studying porosity of a medium, another term, widely used in soil mechanics, is void ratio, e, which is closely related to porosity. The void ratio can be defined as $e = V_v/V_s$, and usually ranges from 0 to 3. The relation between e and n can be expressed as

$$e = \frac{n}{1-n} \quad \text{or} \quad n = \frac{e}{1+e} \tag{2.21}$$

Effective porosity ($\hat{n}$) is another concept of porosity that can be defined as the volume of the continuously connected path for water movement through the soil. In other words, it is the ratio of the continuous fissures over total volume of the medium:

$$\hat{n} = \frac{V_{wm}}{V_T} \tag{2.22}$$

where V_{wm} is the volume of the path of water movement.

2.8.2 SPECIFIC YIELD IN UNCONFINED AQUIFERS

Specific yield is the volume of water that is released from storage in an unconfined aquifer per unit surface area of the aquifer per unit decline in the water table. To determine this coefficient, a given volume of the aquifer is extracted and put on a mesh surface. The ratio of the volume of water discharged from the sample, considering the gravity, over total volume of the sample is its specific yield.

$$S_y = \frac{dV}{Adh} \qquad (2.23)$$

where
  S_y is the specific yield
  dV is the volume of withdrawn water
  h is the level of water in the aquifer
  A is the aquifer area

2.8.3 SPECIFIC RETENTION

Specific retention is described as

$$S_r = n - \hat{n} \qquad (2.24)$$

where
  S_r is the specific retention
  n is the porosity
  $\hat{n}$ is the effective porosity, which is the same as the specific yield for all practical purposes

Different soil particles have different yield and retention. For example, clay has a good retention and low yield.

Therefore, the summation of specific yield and specific retention of an aquifer gives the porosity of this aquifer:

$$n = S_y + S_r \qquad (2.25)$$

Example 2.5

An aquifer with an area of 7 km² experiences a head drop of 0.85 m after 8 years of pumping. If the pumping rate is 5.5 m³/day, determine the specific yield of the aquifer:

Solution

$$S_y = \frac{V_w}{A\Delta h} = \frac{Q\Delta t}{A\Delta h} = \frac{5.5\,\text{m}^3/\text{day} \times 8\,\text{year} \times 365\,\text{day/year}}{7.5\,\text{km}^2 \times 10^6\,\text{m}^2/\text{km}^2 \times 0.85\,\text{m}} = \frac{20,440}{5.525 \times 10^6} = 2.7 \times 10^{-3}$$

Example 2.6

A soil sample has a volume of 180 cm³. The volume of voids in the sample is estimated equal to 67 cm³. Out of the volume of voids, water can move through only 45 cm³. Determine the porosity, specific porosity, specific retention, and specific yield of the soil. What is the area of the aquifer, which the sample was taken from, if pumping at rate 6.0 m³/day causes 1.0 m head drop in the aquifer in 5 years?

Solution

$$n = \frac{V_v}{V_T} = \frac{67}{180} = 0.37$$

$$\hat{n} = \frac{V_{wm}}{V_T} = \frac{45}{180} = 0.25$$

$$S_r = n - \hat{n} = 0.37 - 0.25 = 0.12$$

$$S_y = n - S_r = 0.37 - 0.12 = 0.25$$

$$A = \frac{V_w}{S_y \Delta h} = \frac{Q \Delta t}{A \Delta h} = \frac{6.0 \, \text{m}^3/\text{day} \times 5 \, \text{year} \times 365 \, \text{day/year}}{0.25 \times 1.00 \, \text{m}} = \frac{10,950 \, \text{m}^3}{0.25 \, \text{m}} = 43,800 \, \text{m}^2$$

2.8.4 STORAGE COEFFICIENT AND SPECIFIC STORAGE

By discharging water from, or recharging water to, an aquifer, the storage volume within the aquifer will be changed. This storage volume can be determined for unconfined aquifers as the product of the volume of aquifer, between the beginning and the end of a certain period of time, and the average specific yield of the aquifer. Nonetheless, changes in pressure produce only small changes in storage volume in confined aquifers. When the water is pumped out from an aquifer through a well, the hydrostatic pressure will be reduced and subsequent increases in the aquifer load, which causes compression of the aquifer. Besides, lowering of the pressure results in a small expansion and, in consequence, release of water. The storage coefficient can be used to express the water-yielding capacity of an aquifer.

A storage coefficient, which is also called storativity, is defined as the volume of water released from, or taken into storage of, an aquifer per unit surface area of aquifer over the unit change in the head normal to that surface. Given the value of the storage coefficient, specific storage can be determined by dividing it by unit thickness of the aquifer. Figure 2.9a shows a vertical column of unit area extending through a confined aquifer. The storage coefficient, S, in this aquifer is equal to the volume of water released from the aquifer if the piezometric surface drops a unit distance. This coefficient is dimensionless and expresses the volume of water per volume of aquifer. The common range of storage coefficient in confined aquifers is from 0.00005 to 0.005. Due to this wide range, it can be inferred that there must be a large pressure changes over extensive areas to create substantial water yields.

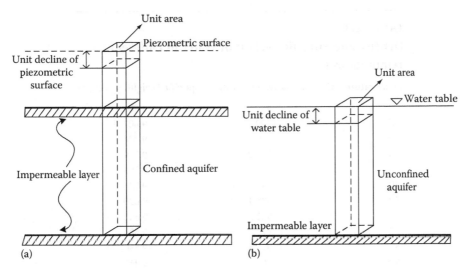

FIGURE 2.9 A schematic for defining storage coefficients of (a) confined and (b) unconfined aquifers.

To determine storage coefficients, pumping tests of wells are usually applied. The relationship between the storage, S, and the saturated aquifer thickness, b, can be expressed as

$$S = 3 \times 10^{-6} b \tag{2.26}$$

where b is in meters.

In an unconfined aquifer, the storage coefficient corresponds to the aquifer's specific yield (Figure 2.9b). Jacob (1940) defined this coefficient as

$$S = \gamma_w b(\alpha + n\beta) \tag{2.27}$$

where
   S is the storage coefficient
   γ_w is the specific weight of water (Table 2.4 presents the specific weight of water in different temperatures)
   b is the thickness of the layer
   α is the compressibility of the bed particles (inverse of the soil particles elasticity module, $\alpha = 1/E_s$)
   β is the compressibility of water (inverse of water elasticity module, $\beta = 1/E_w$)
   n is the porosity of the medium

Table 2.5 presents the elasticity modules of some soils. The water elasticity module is about 2.1×10^9 N/m².

TABLE 2.4

Density and Specific Weight of Water in Different Temperatures

Temperature (°C)	Density, ρ (kg/m³)	Specific Weight, γ_w (kN/m³)
0	999.9	9.806
5	1000	9.807
10	999.7	9.804
20	998.2	9.789
30	995.7	9.765
40	992.2	9.731
50	988.1	9.690
60	983.2	9.642
70	977.8	9.589
80	971.8	9.530
90	965.3	9.467
100	958.4	9.399

Example 2.7

The thickness of the unsaturated layer in a field is 8.2 m. The porosity of the soil was measured equal to 0.28. The soil type is clay with elasticity module of 5.3 × 10⁷. Determine the corresponding storage coefficient for this layer. Assume the temperature is 20°C.

Solution

First compressibility of water and the soil layer are calculated as

TABLE 2.5

Elasticity Modules of Some Soils

Soil Type	E_s (N/m²) × 10⁴
Peat	10–50
Clay	50–1,470
Sand	980–7,850
Gravel	9,806–19,620
Rock	14,710–294,200

$$\beta = \frac{1}{E_w} = \frac{1}{2.1 \times 10^9} = 4.76 \times 10^{-10}\, m^2/N$$

$$\alpha = \frac{1}{E_s} = \frac{1}{5.3 \times 10^7} = 1.89 \times 10^{-8}\, m^2/N$$

$$S = \gamma_w b(\alpha + n\beta) = 9789\,N/m^3 \times 8.2\,m \times [1.89 \times 10^{-8}\, m^2/N + 0.28 \times 4.76 \times 10^{-10}\, m^2/N]$$

$$\Rightarrow S = 1.53 \times 10^{-3}$$

It should be noted that there is a difference between storage coefficient and specific storage, S_S. The difference is that S is the storage coefficient per unit thickness of confined aquifer; while the specific storage, S_S, of a saturated aquifer is the volume of water that a unit volume of aquifer releases from storage due to a unit drop in hydraulic head. Recall that a decrease in hydraulic head h implies a decrease in fluid pressure p and an increase in effective stress σ_e. There are two mechanisms that cause the release of water from storage due to the decrease of hydraulic head. These mechanisms are

1. Increasing σ_e, which results in the compactions of the aquifer. This mechanism controlled by the aquifer compressibility, α.
2. Decreasing p, which causes the expansion of the water. This mechanism controlled by the fluid compressibility β.

In the case of the compaction of the aquifer, the volume of water released from the unit volume of aquifer during compaction and the reduction in volume of the unit volume of aquifer will be equal. The volumetric reduction, dV_T, will be negative, while the amount of water produced dV_w will be positive. Therefore, from Equation 2.13

$$dV_w = -dV_T = \alpha V_T d\sigma_e \qquad (2.28)$$

For a unit volume, $V_T = 1$, if a unit drop in hydraulic head, $dh = -1$, occurs, considering Equation 2.12, $d\sigma_e = -\rho g dh$,

$$dV_w = \alpha \rho g \qquad (2.29)$$

From Equation 2.5, the volume of water produced by the expansion of the water is

$$dV_w = -\beta V_w dp \qquad (2.30)$$

The product of the porosity and the total volume (nV_T) gives the volume of water, V_w. If $V_T = 1$, $dh = -1$, and $dp = \rho g d\psi = \rho g d(h - z) = \rho g dh$, Equation 2.30 can be rewritten as

$$dV_w = \beta n \rho g \qquad (2.31)$$

The specific storage, S_s, can then be calculated by summing up the two terms obtained by Equations 2.30 and 2.31:

$$S_s = \rho g(\alpha + n\beta) \qquad (2.32)$$

Another way to calculate the specific storage coefficient, presented by De Wiest (1965), is

$$S_s = \gamma_w[(1-n)\alpha + n\beta] \qquad (2.33)$$

The dimension of S_s is $[L]^{-1}$. Given the value of S_s, the storage coefficient for confined aquifers can be calculated as

$$S = bS_S \tag{2.34}$$

For unconfined aquifers, the concept of specific yield could be expressed as the difference between storage coefficient and specific storage times the aquifer thickness. In unconfined aquifer, the storage will be reduced by ΔbS or S_y; therefore S and $S_s b$ are not the same:

$$S = S_y + bS_S \tag{2.35}$$

Comparing Equation 2.35 with 2.34, it is perceived that specific yield is so small in confined aquifers that it can be considered negligible. The storage coefficient for confined aquifers is usually equal to or less than 0.005. This value for unconfined aquifers is almost between 0.2 and 0.3.

Example 2.8

The porosity of a confined aquifer is 42%. The thickness of this aquifer is 78 m. The compressibility of water and granular matrix are 4.76×10^{-10} and 6.3×10^{-8}, respectively. Estimate the specific storage and storativity of the aquifer. Also, calculate how much water is released if there is a total head drop of 100 m and the area is 1200 km².

Solution

$$S_s = \rho g(\alpha + n\beta) = (998.2\,kg/m^3) \times (9.81\ m/s^2) \times [(4.76 \times 10^{-10}\ m^2/N)$$

$$+ 0.42 \times (6.3 \times 10^{-8}\ m^2/N)]$$

$$S_s = 2.64 \times 10^{-4}\ (1/m)$$

$$S = b \cdot S_s = 100\,(m) \times 2.64 \times 10^{-4}\,(1/m) = 2.64 \times 10^{-2}$$

The volume of water withdrawn from the area due to the head drop in hydraulic head is

$$V = S\Delta hA = 2.64 \times 10^{-2} \times 100\,m \times 1200\,km^2 \times 10^6 m^2/km^2 = 3.168 \times 10^9 m^3$$

2.8.5 SAFE YIELD OF AQUIFERS

Yield of an aquifer is defined as the rate at which water can be withdrawn without depleting the supply to such an extent that withdrawal a rate higher than safe yield is no longer economically feasible. This concept is applied to the entire aquifer. In other words, the safe yield is the limit to the quantity of water which can be regulatory withdrawn without depletion of aquifer storage reserve.

2.9 STORAGE IN THE UNSATURATED ZONE

In the unsaturated zone, downward (recharge) and upward (evaporation and transpiration) vertical flow, movement and transportation of pollutants from ground surface, and horizontal flow in the capillary zone between the water table and ground surface are important. The amount of water held in partially saturated media is termed as the volumetric water. Figure 2.10a represents a typical distribution of water content above a water table. The capillary zone will be shifted downward by lowering the water table. The volume of water drained from above the water table is shown by the hatched area in Figure 2.10a. Vertical percolation causes this water to be released and in consequence, specific yield becomes an asymptotic function of time (Figure 2.10b). The volumetric water content, θ, which is one of the main properties of unsaturated media, can be calculated as

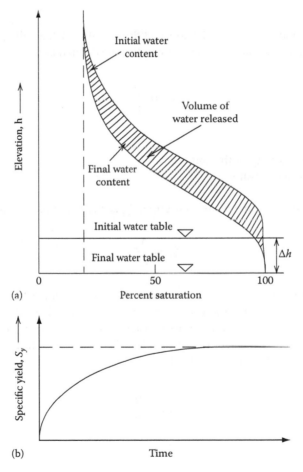

FIGURE 2.10 Movement of water in the zone of aeration by lowering the water table. (a) Water content above the water table. (b) Specific yield as a function of the time of drainage. (From Todd, D.K. and Mays, L.M., *Groundwater Hydrology*, John Wiley & Sons, Inc., New York, 2005.)

$$\theta = \frac{V_w}{V_T} \tag{2.36}$$

To measure this parameter in laboratory, first place the soil is placed and capped tight in a can, to avoid evaporation, and is weighed. Then, it is dried in a forced-draft oven for 10 h or in a convection oven for 24 h. After drying the soil, it is put into a desiccating jar, with desiccant, until the soil cools. The dried and cooled sample is weighed again and the water content, θ_w, is calculated as (Wierenga et al., 1993):

$$\theta_w = \frac{W_{wet\text{-}soil}}{W_{dry\text{-}soil}} - 1 \tag{2.37}$$

where $W_{wet\text{-}soil}$ and $W_{dry\text{-}soil}$ are the wet and dry weights of the soil sample, respectively. The volumetric water content, θ, can then be defined as

$$\theta = \theta_w \frac{\rho_b}{\rho_w} \tag{2.38}$$

where
  ρ_b is the bulk density of the soil
  ρ_w is the density of water

In unsaturated zone, the water saturation, S_w, is a function of the volume of voids:

$$S_w = \frac{V_w}{V_{void}} \tag{2.39}$$

where V_{void} is the volume of the void space in the sample volume V_T.
  In the unsaturated zone, the water content of the sample is usually less than the volume of void space, i.e., in unsaturated condition, θ is less than porosity $(0 < \theta < n)$, and the water saturation is less than 1 $(0 < S_w < 1)$. The water saturation is generally the ratio of porosity over water content:

$$S_w = \frac{\theta}{n} \tag{2.40}$$

Example 2.9

The total volume of a soil sample is 280 cm³. The volume of water and volume of voids were measured equal to 65 and 115 cm³, respectively. Determine the porosity, water content and water saturation of this sample.

Solution

$$n = \frac{V_v}{V_T} = \frac{115}{280} = 0.41 = 41\%$$

$$\theta = \frac{V_w}{V_T} = \frac{65}{280} = 0.23 = 23\%$$

$$S_w = \frac{0.23}{0.41} = 0.56 = 56\%$$

The pressure head, $\psi = h - z$, is negative in the unsaturated zone, positive in the saturated zone and it is zero right at the water table. The reason that water contained in partially saturated soils cannot flow into a borehole is the negative pressure head in the unsaturated zone.

Since air and water fluid phases are present both together in the unsaturated zone, flow in this zone is complicated. Both the volumetric moisture content and the unsaturated hydraulic conductivity depend on the pressure head, or the capillary pressure. The pressure head–moisture content relationship describes how a sample reflects to addition or removal of water. There is a relationship between hydraulic conductivity and pressure head in the unsaturated zone. As a soil dries out, an increasingly large negative pressure head means that the mostly air-filled system has a large resistance to flow passage.

The relationship between the capillary pressure and water content in the unsaturated zone for a particular soil is provided by the soil water characteristic curve. There are three soil water characteristic curves:

- *Water retention or θ_ψ curves*: shows the relationship between soil water content and soil water pressure potential
- *Relative hydraulic conductivity or K_ψ curves*: depicts the relationship between relative hydraulic conductivity and soil water pressure potential
- *Moisture capacity or C_ψ curves*: illustrates the relationship between soil moisture capacity and soil water pressure potential. Soil moisture capacity is defined as the change in moisture content divided by the change in pressure head

For more details please see Schwartz and Zhang (2003).

2.10 WATER-LEVEL FLUCTUATIONS

Changes in groundwater storage due to pumping and recharge result in changes in the water table and the piezometric surface elevations. Besides pumping, other factors such as barometric pressure changes, ocean tides, and use of ground water by plants also influence water levels. The details of water table fluctuations due to pumping and recharge are explained in Chapter 4. For other water table fluctuation factors refer to McWhorter and Sunada (1977).

2.11 GROUNDWATER IN KARST

2.11.1 KARST AQUIFER

"The term *karst* represents terranes with complex geological features and specific hydrogeological characteristics. The karst terranes are composed of soluble rocks, including limestone, dolomite, gypsum, halite, and conglomerates" (Milanovic, 2004); i.e., carbonate rocks (limestone and dolomite) and all their varieties and conglomerates with carbonate matrix represent karst.

Limestones are mostly composed of the mineral calcium carbonate ($CaCO_3$). Dolomites are composed of the mineral dolomite, which is a dual carbonate salt of calcium and magnesium. The chemical composition of dolomite is expressed as $CaMg(CO_3)_2$.

Karstification occurs by the chemical, and sometimes mechanical, action of water in a region of limestone, dolomite, or gypsum bedrock. Therefore, in loess (a sediment formed by the accumulation of wind-blown silt and some sand and clay), clay, lava deposits, and effusive rocks, the voids developed by erosive action of water instead of solution processes cannot be considered as karst.

A nonhomogeneous underground reservoir containing networks of interconnected cracks, caverns, and channels collecting water is called karst aquifer.

2.11.2 TYPES OF KARST

There are different classifications of karst according to its morphological features, structural factors, geographical position, and depositional environment of carbonate rocks, etc. One of the first classifications of karst was introduced by Cvijic (1926) based on the morphological features. He considered three types of karst as

1. *Holokarst* (complete karst) develops in areas included entirely of soluble carbonate rocks. It is characterized by the vast, bare, and rocky land, without arable land and with or without the presence of vegetation.
2. *Merokarst* (incomplete karst) has many properties of nonkarst regions. The karst phenomena in these regions are infrequent and karstifiction occurs in lower depths. Carbonate sediments are covered with arable soil and with vegetation. Merokarsts are also called covered karsts.
3. *The transitional type*: The degree of karstification in this type of karst is between holokarst and merokarst. The transitional type is usually formed in limestone isolated by impermeable and less soluble sediments.

Quinlan (1972) classified karsts based on the soils or sediments covering them. Different types of karst based on this classification are

1. *Subsoil karst*: covered with residual soil
2. *Mantled karmic*: covered with a thin layer of post-karst sediments
3. *Buried karst*: are also called paleokarst and covered by a relatively thick layer of post-karst sediments
4. *Interstratal karst*: covered by pre-karst rocks or sediment

2.11.3 FLUCTUATION OF KARST AQUIFER

One of the distinctive features of karst aquifers is their capability of being filled and emptied by water very fast. This ability is because of the large dimensions of karst channels, their good interconnections, high water-level gradients, and the high permeability of surface zones. In other words, aquifers are formed and drained very fast in karsts. The aquifer reacts quickly to high precipitation in the rainy season. The difference between maximum and minimum levels of a water table can be great in wet seasons.

Milanovic (2004) considered two distinguished periods with different characteristics of fluctuation for karst terrains:

- *The wet season*: with continuous vertical changes in the aquifers water table.
- *The dry season*: with constant lowering of the aquifers water table. This season is also called the recession or depletion period of the aquifer.

2.11.4 RECHARGE OF KARST AQUIFER

There are different categories for karst aquifer recharge. If the karst area recharges itself, it is called autogenic recharge. Allogenic recharge occurs when the adjacent non-karst areas recharge the karst aquifer. Infiltration into shallow holes is called point recharge. The water may also be diffused into fissures in the rock through overlaying soils. Allogenic recharge sometimes infiltrates via shallow holes, while autogenic recharge is often more diffusive. Point recharge via shallow holes is a major pathway for contaminants (Goldsheider and Drew, 2007). Figure 2.11 shows a schematic of karst aquifers and the allogenic and autogenic recharge areas.

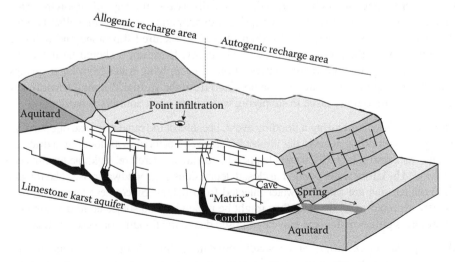

FIGURE 2.11 Heterogeneous karst aquifer system. (From Goldsheider, N. and Drew, D., *Methods in Karst Hydrology*, Taylor & Francis Group, London, U.K., 2007; Hansch, C. and Leo, A., *Substitute Constants for Correlation Analysis in Chemistry and Biology*, John Wiley & Sons, Inc., New York, 1979.)

2.11.5 GROUNDWATER TRACING IN KARST

Tracers are referred to as substances that are added to water to specify its spatial and temporal distribution. They might be soluble or insoluble in water. The most important characteristic of tracers is their sensitivity to accurate concentration detection. Some common tracing technologies are

Dye tracer: Which is the most common technique for investigation of underground links in karst. Na-fluorescein, rhodamine B, rhodamine WT, eozine FE, and salt (NaCl) are some common dye tracers.

Radioactive isotopes: The best advantage of these tracers is that they can be simply detected in a very diluted solution. The most common isotopes for investigation of groundwater in karst are Bromine-82, Iodine-131, Chromium-51, and Tritium.

Proactive isotopes: In this technique, the inorganic salts are used as tracers. Since the contamination by these isotopes is negligible, this technique is recommended for water flows that are used as the source of potable water.

Smoke and gaseous tracers: These tracers are used to evaluate the porosity characteristics of the aeration zone above the aquifer. In this method, smoke is injected to the void above the aquifer and by tracing it; the interconnected karst voids are specified.

For more details of these methods Milanovic (2004) is recommended.

2.11.6 WATER RESOURCES PROBLEMS IN KARST

The main water resources problems in karst zones can be summarized to

Water supply in karst: In karst zones, rapid infiltration causes absence or deficit of surface water to meet supplying demands. Generally, two main ways of supplying water in karst zones are from wells and/or springs. In supplying water from wells, the most important issue is to select the location of wells. Fractured limestone and dolomite aquifers provide the most acceptable water quality. However, it should be noted that wells must be drilled on fracture intersections. In supplying water from karst springs, usually, good quality of water can be obtained from the springs draining fracture systems. The main quality issue in supplying water from karst aquifers is hardness.

Flood flows in karst: During a flooding event, because of the rapid response of open conduit systems, flooding in caves has almost the same characteristics as flooding in drainage basins. As a result, high discharges happen in conduits. Therefore, sinkhole terrains can be flooded by either water ponded in closed depressions or groundwater rise. The flooded sinkhole terrains are filled by water injected from bellow. Since these terrains are not usually expected to be flooded, they might be urbanized and subjected to development. Therefore, significant damage may be imposed due to the flooding of these terrains.

Water pollution in karst: Since water movement through open conduits in karst zones is almost the same as surface water flows, its filtering by soil layers is negligible. Therefore, the pollution of water in open conduits is as susceptible as surface waters. In addition, the calcium, magnesium, and bicarbonate ions existing in karst aquifers, due to the solution of rocks, increase the hardness and alkalinity of water in these aquifers. Fecal material from septic tanks, leachate from landfills, agricultural chemicals, pesticides, herbicides, etc. are some major pollutants that reach ground-water in karst fields without any significant dilution by filtering through soil layers.

Water-quality monitoring in karst: As is discussed in Chapter 9, monitoring networks must be installed under or adjacent to main sources of pollution, such as landfills, to detect pollutants before they spread in the underlying aquifer or move off-site. The monitoring network may be designed for unsaturated or saturated zone or both of them. Because of the heterogeneous distribution of permeability in karst aquifers as well as rapid movement of contaminants through these conduits, there might not be any trace of contaminants in monitoring wells. Therefore, monitoring systems in karst aquifers differ from regular aquifers. The strategy is to excavate trenches to bedrock and inject tracers into them with water flush. Meanwhile, the spring(s) to which the aquifer is drained are determined and these springs are monitored. The monitoring is especially important after flood events.

PROBLEMS

2.1 The porosity and volumetric water content of a sample are 0.40 and 0.18, respectively. Calculate the water saturation of the sample.

2.2 An undisturbed core sample is obtained from an unsaturated aquifer 1.0 m above the water table. The sample is 0.150 m high and 0.050 m in diameter. A dry sample of the aquifer has a specific gravity of 2.65 g/cm³. The weight of the sample is 630 g before drying and 570 g after drying. Calculate the porosity, volumetric water content, and bulk density of aquifer. Assume that the volume of voids is 450 cm³.

2.3 An unconfined aquifer with a specific yield of 0.20 is used as a water supply for irrigation of farm lands. The groundwater is pumped out of this aquifer with the rate of 4.2 m³/day. The area of the aquifer is 1500 m². How long does it take to have one meter drop in water table in this aquifer?

2.4 How much water can be removed from an unconfined aquifer with a specific yield of 0.18 when the water table is lowered 1.0 m? How much water can be removed from a confined aquifer with storage coefficient of 0.0005 when the piezometric surface is lowered 1.0 m? Express your answers in m³/km².

2.5 Determine the total stress acting top on an aquifer, which is underlain a 20 m thick aquitard. The bulk density of the aquitard is 2100 kg/m³. What is the effective stress in the aquifer if the pressure head is 25 m?

2.6 A sample of silty sand has a volume of 215 cm³. It has weight of 514.7 g. After saturating the sample is weighed 594.2 g. The sample is then drained by gravity until it reaches a constant weight of 483.4 g. Finally, the sample is dried in oven for ten hours and it reaches the weight 452.1 g. Assuming the density of water is 1 g/cm³, compute the following:

(a) Water content of the sample.
(b) Volumetric water content of the sample.
(c) Saturation ratio of the sample.
(d) Porosity

    (e) Specific yield
    (f) Specific retention
    (g) Dry bulk density

2.7 The porosity of the unsaturated zone in a field is 0.35. The elasticity module of the medium is 9.5×10^6. Determine the storage coefficient for this layer if the thickness of the unsaturated layer is 10 m.

2.8 What is the specific yield of a soil sample that has the total volume of 250 cm³, void volume of 150 cm³ and flow volume of 100 cm³?

2.9 The average water table elevation has dropped 1.5 m due to the removal of 85 million cubic meters from an unconfined aquifer over an area of 200 km². Determine the specific yield for the aquifer. What is the specific retention of the aquifer if its porosity is 0.42?

2.10 Resolve Example 2.8 for an unconfined aquifer. Assume that specific yield is 2.5%.

2.11 Records of the average rate of precipitation over an area show 680 mm/year rainfall, out of which, 250 mm/year flows overland. Evaluation of time series of the data shows that 300 mm/year is evapotranspirated. The area is irrigated with the rate of 150 mm/year. In this area, 170 mm/year flows through rivers and the rate of outflow to the other basins is 100 mm/year. Determine the influence from seepage. Also, calculate the rate of baseflow. Assume groundwater inflow and outflow remain unchanged and the groundwater is not pumped out.

2.12 Calculate the density of a soil sample if its porosity and a dry bulk density are 0.28 and 1.65 g/cm³, respectively.

2.13 Between bulk density (ρ_b) or particle density (ρ_d), which would be expected to be greater in magnitude? Prove your answer?

2.14 The dry and wet bulk densities of a soil sample are measured equal to 1.65 g/cm³ and 1.98 g/cm³, respectively. Calculate the porosity of the sample. Assume the sample is completely saturated with water.

2.15 Determine the specific yield of an aquifer. The water is pumped with the rate of 6.75 m³/day. There is a 1.2 m drop in water table after 7 years and the area of the aquifer is 8.2×10^6 ha (10000 m²).

REFERENCES

Bishop, A.W. and Blight, G.E. 1963. Some aspects of effective stress in saturated and partly saturated soils. *Geotechnique*, 13, 177–197.

Cvijic, J. 1926. *Geomorfologija (Morfologie Terrestre)*. Tome Secondo, Beograd, Yogoslovia.

Davis, S.N. 1969. Porosity and permeability of natural materials, in *Flow through Porous Media*, ed. R.J.M. De Wiest, Academic Press, New York, pp. 54–89.

De Wiest, R.J.M. 1965. *Geohydrology*. John Wiley & Sons, New York.

Dingman, S.L. 1994. *Physical Hydrology*. Prentice-Hall, Inc., Englewood Cliffs, NJ.

Domenico, P.A. and Mifflin, M.D. 1965. Water from low-permeability sediments and land subsidence. *Water Resources Research*, 1, 563–576.

Goldsheider, N. and Drew, D. 2007. *Methods in Karst Hydrology*. Taylor & Francis Group, London, U.K.

Hansch, C. and Leo, A. 1979. *Substitute Constants for Correlation Analysis in Chemistry and Biology*. John Wiley & Sons Inc., New York.

Jacob, C.E. 1940. On the flow of water in an elastic artesian aquifer. *Transaction. American Geophysical Union*, 2, 574–586.

Johnson, A.I., Moston, R.P., and Morris, D.A. 1968. *Physical and Hydrologic Properties of Water Bearing Deposits in Subsiding Areas in Central California*, U.S. Geological Survey, Professional Paper 497A.

McWhorter, D.B. and Sunada, D.K. 1977. *Ground-Water Hydrology and Hydraulics*. Water Resources Publications, Littleton, Colorado 290 p.

Milanovic, P.T. 2004. *Water Resources Engineering in Karst*. CRC Press LLC, Boca Raton, FL.

Narasimhan, T.N. 1975. A unified numerical model for saturated-unsaturated groundwater flow, PhD dissertation, University of California, Berkeley, CA.

Quinlan, J.F. 1972. Karst related mineral deposits and possible criteria for recognition of paleokarst: A review of preservable characteristics of Holocene and older karstterranes, *24th IGC*, Section 6, Montreal, Canada.

Schwartz, F.W. and Zhang, H. 2003. *Fundamentals of Groundwater*. John Wiley & Sons, Inc., New York.

Terzaghi, K. 1925. *Erdbaumechanic auf bodenphysikalischer grundlage*. Franz Deuticke, Vienna, Austria.

Todd, D.K. 1980. *Groundwater Hydrology*. John Wiley & Sons Inc., New York.

Todd, D.K. and Mays, L.W. 2005. *Groundwater Hydrology*. John Wiley & Sons Inc., New York.

White, W.B. 1988. *Geomorphology and Hydrology of Karst Terrains*. Oxford University Press, New York.

Wierenga, P.J., Young, M.H., Gee, G.W., Hills, R.G., Kincaid, C.T., Nicholson, T.J., and Cady, R.E. 1993. *Soil Characterization Methods for Unsaturated Low Level Waste Sites*, U.S. Nuclear Regulatory Commission, NUREG/CR-5988 (PNL-8480).

3 Groundwater Hydrology

3.1 INTRODUCTION

Hydrology is the study of water and addresses the occurrence, distribution, and movement of water on the Earth. The science of occurrence, distribution, and movement of water under the ground surface is referred to as groundwater hydrology. The hydrologic cycle was introduced in the previous chapter. In hydrologic cycle, groundwater has an extensive role and has been the subject of research for many researchers. It became more important especially during and after the hydraulic mission, the second half of nineteenth century, due to an exponential increase in industrial activities. Industrial improvement created many pollution problems in surface water that caused some of the surface water resources to become unsuitable for certain purposes such as drinking and irrigation. Moreover, population increase and, in consequence, augmentation in need for food production has led to a significant increase in demands of water. Therefore, groundwater is now considered as a clear supplementary source to supply increasing demands of water.

3.2 GROUNDWATER MOVEMENT

The subject of groundwater movement is quite complex. The movement of groundwater will be discussed in this chapter in its simplest forms to illustrate the principles involved. More complex groundwater models will be discussed in the next chapters.

3.2.1 DARCY'S LAW

Darcy's law is the basic empirical relationship governing the movement of groundwater through a porous medium (Bouwer, 1978; De Wiest, 1965; Freeze and Cherry, 1979; Todd, 1980). Imagine water flowing at a rate Q through a cylinder of cross-sectional area A packed with sand and having piezometers at a distance L apart, as shown in Figure 3.1. Henry Darcy in 1856 determined the velocity of water flowing through this cylinder as

$$v = -K \frac{h_1 - h_2}{\Delta l} = Ki \tag{3.1}$$

where
v is the superficial flow velocity, m/s
K is the hydraulic conductivity, m/s
i is the hydraulic gradient, $-dh/dl$, m/m
h_1 is the height of water in the upper piezometer, m
h_2 is the height of water in the lower piezometer, m
Δl is the distance between piezometers, m

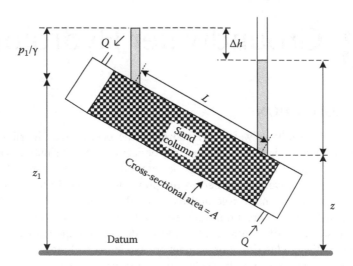

FIGURE 3.1   Pressure distribution and head loss in flow through a sand column.

The minus sign in Darcy's law arises from the fact that the head loss is negative. The discharge through a cross-sectional area of an aquifer is given by

$$\frac{Q}{A} = Ki \Rightarrow Q = KiA \tag{3.2a}$$

or

$$q = Kib \tag{3.2b}$$

where
Q is the hydraulic discharge, m³/s
A is the cross-sectional area, m²
q is the discharge in unit width of the aquifer, m²/s
b is the thickness of the aquifer, m

It should be noted that the application of Equation 3.2 is based on the assumption that the porous medium is homogenous and is saturated with water. The application of Equation 3.2 is illustrated in Example 3.1.

The porosity n has influence on hydraulic conductivity K. In well-sorted sand or fractured rock formations, samples with higher n generally also have higher K. For example, clay-rich soils usually have higher porosities than sandy or gravelly soils but lower hydraulic conductivities.

Using the porosity of the medium and Darcy's law ($Q = Av$), the real velocity of water in a porous medium can be calculated. Regarding the fact that the water moves through the pores in this medium, the value of average velocity (v) in the mentioned relation is not true and the equation should be modified to $Q = A_p v_{real}$. A_p is the area of pores, and v_{real} is the average velocity of water through the pores. Therefore, the average velocity of water through the soil and the average velocity of water through the pores have a direct relation with each other:

$$\begin{aligned} Q &= Av \\ Q &= A_p v_{\text{real}} \end{aligned} \qquad \Rightarrow v = \frac{A_p}{A} v_{\text{real}} \Rightarrow v = n v_{\text{real}} \qquad (3.3)$$

The above relation implies that the velocity in Darcy's law has lower value than the real velocity of water through the soil.

The pure velocity of groundwater is obtained from dividing the superficial flow velocity by the porosity of the porous medium.

Example 3.1

In Figure 3.2, determine the superficial and true pore velocities of flow through the porous medium as well as the time of flow from the upper to the lower reservoir. Assume the following data are applicable:

Cross-sectional area $A = 100\,\text{m}^2$
Head loss $\Delta h = 35\,\text{m}$
Length of flow path $\Delta L = 1500\,\text{m}$
Hydraulic conductivity $K = 7.5 \times 10^{-4}\,\text{m/s}$
Porosity of porous medium $n = 0.45$

Solution

(a) Superficial velocity

$$v = -K\frac{dh}{dl} = -(7.5\times10^{-4}\ \text{m/s})\frac{-35\ \text{m}}{1500\ \text{m}} = 1.75\times10^{-5}\ \text{m/s}$$

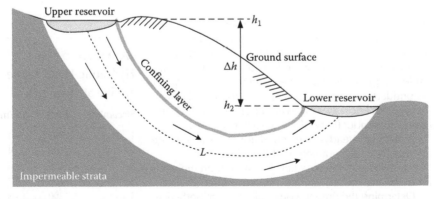

FIGURE 3.2 Two hypothetical reservoirs connected by a subsurface permeable stratum.

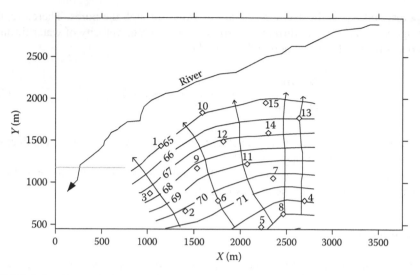

FIGURE 3.3 Contour representation of the potentiometric surface for a confined aquifer being discharged into a river.

(b) Pore velocity

$$v_p = v/n = 1.75 \times 10^{-5}/0.45 = 3.89 \times 10^{-5} \text{ m/s}$$

(c) The time of travel

$$t = \frac{1500 \text{ m}}{3.89 \times 10^{-5} \text{ m/s}} = 3.86 \times 10^7 \text{ s} = 446.4 \text{ days}$$

Example 3.2

Figure 3.3 shows a contour representation of the potentiometric surface for a confined aquifer being discharged into a river. Hydraulic heads in different locations were measured at 15 monitoring wells. Table 3.1 presents these measurements. The aquifer is considered homogeneous and isotropic and has an average thickness of 18 m with a porosity of 0.34 and hydraulic conductivity of 2.6×10^{-4} m/s.

Assuming steady-state conditions, estimate the average hydraulic gradient in the aquifer.

Calculate the rate of water discharge from the aquifer to the stream.

Determine the time it would take for a hypothetical tracer injected in Well #11 to reach the river.

TABLE 3.1
Hydraulic Head Made at Five Streamlines

Streamline	l (m)	Δh (m)
1	900	−4
2	1000	−4
3	1150	−6
4	1300	−7
5	1200	−8

Solution

(a) The hydraulic gradient at each stream line is

Streamline	l (m)	Δh (m)	$\Delta h/l$
1	900	−4	-4.44×10^{-3}
2	1000	−4	-4.00×10^{-3}
3	1150	−6	-5.22×10^{-3}
4	1300	−7	-5.38×10^{-3}
5	1200	−8	-6.67×10^{-3}

The average gradient is therefore: $i = 5.14 \times 10^{-3}$

(b) The rate of water discharge from the aquifer to the stream is

$$q = bKi = 18 \, \text{m} \times 2.6 \times 10^{-4} \, \text{m/s} \times 5.14 \times 10^{-3} \, \text{m/m}$$

$$= 2.41 \times 10^{-5} \frac{\text{m}^3/\text{s}}{\text{m}} = 2.08 \frac{\text{m}^3/\text{day}}{\text{m}}$$

(c) Well #11 is located on line 69. The tracer must pass lines 65, 66, 67, and 68 to reach the river. Therefore, the hydraulic gradient through these lines must be calculated. Regarding Figure 3.3, the coordinates of well 11 are approximately (2050, 1150) and the coordinates of the point that the flow line intersects with line 68 are almost (2050, 1875). Thus, the distance between these two lines is 725 m. The average pore velocity along the flow line intersecting Well # 11 can then be calculated as

$$v_p = -\frac{K}{n} \cdot \frac{\Delta h}{\Delta l} = \frac{2.6 \times 10^{-4}}{0.34} \times \frac{4}{725} = 4.22 \times 10^{-6} \, \text{m/s}$$

To reach the river the tracer should travel approximately 1100 m, which is the distance between well 11 and the intersection point of the flow line and river (2050, 2250). The travel time can thus be calculated as

$$t_{tr} = \frac{d}{v} = \frac{1100 \, \text{m}}{4.22 \times 10^{-6} \, \text{m/s}} \cong 2.61 \times 10^8 \, \text{s} \cong 8.27 \, \text{years}$$

3.2.1.1 Validity of Darcy's Law

In laminar flow, the velocity is proportional to the first power of the hydraulic gradient. Hence, it is acceptable to believe that Darcy's law can be applied to laminar flow in porous media. A reasonable criterion to determine whether the flow is laminar or turbulent is the Reynold's number. Therefore, this number can be considered as a factor to establish the limits of flow described by Darcy's law. Reynold's number is expressed as

$$N_R = \frac{\rho v D}{\mu} \tag{3.4}$$

where

 ρ is the fluid density
 v is the velocity
 D is the diameter (of pipe)
 μ is the viscosity of the fluid

It can be inferred that the above equation has been adopted for flow in pipe. To be able to apply this equation to a porous medium, Darcy velocity and an effective grain size d_{10} are substituted for v and D, respectively.

It has been proven experimentally that Darcy's law is valid for $N_R < 1$ and it is not that far away for up to $N_R = 10$ (Ahmed and Sunada, 1969). Considering most natural groundwater flow occurs with Reynold's number less than 1, Darcy's law is applicable to groundwater movement.

3.2.2 HYDRAULIC HEAD

In a groundwater system in saturated zone, hydraulic head at a point (Figure 3.4) can be expressed as

$$h = \frac{z + p}{\rho g} \tag{3.5}$$

where

 z is the elevation head above a datum plane (m)
 p is the fluid pressure at the point applied by the column of water above the point (Pa)
 ρ is the water density (kg/m³)
 g is the gravity constant (m/s²)
 ρg expresses the specific weight of water (γ)

In groundwater studies, fluid potential and hydraulic head are equivalent; however, hydraulic head is most commonly used. Water can flow from a region of lower pressure to a region of higher pressure if the total head at the starting point is greater than that at the ending point.

In the field of petroleum and natural gas engineering, pressure is generally used in place of head because pressures at great depths are normally so great and elevation heads are often insignificant.

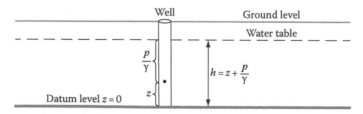

FIGURE 3.4 Illustration of hydraulic head.

3.2.3 HYDRAULIC CONDUCTIVITY

The hydraulic conductivity is defined in two separate parts considering the saturation or unsaturation status of the media.

3.2.3.1 Hydraulic Conductivity in Saturated Media

Hydraulic conductivity, K, contains the properties for both medium and fluid. It can be used for evaluating water transmissivity in a porous medium. Hydraulic conductivity of a saturated media is defined as

$$K_s = k \frac{\gamma}{\mu} \tag{3.6}$$

where
K_s is the saturated hydraulic conductivity, cm/s or ft/day
$k = Cd^2$ is the specific or intrinsic permeability
C is the constant of proportionality, unitless
d is the grain size of porous medium, m
γ is the specific weight of water, kN/m^3
μ is the dynamic viscosity of water, N s/m^2

The physical meaning of hydraulic conductivity is stated as "the volume of liquid flowing perpendicular to a unit area of porous medium per unit time under the influence of a hydraulic gradient of unity." Earlier literature described this phenomenon as field coefficient of permeability with units of gallons per day per square feet. This name and definition are now rarely used. Table 3.2 lists the hydraulic conductivities of several lithologies. It can be also obtained by Table 3.3.

Example 3.3

The saturated hydraulic conductivity of a coarse sand measured in a laboratory at 20°C is 2.74×10^{-4} m/s. What is the intrinsic permeability?

At 20°C: $\rho_w = 998.29$ kg/m^3, $\mu_w = 1.003 \times 10^{-3}$ kg/ms

$$k = \frac{\mu_w}{\rho_w g} K_s = \frac{1.003 \times 10^{-3} \text{ kg/ms}}{998.29 \text{ kg/m}^3 \times 9.81 \text{ m/s}^2} \times 2.74 \times 10^{-4} \text{ m/s}$$

$$= 2.81 \times 10^{-11} \text{ m}^2 = 2.81 \times 10^{-7} \text{ cm}^2$$

3.2.3.2 Hydraulic Conductivity in Unsaturated Media

In unsaturated media, the hydraulic gradient, molecular attraction, and surface tension influence water. Therefore, unsaturated hydraulic conductivity is a function of the pressure head, Ψ (always negative in unsaturated media). The unsaturated hydraulic conductivity is shown as $K(\Psi)$.

The unsaturated hydraulic conductivity is at its maximum when the infiltration capacity is reached. In dry condition, water will not flow any significant distance

TABLE 3.2
Important Physical Properties of Soil and Rock

Lithology	Hydraulic Conductivity (cm/s)	Compressibility, α (m²/N or Pa⁻¹)
Unconsolidated		
Gravel	10^{-2}–10^2	10^{-8}–10^{-10}
Sand	10^{-4}–1.0	10^{-7}–10^{-9}
Sat	10^{-7}–10^{-3}	No data
Clay	10^{-10}–10^{-7}	10^{-6}–10^{-8}
Glacial till	10^{-10}–10^{-4}	10^{-6}–10^{-8}
Indurated fractured basalt	10^{-5}–1.0	10^{-8}–10^{-10}
Karst limestone	10^{-4}–10.0	Not applicable
Sandstone	10^{-8}–10^{-4}	10^{-11}–10^{-10}
Limestone, dolomite	10^{-7}–10^{-4}	$<10^{-10}$
Shale	10^{-11}–10^{-7}	10^{-7}–10^{-8}
Fractured crystalline rock	10^{-7}–10^{-2}	10.0^{-10}
Dense crystalline rock	10^{-12}–10^{-8}	10^{-9}–10^{-11}

Source: Adapted from Domenico, P.A. and Schwartz, F.W., *Physical and Chemical Hydrogeology*, John Wiley & Sons, New York, 1990; Freeze, R.A. and Cherry, J.A., *Groundwater*, Prentice-Hall, Englewood Cliffs, NJ, 1979; Fetler, C.W. *Applied Hydrogeology*, 3rd ed., Mac millan College Publishing Co. Inc., New York, 1994; Narasimhan, T.N. and Goyal, K.E., *Subsidence Due to Geothermal Fluid Withdrawal*, in Man-Induced Land Subsidence, Reviews in Engineering Geology, V. VI, Geological Society of America, 35–66 p., 1984.

unless the water content in the soil is sufficient and the pressure head becomes less negative. Then, the specific discharge is

$$q = -K(\Psi)h\frac{\partial h}{\partial l} \tag{3.7}$$

or

$$q = -K(\theta)h\frac{\partial h}{\partial l} \tag{3.8}$$

where θ is the water content of the soil. There are three methods for computing unsaturated hydraulic conductivity as a function of water content. These methods have been summarized in Rawls et al. (1993). The three methods based on dimensionless equations, which makes them applicable to any set of units.

Hydraulic conductivity in unsaturated zone can be estimated by the formula introduced by Brooks and Corey (1964) as

$$\frac{K(\theta)}{K_s} = \left[\frac{(\theta - \theta_r)}{(n - \theta_r)}\right]^m \tag{3.9}$$

TABLE 3.3

Range of Values of Hydraulic Conductivity and Permeability

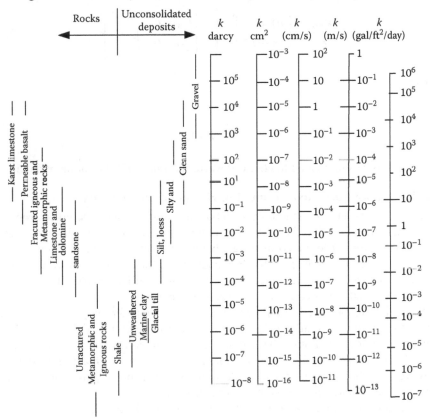

Source: Freeze, R. and Cherry, J., *Groundwater*, Prentice-Hall, Englewood Cliffs, NJ, 604 p., 1979.

where

K_s is the saturated hydraulic conductivity

θ_r is the residual water content

n is the porosity

and

$$m = 2 + 3\lambda \tag{3.10}$$

where λ is the pore-size index (Brooks and Corey, 1964). Equation 3.9 is valid when $(n - \theta_r) > (\theta - \theta_r)$. If $(n - \theta_r) \leq (\theta - \theta_r)$, $K(\theta)/K_s = 1$.

Campbell (1974) presents the following formula for hydraulic conductivity in unsaturated zone:

$$\frac{K(\theta)}{K_s} = \left(\frac{\theta}{n}\right)^m \tag{3.11}$$

and the following formula was presented by Van Genuchten (1980) for hydraulic conductivity in unsaturated zone:

$$\frac{K(\theta)}{K_s} = \left[\frac{(\theta-\theta_r)}{(n-\theta_r)}\right]^{1/2}\left\{1-\left[\left(\frac{(\theta-\theta_r)}{(n-\theta_r)}\right)^{1/m}\right]^m\right\}^2 \tag{3.12}$$

where

$$m = \frac{\lambda}{(\lambda+1)} \tag{3.13}$$

The reader is recommended to check Shwartz and Zhang (2003) for more details.

3.2.3.3 Laboratory Measurement of Hydraulic Conductivity

To measure hydraulic conductivity in laboratory, samples must be taken from the aquifer and transferred to the laboratory in undisturbed condition, which is, in general, possible for consolidated materials but usually not for unconsolidated materials. It is also rarely possible for fissured aquifers. Cylinders or cubes with diameter and length of 25–50 mm are typical for consolidated aquifer samples. To determine permeability in three directions, cubic samples are used.

Since the samples need to be representative of the aquifer, they can be obtained where the stratum of aquifer is horizontal. They can be collected from bore holes that intersect several strata. The samples can also be taken from cliffs or quarry faces. If the stratum is inclined, the samples are taken from bore holes or the outcrop. However, bore hole samples are more desirable, since the outcrop material are likely to be weathered and unrepresentative of the aquifer.

In laboratory, permeameter is used to determine the hydraulic conductivity with application of Darcy's law. There are two types of permeameter: the constant head permeameter (Figure 3.5a) and the falling head permeameter (Figure 3.5b). The constant

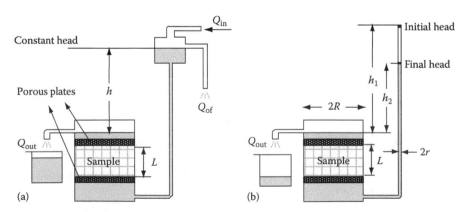

FIGURE 3.5 Permeameters for measuring hydraulic conductivity under: (a) constant head; (b) falling head.

head permeameter is used for noncohesive soils (e.g., sands and gravels). The falling head permeameter is used for materials which have lower hydraulic conductivity.

In the permeameter test, groundwater from the same aquifer must be used. This is because some materials are sensitive to changes in water chemistry, such as clays that may swell or shrink, but this water is in chemical equilibrium with the aquifer material.

To obtain the hydraulic conductivity for the constant head permeameter (Figure 3.5a):

Q_{of} is the rate of overflow. From Darcy's law we have

$$Q = -KA\frac{h_2 - h_1}{L}$$

Therefore

$$K = \frac{QL}{Ah} \qquad (3.14)$$

where
   Q is the flow rate
   A is the cross-sectional area
   L is the length of the sample
   h is the constant head

For the falling head permeameter, the hydraulic conductivity is obtained as following:

$$Q_{in} = -A_t \frac{dh}{dt}$$

$$Q_{out} = KA_c \frac{h}{l}$$

where A_t and A_c are the cross-sectional area of the tube and the container, respectively. In steady state condition, $Q_{in} = Q_{out}$:

$$-A_t \frac{dh}{dt} = KA_c \frac{h}{l}$$

$$\int_{h_1}^{h_2} \frac{dh}{h} = \frac{-K}{L}\frac{A_c}{A_t}\int_{t_1}^{t_2} dt$$

$$\left[\ln h\right]_{h_1}^{h_2} = \frac{-K}{L}\frac{A_c}{A_t}(t_2 - t_1)$$

$$\ln\frac{h_2}{h_1} = \frac{-K}{L}\frac{A_c}{A_t}(t_2 - t_1)$$

$$K = \frac{A_t L}{A_c(t_2 - t)} \ln \frac{h_1}{h_2} \qquad (3.15)$$

where
  L is the length of the sample
  h_1 and h_2 are the heads at the beginning, t_1, and at time t_2 later

3.2.3.4   Field Measurement of Hydraulic Conductivity

Pumping tests are common ways for measuring hydraulic conductivity at field. The hydraulic conductivity is estimated from observations of the water levels near pumping wells. The value of hydraulic conductivity obtained from pumping tests is not as accurate as the punctual information from laboratory tests. However, pumping tests have the advantages that the aquifer is not disturbed. The pumping tests are described in detail in Chapter 4.

Tracer tests can also be another alternative for hydraulic conductivity measurements at field. In tracer tests, a dye such as fluoresce or a salt such as calcium chloride can be used. If there is a drop of water table, h, between the injection well and the observation bore hole that are L meters apart from each other, the pore velocity obtained by Darcy's law can be equated to that obtained by dividing the distance by the travel time between the two bore holes. Rearranging the relationship gives

$$K = \frac{nL^2}{ht} \qquad (3.16)$$

where
  n is the porosity
  t is the travel time between the two bore holes

Since there might be some uncertainties about the flow direction in actual conditions, this test is hard to be accomplished in practice. Besides, if the distance between the bore holes is long, the travel time becomes too long, which makes the test unreasonable.

There are some other commonly used tests for determination of hydraulic conductivity. These tests are the slug test, the auger-hole test, and the piezometer test. In these tests are the rate of rise of the water in a bore hole is measured after the water level has been lowered by removal of water with a bailer or bucket and the hydraulic conductivity is determined based on the this rate. Although these tests give more localized values for the hydraulic conductivity, but they are less expensive to conduct. More detail about these tests can be found in Bouwer (1978).

3.3 HOMOGENEOUS AND ISOTROPIC SYSTEMS

A homogeneous formation has uniform hydraulic conductivity at all points in the aquifer. In contrast, if the hydraulic conductivity varies with location, the formation is called heterogeneous. Figure 3.6 shows three different ways that hydraulic conductivity changes in heterogeneous formations.

In a geologic formation, if at a given point the hydraulic conductivity is not changed in all directions, the formation is isotropic. In anisotropic formations, the hydraulic conductivity varies with direction. The system depicted in Figure 3.7a has the same consolidation in different directions and shows an isotropic formation. In Figure 3.7, the system (b) has further consolidation in vertical direction. This system shows an anisotropic formation.

If the coefficient of permeability is independent of the direction of the velocity, the soil is said to be an isotropic flow medium. Moreover, if the soil has the same coefficient of permeability at all points within the region of flow, the soil is said to be homogeneous and isotropic. If the coefficient of permeability is dependent on the direction of the velocity and if this directional dependence is the same at all points of the flow region, the soil is said to be homogeneous and anisotropic. In homogeneous and anisotropic soils, the coefficient of permeability is dependent on the direction of the velocity but independent of the space coordinates.

Most soils are anisotropic to some degree. Sedimentary soils often exhibit thin alternating layers. In general, in homogeneous natural deposits, the coefficient of permeability in the horizontal direction is greater than that in the vertical (Snow, 1969).

Although idealized aquifers that are homogeneous and isotropic do not exist, for mathematical calculations of the storage and flow of groundwater, the aquifers are assumed to be homogeneous and isotropic. However, these assumptions give acceptable quantitative approximations, particularly in large-scale studies.

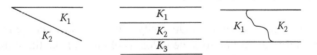

FIGURE 3.6 Hydraulic conductivity formations in three types of heterogeneous systems.

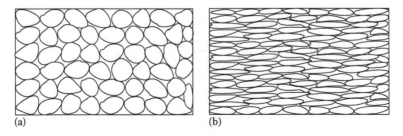

(a) (b)

FIGURE 3.7 (a) Isotropic and (b) anisotropic sediment deposits.

3.3.1 HYDRAULIC CONDUCTIVITY IN MULTILAYER STRUCTURES

In geology, if a structure has various characteristics (soil type, permeability, compressibility, etc.) in different depths, it is called multilayer structures. The hydraulic conductivity of this type of structure differs in vertical and horizontal directions and obtains as follows:

(a) Average horizontal hydraulic conductivity: In this status, the hydraulic gradient is equal in all layers and total discharge is the summation of the discharge of all the layers.

$$i = i_1 = i_2 = \cdots = i_n$$

$$Q = Q_1 + Q_2 + \cdots + Q_n$$

$$K_H iA = K_1 i_1 A_1 + K_2 i_2 A_2 + \cdots + K_n i_n A_n$$

$$K_H ib = K_1 i_1 b_1 + K_2 i_2 b_2 + \cdots + K_n i_n b_n$$

$$K_H = \frac{\sum_{i=1}^{n} K_i b_i}{b} \qquad (3.17)$$

where
K_H is the horizontal hydraulic conductivity
i_i is the hydraulic gradient of the layer
Q_i is the discharge of the layer
A_i is the area of the layer
b_i is the thickness of the layer

(b) Average vertical hydraulic conductivity: The rate of discharge in all layers is equal in this status but, total head loss is obtained from the summation of the head losses of all layers:

$$dh = dh_1 + dh_2 + \cdots + dh_n$$

$$Q = Q_1 = Q_2 = \cdots = Q_n$$

$$\frac{b}{K}Q = \frac{b_1}{K_1}Q_1 + \frac{b_2}{K_2}Q_2 + \cdots + \frac{b_n}{K_n}Q_n$$

$$K_V = \frac{b}{\sum_{i=1}^{n} b_i/K_i} \qquad (3.18)$$

The horizontal hydraulic conductivity in alluvium is normally greater than that in the vertical direction:

$$K_H > K_V \tag{3.19}$$

Imagine a two-layer case. The above status can be proven as

$$\frac{K_1 b_1 + K_2 b_2}{b_1 + b_2} > \frac{b_1 + b_2}{b_1/K_1 + b_2/K_2} \tag{3.20}$$

which reduces to

$$\frac{b_1}{b_2}(K_1 - K_2)^2 > 0 \tag{3.21}$$

Because the left side is always positive, it must be greater than zero, thereby Equation 3.19 is confirmed.

Ratios of K_H/K_V, usually fall in the range of 2–10 for alluvium, but values up to 100 or more occur where clay layers are present. For consolidated geologic materials, anisotropic conditions are governed by the orientation of strata, fractures, solution openings, or other structural conditions, which do not necessarily possess a horizontal alignment.

In applying Darcy's law to a two-dimensional flow in an anisotropic media, the appropriate value of K must be selected for the direction of flow. For diagonal directions, the K value can be obtained from

$$\frac{1}{K_\beta} = \frac{\cos^2 \beta}{K_H} + \frac{\sin^2 \beta}{K_V} \tag{3.22}$$

where K_β is the hydraulic conductivity in the direction making an angle β with the horizontal.

This equation relates the principal conductivity components in horizontal and vertical directions to the resultant K_β in the angular direction β. Assuming $H = r\cos \beta$ and $V = r\sin \beta$, we can rewrite Equation 3.22 as

$$\frac{1}{K_\beta} = \frac{H^2}{K_H} + \frac{V^2}{K_V} \tag{3.23}$$

For cases where flow is at some angle to a geological boundary between layers (Figure 3.8), the tangent refraction law affects the groundwater flow. The extent of refraction of the flow lines will depend on the hydraulic conductivity ratio of the two layers. This ratio is equal to the ratio of the tangents of the angles that the flow lines make with the normal to the boundary, or

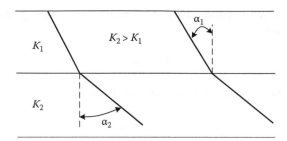

FIGURE 3.8 Diagram of flow line refraction at the boundary between two layers with different permeability.

$$\frac{K_1}{K_2} = \frac{\tan(\alpha_1)}{\tan(\alpha_2)} \tag{3.24}$$

where
K_1 and K_2 are the hydraulic conductivity of layers 1 and 2, respectively
α is the refraction angle

Example 3.4

A constant-head permeameter has a cross-sectional area of 150 cm². At a head of 20 cm, the permeameter discharges 57 cm³ in 10 min through a sample of 40 cm length. Calculate the hydraulic conductivity in cm/s. From the hydraulic conductivity value, name the type of soil. Compute the intrinsic permeability if the test was carried out at 15°C?

Solution

$$K = \frac{VL}{Ath} = \frac{57 \text{ cm}^3 \times 40 \text{ cm}}{150 \text{ cm}^2 \times 360 \text{ s} \times 20 \text{ cm}} = 2.11 \times 10^{-3} \text{ cm/s}$$

$K = 2.11 \times 10^{-3}$ cm/s $\Rightarrow$ using Table 3.3, the soil must be silty sand or clean sand.

at 15°C: $\rho_w = 999.19 \text{ kg/m}^3, \quad \mu_w = 1.139 \times 10^{-3} \text{ kg/ms}$

$$k = \frac{\mu_w}{\rho_w g} K = \frac{1.139 \times 10^{-3} \text{ kg/ms}}{999.19 \text{ kg/m}^3 \times 9.81 \text{ m/s}^2} \times 2.11 \times 10^{-5} \text{ m/s} = 2.45 \times 10^{-12} \text{ m}^2$$

$k = 2.45 \times 10^{-8} \text{ cm}^2$

Example 3.5

In a falling-head test, the initial head at $t = 0$ is 72 cm. At $t = 30$ min, the head is 66 cm. The diameters of the standpipe and the specimen are 1.5 and 25 cm, respectively. The length of the specimen is 25 cm. Calculate the hydraulic conductivity and intrinsic permeability of the specimen.

$$A_t = \frac{\pi \times 1.5^2}{4} = 1.77 \text{ cm}^2$$

$$A_c = \frac{\pi \times 25^2}{4} = 490.86 \text{ cm}^2$$

$$h_0 = h(t = 0) = 72 \text{ cm}$$

$$h_t = h(t = 30 \text{min}) = 66 \text{ cm}$$

$$K = \frac{A_t}{A_c} \frac{L}{t} \ln\left[\frac{h_0}{h_1}\right] = \frac{1.77 \text{ cm}^2}{490.86 \text{ cm}^2} \times \frac{25 \text{ cm}}{30 \text{ min}} \times \ln\left(\frac{72 \text{ cm}}{66 \text{ cm}}\right)$$

$$= 2.61 \times 10^{-4} \text{ cm/min} = 4.35 \times 10^{-6} \text{ cm/s} = 4.35 \times 10^{-8} \text{ m/s}$$

Assuming the experiment was conducted at 20°C:

$$\rho_w = 998.29 \text{ kg/m}^3, \quad \mu_w = 1.003 \times 10^{-3} \text{ kg/ms}$$

$$k = \frac{\mu_w}{\rho_w g} K = \frac{1.003 \times 10^{-3} \text{ kg/ms}}{998.29 \text{ kg/m}^3 \times 9.81 \text{m/s}^2} \times 4.35 \times 10^{-8} \text{ m/s} = 4.46 \times 10^{-15} \text{ m}^2$$

Example 3.6

Five horizontal formations with an equal thickness of 1 m for each of them exists in a hydrogeological system. The hydraulic conductivities of the formations are 40, 20, 30, 100, and 10 cm/day, respectively. Calculate equivalent horizontal and vertical hydraulic conductivities. Also, if the flow in the uppermost layer is at an angle of 30° away from the horizontal direction relative to the boundary, calculate flow directions in all of the formations.

$$d_i = d = 1\text{m}, \quad i = 1...5$$

$$K_h = \frac{\sum_{i=1}^{5} K_i d}{\sum_{i=1}^{5} d} = \frac{\sum_{i=1}^{5} K_i}{5} = \frac{40 + 20 + 30 + 100 + 10}{5} = 40.0 \text{ cm/day}$$

$$K_v = \frac{\sum_{i=1}^{5} d}{\sum_{i=1}^{5} d/K_i} = \frac{5}{\sum_{i=1}^{5} 1/K_i} = \frac{5}{(1/40) + (1/20) + (1/30) + (1/100) + (1/10)} = 22.9 \text{ cm/day}$$

Angle of incidence: α_1 Angle of refraction: α_2

$$\frac{k_i}{k_r} = \frac{\tan\alpha_i}{\tan\alpha_r} \Rightarrow \alpha_i = \tan^{-1}\left(\frac{K_i}{K_r} \tan\alpha_r\right)$$

The angle of refraction in one layer constitutes the angle of incidence to the next one:

$$\frac{20}{40} = \frac{\tan\alpha_i}{\tan(30)} \Rightarrow \alpha_i = \tan^{-1}\left(\frac{20}{40}\tan(30)\right) = 16.1°$$

and considering the above relationship, the direction of all formations are

$$\alpha_1 = 30°, \quad \alpha_2 = 16.1°, \quad \alpha_3 = 23.41°, \quad \alpha_4 = 55.28°, \quad \alpha_5 = 8.21°$$

3.4 TRANSMISSIVITY

Transmissivity is the amount of water that moves horizontally through a unit width of a saturated aquifer in result to a unit change in gradient.

$$T = Kb \tag{3.25}$$

where
K is the hydraulic conductivity
b is the depth of the saturated aquifer

This dimensional characteristic derives from the definition of transmissivity as the flow rate of water passing through a unit width of aquifer perpendicular to flow direction over the entire thickness of the aquifer (Figure 3.9).

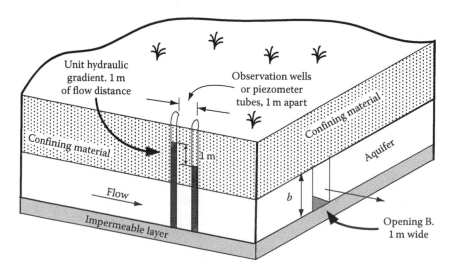

FIGURE 3.9 A schematic for defining transmissivity. (Adapted from Ferris, J.G. et al., *Theory of Aquifer Tests*, U.S. Geological Survey, Water-Supply Paper 1536E, 174p., 1962. With permission.)

Transmissivity is usually reported in units of square feet per day or square meters per day. The total rate of flow (Q) through any area (A) of the aquifer perpendicular to the flow direction under the gradient (i) is then given as

$$Q = WT_i \tag{3.26}$$

where
W is the width of aquifer
i is the hydraulic gradient
T is transmissivity in (m²/s)
Q is the discharge rate in (m³/s)

For multi-layer aquifers, the transmissivity is calculated as follows:

$$T = \sum_{i=1}^{n} T_i \tag{3.27}$$

By multiplying the pertinent K values of hydraulic conductivity from Table 3.2 by the range of reasonable aquifer thicknesses, say 5–100 m, a range of values of T can be calculated. Transmissivities greater than 0.015 m²/s (or 0.16 ft²/s or 100,000 gal/day/ft) represent good aquifers for water well exploitation.

In an unconfined aquifer, the transmissivity is not as well defined as in a confined aquifer. But still Equation 3.25 can be used for unconfined aquifer. However, in this case, b is now the saturated thickness of the aquifer or the height of the water table above the top of the underlying aquitard that bounds the aquifer.

3.5 DUPUIT–FORCHHEIMER THEORY OF FREE-SURFACE FLOW

The movement of groundwater in unconfined systems bounded by a free surface is usually treated as a steady-state flow problem, and a simple Dupuit–Forchheimer analysis is often used. The Dupuit–Forchheimer analysis is based on the assumption that streamlines are virtually horizontal (potential lines are vertical) and that the slope of the hydraulic grade line is the same as the free-surface slope (Bouwer, 1978; De Wiest, 1965; Hansch and Leo, 1979; Todd, 1980). Although the model is for unconfined aquifers, with suitable assumptions it can be extended to confined aquifers.

Example 3.7

In Figure 3.10, the irrigation water is being infiltrated with uniform intensity into the ground. If the intensity of infiltration is f, suggest a relationship for calculating the highest level of the water table.

Solution
Using the Dupuit–Forchheimer assumptions and Darcy's law, the velocity of water in x-direction at the highest point of the water table is

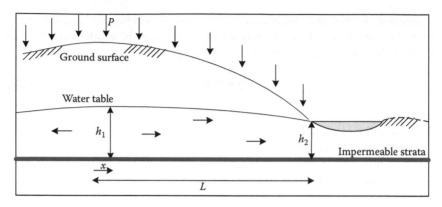

FIGURE 3.10 Definition sketch for Example 3.7.

$$v_x = -K\frac{dh}{dx}$$

dh/dx shows the slope of water table. The rate of discharge in unit width at distance x is

$$q_x = -Kh\frac{dh}{dx}$$

On the other hand, the rate of discharge due to the infiltration, f_x, is

$$f_x = -Kh\frac{dh}{dx} \Rightarrow -f_x dx = Khdh$$

Integrating the above equation from the point of peak (P) to the stream flow,

$$K(h_1^2 - h_2^2) = PL^2$$

So,

$$h_1 = \sqrt{h_2^2 + \frac{PL^2}{K}}$$

Example 3.8

A wastewater treatment plant is supposed to release its effluent over land. This waste water should be infiltrated into the land to avoid overland flow. The slope is 1% and the soil type is sandy clay with hydraulic conductivity of 4.2 m/day. The depth of permeable layer is 10 m (Figure 3.11). If the effluent rate is 3 cm/day, determine the length at which the effluent must be released to avoid overland flow?

Solution

The maximum rate of groundwater flow occurs when the soil layer between the permeable strata and ground surface is totally saturated. In this condition, Using

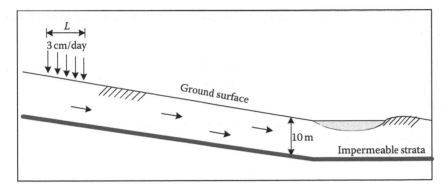

FIGURE 3.11 Definition sketch for Example 3.8.

the Dupuit–Forchheimer assumptions and Darcy's law, the rate of discharge in unit width is

$$q = Kh\frac{dh}{dx} = 4.2 \times 10 \times 0.01 = 0.42\ \text{m}^2/\text{day}$$

The effluent rate is 3 cm/day, which is released over the land in length L. Hence, the rate of discharge in unit width, q, is $0.03L$ (m²/day). The length required to release the wastewater in order to avoid overland flow is

$$0.03L = 0.42 \Rightarrow L = \frac{0.42}{0.03} = 14\ \text{m}$$

3.5.1 GROUNDWATER FLOW IN UNSATURATED ZONE

In the unsaturated zone, the Darcy equation is written as

$$q = -K(\psi)\nabla h \qquad (3.28)$$

where
  q is the Darcy velocity vector [L/T]
  $K(\psi)$ is the hydraulic conductivity tensor [L/T]
  h is the hydraulic head [L]

The hydraulic-conductivity tensor in the unsaturated zone is expressed as

$$K(\psi) = K_r(\psi)K_s \qquad (3.29)$$

where
  K_s is the saturated hydraulic conductivity tensor
  $K_r(\psi)$ is the relative hydraulic conductivity
  $K_r(\psi)$ is dimensionless and varies between 0 and 1

The linear groundwater velocity in the unsaturated zone is related to the Darcy velocity by

$$v = \frac{q}{\theta} = \frac{q}{sn}$$

(3.30)

where
v is the linear groundwater velocity or groundwater pore velocity vector
s is the saturation
n is the porosity

Example 3.9

Under steady-state infiltration, the average saturation of soil is 0.80 in an aquifer having a porosity of 0.42. The Darcy velocity in the unsaturated zone is 3.75 cm/h. If the thickness of the unsaturated zone is 18 m, calculate the time required for a drop of water at the ground surface to travel to the water table.

Solution

$$v = \frac{3.75\,\text{cm/h}}{(0.42)(0.80)} = 11.16\ \text{cm/h}$$

$$t = \frac{1800\ \text{cm}}{11.16\ \text{cm/h}} = 161.29\,\text{h} = 6.72\ \text{days}$$

3.6   FLOWNETS

One of the simplest procedures to solve a flow equation is a graphical approach. In graphical approach, a unique set of streamlines and equipotential lines that describe flow within a domain are sketched. The streamlines (or flow lines) indicate the path through which water moves in the aquifer. The equipotential lines represent the contours of equal head in the aquifer intersect the streamlines. In this section, flownets in isotropic and homogeneous media as well as heterogeneous and anisotropic media are discussed and evaluated.

3.6.1   ISOTROPIC AND HOMOGENEOUS MEDIA

For steady-state conditions, the two-dimensional groundwater flow equation for an isotropic and homogeneous media is described by the Laplace equation as

$$\frac{\partial^2 h}{\partial x^2} + \frac{\partial^2 h}{\partial y^2} = 0$$

(3.31)

Laplace equation will be discussed in detail in Chapter 4. A flownet is a graphical solution to the above equation. It depicts a selected number of flowline and equipotential lines in the flow system. A quantitative and graphical solution to the groundwater flow problem is obtained from the flownet. A flownet may give distribution

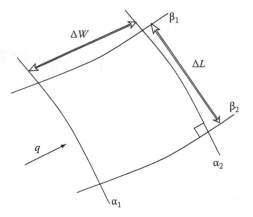

FIGURE 3.12 A flownet element of two-dimensional flow.

of heads, discharges, areas of high (or low) velocities, and the general flow pattern. The flow channel between adjacent streamlines is called a streamtube. The equipotential lines are perpendicular to the streamtubes (Figure 3.12). The discharge in the streamtube per unit width perpendicular to the plane of the figure is

$$\Delta Q = \alpha_2 - \alpha_1 = \Delta\alpha \tag{3.32}$$

$$q = \frac{\Delta Q}{\Delta W} = \frac{\Delta\alpha}{\Delta W} \tag{3.33}$$

Darcy's law gives

$$q = -\frac{K\Delta h}{\Delta L} = \frac{\beta_2 - \beta_1}{\Delta L} = \frac{\Delta\beta}{\Delta L} \tag{3.34}$$

Also, taking the continuity into account:

$$q = \frac{\Delta\alpha}{\Delta W} = \frac{\Delta\beta}{\Delta L} \tag{3.35}$$

In the solution of a flow problem by flownet construction, the first step is to draw the two-dimensional flow domain in a way that all boundary locations, wells, etc. are considered relative to one another. A trial and error procedure is used to sketch a flownet.

To construct a flownet for an isotropic and homogeneous system, we need to sketch the streamlines and equipotential lines. Equipotential lines express the hydraulic head in the domain. These lines are perpendicular to the equipotential lines. The same hydraulic head drops between the equipotential lines form curvilinear squares formed by the streamlines and equipotential lines. These squares are described as the curved side squares that are tangent to an inscribed circle (Figure 3.13).

Assuming no inflow to/outflow from the region in the internal part of the net, the same quantity of ground water flows between adjacent pairs of flow lines. In addition, the hydraulic head drop between two adjacent equipotential lines is the same.

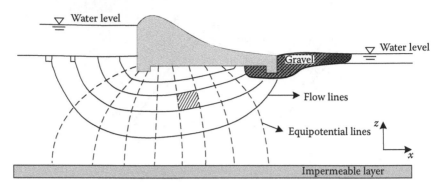

FIGURE 3.13   A flownet showing seepage under a dam.

Figure 3.13 shows a flownet depicting seepage under a dam in an x-z plane. In this figure, the solid lines show stream lines and dashed lines express the equipotential lines. Flow occurs because the hydraulic head of water in the pool above the dam is higher than in the pool below the dam. The flownet is a theoretical representation of flow beneath the dam. The bottom surface of the reservoir is considered as an equipotential line, which has a constant head boundary across which the flow is directed downward.

The base of the dam is a flow line. The pool below the dam receives discharge and provides another constant-head boundary or another equipotential line. Because the system is isotropic and homogeneous, flow lines and equipotential lines intersect to form curvilinear squares.

Some strategies for sketching a flownet suggested by Bennett (1962) as follows:

1. Use only four or five flow channels in a first attempt at sketching.
2. Observe the appearance of the entire flownet; do not try to adjust details until the entire net is approximately correct.
3. Take it into account that frequently parts of a flownet consist of straight and parallel lines, which result in uniformly sized true squares.
4. In a flow system that has symmetry, only a section of the net needs to be constructed because the other parts are images of that section.
5. During the sketching of the net, consider that the size of the rectangle changes gradually; all transitions are smooth, and where the paths are curved they are of elliptical or parabolic shape.

It is very useful to keep several simple flow cases in mind before you begin studying complex flow systems. In preparing a flownet, the following rules must be taken into account:

1. A no-flow boundary is a streamline.
2. If there is no recharge or evapotranspiration, the water table is a streamline. When there is recharge, the water table is neither a flow line nor an equipotential line.

3. Streamlines end at extraction wells, drains, and gaining streams and they start from injection wells and losing streams.
4. Streamlines divide a flow system into two symmetric parts.
5. Streamlines begin and end at the water table in areas of ground-water recharge and discharge, respectively.

Using Darcy's equation for a flow channel in the two-dimensional space,

$$\Delta Q = T\Delta h \frac{\Delta W}{\Delta L} \tag{3.36}$$

where
ΔQ is the flow through a streamtube
T is the transmissivity of the aquifer
Δh is the hydraulic-head drop between two equipotential lines
ΔW is the distance between two streamlines
ΔL is the distance between two equipotential lines (Figure 3.12)

In sketching a flownet, we always try to provide a mesh of squares. Therefore, ΔW is equal to ΔL and $\Delta W/\Delta L$ is equal to 1. If there are a total number of n_f flow channels and n_d hydraulic-head drops (i.e., equipotential lines), Darcy's equation is written as

$$Q = \frac{n_f}{n_d} T\Delta H \tag{3.37}$$

or

$$q = \frac{n_f}{n_d} K\Delta H \tag{3.38}$$

where
Q is the total flow rate and is equal to $n_f \Delta Q$
ΔH is the total hydraulic head drop and is equal to $n_d \Delta h$

Example 3.10

The groundwater is withdrawn from an aquifer with rate 20,000 m³/day. Due to this withdrawal, the hydraulic head drops at four equipotential lines. After sketching the flownet for this formation, it is determined that the groundwater is flowing through eight tubes. Assuming the transmissivity is 700 m²/day, determine the total hydraulic head drop.

Solution

$$\Delta H = \frac{n_d Q}{T n_f} = \frac{(4)(20,000 \text{ m}^3/\text{day})}{(700 \text{ m}^2/\text{day})(8)} = 14.28 \text{ m}$$

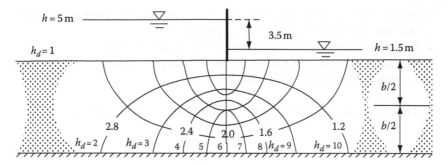

FIGURE 3.14   Flownet of seepage under a sheet pile.

Example 3.11

Estimate the discharge under the sheet piling shown in Figure 3.14 using the flownet method. Assume that the sheet piling penetrates to 1/2 of the aquifer thickness $K = 3.2 \times 10.2$ cm/s.

Solution

The flownet construction is started with a vertical equipotential line, directly below the sheet piling. The upper streamline is located right in the vicinity of the sheet piling. Since it is assumed that the Darcy velocity will decrease with depth, the spacing between the streamlines is increased with depth. The equipotential line immediately adjacent to the vertical one is sketched next. In drawing these equipotential lines, the condition of orthogonality as well as the condition that each element forms a square must be taken into account. The process of drawing streamlines and equipotential lines is continued until we get to the vicinity of the terminal point. Since sketching a flownet is based on trial and error, some readjustments of both the streamlines and the equipotential lines are usually required.

From Figure 3.14, The number of streamtubes is 5 and the number of equipotential drops is 10. Thus, the discharge per meter of sheet pile length is

$$Q = K \frac{n_f}{n_d} \Delta H = \frac{5}{10} \times 3.2 \times 10^{-4} \text{ m/day} \times 3.5 \text{ m} = 5.6 \times 10^{-4} \text{ m}^2/\text{s}$$

3.6.2 HETEROGENEOUS MEDIA

In a heterogeneous media, the equipotential lines and the flow lines do not necessarily intersect to form squares. A constant discharge in a stream tube for two adjacent media with differing hydraulic conductivity is estimated as

$$\Delta Q = T_1 \Delta h_1 \frac{\Delta W_1}{\Delta L_1} = T_2 \Delta h_2 \frac{\Delta W_2}{\Delta L_2} \tag{3.39}$$

If $\Delta h_1 = \Delta h_2$:

$$\frac{T_1}{T_2} = \frac{\Delta L_1 \Delta W_2}{\Delta L_2 \Delta W_1} \tag{3.40}$$

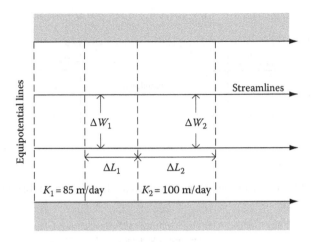

FIGURE 3.15 Definition sketch for Example 3.12.

Example 3.12

Calculate the flow rate and the length ΔL_2 for the heterogeneous aquifer shown in Figure 3.15:

Assume that $\Delta W_1 = \Delta W_2 = 8\,\text{m}$ and $\Delta L_1 = 10\,\text{m}$ and that the thickness of the aquifer is 150 m.

Solution

$$\Delta Q = T_1 \Delta h_1 \frac{\Delta W_1}{\Delta L_1} = (85\ \text{m/day})(150\ \text{m})(1\ \text{m})\frac{8\ \text{m}}{10\ \text{m}} = 10,200\ \text{m}^3/\text{day}$$

The segment length ΔL_2 is

$$\Delta L_2 = \frac{T_2}{T_1}\Delta L_1 = \frac{(100\ \text{m/day})(150\ \text{m})}{(85\ \text{m})(150\ \text{m})}(10\ \text{m}) = (11.76\ \text{m})$$

3.6.3 ANISOTROPIC MEDIA

In anisotropic media, there are not right angles at the interception of streamlines and equipotential lines except when flow is aligned with one of the principal directions of hydraulic conductivity or transitivity. Therefore, in an anisotropy medium, the flownet may be sketched by transforming the flow field. This procedure can be summarized as follows (Shwartz and Zhang, 2003):

1. Determine the direction of maximum transmissivity in the x-direction and the direction of minimum transmissivity in the y-direction.
2. Multiply the dimension in the x-direction by a factor of $(K_y/K_x)^{1/2}$. Sketch the flownet in the transformed flow domain.
3. Project the flownet back to the original dimension by dividing the x-coordinates of the flownet by a factor of $(K_y/K_x)^{1/2}$.

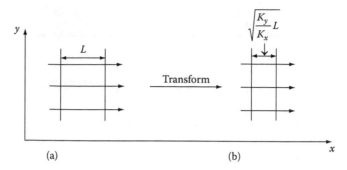

FIGURE 3.16 Flownet in anisotropic media (a) original cross section; (b) transformed cross section.

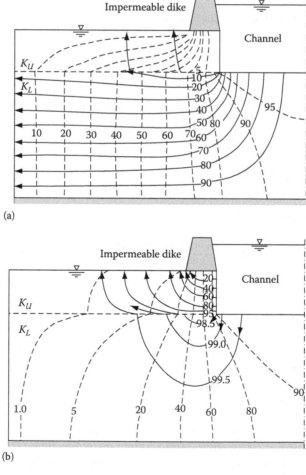

FIGURE 3.17 Flownets for seepage through two different anisotropic two-layer systems: (a) $K_U/K_L = 1/50$; (b) $K_U/K_L = 50$. (From Todd, D.K. and Mays, L.M., *Groundwater Hydrology*, John Wiley & Sons Inc., New York, 2005.)

Figure 3.16 shows the flownets in original and transformed coordinate systems. As depicted in this figure, in the transformed system, the flownet consists of squares while in the original system, it consists of rectangles. Flownets for seepage from one side of a channel through two different anisotropic two-layer systems are depicted in Figure 3.17.

3.7 STATISTICAL METHODS IN GROUNDWATER HYDROLOGY

Groundwater flows, the rate of discharges from and recharges to aquifers, and groundwater quality parameters are probabilistic time series that must be analyzed and interpreted by statistical time series analyses. A time series contains a sequence of measurements (e.g., daily water withdrawal from a well). A statistical sample contains measurements that were taken at irregular time intervals. For example, the sample may only contain data for a summer season. The simplest analyses assume that there is no correlation between individual members of the time series. In this case it does not matter whether the data collected are a sample or a time series. All measurements that constitute to a time series or a sample of a measured parameter are fitted to a probability distribution function. The most simple probability distribution function is the normal distribution, which defines the symmetric distribution of the individual members of the time series around the series mean value.

Normal distribution includes negative values, whereas the great majority of real-time series of statistical samples contains only positive numbers as well as zeros. In other words, the probability function is defined from 0 to ∞ and not from −∞ to +∞. Some series are skewed and not symmetrical. This is especially true for water quality and hydrologic data. Therefore, the time series data is transformed to their logarithmic values, $x = \log X$, which will yield the log-normal distribution. This transformation results in obtaining a probability distribution from zero to infinity and also resolves the problem of skewness of the data. There are some other asymmetrical probability distributions for analyzing the hydrologic data such as log Pearson type III distribution and Gumbel distribution. Normal and lognormal distributions are explained later in this section.

3.7.1 NORMAL DISTRIBUTION

The normal distribution is a bell-shaped, symmetric distribution. It is described mathematically by its probability density function

$$f(x) = \frac{1}{\sigma\sqrt{2\pi}} \exp\left[-\frac{1}{2\sigma^2}(x-\mu^2)\right], \quad -\infty < x < \infty, -\infty < \mu < \infty, \sigma > 0 \quad (3.41)$$

where $f(x)$ is the height (ordinate) of the curve at the value x. The density function is completely specified by two parameters, μ and σ^2, which are the mean and variance, respectively, of the distribution. The notation $N(\mu, \sigma^2)$ is used to denote a normal probability density function (in short, normal distribution) with mean μ and variance σ^2. The series mean is defined as

$$\mu = \sum_{1}^{N} \frac{X_i}{N} \qquad (3.42)$$

where
  X_i is an individual measurement of the time series
  N is the number of members in the time series or in the sample data

The variance is then defined as

$$\sigma^2 = \sum_{1}^{N} \frac{(X_i - \mu)^2}{(N-1)} \qquad (3.43)$$

The standard deviation is the square root of the variance, $s = \sqrt{\sigma^2}$.

Figure 3.18 shows two normal distributions. The solid curve is $N(0,1)$ and the dashed curve is $N(2,1.5)$. There is a different normal distribution for each combination of μ and σ^2. However, they can all be transformed to the $N(0,1)$ distribution by the transformation:

$$Z = \frac{X - \mu}{\sigma} \qquad (3.44)$$

Z is called a standard normal deviate.

The density function $f(x)$ and the cumulative distribution function (CDT) for $aN(\mu,\sigma^2)$ distribution are depicted in Figure 3.19. The CDT is the probability that the random variable X will take on a value less than or equal to a specified value x and is denoted by $F(x)$:

$$F(x) = \text{prob}[X \le x] \qquad (3.45)$$

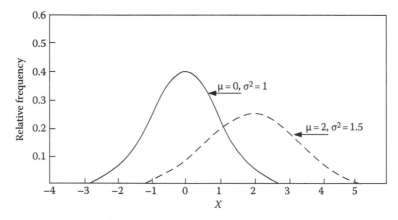

FIGURE 3.18   Two normal distributions: $N(0,1)$ and $N(2,1.5)$.

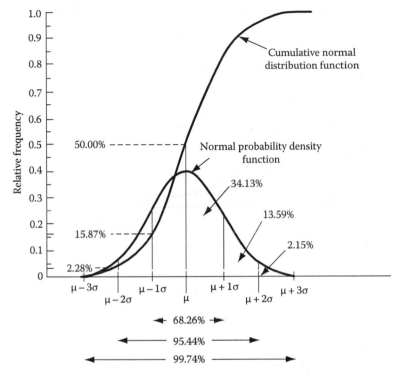

FIGURE 3.19 Areas under the normal probability density function and the cumulative normal distribution function. (From Sokal, R.R. and Rohlf, R.J., *Biometry*, W.H. Freeman, San Francisco, CA, 133 p., 1981.)

In other words, $F(x)$ gives the cumulative percentage of the normal density function that lies between $-\infty$ and the point x on the abscissa (Gilbert, 1987).

As shown in Figure 3.19, 2.15% of the density function lies above/under $\mu \pm 2\sigma^2$. Also, 95.44% of the density function lies between $\mu - 2\sigma^2$ and $\mu + 2\sigma^2$.

3.7.2 LOGNORMAL DISTRIBUTION

The lognormal distribution is used to model many kinds of hydrological and environmental data. Two-, three-, and four-parameter lognormal distributions can be defined. The two-parameter lognormal distribution is discussed in this section. For other lognormal distributions, please refer to statistics references. The two parameter lognormal density function is

$$f(x) = \frac{1}{x\sigma_y\sqrt{2\pi}} \exp\left[-\frac{1}{2\sigma_y^2}(\ln x - \mu_y)^2\right], \quad x > 0, \ -\infty < \mu_y < \infty, \ \sigma_y > 0 \quad (3.46)$$

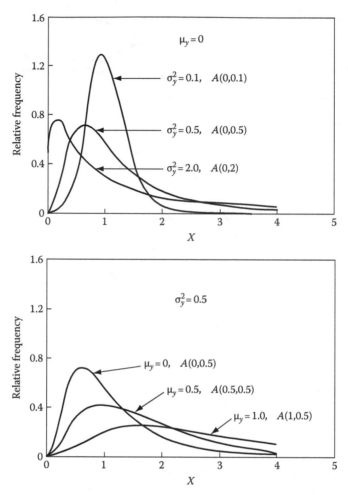

FIGURE 3.20 Lognormal distributions for different values of the parameters μ_y and σ_y^2. (From Aitchison, J. and Brown, J.A.C., *The Lognormal Distribution*, Cambridge University Press, Cambridge, U.K., 176 p., 1969.)

where μ_y and σ_y^2, the two parameters of the distribution, are the true mean and variance, respectively, of the transformed random variable Y which is equal to $\ln X$. A two parameter lognormal distribution with parameters μ_y and σ_y^2 is expressed as $\Lambda(\mu_y, \sigma_y^2)$. Figure 3.20 shows some two-parameter lognormal distributions.

One of the most preferred methods to estimate the parameters μ_y and σ_y^2 is maximum likelihood. This method uses the following estimators:

$$\bar{y} = \frac{1}{n} \sum_{i=1}^{n} y_i \tag{3.47}$$

and

$$s_y^2 = \frac{1}{n} \sum_{i=1}^{n} (y_i - \bar{y})^2 \qquad (3.48)$$

where $y_i = \ln x_i$. The standard deviation is then, $s_y = \sqrt{S_y}$.

3.7.3 t-DISTRIBUTION

The form of t-distribution is determined by a quantity called the degrees of freedom (DF), which can be calculated from the sample size, N, as $DF = N - 1$.

If the parent distribution of X is normal, the following quantity has a t-distribution with $DF = N - 1$:

$$T = \frac{N^{1/2} [m_X - \mu_x]}{S_X} \qquad (3.49)$$

The t-distribution is symmetrical about its mean value, $\mu_T = 0$, and it has a variance $\sigma_T^2 = DF/DF - 2$. The quintiles of the distribution can be calculated if DF is given. Table 3.4 gives the values for $t_{1-\alpha/2,N-1}$ for various DF. $t_{1-\alpha/2,N-1}$ is $(1-\alpha/2)$ quantile of the t-distribution with $N - 1$ degrees of freedom.

As DF gets large, the t distribution approaches the standard normal distribution. If N is equal to or greater than 30, t-distribution is a good approximation even for skewed parent distributions, although T follows the t-distribution only when the parent distribution is normal (Barrette and Goldsmith, 1976).

3.7.4 CHI-SQUARE (χ^2) DISTRIBUTION

The form of the χ^2 distribution also depends on the degrees of freedom. If the parent distribution of X is normal, the following quantity has a χ^2 distribution with $DF = N - 1$.

$$\chi^2 = \frac{(N-1)s_X^2}{\sigma_X^2} \qquad (3.50)$$

The mean of the distribution $\mu_X^2 = DF$, and its variance $\sigma_{\chi^2}^2 = 2DF$. This distribution is asymmetrical. As N gets large, the χ^2 distribution approaches the normal distribution (Yevjevich, 1972). The values of $\chi_{1-\alpha/2,N-1}^2$ for various DF are presented in Table 3.4.

3.7.5 ERRORS

In general, the true values of the quintiles of hydrologic quantity populations and correlation coefficients between them are unknown. However, the true values can be estimated by taking samples from the populations and calculating the sample quintiles and correlation coefficients.

TABLE 3.4

0.025 and 0.975 Quantiles of t-Distribution and χ^2 Distribution for Different Values of Degrees of Freedom

DF	t-Distribution		χ^2 Distribution	
	$t_{0.025,DF}$	$t_{0.975,DF}$	$\chi^2_{0.025,DF}$	$\chi^2_{0.975,DF}$
10	−2.23	2.23	3.25	20.5
11	−2.20	2.20	3.82	21.9
12	−2.18	2.18	4.40	23.3
13	−2.16	2.16	5.01	24.7
14	−2.14	2.14	5.63	26.1
15	−2.13	2.13	6.26	27.5
16	−2.12	2.12	6.91	28.8
17	−2.11	2.11	7.56	30.2
18	−2.10	2.10	8.23	31.5
19	−2.09	2.09	8.91	32.9
20	−2.09	2.09	9.59	34.2
21	−2.08	2.08	10.3	35.5
22	−2.07	2.07	11.0	36.8
23	−2.07	2.07	11.7	38.1
24	−2.06	2.06	12.4	39.4
25	−2.06	2.06	13.1	40.6
30	−2.04	2.04	16.8	47.0
40	−2.02	2.02	24.4	59.3
50	−2.01	2.01	32.4	71.4
60	−2.00	2.00	40.5	83.3
100	−1.98	1.98	74.2	129.6

3.7.5.1 Sampling Error

Sampling error is the uncertainty inherent in these sample estimates of population quantities. It refers to the uncertainty in estimates of the population value based on the sample of finite size.

Assume that an infinite number of samples of size N are taken from a population and the desired statistics for each sample is calculated. The value of each statistic are functions of random variables and thus can be considered populations of random variables. Random variables are the random measured values in the sample. Each such population has a probability distribution, with its own quintiles and other statistics. These distributions, which are derived by sampling, are known as sampling distributions. Parent distribution is the underlying population from which the sample values are taken.

3.7.5.2 Standard Errors

The standard deviations of sampling distributions are called standard errors of the sample estimates. If the form of the sampling distribution is known, this information, along with the standard error, can be used to compute absolute measures of the uncertainty in the form of confidence intervals.

If the parent distribution is normal, formulas for computing standard errors can be developed from statistical theory. The theoretical standard errors are functions of the population and are inversely related to the sample size.

If the population of the sample mean, $\bar{x}$, has standard deviation, s_X, the standard error of the mean is calculated as

$$s_X = \frac{s}{N^{1/2}} \tag{3.51}$$

If the data are approximately normally distributed, the standard error of the sample standard deviation, s_s, can be calculated as (Ycvjevich, 1972)

$$s_s = \frac{s}{(2N)^{1/2}} \tag{3.52}$$

3.7.6 ESTIMATING QUANTILES (PERCENTILES)

Quintiles of distributions are estimated to determine whether hydrologic data or environmental pollution levels exceed specified limits. For instance, a regulation may require that the true 0.98 quintile of the maximum discharge from a well, $\times 0.98$, must not exceed $2\,m^3/day$. In practice, xp must be estimated from data. There are two methods for estimating xp when a normal or lognormal distribution is underlying the data. These methods are probability plotting method and the parametric method. For the procedure of the probability plotting method, the reader is referred to statistical methods books such as Gilbert (1987), Price et al. (1981), etc. We discuss the parametric and nonparametric methods in this section.

In nonparametric method, there is no need for an underlying distribution. In this method, the data is first ranked from the smallest to the largest. The rank of $(100p)$ th percentile is then

$$M = p(N+1) \tag{3.53}$$

where
   M is the ascending order of magnitude
   N is the number of data points
   p is the fraction of the population below percentile of interest

In other words, p is the nonexceedance probability p can then be calculated as

$$p(\%) = M(N+1) \tag{3.54}$$

The median corresponds to 50% value.

Example 3.13

What is the rank of the 90th percentile of 10 observations? What is the 95th percentile of these observations?

Solution

For 90th percentile, $p = 0.9$, $M = (0.9)(10 + 1) = 9.9$. The 90th percentile is then calculated by interpolation between the 9th and 10th largest observations.
   For 95th percentile, $p = 0.95$, $M = (0.95)(10 + 1) = 10.45!!!$
   We can not estimate the 95th percentile with only 10 observations using these estimators. For extreme percentiles, we need to have sufficiently large sample size. An alternative estimator is $M = pN$; but it will not provide estimates as good as those from the other estimator.

Quantiles of a normal distribution, $\hat{x}_p$, can be estimated by using the sample mean, $\bar{x}$, and standard deviation, s, as

$$\hat{x}_p = \bar{x} + Z_p s \tag{3.55}$$

where Z_p is the pth quintile of the standard normal distribution. Appendix A, Table A1 gives values of p that correspond to Z_p.
   Quantiles of a two-parameter lognormal distribution is also computed by

$$\hat{x}_p = \exp(\bar{y} + Z_p s_y) \tag{3.56}$$

where $\bar{y}$ and s_y are the estimates of μ_y and σ_y, respectively.

Example 3.14

In an aquifer, the mean of the logs of the annual rate of base flow from the aquifer through a surface stream is 0.28 (log of m^3/day). The standard deviation of the logs is 0.075 (log of m^3/day). Calculate the 90th percentile if a lognormal distribution is underlying the data.

Solution

From Table A1, $Z_p = 1.282$:

$$Q_{90\%} = \exp(0.25 + 1.282 \times 0.075) = 1.41\,m^3/day$$

Example 3.15

The daily water withdrawal from a well is presented in the second column of the following table. Calculate the arithmetic and logarithmic mean of the data. Also, determine a value that would have a 95% probability of not being exceeded ($p[X \le X_n] = 0.95$).

Solution

In the following table, columns (1) and (2) are given. We calculate columns (3) through (5).

Using Table A1, $Z_{0.95} = 1.645$. The 95% probability of not being exceeded is then calculated as

$$X_{0.95} = 10^{\mu_y + Z_p s_y} = 10^{0.2516 + 1.645 \times 0.269} = 4.94 \ \text{m}^3/\text{day}$$

3.7.7 PROBABILITY/FREQUENCY/RECURRENCE INTERVAL

Hydrologic and water quality analysis reports express the extreme values in terms of probabilities of excellence, frequency of occurrence, and recurrence interval/return period. All three terms have the same meaning. Since we deal with time series, the probability can be converted to an average time between two occurrences of the extreme value, called the recurrence interval (Novotny, 2003).

Assume that we have a 10-year time series of a hydrologic parameter, which is measured based on a monthly basis. The total number of data is therefore $10 \times 12 = 120$.

Analyses show that the mean and standard deviation of the data are 150 and 48, respectively. A value of the series that would have a probability of 5% of being exceeded (95% of cumulative probability of being less than or equal to) is then

$$X(p[X \geq X_n] = 5\%) = \mu + Z_p \sigma = 150 + 1.64 \times 48 = 228.7$$

The recurrence interval between two occurrences of measurements that would equal or exceed 228.7 is then a reciprocal of the probability times the sampling interval:

$$T_r = \frac{100}{p(\%)}(\Delta t) = \frac{100}{10}(1 \ \text{month}) = 10 \ \text{months}$$

The frequency is then simply a reciprocal of the recurrence interval—the measurements that would equal or exceed 228.7 will occur approximately once in 10 months.

3.8 TIME SERIES ANALYSIS

To develop appropriate operation policies for water resources systems that are applicable to real-time decision making, forecasting the future state of the resources is necessary. For instance, consider a well field, which supplies water for agricultural and domestic needs in an area. The amount of scheduled withdrawal depends on the probable range of inflow to the aquifer. Considering the fact that there is a sound lack of adequate knowledge about physical processes in the hydrologic cycle, the application of statistical models in forecasting and generating synthetic data have been expanded by researchers.

In this section, basic principles of hydrologic time series modeling and different types of statistical models are discussed. More details on statistical modeling can be found in Salas et al. (1988) and Brockwell and Davis (1987).

3.8.1 NONSTATIONARY HYDROLOGIC VARIABLES

Due to variations that are the result of natural and human activities, the hydrologic variables are mostly nonstationary. These variations in hydrologic modeling are classified as follows (Karamouz et al., 2003):

1. *Trend* is a gradual change (increasing or decreasing) in the average value of the variable. Trend is unidirectional and is mainly caused by anthropogenic changes in nature (changes in nature in result of human activities). A trend, T_t, is represented by a continuous and differentiable function of time. It can be modeled by different linear and polynomial functions such as
   - Linear function: $T_t = a + bt$
   - Polynomial function: $T_t = a + bt + ct^2 + \cdots + dt^m$
   - Power functions: $T_t = a - br^t$ $(0 < r < 1)$ and $T_t = 1/(a + br^t)(0 < r < 1)$
2. *Jump* is a sudden change in the observed values that can be positive or negative. The main reasons for jumps in hydrologic series are natural disruptions and human activities.
3. *Periodicity* can be defined as the cyclic variations in the hydrologic time series. These variations are usually repeated over the same intervals of time.
4. *Randomness* is the variation due to uncertainties of natural processes. The random component of the hydrologic time series is classified as autoregressive or purely random.

In statistical modeling of time series, it is assumed that the time series is purely random. Therefore, all trends, jumps, and periodicities must be removed from the time series.

3.8.2 HYDROLOGIC TIME SERIES MODELING

The following steps are involved in a systematic approach to hydrologic time series modeling (Salas et al., 1988):

- *Data preparation*: Removing trends, periodicity, outlying observations, and fitting the data to a proper distribution.
- *Identification of model composition*: Different compositions for model selection are univariate model, multivariate model, and a combination of each of these models with disaggregation models. Based on the characteristics of the water resources system and existing information, it is identified that which composition is better to be used.
- *Identification of model type*: Different types of models mostly used in hydrologic studies are AR (autoregressive), ARMA (autoregressive moving average), and ARIMA (autoregressive integrated moving average). These models will be discussed in detail later in this chapter. The appropriate type of model is selected due to statistical characteristics of the time series and the modeler input and knowledge about different types of models.

- *Identification of model form*: The statistical characteristics of the time series determine the form of the selected model. The periodicity of the data and how it can be considered in the model is the main issue in this step.
- *Estimation of model parameters*: There are different methods for parameter estimation, such as method of moments and method of maximum likelihood. These methods will be discussed later in this chapter.
- *Testing the goodness of fit of the model*: In this step, different assumptions such as independence and normality of residuals should be checked.
- *Evaluation of uncertainties*: The uncertainties in model and parameters as well as natural uncertainties in data should be evaluated separately.

3.8.3 DATA PREPARATION

As mentioned before, data preparation involves removing trend, removing outlying observations, removing periodicity, and fitting the data to a proper distribution. In using statistical models for hydrologic time series, considering the variable as a pure random variable is the principal assumption. Therefore, if the series contains unnatural components such as synthetic recharge of an aquifer, they should be first removed. Time sequence plot of the series is usually used to detect trend and seasonality. For instance, consider a nonseasonal series, x_t, which contains trend component as follows:

$$x_t = T_t + z_t \tag{3.57}$$

where T_t and z_t are trend and random components, respectively. The trend components are defined as a function of time as

$$T_t = f(t, \ldots, t^m, \beta_1, \ldots, \beta_m) \tag{3.58}$$

where $\beta_1, \ldots, \beta_m$ are the parameters to be estimated. These parameters can be estimated based on least-square method by minimizing the following sum:

$$\sum_t (x_t - T_t)^2 \tag{3.59}$$

To estimate $\beta_1, \ldots, \beta_m$, the partial derivatives of the above sum with respect to each unknown parameter, $\beta_1, \ldots, \beta_m$, is equated to zero. The random series is then generated using Equation 3.57.

Example 3.16

A series of annual water withdrawal from a well to supply agricultural demands for a period of 10 years is shown in Table 3.5. Estimate a linear trend function and generate the random series of water consumption by removing the trend.

TABLE 3.5
Annual Water Consumption in a City in Example 3.15

Year	Water Consumption (MCM/Year)
1	44
2	56
3	52
4	48
5	61
6	55
7	65
8	43
9	63
10	49

Solution

The linear trend function is

$$T_t = \alpha t + \beta \tag{3.60}$$

In least square method the following sum should be minimized:

$$\sum_{t=1}^{10} (x_t - T_t)^2 = \sum_{t=1}^{10} (x_t - \alpha t - \beta)^2$$

α and β can be calculated as

$$\alpha = \frac{10\sum_{t=1}^{10} tx_t - \sum_{t=1}^{10} t \sum_{t=1}^{10} x_t}{10\sum_{t=1}^{10} t^2 - \left(\sum_{t=1}^{10} t\right)^2} = 0.57$$

$$\beta = \bar{x} - \alpha\bar{t} = 50.47$$

The random component after removing trend is shown in Table 3.6.

The above method is called the classical decomposition of the series into a trend component, a seasonal component, and a random component. There is another method for removing trend and seasonality, which is widely used in modeling of hydrologic time series. This method is called differencing method. In this method, the first differencing operator, ∇, is

$$\nabla(x_t) = x_t - x_{t-1} = (1 - B)x_t \tag{3.61}$$

TABLE 3.6
Random Components of Annual Water Consumption—Example 3.15

Year	Water Consumption (MCM/Year)	Trend Component	Random Component
1	44	51.04	44 − 51.04 = −7.04
2	56	51.61	4.39
3	52	52.18	−0.18
4	48	52.75	−4.75
5	61	53.32	7.68
6	55	53.88	1.12
7	65	54.45	10.55
8	43	55.02	−12.02
9	63	55.59	7.41
10	49	56.16	−7.16

where B is the backward shift operator and is defined as

$$B(x_t) = x_{t-1} \tag{3.62}$$

Higher orders of the operators B and ∇ can be defined as

$$B^i(x_t) = x_{t-j} \quad \text{and} \quad \nabla^j(x_t) = \nabla(\nabla^{j-1}(x_t)), \quad j \geq 1 \tag{3.63}$$

Applying the first-order difference operator to the linear trend function gives the constant function $\nabla(T_t) = \alpha$. Similarly, any polynomial trend of degree k can be reduced to a constant by application of the operator ∇^k. Hence, if a hydrologic time series has a polynomial trend such as

$$T_t = \sum_{i=0}^{m} a_i t^i \tag{3.64}$$

and its operator can be written as

$$\nabla^m(x_t) = m! a_m + \nabla^m z_t \tag{3.65}$$

In most of the parameter estimation methods, it is assumed that the time series fits a normal distribution. But in many cases, the hydrologic time series do not follow normal distribution and so it is necessary to transform those variables to normal or log-normal before statistical modeling. The mostly used variable transformation is the two-parameter transformation developed by Box and Cox (1964). This transformation is

$$z_t = \frac{(x_t + \lambda_2)^{\lambda_1} - 1}{\lambda_1 g^{(\lambda_1 - 1)}}, \quad \lambda_1 > 0 \tag{3.66}$$

and

$$z_t = g \ln(x_t + \lambda_2), \quad \lambda_1 = 0 \tag{3.67}$$

where
λ_1 and λ_2 are the model parameters
g is the geometric mean of $x_t + \lambda_2$
x_t and z_t are the original and transformed series, respectively

The parameter λ_1 controls the strength of the transformation. $\lambda_1 = 1$ corresponds to the original data and $\lambda_1 = 0$ to a logarithm. Since the values are scaled by the geometric mean to keep the variance constant in this transformation, mean squared errors between two different transformations can be compared.

Hashino and Delleur (1981) developed another transformation similar to Box–Cox as

$$z_t^{(\lambda)} = \frac{(x_t + c) - 1}{\lambda}, \quad \lambda \neq 0 \tag{3.68}$$

and

$$z_t^{(\lambda)} = \ln(x_t + c), \quad \lambda = 0 \tag{3.69}$$

where λ and c are λ_1 and λ_2 in Box–Cox transformation. In this transformation, the values are not scaled and therefore is not allowing for the direct comparison of mean squared errors between two different transformations.

3.8.4 Parameter Estimation

The basic methods for estimating model parameters are methods of moments, least squares, and maximum likelihood. These methods are discussed in the following.

3.8.4.1 Method of Moments

Consider a sample $x_1, \ldots, x_N$. The nth sample moment is defined as follows:

$$M_n = \left(\frac{1}{N}\right) \sum_{i=1}^{N} x_i^n \tag{3.70}$$

When a model is fitted to a sample, the moment parameter estimates can be obtained by equating sample moments and population moments.

Example 3.17

Assume that the following model is fitted to the sample $x_1, \ldots, x_N$:

$$x_t = \alpha z_t + \beta$$

where
  z_t is an independent random variable with mean zero and variance one
  α and β are the model parameters

Estimate the model parameters with the method of moments.

Solution

The first and second sample moments can be estimated as

$$M_1 = \left(\frac{1}{N}\right) \sum_{i=1}^{N} x_i \quad \text{and} \quad M_2 = \left(\frac{1}{N}\right) \sum_{i=1}^{N} x_i^2$$

The first two population moments of the model are

$$E[X^1] = \beta \quad \text{and} \quad E[X^2] = \alpha^2 + \beta^2$$

By equating the first population and sample moments

$$\hat{\beta} = \left(\frac{1}{N}\right) \sum_{i=1}^{N} x_i$$

which is an estimate of the parameter β. And the estimate of parameter α is obtained by equating the second moment as

$$\alpha = \sqrt{\frac{1}{N} \sum_{i=1}^{N} x_i^2 - \hat{a}^2}$$

3.8.4.2 Method of Least Squares

Consider that the following model is fitted to the sample $x_1, \ldots, x_N$:

$$\hat{x}_t = f(x_{t-1}, x_{t-2}, \ldots, \alpha_1, \alpha_2, \ldots, \alpha_m) + \varepsilon_t \tag{3.71}$$

where
ε_t is the model residual
$\alpha_1, \alpha_2, \ldots, \alpha_m$ are the model parameters

In the least square method, the following must be minimized:

$$\sum_{t=1}^{N} (x_t - \hat{x}_t)^2 = \sum_{t=1}^{N} (x_t - f(x_{t-1}, x_{t-2}, \ldots, \hat{\alpha}_1, \hat{\alpha}_2, \ldots, \hat{\alpha}_m))^2 \tag{3.72}$$

In this method, all partial derivatives with respect to estimated values of the parameters, $\hat{\alpha}_1, \hat{\alpha}_2, \ldots, \hat{\alpha}_m$, should be equated to zero. Therefore,

$$\frac{\partial \sum_{t=1}^{N} (x_t - \hat{x}_t)^2}{\partial \hat{\alpha}_1} = 0, \ldots, \frac{\partial \sum_{t=1}^{N} (x_t - \hat{x}_t)^2}{\partial \hat{\alpha}_m} = 0 \tag{3.73}$$

These equations are solved simultaneously and, in consequence, model parameters, $\hat{\alpha}_1, \hat{\alpha}_2, \ldots, \hat{\alpha}_m$, are estimated.

Example 3.18

Find the model parameter, α, using least square method for the following model which is used for the sample $x_1, \ldots, x_N$:

$$x_t = \alpha x_{t-1} + \varepsilon_t$$

Solution

The square of errors is

$$\sum \varepsilon_t^2 = \sum (x_t - \alpha x_{t-1})^2$$

The partial derivatives of the sum with respect to α is

$$\frac{\partial \sum (x_t - \alpha x_{t-1})^2}{\partial \alpha} = \sum (x_{t-1}(\alpha x_{t-1} - x_t))$$

Therefore, the parameter is estimated as

$$\alpha = \frac{\sum x_t x_{t-1}}{\sum x_{t-1}^2}$$

3.8.4.3   Method of Maximum Likelihood

The likelihood function is expressed as

$$L(\varepsilon) = \prod_{t=1}^{N} f(\varepsilon_t, \alpha_1, \ldots, \alpha_m) \tag{3.74}$$

To simplify the calculations, it is desirable to transform the multiplication operation to addition operation. In this method, log of the function is taken to convert the operations. The log-likelihood function will then be maximized. By maximizing the function $L(\varepsilon)$, the maximum likelihood of the parameters, $\alpha_1, \ldots, \alpha_m$, is obtained:

$$\log(L(\varepsilon)) = \ln \prod_{t=1}^{N} f(\varepsilon_t, \alpha_1, \ldots, \alpha_m) = \sum_{t=1}^{N} \ln \left[f(\varepsilon_t, \alpha_1, \ldots, \alpha_m) \right] \tag{3.75}$$

Partial derivatives with respect to the parameters, $\alpha_1, \ldots, \alpha_m$, will then be equated to zero to maximize the above sum:

$$\frac{\partial \text{Log}(L(\varepsilon))}{\partial \alpha_1} = 0, \ldots, \frac{\partial \text{Log}(L(\varepsilon))}{\partial \alpha_m} = 0 \tag{3.76}$$

The maximum likelihood estimate of the parameters can be obtained by simultaneously solving the above equations.

3.8.5   Goodness of Fit Tests

Two of the most common test statistics for the goodness of fit are the chi-square, χ^2, test for testing the hypothesis of goodness of fit of a given distribution, and the

Smirnov–Kolmogorov statistic for estimating maximum absolute difference between the cumulative frequency curve of sample data and the fitted distribution function. The chi-square test is explained here and the reader is encouraged to check Salas et al. (1988) for the other test.

Lapin (1990) suggested the following steps for hypothesis testing:

1. Formulate the null hypothesis, H_0, which usually is that the data represent a specific distribution.
2. Select the test procedure, such as chi-square, and test statistic.
3. Select the significance/acceptance level.
4. Compute the value of the test statistic for the sample data.
5. Find the corresponding critical value for χ^2 from the χ^2 distribution table.
6. Compare the obtained value of χ^2 with the critical value. If the former is smaller than the latter one, the null hypothesis of fitness of the selected distribution is accepted. Otherwise, it is rejected.

3.8.5.1 Chi-Square Goodness of Fit Test

This test is based on the comparison of two sets of frequencies. To accomplish this test, the data are first divided into a number of categories. Then the actual frequencies (f_i) in each category (class) is compared with the expected frequencies, $(\hat{f}_i)$, which are estimated based on the fitted distribution. The test statistic is estimated as

$$\chi^2 = \sum_{i=1}^{NC} \frac{(f_i - \hat{f}_i)^2}{\hat{f}_i}$$ (3.77)

where
 i is the number of classes
 NC is the total number of categories

This estimation is close enough for testing the goodness-of-fit whenever the expected frequency in any category is equal to or greater than 5. χ^2 is chi-square distributed with $NC - NP - 1$ degree of freedom, where NP is the number of parameter to be estimated.

Example 3.19

A well field is used as a supplementary source of water to supply the water demands of an urban area in dry periods, i.e., when there is the deficit of surface water. The classified monthly withdrawal data, which is divided into nine categories is presented in Table 3.7. Previous studies show that the withdrawal data in this well field is exponentially distributed. Therefore, the probability of the monthly withdrawal, D_w, is equal to or below x and is expressed as

$$\text{Prob}[D_w \le x] = 1 - e^{-\lambda x}$$

TABLE 3.7

Withdrawal Data in Example 3.18

Interval (*i*)	Withdrawal Data Classes (MCM)	Actual Frequency (*f_i*)
1	$0 \leq D_w < 5$	38
2	$5 \leq D_w < 10$	26
3	$10 \leq D_w < 15$	12
4	$15 \leq D_w < 20$	10
5	$20 \leq D_w < 25$	8
6	$25 \leq D_w < 30$	3
7	$30 \leq D_w < 35$	2
8	$35 \leq D_w < 40$	1
9	$40 \leq D_w$	0

where λ is estimated equal to 0.105. Use the chi-square test with 5% significance level to test it whether the selected distribution fits the sample data.

Solution

Using the given exponential distribution, the expected frequencies are computed and presented in the third column of Table 3.8. For example, for the first class of the data, the probability of withdrawing less than 5 MCM water from the well is

$$\text{Prob}[D_w < 5] = 1 - e^{-0.105 \times 5} = 0.41$$

TABLE 3.8

Estimated Expected Frequencies and Chi-Square Statistic for the Withdrawal Data—Example 3.18

Interval (*i*)	Actual Frequency (*f_i*)	Exponential Cumulative Probability at Upper Limit	Expected Frequencies ($\hat{f_i}$)	Actual Frequency (*f_i*)	$f_i - \hat{f_i}$	χ^2
1	38	0.41	41	38	−3	0.22
2	26	0.65	24	26	2	0.17
3	12	0.79	14	12	−2	0.29
4	10	0.88	9	10	1	0.11
5	8	0.93	5	8	3	1.8
6	3	0.96	3	3	−1	0.14
7	2	0.97	1	2		
8	1	0.99	2	1		
9	0	1	1	0		
Sum	100	—	100		—	2.73

This value is then multiplied by 100 for the first interval. The resulting value is presented in the forth column of Table 3.8. From the second interval on, the expected frequencies are estimated based on the difference between cumulative probabilities multiplied by 100. For example, for the second interval,

$$\text{Expected frequency} = (0.65 - 0.41) \times 100 = 24$$

The actual frequency for the intervals six to nine in Table 3.8 is less than 5. Therefore, these classes can be grouped together. The chi-square statistic is then estimated as 2.73. The degree of freedom for the test is $6 - 1 - 1 = 4$. From Appendix A, Table A2, for 5% significance level and four degrees of freedom, the critical value is $\chi^2_{0.05(4)} = 9.488$. Since, the estimated value of the test statistics, 2.73, is smaller than the critical value, 9.488, the null hypothesis that the monthly precipitation is exponentially distributed is accepted.

3.8.5.2 Tests of Normality

There are different ways to test the normality of a time series. The graphical test is the most widely used test to check the hypothesis that the time series of interest is normal. In this test, the empirical distribution of the series is plotted on a normal probability paper. Then it is checked if the plotted points fit a straight line. The chi-square test is another widely used method for testing the normality of data. Another common normality test is the skewness test. The skewness coefficient of a time series x_t is estimated as follows:

$$\hat{\gamma} = \frac{1/N \sum_{t=1}^{N} (x_t - \bar{x})^3}{\left[1/N \sum_{t=1}^{N} (x_t - \bar{x})^2 \right]^{3/2}} \tag{3.78}$$

where
  $\hat{\gamma}$ is the skewness coefficient
  N is the total number of sample data
  $\bar{x}$ is the sample mean

This test is based on the fact that the skewness coefficient of a normal variable is zero. If the series is normally distributed, the mean and variance of the skewness coefficient are zero and $6/N$, respectively (Snedecor and Cochran, 1967). $(1-\alpha)$ probability limits on γ could be defined as

$$\left[-z_{1-\alpha/2} \sqrt{\frac{6}{N}}, \quad z_{1-\alpha/2} \sqrt{\frac{6}{N}} \right] \tag{3.79}$$

where $z_{1-\alpha/2}$ is the $1 - \alpha/2$ quantile of the standard normal distribution. Therefore, if $\hat{\gamma}$ falls within the limits of Equation 3.79, the hypothesis of normality is accepted. Otherwise it is rejected. This procedure is recommended for the sample sizes greater than 150. For smaller sample sizes, the computed coefficient of skewness is compared with the tabulated values of skewness coefficient, $\gamma_\alpha(N)$, presented in Table 3.9. If $|\hat{\gamma}| < \gamma_\alpha(N)$, the hypothesis of normality is accepted.

TABLE 3.9

Values of Skewness Coefficient for Normality for Sample Sizes Less Than 150

N	$\alpha = 0.02$	$\alpha = 0.01$
25	1.061	0.711
30	0.986	0.662
35	0.923	0.621
40	0.870	0.587
45	0.825	0.558
50	0.787	0.534
60	0.723	0.492
70	0.673	0.459
80	0.631	0.432
90	0.596	0.409
100	0.567	0.389
125	0.508	0.350
150	0.464	0.321
175	0.430	0.298

Source: Snedecor, G.W. and Cochran, W.G., *Statistical Methods*, Iowa State University Press, Ames, IA, 1967.

Example 3.20

The following logarithmic transformation is applied to the monthly rates of recharge to an aquifer:

$$y_t = \log_{10}(x_t)$$

where x_t and y_t are the original and the transformed rate of recharge in month t, respectively. The skewness coefficient of transformed series is estimated to be 0.8. Use the skewness test to check on the normality of the transformed data. Consider a 5% significance level for this test. The number of data in the time series is 1020.

Solution

Using Equation 3.79, the 95% probability limits on skewness coefficient is

$$\left[-z_{0.975}\sqrt{\frac{6}{N}}, \ z_{0.975}\sqrt{\frac{6}{N}} \right] = \left[-1.96\sqrt{\frac{6}{1020}}, \ 1.96\sqrt{\frac{6}{1020}} \right] = [-0.15, \ 0.15]$$

The value of $z_{0.975}$ can be read from the table of normal distribution in Appendix A. Because 0.8 does not fall within the above limits, the series is not normally distributed.

3.8.6 AKAIKE'S INFORMATION CRITERION

Akaike's information criterion (AIC), which was first proposed by Akaike (1974), is usually used as the primary criterion for model selection. It considers the parsimony in model building. He defined the AIC among competing ARMA (p,q) models as follows:

$$AIC(p,q) = N \ln(\hat{\sigma}^2(\varepsilon)) + 2(p+q) \qquad (3.80)$$

where
  N is the sample size
  $\hat{\sigma}^2(\varepsilon)$ is the maximum likelihood estimate of the residual variance

The model with minimum AIC is selected.

The goodness-of-fit tests should be applied to the model with minimum AIC to make sure the residuals are consistent with their expected behavior. If the model residuals do not pass the test, the models with higher values of AIC should be checked.

3.8.7 AUTOREGRESSIVE MODELING

The autoregressive (AR) models are effective tools for modeling hydrologic time series such as streamflow in low flow season, which is mainly supplied from groundwater and has low variations (Salas et al., 1980). These models incorporate the correlation between time sequences of variables. They have been widely used in hydrologic time series modeling. AR models are classified into the following subsets:

- AR models with constant parameters, which can be used to model annual series
- AR models with timely variable parameters, which can be used to model seasonal (periodic) series

The basic form of the AR model of order p, AR(p), with constant parameters is

$$z_t = \sum_{i=1}^{p} \varphi_i z_{t-i} + \varepsilon_t \qquad (3.80a)$$

where
  z_t is the time-dependent normal and standardized series $\{N(0,1)\}$
  φ_i is the autoregressive coefficient
  ε_t is the time independent variable (white noise)
  p is the order of autoregressive model

Equation 3.80 can be applied to the standardized series. To obtain a standardized series, the following equation is used:

$$z_t = \frac{x_t - \mu}{\sigma} \qquad (3.81)$$

where μ and σ are the mean and standard deviation of the series x_t. The parameter set of the model is

$$\{\mu, \sigma, \varphi_1, \ldots, \varphi_p, \sigma^2(\varepsilon)\} \qquad (3.82)$$

where $\sigma^2(\varepsilon)$ is the variance of time independent series. The Yule–Walker linear equations are solved simultaneously to estimate the model parameters. These equations can be expressed as

$$r_i = \hat{\varphi}_1 r_{i-1} + \hat{\varphi}_2 r_{i-2} + \cdots + \hat{\varphi}_p r_{i-p} \quad (i > 0) \qquad (3.83)$$

where r_i is the sample correlation coefficients of lag i. The parameter $\sigma^2(\varepsilon)$ is estimated as

$$\hat{\sigma}^2(\varepsilon) = \frac{N\hat{\sigma}^2}{(N-p)}\left(1 - \sum_{i=1}^{p} \hat{\varphi}_i r_i\right) \qquad (3.84)$$

where
  N is the number of the data
  $\hat{\sigma}^2$ is the sample variance

To meet the stationary condition by the model parameters, the roots of the following equation should lie inside the unit circle (Yevjevich, 1972):

$$u^p - \hat{\varphi}_1 u^{p-1} - \hat{\varphi}_2 u^{p-2} - \cdots - \hat{\varphi}_p = 0 \qquad (3.85)$$

In other words, $|u_i| < 1$. To forecast or generate annual AR models, the following relation is used:

$$\hat{z}_t = \hat{\varphi}_1 \hat{z}_{t-1} + \cdots + \hat{\varphi}_p \hat{z}_{t-p} + \hat{\sigma}(\varepsilon)\zeta_t \qquad (3.86)$$

where ζ_t is the standardized normal variable.

Example 3.21

For an AR(2) model, the parameters have been estimated as $\varphi_1 = 0.5$ and $\varphi_2 = 0.3$. Check the parameters stationary condition.

Solution

Using Equation 3.85,

$$u^2 - 0.5u - 0.3 = 0 \Rightarrow \begin{cases} u_1 = 0.85 \\ u_2 = -0.35 \end{cases}$$

The roots lie within the unit circle; therefore, the parameters pass the stationary condition.

Example 3.22

An AR(2) model is selected for a 20-year normal and standardized annual quality data of an aquifer. The first and second correlation coefficients are estimated as $r_1 = 0.6$ and $r_2 = 0.35$. Estimate the model parameters if the variance of normal and standardized inflow series is estimated as 1.5.

Solution

Using Equation 3.83, we have

$$\begin{cases} r_1 = \varphi_1 + \varphi_2 r_1 \\ r_2 = \varphi_2 + \varphi_1 r_1 \end{cases}$$

By solving the above equations,

$$\varphi_1 = \frac{r_1(1 - r_2)}{1 - r_1^2} \quad \text{and} \quad \varphi_2 = \frac{r_2 - r_1^2}{1 - r_1^2}$$

Therefore,

$$\varphi_1 = 0.61 \quad \text{and} \quad \varphi_2 = -0.02$$

The variance of model residuals is now estimated using Equation 3.84:

$$\hat{\sigma}^2(\varepsilon) = \frac{20 \times 1.5}{20 - 2}(1 - 0.6 \times 0.61 + 0.35 \times 0.02) = 1.173$$

In order to check whether an AR(p) is an appropriate model for a specific time series, the behavior of the partial autocorrelation function (PACF) of the series is estimated and investigated. PACF shows the time dependence structure of the series in a different way. The partial autocorrelation lag k represents the correlation between z_1 and z_{k+1}, adjusted for the intervening observations $z_2, ..., z_k$. For an AR(p) process, the partial autocorrelation function is the last autoregressive coefficient of the model, $\varphi_k(k)$, which is estimated by rewriting equation 3.83 as following and successively solving it:

$$r_i = \hat{\varphi}_1(k)\, r_{i-1} + \hat{\varphi}_2(k)\, r_{i-2} + \cdots + \hat{\varphi}_k(k)\, r_{i-k} \tag{3.87}$$

A set of linear equations are then resulted from the above equation. These equations are

$$\hat{\varphi}_1(k)\, r_0 + \hat{\varphi}_2(k)\, r_1 + \cdots + \hat{\varphi}_k(k)\, r_{k-1} = r_1$$

$$\hat{\varphi}_1(k)\, r_1 + \hat{\varphi}_2(k)\, r_0 + \cdots + \hat{\varphi}_k(k)\, r_{k-1} = r_2$$

$$\vdots$$

$$\hat{\varphi}_1(k)\, r_{k-1} + \hat{\varphi}_2(k)\, r_{k-2} + \cdots + \hat{\varphi}_k(k)\, r_0 = r_k$$

The above equations can be written as

$$
\begin{bmatrix}
1 & r_1 & r_2 & \cdots & r_{k-1} \\
r_1 & 1 & r_1 & \cdots & r_{k-2} \\
r_2 & r_1 & 1 & \cdots & r_{k-3} \\
\vdots & \vdots & \vdots & & \vdots \\
r_{k-1} & r_{k-2} & r_{k-3} & & 1
\end{bmatrix}
\begin{bmatrix}
\phi_1(k) \\
\phi_2(k) \\
\phi_3(k) \\
\vdots \\
\phi_k(k)
\end{bmatrix}
=
\begin{bmatrix}
r_1 \\
r_2 \\
r_3 \\
\vdots \\
r_k
\end{bmatrix}
\tag{3.88}
$$

The partial autocorrelation function, $\varphi_k(k)$, can be determined by successively applying above relation. The partial correlogram of an autoregressive process of order p has peaks at lags 1 through p and then cuts off. Hence, the partial autocorrelation function is used to identify the p of an AR(p) model.

The basic form of the of the AR model with periodic parameters is

$$z_{v,\tau} = \sum_{i=1}^{P} \varphi_{i,\tau} z_{v,\tau-i} + \sigma_\tau(\varepsilon) \zeta_{v,\tau} \tag{3.89}$$

where
  $z_{v,\tau}$ is the normal and standardized value in year v and season τ
  $\varphi_{i,\tau}$ is the periodic autoregressive coefficients
  $\sigma_\tau(\varepsilon)$ is the periodic standard deviation of residuals

$z_{v,\tau}$ is estimated using seasonal mean and variance as

$$z_{v,\tau} = \frac{x_{v,\tau} - \mu_\tau}{\sigma_\tau} \tag{3.90}$$

where μ_τ and σ_τ are the mean and standard deviation of x in season τ. The parameter set of the model can be summarized as

$$\{\mu_\tau, \sigma_\tau, \varphi_{1,\tau}, \ldots, \varphi_{P,\tau}, \sigma_\tau^2(\varepsilon), \tau = 1, \ldots, \eta\} \tag{3.91}$$

where η is the total number of seasons. Equation 3.83 can be repeatedly used for each season. Therefore,

$$r_k = \sum_{i=1}^{P} \hat{\varphi}_{i,\tau} r_{|k-i|,\tau-\min(k,i)} \quad (k > 0) \tag{3.92}$$

The residual variance can be estimated using the following relation:

$$\hat{\sigma}_{\tau}^2(\varepsilon) = 1 - \sum_{j=1}^{P} \hat{\varphi}_{j,\tau} \hat{r}_{j,\tau} \tag{3.93}$$

3.8.8 MOVING AVERAGE PROCESS

If the series, z_t, is dependent only on a finite number of previous values of a random variable, ε_t, then the process can be called moving average process. The moving average model of order q, MA(q), can be expressed as

$$z_t = \varepsilon_t - \theta_1 \varepsilon_{t-1} - \theta_2 \varepsilon_{t-2} - \theta_3 \varepsilon_{t-3} - \cdots - \theta_q \varepsilon_{t-q} \tag{3.94}$$

or

$$z_t = -\sum_{j=0}^{q} \theta_j E_{t-j} \quad (\theta_0 = -1) \tag{3.95}$$

where $\theta_1, \ldots, \theta_q$ are q orders of MA(q) model parameters. The parameter set of the model is summarized as

$$\{\mu, \theta_1, \ldots, \theta_q, \sigma^2(\varepsilon)\} \tag{3.96}$$

The parameters of the model should satisfy the invertibility condition. For this purpose, the roots of the following polynomial should lie inside the unit circle:

$$u^q - \hat{\theta}_1 u^{q-1} - \hat{\theta}_2 u^{q-2} - \cdots - \hat{\theta}_q = 0 \tag{3.97}$$

Example 3.23

For an MA(2) model, the parameters have been estimated as $\theta_1 = 0.65$ and $\varphi_2 = 0.3$. Check whether the parameters pass the invertibility condition.

Solution

Using Equation 3.97,

$$u^2 - 0.65u - 0.3 = 0 \Rightarrow \begin{cases} u_1 = 0.96 \\ u_2 = -0.31 \end{cases}$$

The roots lie within the unit circle; therefore, the parameters pass the stationary condition.

3.8.9 Autoregressive Moving Average Modeling

The autoregressive-moving average (ARMA) model of order (p,q) can be defined by combining an autoregressive model of order p and a moving average model of order q as follows:

$$z_t - \varphi_1 z_{t-1} - \cdots - \varphi_p z_{t-p} = \varepsilon_t + \theta_1 \varepsilon_{t-1} + \cdots + \theta_q \varepsilon_{t-q} \tag{3.98}$$

The ARMA(p,q) model is expressed as

$$\varphi(B)z_t = \theta(B)\varepsilon_t \tag{3.99}$$

where $\varphi(z)$ and $\theta(z)$ are the pth and qth degree polynomials:

$$\varphi(z) = 1 - \varphi_1 z - \cdots - \varphi_p z^p \tag{3.100}$$

$$\theta(z) = 1 + \theta_1 z + \cdots + \theta_q z^q \tag{3.101}$$

The parameter set of ARMA(p,q) model is $\{\mu, \theta_1, \ldots, \theta_q, \varphi_1, \ldots, \varphi_p, \sigma^2(\varepsilon)\}$. These parameters should satisfy both the conditions of invertibility and stationarity.

3.8.9.1 Generation and Forecasting Using ARMA Models

Two major uses of the time series modeling are generating synthetic values and forecasting future events. The generation of synthetic time series often consists of the generation of the independent normal variables in a series, which is defined as random process.

Once the ARMA model is fitted to a time series, the following procedures are accomplished to generate or forecast the values of that time series. For the generation of the values of the times series using the ARMA(p,q) model (Equation 3.98), it is necessary to give p initial Z values to the model. The Equation 3.98 may be used recursively to generate desired numbers of Z_t. By generating an adequately long series and deleting 50 or 100 initial terms, the transient effect of the initial values is negligible (Salas et al., 1988). The synthetic generated values conserve the statistical properties of the historical data.

To use ARMA(p,q) for forecasting Z values in a lead time L, the following equations are used if $Z_t(L)$ denotes the value of Z_t at lead time L:

$$Z_t(L) = \varphi_1 Z_{t+L-1} + \varphi_2 Z_{t+L-2} + \cdots + \varphi_p Z_{t+L-p} - \theta_1 \varepsilon_{t+L-1} - \cdots - \theta_q \varepsilon_{t+L-q} \quad \text{for } L \leq q \tag{3.102}$$

$$Z_t(L) = \varphi_1 Z_{t+L-1} + \varphi_2 Z_{t+L-2} + \cdots + \varphi_p Z_{t+L-p} \quad \text{for } L > q \tag{3.103}$$

3.8.10 Autoregressive Integrated Moving Average Modeling

A time series should have the following characteristics to be eligible for applying the ARMA models to it:

- There is no apparent deviation from stationarity in the series.
- Autocorrelation function is rapidly decreasing.

Considering the fact that a time series does not necessarily meet these conditions, a proper transformation is usually performed to generate the time series with the two above conditions. This has been usually achieved by differencing, which is necessary for ARIMA models, which are very powerful for describing stationary and nonstationary time series and is called an autoregressive integrated moving average process. The nonseasonal form of ARIMA models of order (p, d, q) is expressed as

$$\varphi(B)(1-B)^d z_t = \theta(B)\varepsilon_t \tag{3.104}$$

where $\varphi(B)$ and $\theta(B)$ are polynomials of degree p and q, respectively:

$$\varphi(B) = 1 - \varphi_1 B - \varphi_2 B^2 - \cdots - \varphi_p B^p \tag{3.105}$$

$$\theta(B) = 1 - \theta_1 B - \theta_2 B^2 - \cdots - \theta_q B^q \tag{3.106}$$

The model has $p + q + 1$ parameters, which are $\{\varphi_1, \ldots, \varphi_p, \theta_1, \ldots, \theta_q, \sigma^2(\varepsilon)\}$.

The seasonal form of ARIMA models of nonseasonal order (p, d, q), and of seasonal order $(P, D, Q)_w$ with seasonality w has also been developed to model the seasonal hydrologic time series. The general form of multiplicative ARIMA(p, d, q) $(P, D, Q)_w$ is

$$(1 - \Phi_1 B^w - \Phi_2 B^{2w} - \cdots - \Phi_p B^{Pw})(1 - \varphi_1 B - \varphi_2 B^2 - \cdots - \varphi_p B^p)(1 - B^w)^D (1 - B)^d z_t$$

$$= (1 - \Theta_1 B^w - \Theta_2 B^{2w} - \cdots - \Theta_Q B^{Qw})(1 - \theta_1 B - \theta_2 B^2 - \cdots - \theta_q B^q)\varepsilon_t \tag{3.107}$$

where
ε_t is the independently distributed random variable
B is the backward operator as $B(z_t) = z_{t-1}$
$(1 - B^w)^D$ is the Dth seasonal difference of season w
$(1 - B)^d$ is the dth nonseasonal difference
p is the order of nonseasonal autoregressive model
q is the order of nonseasonal moving average model
P is the order of seasonal autoregressive model
Q is the order of seasonal moving average model
Φ is the seasonal autoregressive parameter
Θ is the seasonal moving average parameter
φ is the nonseasonal autoregressive parameter
θ is the nonseasonal moving average parameter

Equation 3.107 can also be written as

$$\Phi(B^w)\varphi(B)(1 - B^w)^D(1 - B)^d z_t = \Theta(B^w)\theta(B)\varepsilon_t \tag{3.108}$$

Example 3.24

Formulate the model ARIMA$(1,0,1)(1,0,1)_{12}$.

Solution
Using Equation 3.107, we have

$$(1 - \Phi_1 B^w)(1 - \varphi_1 B)z_t = (1 - \Theta_1 B^w)(1 - \theta_1 B)\varepsilon_t$$

or

$$z_t - \varphi_1 z_{t-1} - \Phi_1 z_{t-12} + \varphi_1 \Phi_1 z_{t-13} = \varepsilon_t - \theta_1 \varepsilon_{t-1} - \Theta_1 \varepsilon_{t-12} - \theta_1 \Theta_1 \varepsilon_{t-13}$$

Therefore, the model has non-linear parameters because of the terms $\theta\Theta$ and $\varphi\Phi$.

3.8.10.1   Time Series Forecasting Using ARIMA Models

The ARIMA models are nonstationary. Therefore, it is not reasonable to use them for synthetic generation of stationary time series. However, they can be useful for forecasting. The equations of ARIMA models, mentioned before in this section, could be used for data forecasting. As an example, the forecasting equations for ARIMA $(1,0,1)(1,0,1)_{12}$ could be represented as follows:

$$z_t(L) = \varphi_1 z_{t-1+L} + \Phi_1 z_{t-12+L} - \varphi_1 \Phi_1 z_{t-13+L} - \theta_1 \varepsilon_{t-1+L}$$

$$- \Theta_1 \varepsilon_{t-12+L} - \theta_1 \Theta_1 \varepsilon_{t-13+L} \quad \text{for } L \le 12 \tag{3.109}$$

$$z_t(L) = \varphi_1 z_{t-1+L} + \Phi_1 z_{t-12+L} - \varphi_1 \Phi_1 z_{t-13+L} \quad \text{for } L > 12 \tag{3.110}$$

Example 3.25

Table 3.10 presents a 16 year monthly time series of rates of recharge to an aquifer. Determine the best autoregressive model for these data and forecast the rate of recharge for the first half of water year 2007.

Solution
In data preparation for forecasting, first the stationary condition of the time series must be checked. In other words, the time series must meet the stationary condition in order to forecast newer data. Otherwise, the forecasted data are not accurate.

In this example, evaluation of the time series shows that there are no trend, outlying observations, and periodicity in the data. The data is plotted on a semilogarithmic graph (Figure 3.21). As demonstrated in this figure, the data does not fit the normal distribution. Therefore, the data are normalized and standardized. Figure 3.22 depicts the normalized and standardized data.

TABLE 3.10
Monthly Time Series of Rates of Recharge to an Aquifer (in./Month)

	October	November	December	January	February	March	April	May	June	July	August	September
1991	1.71	1.72	1.45	1.41	1.66	8.74	8.09	4.73	0.67	0.11	0.01	0.86
1992	2.68	2.28	2.68	2.42	3.16	3.67	9.87	6.07	3.20	0.72	0.05	0.52
1993	1.26	1.93	1.86	1.61	1.87	2.46	6.07	4.83	0.38	0.02	0.13	0.32
1994	1.06	1.95	1.87	1.53	1.90	5.03	7.98	8.41	4.25	1.38	1.14	0.88
1995	1.13	1.63	1.82	1.85	2.66	3.44	4.89	0.92	0.22	0.05	0.03	0.20
1996	0.92	2.05	1.72	1.46	1.92	3.25	10.81	8.33	0.65	0.02	0.00	0.67
1997	0.92	1.37	1.78	2.23	2.02	2.38	5.84	1.82	1.39	0.62	0.03	0.38
1998	0.82	1.38	1.49	1.57	1.36	2.45	6.17	8.14	2.72	0.70	0.13	0.23
1999	0.90	1.56	1.36	1.48	1.90	2.31	14.14	13.27	5.18	0.83	0.45	0.49
2000	1.25	1.45	1.89	2.05	2.48	4.50	12.89	10.02	3.41	1.22	0.45	0.79
2001	2.10	2.10	2.69	2.71	3.34	4.24	5.74	3.78	1.99	0.94	0.07	0.84
2002	1.08	1.55	1.36	1.27	1.28	1.35	9.15	9.76	1.29	0.86	0.00	0.98
2003	1.07	1.35	1.38	1.57	2.88	2.55	4.72	3.31	2.05	0.02	0.42	0.01
2004	0.30	0.61	0.66	0.90	1.06	1.94	8.39	6.85	3.20	0.57	0.07	0.18
2005	0.78	1.05	1.10	1.32	4.11	3.07	3.35	5.79	1.85	0.59	1.37	1.87
2006	0.52	1.05	1.14	1.42	1.65	2.35	2.73	2.61	0.70	0.04	0.03	0.03

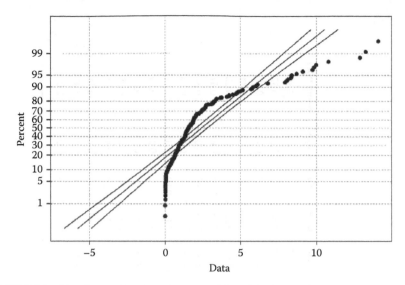

FIGURE 3.21   The time series plotted on a semi-logarithmic graph.

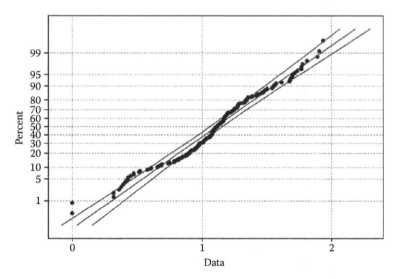

FIGURE 3.22   Normalized and standardized data plotted on a semi-logarithmic graph.

To determine the parameters of an ARIMA model, autocorrelation function (ACF) graph and partial autocorrelation function (PACF) graph can be used. An ACF graph evaluates the correlation between the data of a time series and a PACF graph shows the correlation between autoregressive coefficients in different time lags. For each of these graphs, a confidence limit is set. Only if all of the correlation values are within this limit, the model can be selected and the forecasting is

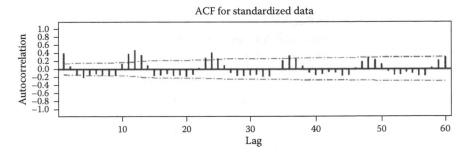

FIGURE 3.23   ACF graph for different lags.

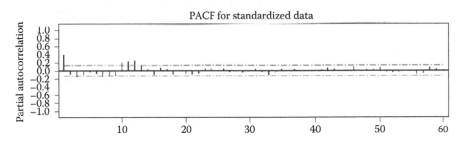

FIGURE 3.24   PACF graph for different lags.

acceptable. Using ACF and PACF graphs of normalized data, the values of parameters p and q of the ARIMA model are determined.

The time lags that violate the confidence limits of ACF and PACF graphs represent the values of parameters p and q, respectively.

In this example, after normalizing and standardizing the data, their corresponding ACF and PACF graphs are plotted. Figures 3.23 and 3.24 show these graphs. A 95% confidence limit is considered for these graphs. Using ACF and PACF graphs, the parameters p and q are approximated. According to these figures, values of both p and q are less than or equal to 1.

The final structure of the forecasting model is determined by trial and error. Then, Akaike test is used to compare different ARIMA(p,q) models and select the best one. For this purpose we have

$$AIC(p,q) = N \ln(\sigma_\varepsilon^2) + 2(p+q)$$

The model with less AIC is selected as the best one. Table 3.11 presents the results for different ARIMA models. According to this table, the model ARIMA(1,1,1)(1,1,1) has the least AIC and is selected for our forecasting purpose.

Figures 3.25 and 3.26 show the ACF and PACF graphs for the residuals. These figures illustrate that the values of errors in both of the graphs are within the confidence limit. Therefore, the selected forecasting model, ARIMA(1,1,1)(1,1,1), is confirmed.

TABLE 3.11
Values of AIC for Different ARIMA Models

ARIMA Model	AIC	Residual Error
(1,0,0)(1,0,1)	−125	0.2975
(1,1,1)(1,1,1)	−131	0.2886
(1,0,0)(1,0,1)	−129	0.2868

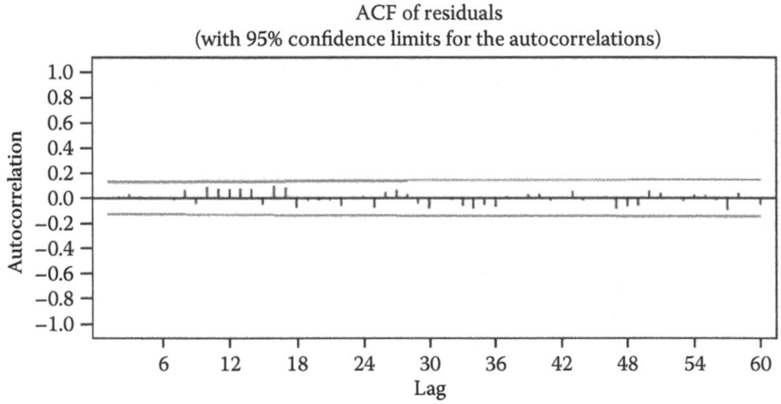

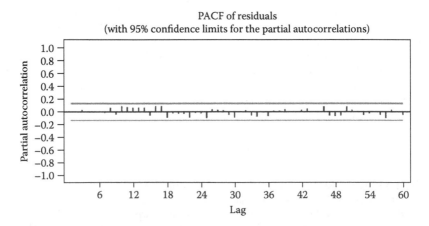

ACF of residuals
(with 95% confidence limits for the autocorrelations)
Autocorrelation
1.0
0.8
0.6
0.4
0.2
0.0
−0.2
−0.4
−0.6
−0.8
−1.0
6
12
18
24
30
36
42
48
54
60
Lag

FIGURE 3.25 ACF graph for residuals of the data.

PACF of residuals
(with 95% confidence limits for the partial autocorrelations)
Partial autocorrelation
1.0
0.8
0.6
0.4
0.2
0.0
−0.2
−0.4
−0.6
−0.8
−1.0
6
12
18
24
30
36
42
48
54
60
Lag

FIGURE 3.26 PACF graph for residuals of the data.

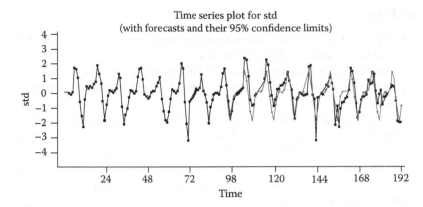

FIGURE 3.27 Comparison of observed and forecasted values for rates of recharge to the aquifer using ARIMA(1, 1, 1)(1, 1, 1) model.

The goodness-of-fit tests are usually used to evaluate the performance of forecasting models. However, this can be evaluated more efficiently by developing (forecasting) a long-term time series of the data and comparing this series with the corresponding historical time series. For this purpose, the first 36–48 months are usually used to estimate the parameters of forecasting model. Then, the forecasting process is done for the rest of the historical time series. In this process, after forecasting the value of each month, the actual value of the time series is substituted with the forecasted value to forecast the next month. This process continues until the last month of the time series.

In this example, the first 92 months of the recharge rate time series were used to train the forecasting model. Then, the remaining 100 months were forecasted. Figure 3.27 expresses the actual and forecasted values.

This figure illustrates the acceptable performance of the selected forecasting model. Therefore, this model is used to generate 6 months of recharge rates for the first half of 2007. Since the generated values are standardized and normalized, the forecasted values must be converted to the regular format. It should be noted that if the main time series has a trend, this trend must also be added to the forecasted values. Table 3.12 presents the forecasted values for rates of recharge to the aquifer.

TABLE 3.12

Forecasted Values for Rates of Recharge to the Aquifer for the First Half of Water Year 2007

Month	1	2	3	4	5	6
Recharge rate	0.39	0.70	0.78	0.85	1.31	1.73

PROBLEMS

3.1 Which property is a better measure of the productivity of an aquifer: porosity or hydraulic conductivity? Explain why.

3.2 A confined aquifer has a transitivity of $100\,m^2$/day. If the slope of the piezometric surface is 0.0005, compute the flow rate of water through the aquifer per km of width.

3.3 An aquifer has a hydraulic conductivity of 10.0 m/day. Compute the velocity of water through the aquifer if the slope of the water table is 0.002 and the porosity is 0.2. How far will the ground water travel in 1.0 year?

3.4 Two piezometers were installed at two wells one km apart. Well's information is listed below. If the intrinsic permeability of the aquitard is $0.005\,m^2$, determine the Darcy's velocity.

	Observation Well	
	A	B
Depth to bottom of well	114.7	125.8
Depth of water surface	48.4	51.2

3.5 Piezometers A and B are installed in a confined aquifer with an estimated hydraulic conductivity of 12 m/day and a porosity of 0.25 and one km apart. Piezometer A is at the true north and piezometer B is located at an angle of 45° clockwise from A. The water level at B is 15 cm below A. What is the Darcy velocity along line AB. Explain why this is not the actual seepage velocity in the aquifer. Piezometer C is also located in the aquifer 0.5 km from A at an angle of 90° clockwise from it. The water level at C is 10 cm below A. Calculate the Darcy velocity and real seepage velocity in the aquifer.

3.6 An aquifer formation has a hydraulic conductivity of 18 m/day when the aquifer is at a temperature of 25°C. Determine the intrinsic permeability of the aquifer. If the fluid in the aquifer consists of spilled tetrachloroethylene, then determine the hydraulic conductivity of tetrachloroethylene movement. Under the same piezometric gradient, which fluid moves faster?

3.7 An anisotropic aquifer system has horizontally parallel layers with hydraulic conductivities of $K_{11} = 15$ m/day, $K_{22} = 5$ m/day. The thicknesses of the first and second layers are 2 m and 4 m, respectively. If the water table drops 1.2 m in 800 m in the aquifer, how much will the seepage velocity be? Compare this velocity with the seepage velocity when the aquifer is homogeneous with a hydraulic conductivity equal to the average of those two layers.

3.8 In an aquifer, $K_{xx} = 100$ m/day, $K_{yy} = 10$ m/day, and K_{xy} can be ignored. Where the x and y axes are along the east/west and north/south directions, receptivity. Groundwater flows toward east direction in this aquifer. The development of a new well field will cause the mean head gradient to shift from the east to southeast. Compare the specific discharge in the direction of the head gradient for a gradient of 0.01 in the east direction with the specific discharge in the direction of the head gradient of 0.01 in the southeast direction. In either of these cases, does the groundwater flow have the direction of the head gradient?

3.9 Four horizontal, homogeneous, isotropic geologic formations each of them 5 m thick, overlie one another. If the hydraulic conductivities are 10^{-4}, 10^{-6}, 10^{-4}, and 10^{-6} m/s, respectively, calculate the horizontal and vertical components of hydraulic conductivity.

3.10 Table 3.13 shows the rates of recharge to an aquifer. Test the stationarity of the time series in its mean. Define a trend model and the residuals of the series.

3.11 The monthly discharges for 80 months from an aquifer are divided into 8 categories as shown in the Table 3.14. Previous studies show that the precipitation data in this station is normally distributed. If the mean and standard deviation for the data are 20.48 (mm) and 11.84 (mm), respectively, use the chi-square test with 5% significance level and comment on the selected distribution for the sample data.

TABLE 3.13

The Rates of Recharge to an Aquifer in Problem 3.10

Time Sequences	Rate of Recharge (m³/hr)	Time Sequences	Rate of Recharge (m³/hr)
1	200	11	172
2	212	12	180
3	302	13	185
4	212	14	190
5	192	15	302
6	186	16	215
7	148	17	218
8	312	18	158
9	220	19	142
10	195	20	175

TABLE 3.14

The Monthly Discharges from an Aquifer in Problem 3.14

Interval (i)	Discharges from the Aquifer (m³/d)	Actual Frequency (f_i)
1	$0 \leq P < 5$	9
2	$5 \leq P < 10$	13
3	$10 \leq P < 15$	5
4	$15 \leq P < 20$	10
5	$20 \leq P < 25$	11
6	$25 \leq P < 30$	12
7	$30 \leq P < 35$	8
8	$35 \leq P < 40$	12

TABLE 3.15

The Monthly Discharges from an Aquifer in Problem 3.12

Interval (i)	Discharges from the Aquifer (m^3/d)	Actual Frequency (f_i)
1	$0 \leq P < 5$	27
2	$5 \leq P < 10$	24
3	$10 \leq P < 15$	9
4	$15 \leq P < 20$	8
5	$20 \leq P < 25$	5
6	$25 \leq P < 30$	4
7	$30 \leq P < 35$	2
8	$35 \leq P < 40$	1

3.12 The classified monthly discharges from an aquifer are divided into 8 categories as shown in the Table 3.15. Previous studies show that the precipitation data in this station is normally distributed. Use the chi-square test with 5% significance level and comment on the selected distribution for the sample data.

3.13 The time series of inflow to an aquifer could be shown in the form of the following equation. Categorize this model as a multiplicative ARIMA model.

$$(1 - B^{10})Z_t = (1 + 0.4B)(1 - 0.7B^{10})\varepsilon_t.$$

3.14 Write the mathematical expression of order 1 and order 2 non-seasonal ARIMA. Why is the seasonal differencing done? Write the mathematical expression for the second order seasonal differencing for monthly time series. (Assume 12 for the number of seasons)

3.15 ARMA(1,1) model is fitted to the time series of groundwater quality data in an aquifer with parameters $\varphi = -0.6$ and $\theta = -0.4$. Plot the auto correlation functions of the time series from lag-1 to lag-5.

APPENDIX A

TABLE A1
Normal Distribution

z	0	0.01	0.02	0.03	0.04	0.05	0.06	0.07	0.08	0.09
0	0.0000	0.0040	0.0080	0.0120	0.0160	0.0199	0.0239	0.0279	0.0319	0.0359
0.1	0.0398	0.0438	0.0478	0.0517	0.0557	0.0596	0.0636	0.0675	0.0714	0.0753
0.2	0.0793	0.0832	0.0871	0.0910	0.0948	0.0987	0.1026	0.1064	0.1103	0.1141
0.3	0.1179	0.1217	0.1255	0.1293	0.1331	0.1368	0.1406	0.1443	0.1480	0.1517
0.4	0.1554	0.1591	0.1628	0.1664	0.1700	0.1736	0.1772	0.1808	0.1844	0.1879
0.5	0.1915	0.1950	0.1985	0.2019	0.2054	0.2088	0.2123	0.2157	0.2190	0.2224
0.6	0.2257	0.2291	0.2324	0.2357	0.2389	0.2422	0.2454	0.2486	0.2517	0.2549
0.7	0.2580	0.2611	0.2642	0.2673	0.2704	0.2734	0.2764	0.2794	0.2823	0.2852
0.8	0.2881	0.2910	0.2939	0.2967	0.2995	0.3023	0.3051	0.3078	0.3106	0.3133
0.9	0.3159	0.3186	0.3212	0.3238	0.3264	0.3289	0.3315	0.3340	0.3365	0.3389
1	0.3413	0.3438	0.3461	0.3485	0.3508	0.3531	0.3554	0.3577	0.3599	0.3621
1.1	0.3643	0.3665	0.3686	0.3708	0.3729	0.3749	0.3770	0.3790	0.3810	0.3830
1.2	0.3849	0.3869	0.3888	0.3907	0.3925	0.3944	0.3962	0.3980	0.3997	0.4015
1.3	0.4032	0.4049	0.4066	0.4082	0.4099	0.4115	0.4131	0.4147	0.4162	0.4177
1.4	0.4192	0.4207	0.4222	0.4236	0.4251	0.4265	0.4279	0.4292	0.4306	0.4319
1.5	0.4332	0.4345	0.4357	0.4370	0.4382	0.4394	0.4406	0.4418	0.4429	0.4441
1.6	0.4452	0.4463	0.4474	0.4484	0.4495	0.4505	0.4515	0.4525	0.4535	0.4545
1.7	0.4554	0.4564	0.4573	0.4582	0.4591	0.4599	0.4608	0.4616	0.4625	0.4633
1.8	0.4641	0.4649	0.4656	0.4664	0.4671	0.4678	0.4686	0.4693	0.4699	0.4706
1.9	0.4713	0.4719	0.4726	0.4732	0.4738	0.4744	0.4750	0.4756	0.4761	0.4767
2	0.4772	0.4778	0.4783	0.4788	0.4793	0.4798	0.4803	0.4808	0.4812	0.4817
2.1	0.4821	0.4826	0.4830	0.4834	0.4838	0.4842	0.4846	0.4850	0.4854	0.4857
2.2	0.4861	0.4864	0.4868	0.4871	0.4875	0.4878	0.4881	0.4884	0.4887	0.4890
2.3	0.4893	0.4896	0.4898	0.4901	0.4904	0.4906	0.4909	0.4911	0.4913	0.4916
2.4	0.4918	0.4920	0.4922	0.4925	0.4927	0.4929	0.4931	0.4932	0.4934	0.4936
2.5	0.4938	0.4940	0.4941	0.4943	0.4945	0.4946	0.4948	0.4949	0.4951	0.4952
2.6	0.4953	0.4955	0.4956	0.4957	0.4959	0.4960	0.4961	0.4962	0.4963	0.4964
2.7	0.4965	0.4966	0.4967	0.4968	0.4969	0.4970	0.4971	0.4972	0.4973	0.4974
2.8	0.4974	0.4975	0.4976	0.4977	0.4977	0.4978	0.4979	0.4979	0.4980	0.4981
2.9	0.4981	0.4982	0.4982	0.4983	0.4984	0.4984	0.4985	0.4985	0.4986	0.4986
3	0.4987	0.4987	0.4987	0.4988	0.4988	0.4989	0.4989	0.4989	0.4990	0.4990

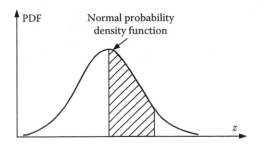

PDF

Normal probability density function

z

For example, in order to calculate $Z_{0.9}$, the Z value corresponding to $(0.9 - 0.5) = 0.4$ should be find in the Normal distribution table. $Z_{0.9} = 1.282$.

TABLE A2

Chi-Square Distribution (The Following Table Provides the Values of χ_α^2 that Correspond to a given Upper-Tail Area α and a Specified Number of Degrees of Freedom)

Degrees of Freedom	Upper-Tail Area α						
	0.99	0.98	0.95	0.90	0.80	0.70	0.50
1	0.000157	0.000628	0.00393	0.0158	0.0642	0.148	0.455
2	0.0201	0.0404	0.103	0.211	0.446	0.713	1.386
3	0.115	0.185	0.352	0.584	1.005	1.424	2.366
4	0.297	0.429	0.711	1.064	1.649	2.195	3.357
5	0.554	0.752	1.145	1.610	2.343	3.000	4.351
6	0.872	1.134	1.635	2.204	3.070	3.828	5.348
7	1.239	1.564	2.167	2.833	3.822	4.671	6.346
8	1.646	2.032	2.733	3.490	4.594	5.527	7.344
9	2.088	2.532	3.325	4.168	5.380	6.393	8.343
10	2.558	3.059	3.940	4.865	6.179	7.267	9.342
11	3.053	3.609	4.575	5.578	6.989	8.148	10.341
12	3.571	4.178	5.226	6.304	7.807	9.034	11.340
13	4.107	4.765	5.892	7.042	8.634	9.926	12.340
14	4.660	5.368	6.571	7.790	9.467	10.821	13.339
15	5.229	5.985	7.261	8.547	10.307	11.721	14.339
16	5.812	6.614	7.962	9.312	11.152	12.624	15.338
17	6.408	7.255	8.672	10.085	12.002	13.531	16.338
18	7.015	7.906	9.390	10.865	12.857	14.440	17.338
19	7.633	8.567	10.117	11.651	13.716	15.352	18.338
20	8.260	9.237	10.851	12.443	14.578	16.266	19.337
21	8.897	9.915	11.591	13.240	15.445	17.182	20.337
22	9.542	10.600	12.338	14.041	16.314	18.101	21.337
23	10.196	11.293	13.091	14.848	17.187	19.021	22.337
24	10.856	11.992	13.848	15.659	18.062	19.943	23.337
25	11.524	12.697	14.611	16.473	18.940	20.867	24.337
26	12.189	13.409	15.379	17.292	19.820	21.792	25.336
27	12.879	14.125	16.151	18.114	20.703	22.719	26.336
28	13.565	14.847	16.928	18.939	21.588	23.647	27.336
29	14.256	15.574	17.708	19.768	22.475	24.577	28.336
30	14.953	16.306	18.493	20.599	23.364	25.508	26.336

TABLE A2 (continued)

Chi-Square Distribution (The Following Table Provides the Values of χ^2_α that Correspond to a given Upper-Tail Area α and a Specified Number of Degrees of Freedom)

Degrees of Freedom	Upper-Tail Area α						
	0.30	0.20	0.10	0.05	0.02	0.01	0.001
1	1.074	1.642	2.706	3.841	5.412	6.635	10.827
2	2.408	3.219	4.605	5.991	7.824	9.210	13.815
3	3.665	4.642	6.251	7.815	9.837	11.345	16.268
4	4.878	5.989	7.779	9.488	11.668	13.277	18.465
5	6.064	7.289	9.236	11.070	13.388	15.086	20.517
6	7.231	8.558	10.645	12.592	15.033	16.812	22.457
7	8.383	9.803	12.017	14.067	16.622	18.475	24.322
8	9.524	11.030	13.362	15.507	18.168	20.090	26.125
9	10.656	12.242	14.684	16.919	19.679	21.666	27.877
10	11.781	13.442	15.987	18.307	21.161	23.209	29.588
11	12.899	14.631	17.275	19.675	22.618	24.725	31.264
12	14.011	15.812	18.549	21.026	24.054	26.217	32.909
13	15.119	16.985	19.812	22.362	25.472	27.688	34.528
14	16.222	18.151	21.064	23.685	26.873	29.141	36.123
15	17.322	19.311	22.307	24.996	28.259	30.578	37.697
16	18.418	20.465	23.542	26.296	29.633	32.000	39.252
17	19.511	21.615	24.769	27.587	30.995	33.409	40.790
18	20.601	22.760	25.989	28.869	32.346	34.805	42.312
19	21.689	23.900	27.204	30.144	33.687	36.191	43.820
20	22.775	25.038	28.412	31.410	35.020	37.566	45.315
21	23.858	26.171	29.615	32.671	36.343	38.932	46.797
22	24.939	27.301	30.813	33.924	37.659	40.289	48.268
23	26.018	28.429	32.007	35.172	38.968	41.638	49.728
24	27.096	29.553	33.196	36.415	40.270	42.980	51.179
25	28.172	30.675	34.382	37.652	41.566	44.314	52.620
26	29.246	31.795	35.563	38.885	42.856	45.642	54.052
27	30.319	32.912	36.741	40.113	44.140	46.963	55.476
28	31.391	34.027	37.916	41.337	45.419	48.278	56.893
29	32.461	35.139	39.087	42.557	46.693	49.588	58.302
30	33.530	36.250	40.256	43.773	47.962	50.892	59.703

REFERENCES

Ahmed, N. and Sunada, D.K. 1969. Nonlinear flow in porous media. *Journal of Hydraulics Division, ASCE*, 95(HY6), 1847–1857.

Aitchison, J. and Brown, J.A.C. 1969. *The Lognormal Distribution*. Cambridge University Press, Cambridge, U.K., 176 p.

Akaike, H. 1974. A new look at the statistical model identification. *IEEE Transactions on Automatic Control*, 19(6), 716–723.

Barrette, J.P. and Goldsmith, L. 1976. When is n large enough? *The American Statistician*, 30, 67–71.

Bennett, R.R. 1962. Flow net analysis, in *Theory of Aquifer Tests*. J.G. Ferris, D.B. Knowlers, R.H. Brown, and R.W. Stallman, eds., US Geol. Serv. Prof. Paper 708, 70 p.

Bouwer, H. 1978. *Groundwater Hydrology*. McGraw-Hill Book Company, New York, 480 p.

Box, G.E.P. and Cox, D.R. 1964. An analysis of transformation. *Journal of the Royal Statistical Society*, B26, 211–252.

Brockwell, P.J. and Davis, R.A. 1987. *Time Series Theory and Methods*. Library of Congress Cataloging in Publication Data.

Brooks, R.H. and Corey, A.T. 1964. *Hydraulics Properties of Porous Media*. Hydrology Paper 3, Colorado State University, Fort Collins, CO, 71 p.

Campbell, G.S. 1974. A simple method for determining unsaturated conductivity from moisture retention data. *Soil Science*, 117, 311–314.

De Wiest, R.J.M. 1965. *Geohydrology*. John Wiley & Sons, New York, 463 p.

Ferris, J.G., Knowles, D.B., Browne, R.H., and Stallman, R.W. 1962. *Theory of Aquifer Tests*, U.S. Geological Survey, Water-Supply Paper 1536E, 174 p.

Freeze, R. and Cherry, J. 1979. *Groundwater*. Prentice-Hall, Englewood Cliffs, NJ, 604 p.

Gilbert, R.O. 1987. *Statistical Methods for Environmental Pollution Monitoring*. John Wiley & Sons Inc., New York.

Hansch, C. and Leo, A. 1979. *Substitute Constants for Correlation Analysis in Chemistry and Biology*. John Wiley & Sons Inc., New York, pp. 841–847.

Hashino, M. and Delleur, J.W. 1981. *Investigation of the Hurst Coefficient and Optimization of ARMA Models for Annual River Flow*. Report CE-HSE-81-1, Civil Engng, Purdue Univ., West Lafayette. Indiana, USA.

Jacob, C.E. 1940. On the flow of water in an elastic artesian aquifer, *Transactions. American Geophysical Union*, 2, 574–586.

Karamouz, M., Szidarovszky, F., and Zahraie, B. 2003. *Water Resources Systems Analysis*. Lewis Publishers, CRC Press, Boca Raton, FL, 590 p.

Lapin, L.L., 1990. *Probability and Statistics for Modern Engineering*. PWS-KENT Publishing Company, Boston, MA, 810 p.

Novotny, V.N. 2003. *Water Quality—Diffuse Pollution and Watershed Management*, 2nd edn. John Wiley & Sons Inc., Hoboken, NJ.

Price, K.R., Gilbert, R.O., and Gano, K.A. 1981. *Americium-241 in Surface Soil Associated with the Hanford Site and Vicinity*. Pacific Northwest Laboratory, Richland, WA, PNL-3731.

Rawls, W.J., Ahuja, L.R., Brakensiak, D.L., and Shirmohammadi, A. 1993. Infiltration and soil water movement, in *Handbook of Hydrology*, D.R. Maidment, ed., McGraw-Hill, New York, pp. 5.1–5.51.

Salas, J.D., Delleur, J.W., Yevjevich, V., and Lane, W.L. 1988. *Applied Modeling of Hydrological Time Series*. Water Resources Publications, Littleton, CO, 484 p.

Schwartz, F.W. and Zhang, H. 2003. *Fundamentals of Groundwater*. John Wiley & Sons, Inc., New York, 583 p.

Snedecor, G.W. and Cochran, W.G. 1967. *Statistical Methods*. 6th edn. Iowa State University Press, Ames, IA.

Snow, D.T. 1969. Anisotropic permeability of fractured media, *Water Resources Research*, 5, 1273–1289.

Sokal, R.R. and Rohlf, R.J. 1981. *Biometry*, 2nd edn. W.H. Freeman, San Francisco, CA, 133 p.

Todd, D.K. 1980. *Groundwater Hydrology*. John Wiley & Sons Inc., New York, 621 p.

Todd, D.K. and Mays, L.M., 2005. *Groundwater Hydrology*. John Wiley & Sons Inc., New York.

Van Genuchten, M.Th. 1980. A closed form equation for predicting the hydraulic conductivity of unsaturated soils, *Soil Society of America Journal*, 44, 892–898.

Yevjevich, V. 1972. *Probability and Statistics in Hydrology*. Water Resources Publications, Fort Collins, CO, 302 p.

4 Hydraulics of Groundwater

4.1 INTRODUCTION

Groundwater in its natural state moves due to hydraulic forces. Because hydraulic head represents the energy of water, groundwater flows from locations of higher head, usually upland areas, to locations of lower head, such as lowland areas, marshes, springs, and rivers. This is a principle to use water-level data obtained from wells, springs, and surface water features to determine the horizontal and vertical direction of groundwater movement and to estimate the rate of groundwater flow. In the subsurface, water occurs in four phases: water vapor from transpiration by plants and direct evaporation from the water table, condensed water absorbed by dry soil particles, water hold around soil particles by molecular attraction, and water under the influence of gravity.

In the saturated zone, groundwater flows through interconnected voids in response to the difference in fluid pressure and elevation. The change in hydraulic head over a specified distance in a given direction is called the hydraulic gradient. The quantity of groundwater moving through a volume of rock can be estimated using Darcy's law which is a function of hydraulic gradient. Darcy's law determines the rate of groundwater flow through an area.

The groundwater flow equation is used in hydrogeology and it is a mathematical relationship representing groundwater flow through an aquifer. The mathematical description of groundwater flow is based on the principles of conservation of mass, energy, and momentum. In this chapter, groundwater flow equations are developed using these principles in confined and unconfined aquifers for steady- and unsteady-state conditions. Groundwater flow equations in an aquifer subjected to the impacts of a pumping well are also discussed in this chapter. Generally, groundwater flow equations are expressed in terms of partial differential equations where spatial coordinates and time are independent variables. In addition, the different methods for artificial recharge are explained as a tool to store surplus surface water underground.

4.2 CONTINUITY EQUATION

The conservation of mass states that for a given increment of time the difference between the mass flowing in across the boundaries, the mass flowing out across the boundaries, and the sources within the volume, is the change in storage. The continuity

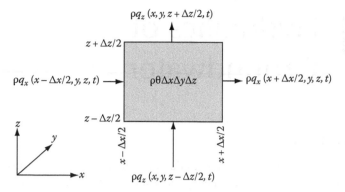

FIGURE 4.1   Mass conservation in elementary control volume.

equation, as a fundamental law of groundwater flow, expresses the principle of mass conservation as follows:

[Mass inflow rate − mass outflow rate = Rate of change of mass storage with time]

(4.1)

Considering an elementary control volume of soil (Figure 4.1), which has the volume of ($\Delta x \Delta y \Delta z$), the mass of groundwater, M, in this control volume is as

$$M = \rho \theta \Delta x \Delta y \Delta z \tag{4.2}$$

where θ is the moisture content of the porous medium. The Equation 4.1 can be written as follows:

$$\frac{\partial M}{\partial t} = \text{inflow} - \text{outflow} \tag{4.3}$$

The inflow and outflow can be calculated for each side of element. For example, the mass of groundwater inflow to the left side is as

$$\rho q_x \left(\frac{x - \Delta x}{2}, y, z, t \right) \Delta y \Delta z \approx \left(\rho q_x - \frac{\Delta x}{2} \frac{\partial \rho q_x}{\partial x} \right) \Delta y \Delta z \tag{4.4}$$

A similar equation can be derived for other sides. Considering these equations, the total inflow minus outflow can be derived as follows:

$$\frac{\partial M}{\partial t} = -\left(\frac{\partial \rho q_x}{\partial x} + \frac{\partial \rho q_y}{\partial y} + \frac{\partial \rho q_z}{\partial z} \right) \Delta x \Delta y \Delta z \tag{4.5}$$

Based on Equation 4.3, the change in storage is calculated by

$$\frac{\partial M}{\partial t} = \frac{\partial}{\partial t}(\rho\theta\Delta x\Delta y\Delta z) = \rho\left(\frac{\theta}{\rho}\frac{\partial\rho}{\partial t} + \frac{\partial\theta}{\partial t} + \frac{\theta}{\Delta z}\frac{\partial\Delta z}{\partial t}\right)\Delta x\Delta y\Delta z \qquad (4.6)$$

Compression of porous medium and water can be considered using the following equations, respectively:

$$\frac{1}{\Delta z}\frac{\partial\Delta z}{\partial t} = \alpha\frac{\partial p}{\partial t} \qquad (4.7)$$

$$\frac{1}{\Delta z}\frac{\partial\Delta p}{\partial t} = \beta\frac{\partial p}{\partial t} \qquad (4.8)$$

where
   α is the elastic compressibility coefficient of a porous medium
   p is the water pressure
   β is the compressibility coefficient of water

Substitution of these equations in Equation 4.6 gives

$$\frac{\partial M}{\partial t} = \rho\left(\theta(\alpha+\beta)\frac{\partial p}{\partial t} + \frac{\partial\theta}{\partial t}\right)\Delta x\,\Delta y\,\Delta z \qquad (4.9)$$

Therefore, using Equations 4.2 and 4.5, the continuity equation can be derived as

$$\rho\left(\theta(\alpha+\beta)\frac{\partial p}{\partial t} + \frac{\partial\theta}{\partial t}\right) = -\left(\frac{\partial\rho q_x}{\partial x} + \frac{\partial\rho q_y}{\partial y} + \frac{\partial\rho q_z}{\partial z}\right) \qquad (4.10)$$

The left side of this equation describes the change in volume of water in porous medium due to change in water content or compression of the water and the medium. Considering ρ is constant and no change in moisture content, Equation 4.10 can be summarized in two-dimensional as expressed in Example 4.1.

Example 4.1

Derive the steady-state flow equation in an aquifer in terms of hydraulic head h with the following hydraulic conductivity function:

$$K = K_0(a + be^{cx} + de^{fy})$$

where a, b, c, d, and f are system's parameters.

Solution

The continuity equation (Equation 4.10) in steady-state condition can be simplified as follows:

$$\frac{\partial q_x}{\partial x} + \frac{\partial q_y}{\partial y} = 0$$

Applying Darcy's law, we have

$$\frac{\partial}{\partial x}\left(K\frac{\partial h}{\partial x}\right) + \frac{\partial}{\partial y}\left(K\frac{\partial h}{\partial y}\right) = 0$$

or

$$\frac{\partial}{\partial x}\left(K_0(a+be^{cx}+de^{fy})\frac{\partial h}{\partial x}\right) + \frac{\partial}{\partial y}\left(K_0(a+be^{cx}+de^{fy})\frac{\partial h}{\partial y}\right) = 0$$

Finally,

$$(a+be^{cx}+de^{fy})\frac{\partial^2 h}{\partial x^2} + cbe^{cx}\frac{\partial^2 h}{\partial x^2} + (a+be^{cx}+de^{fy})\frac{\partial^2 h}{\partial y^2} + fde^{cx}\frac{\partial^2 h}{\partial y^2} = 0$$

4.3 EQUATION OF MOTION IN GROUNDWATER

The equation of motion can be derived using conservation of momentum. Considering the elementary control volume (Figure 4.1), the forces that usually act on the water in the control volume are

- Pressure forces
- Gravity forces
- Reaction forces of solids

For example, the pressure force on the left-hand side of control volume in Figure 4.1 is as follows:

$$\theta p\left(\frac{x-\Delta x}{2, y, z, t}\right)\Delta y\Delta z \tag{4.11}$$

The left-hand side also has the same force, but in the opposite direction. Therefore, the resulting pressure force component in the x direction is as follows (Delluer, 1999):

$$-\frac{\partial \theta p}{\partial x}\Delta x\Delta y\Delta z \tag{4.12}$$

The other components of pressure force on control volume, in y and z directions can be obtained using a similar method.

The gravity force, which is equal to the total weight of water in the control volume and acts in the negative side of z direction, is

$$-\rho g \theta \Delta x \Delta y \Delta z \qquad (4.13)$$

where g is the gravity constant. The reaction forces are usually defined as average body forces per water volume. It consists of the friction forces due to water movement and the forces that act against the water pressure. The friction and reaction forces are denoted as $r = (r_x, r_y, r_z)$ and $f = (f_x, f_y, f_z)$, respectively. The x component of these forces is as follows:

$$(r_x + f_x)\theta \Delta x \Delta y \Delta z \qquad (4.14)$$

It can have similar components in y and z directions.

Using del operator $\nabla = \left(\dfrac{\partial}{\partial x}, \dfrac{\partial}{\partial y}, \dfrac{\partial}{\partial z} \right)$, the effects of all forces can be combined in one vector as follows:

$$[-\nabla(\theta p) - \rho g \nabla z + (r + f)\theta]\Delta x \Delta y \Delta z \qquad (4.15)$$

Considering that when the fluid is at rest, the friction f is zero and the water pressure is hydrostatic, the overall reaction force can be evaluated as follows (Delluer, 1999):

$$r = \frac{p}{\theta}\nabla 0 \qquad (4.16)$$

In the case of groundwater motion, the sum of forces is equal to the changes of fluid momentum and the friction force is not zero. As the groundwater flow is generally very slow, the changes in momentum are negligible and, therefore, the forces that act on the fluid in control volume are approximately in equilibrium.

$$-\nabla p - \rho g \nabla z + f \approx 0 \qquad (4.17)$$

The friction force can be represented as

$$f = -\frac{\mu}{k} q \qquad (4.18)$$

where
   μ is the dynamic viscosity of fluid
   k is the intrinsic permeability
   q is the groundwater flux

The equation of motion can be expressed using Equations 4.12 and 4.13 as follows:

$$q = -\frac{k}{\mu}(\nabla p + \rho g \nabla z) \tag{4.19}$$

where the gradients of density are negligible, the motion equation can be simplified as

$$q = -\frac{k\rho g}{\mu}(\nabla \psi + z) = -K\nabla h \tag{4.20}$$

where $\psi = \int \frac{dp}{\rho g}$ is the pressure potential and the other variables have been defined in previous sections. The above equation clarifies the principles and assumptions that result in Darcy's law. One of the most important assumptions of Darcy's law is that the flow of water in porous medium is slow and the large values of frictional forces balance the driving forces. In the anisotropic porous media, Darcy's law in Cartesian coordinates becomes

$$q_x = -K_x \frac{\partial h}{\partial x} \tag{4.21}$$

$$q_y = -K_y \frac{\partial h}{\partial y} \tag{4.22}$$

$$q_z = -K_z \frac{\partial h}{\partial z} \tag{4.23}$$

4.3.1 GROUNDWATER FLOW EQUATION

The groundwater equation can be derived by a combination of the continuity equation and the equation of motion as expressed in Equations 4.10 and 4.20, as follows:

$$\theta(\alpha + \beta)\frac{\partial p}{\partial t} + \frac{\partial \theta}{\partial t} = \nabla\left[\frac{k}{\mu}(\nabla p + \rho g \nabla z)\right] \tag{4.24}$$

This equation can be written as follows:

$$S\frac{\partial p}{\partial t} = \nabla\left[\frac{k}{\mu}(\nabla p + \rho g \nabla z)\right] \tag{4.25}$$

where S is the storage coefficient and is equal to

$$S = \theta(\alpha + \beta) + \frac{\partial \theta}{\partial p} \tag{4.26}$$

S is related to water and soil characteristics such as saturated and unsaturated conditions of soil, and the water pressure in a confined aquifer. Therefore, the three-dimensional equation of groundwater movement as a function of water pressure (not as a function of groundwater potential) can be derived as follows:

$$S\frac{\partial p}{\partial t} = \frac{\partial}{\partial x}\left(\frac{k_x}{\mu}\frac{\partial p}{\partial x}\right) + \frac{\partial}{\partial y}\left(\frac{k_y}{\mu}\frac{\partial p}{\partial y}\right) + \frac{\partial}{\partial z}\left(\frac{k_z}{\mu}\frac{\partial p}{\partial z}\right) \tag{4.27}$$

The general form of the equation of groundwater flow is usually simplified in practice. Ignoring density effects, Equation 4.24 is simplified as follows:

$$\theta(\alpha+\beta)\frac{\partial p}{\partial t} + \frac{\partial\theta}{\partial t} = \nabla(K\nabla h) \tag{4.28}$$

In groundwater, the density is usually considered to be constant; therefore, the temporal variation of water pressure and groundwater potential are related as follows:

$$\frac{\partial p}{\partial t} = \rho g\frac{\partial h}{\partial t} \tag{4.29}$$

Therefore, the following basic groundwater flow equation is derived:

$$\rho g\theta(\alpha+\beta)\frac{\partial h}{\partial t} + \frac{\partial\theta}{\partial t} = S_s\frac{\partial h}{\partial t} = \nabla(K\nabla h) \tag{4.30}$$

where S_s is the specific storage coefficient. Written in the Cartesian coordinates, the equation of saturated groundwater flow becomes

$$S_s\frac{\partial h}{\partial t} = \frac{\partial}{\partial x}\left(K_x\frac{\partial h}{\partial x}\right) + \frac{\partial}{\partial y}\left(K_y\frac{\partial h}{\partial y}\right) + \frac{\partial}{\partial z}\left(K_z\frac{\partial h}{\partial z}\right) \tag{4.31}$$

In steady-state condition, the Equation 4.31 can be simplified as follows:

$$\frac{\partial}{\partial x}\left(K_x\frac{\partial h}{\partial x}\right) + \frac{\partial}{\partial y}\left(K_y\frac{\partial h}{\partial y}\right) + \frac{\partial}{\partial z}\left(K_z\frac{\partial h}{\partial z}\right) = 0 \tag{4.32}$$

This equation shows that the difference in groundwater potential causes the movement of water in porous media and the fluxes depend upon the hydraulic conductivity of the medium in different directions (Karamouz et al., 2003).

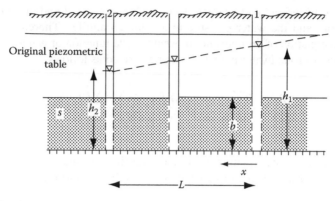

FIGURE 4.2   A confined aquifer with steady-flow conditions.

Example 4.2

Develop the groundwater flow equations for a confined aquifer shown in Figure 4.2. Assume the aquifer is homogeneous with isotropic and steady-flow conditions. If the hydraulic gradient between two wells in the distance of 250 m is equal to 0.45, find the flow velocity and aquifer discharge. Assume $k = 5 \times 10^{-6}$ m/s and $b = 30$ m.

Solution

For the one-dimensional confined aquifer system, the governing equation is given by the following equation:

$$\frac{d}{dx}\left(\frac{dh}{dx}\right) = 0$$

or

$$\frac{dh}{dx} = C_1$$

Applying Darcy's law, we have

$$\frac{dh}{dx} = -\frac{V}{K}$$

So,

$$h = -\frac{V}{K}x + C_2$$

with the boundary conditions,

$$h(0) = h_1, \quad h(L) = h_2$$

$$V = -\frac{K}{L}(h_2 - h_1) \quad \text{and} \quad q = -\frac{Kb}{L}(h_2 - h_1)$$

$$V = -K \frac{(h_2 - h_1)}{L} = 5 \times 10^{-6} \times 0.45 = 2.25 \times 10^{-6} \text{ m/s}$$

$$q = Vb = 2.25 \times 10^{-6} \times 30 = 6.75 \times 10^{-5} \text{ m}^2/\text{s}$$

Example 4.3

Develop the groundwater flow equations for a two-dimensional horizontal semi-confined or leaky aquifer, which has the thickness $b(x, y)$ at each point (x, y).

Solution

The average value of the hydraulic head at each point is as follows:

$$\bar{h} = \frac{1}{b} \int_0^b h \, dz$$

Using Equation 4.31 and considering $T = kb$ the flow equation is

$$S_s b \frac{\partial \bar{h}}{\partial t} = \frac{\partial}{\partial x} \left(T_x \frac{\partial \bar{h}}{\partial x} \right) + \frac{\partial}{\partial y} \left(T_y \frac{\partial \bar{h}}{\partial y} \right) + q_z(b)$$

where $q_z(b)$ is the vertical recharge or discharge such as leakage, pumping, or injection. The vertical leakage from the upper semi-confining layer can be calculated using Darcy's law as follows:

$$q_z(b) = \frac{K_a(H_a - \bar{h})}{b_a}$$

where
 K_a is the hydraulic conductivity of upper confining layer
 H_a is the head in upper boundary of confining layer
 b_a is the thickness of the semi-confined layer

Considering the point injection and pumping wells in the system, the flow equation can be written as follows (Hantush and Jacob, 1955; Willis and Yeh, 1987):

$$S_s b \frac{\partial \bar{h}}{\partial t} = \frac{\partial}{\partial x} \left(T_x \frac{\partial \bar{h}}{\partial x} \right) + \frac{\partial}{\partial y} \left(T_y \frac{\partial \bar{h}}{\partial y} \right) + \frac{K_a(H_a - \bar{h})}{b_a} \pm \sum_{w \in W} Q_w \delta_w$$

where
 $-Q_w$ is the discharge ($+Q_w$ is the recharge) from the pumping (injection) well w
 δ_w is a zero-one integer variable

The variable δ_w is one if there is pumping or injection at well site w.

Example 4.4

Determine the hydraulic head distribution in the homogeneous and isotropic confined aquifer shown in steady-flow conditions as shown in Figure 4.3 in general form. Then, consider two channels with water depth as 80 and 70 m in two sides of the confined aquifer with varied thickness from 50 to 10 m. Find the hydraulic head in $L/2$, $L/3$ from the datum.

Solution

For the one-dimensional aquifer system, the Equation 4.32 can be written as

$$\frac{d}{dx}\left(Kb\frac{dh}{dx}\right) = 0$$

with the boundary conditions,

$$h(0) = h_1, \quad h(L) = h_2$$

and

$$b = b_0 e^{-cx}$$

Simplifying the equation,

$$Kb\frac{d^2h}{dx^2} + K\frac{db}{dx}\frac{dh}{dx} = 0$$

or

$$\frac{d^2h}{dx^2} - c\frac{dh}{dx} = 0$$

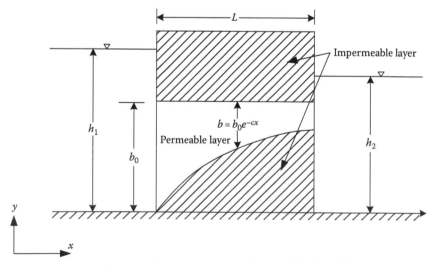

FIGURE 4.3 A confined aquifer between two impermeable layers. (From Willis, R. and Yeh, W.W.-G., *Groundwater Systems Planning and Management*, Prentice Hall, Englewood Cliffs, NJ, 1987.)

The solution of this second-order, linear ordinary differential equation is

$$h = C_1 e^{cx} + C_2$$

where C_1 and C_2 are constants of integration. Using the boundary conditions to evaluate these constants, the steady-state head distribution is described by the equation (Willis and Yeh, 1987)

$$h = h_1 - \left[\frac{(h_1 - h_2)(e^{cx} - 1)}{(e^{cL} - 1)} \right]$$

$$b = b_0 e^{-cx} \quad 50 = 10 e^{-c \times 150} \quad c = -0.011$$

$$L/2 = 75\,\text{m} \quad h = 80 - \left[\frac{(80 - 70)(e^{-0.011 \times 75} - 1)}{(e^{-0.011 \times 150} - 1)} \right] = 73.06\,\text{m}$$

$$L/3 = 50\,\text{m} \quad h = 80 - \left[\frac{(80 - 70)(e^{-0.011 \times 50} - 1)}{(e^{-0.011 \times 150} - 1)} \right] = 74.77\,\text{m}$$

Example 4.5

Develop the groundwater flow equations for a homogeneous and isotropic unconfined aquifer shown in steady-flow condition as shown in Figure 4.4. If the water table in two wells are equal to 60 and 43 m, find the water table in the observation well in middle distance between the two wells. With the assumption of a linear variation of water table between two wells, find the water table in the observation well.

Solution

Applying Darcy's law, we have

$$q = -KA \frac{dh}{dx}$$

and $A = h \times 1$

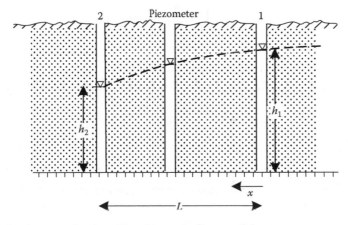

FIGURE 4.4 An unconfined aquifer with steady-flow conditions.

with the boundary conditions,

$$h(0) = h_1, \quad h(L) = h_2$$

Direct integration of the equation produces

$$\int_0^L q\,dx = -K \int_{h_1}^{h_2} h\,dh$$

So, $q = -\dfrac{K}{2L}(h_2^2 - h_1^2)$ and $h = \sqrt{h_1^2 - (h_1^2 - h_2^2)\dfrac{X}{L}}$

$$h = \sqrt{h_1^2 - (h_1^2 - h_2^2)\dfrac{X}{L}} = \sqrt{60^2 - (60^2 - 43^2) \times \dfrac{L/2}{L}} = 52.2\,\text{m}$$

With the assumption of a linear variation of water table:

$$h = 60 - \frac{(60 - 43)}{2} = 51.5\,\text{m}$$

It shows with a parabolic water table assumption, water table is observed more than linear variation.

Example 4.6

Determine the hydraulic head for the homogeneous and isotropic unconfined aquifer shown in Figure 4.5, in steady-flow condition. Assume that the aquifer has a uniform recharge rate, $R(L/T)$.

Solution

For the one-dimensional aquifer system, the governing equation can be written as

$$\frac{d}{dx}\left[Kh\frac{dh}{dx} \right] + R = 0$$

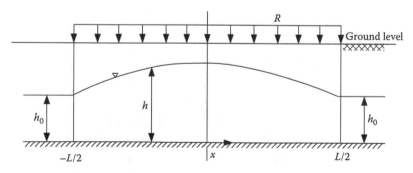

FIGURE 4.5   An unconfined flow with uniform recharge rate in Example 4.6.

and the boundary conditions

$$h(-L/2) = h_0$$

$$\left.\frac{dh}{dx}\right|_{x=0} = 0$$

Direct integration of the equation produces

$$h^2 = -\frac{Rx^2}{K} + C_1 x + C_2$$

where C_1 and C_2 are the constants of integration. Evaluating these constants from the known boundary condition information, the steady-state head distribution is described by the ellipse equation, or

$$h^2 = h_0^2 - \frac{R}{K}\left[x^2 - \left(\frac{L}{2}\right)^2\right]$$

Example 4.7

Artificial recharge of groundwater with spreading basins is used to mitigate or control saltwater intrusion into coastal aquifers. For the recharge system shown in Figure 4.6, determine the steady-state head distribution. Assume that the aquifer is homogeneous and isotropic and that for $x > x_e$ (the effective radius) $h = d$, the initial head in the system.

Solution

For the one-dimensional aquifer system, the governing equation can be written as

$$\frac{d}{dx}\left[Kh\frac{dh}{dx}\right] + f(x) = 0$$

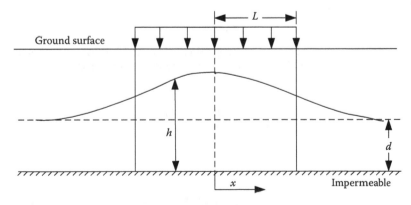

FIGURE 4.6 An unconfined aquifer system with uniform recharge rate in Example 4.7.

where $f(x)$ is the source term that is defined as

$$f(x) = \begin{cases} R, & |x| \le L \\ 0, & |x| > L \end{cases}$$

The boundary conditions of the problem require that

$$h(x_e) = d, \quad \left.\frac{dh}{dx}\right|_{x=0} = 0$$

and that there be continuity of the head and the velocity at the boundary, $x = L$, or

$$h|_{L^-} = h|_{L^+}, \quad \left.\frac{dh}{dx}\right|_{L^-} = \left.\frac{dh}{dx}\right|_{L^+}$$

For the flow region, $|h| \le L$, the governing equation is

$$\frac{d}{dx}\left[h\frac{dh}{dx}\right] + \frac{R}{K} = 0$$

Direct integration of the equation produces

$$h^2 = -\frac{R}{K}x^2 + C_2$$

where C_2 is a constant of integration and we have used the gradient condition at $x = 0$ to evaluate the first constant of integration, $C_1 = 0$. For the flow domain, $|x| > L$, the governing equation is simply,

$$\frac{d}{dx}\left[h\frac{dh}{dx}\right] = 0$$

The solution is given as

$$h^2 = C_3x + C_4$$

where C_3 and C_4 are constants of integration. We evaluate these constants from the given boundary conditions. For example, continuity of h and dh/dx at $x = L$ requires that

$$2h\left.\frac{dh}{dx}\right|_{L^-} = -\frac{2RL}{K} = 2h\left.\frac{dh}{dx}\right|_{L^+} = C_3$$

and

$$C_3 = -\frac{2RL}{K}$$

At the effective distance, x_e, we also have

$$h(x_e) = d \Rightarrow d^2 = \frac{2(-RL)}{K} x_e + C_4$$

or

$$C_4 = d^2 + \frac{2RLx_e}{K}$$

Also at $x = L$,

$$h^2 = \frac{-RL^2}{K} + C_2 = \frac{2(-Rl)}{K} L + d^2 + \frac{2Rl}{K} x_e$$

or

$$C_2 = \frac{-RL^2}{K} + d^2 + \frac{2RL}{K} x_e$$

The head distribution in the aquifer system is then given by the equations (Willis and Yeh, 1987)

$$h^2 = \frac{-R}{K} x^2 - \frac{RL^2}{K} + d^2 + \frac{2RLx_e}{K}, \quad |x| \le L$$

$$h^2 = \frac{-rRL}{K} x + d^2 + \frac{2RLx_e}{K}, \quad L < |x| \le x_e$$

$$h = d, \quad |x| \ge x_e$$

4.4 WELLS

Well drilling aids groundwater exploration with an objective to discover aquifers in different hydrogeological conditions and determination of hydraulic parameters. It leads to decline in the water level that limits the yield of the basin. Water is removed from the aquifer surrounding the well during water pumping and the piezometric surface decreases. Instead of moving toward the natural discharge area, the groundwater within the influence of the pump flows toward the well from every direction. Drawdown is the distance by which the piezometric surface is lowered. Hence, prediction of hydraulic-head drawdowns in aquifers under proposed pumping schemes is one of the goals of groundwater resource study.

The pumping well creates an artificial discharge area by drawing down (lowering) the water table around the well.

The conical-shaped depression of the water table around a pumping well is caused by the withdrawal of water; a valley in the water table. This area of drawdown is

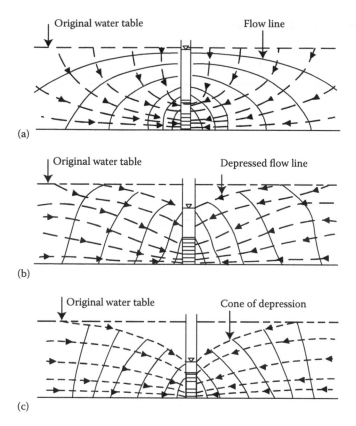

FIGURE 4.7   Development of flow distribution about a discharging well in an unconfined aquifer: (a) initial stage, (b) intermediate stage, (c) steady-state stage. (From U.S. Bureau of Reclamation, *Groundwater Manual*, U.S. Government Printing Office, Denver, CO, 1981.)

called the cone of depression. Figure 4.7 shows the initial discharge derived from casing storage and aquifer storage immediately surrounding the well when water is pumped from a well. At the initial stage in pumping an unconfined aquifer, water begins to flow toward the well screen (Figure 4.7a). At the intermediate stage in pumping an unconfined aquifer, although dewatering of the aquifer materials near the well bore continues, the radial component of the flow becomes more pronounced (Figure 4.7b). At the approximate steady-state stage in pumping an unconfined aquifer, the profile of the cone of depression is established. Nearly all water originates near the outer edge of the area of influence, and a stable, mainly radial flow pattern is established as shown in Figure 4.7c.

If pumping continues, more water must be derived from the aquifer storage at greater distances from the bore of the well and the following items are observed:

1. The cone of depression is expanded.
2. The radius of influence of the well increases due to the expansion of the cone.

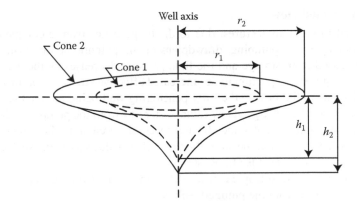

FIGURE 4.8   Changes in radius and depth of cone of depression.

3. The drawdown is incremented at any point with the increase in the depth of the cone to provide the additional head required to move the water from greater distance.
4. The cone expands and deepens more slowly with time since an increasing volume of stored water is available with horizontal expansion of the cone.

The expansion of the cone of depression during equal intervals of time is much like that shown in Figure 4.8. Calculations of the volume of each cone would show that cone 2 has twice the volume as that of cone 1, and cone 3 has three times the volume as that of cone 1. It is because of a constant pumping rate that the same volume of water is discharged from the well. Thus, if the aquifer is homogeneous and the well is being pumped at a constant rate, the increase in the volume of the cone of depression is constant over time. Water table, shaped now in the form of a cone, is steepest where it meets the well. Farther away from the well, the surface is flatter and beyond a certain distance, called the radius of influence, the surface of the cone is almost as flat as the original water table.

After sometime, deepening or expansion of the cone during short intervals of pumping is barely visible and it has been stabilized. This often misleads observers to conclude that the cone will not expand or deepen as pumping continues. The cone of depression continues to enlarge, until the flow in the aquifer with the source of surface water, vertical recharge from precipitation is intercepted to equal the pumping rate and the leakage occurs through overlying or underlying formations occur.

Equilibrium occurs when continued pumping results in no further drawdown and the cone stops expanding. It might occur within a few hours after pumping begins; or never occurs even though the pumping period is extended for years. Well discharge to drawdown can be related by deriving a radial flow equation.

4.4.1   Steady Flow into a Well

Steady-state groundwater problems are relatively simpler. Expression for steady-state radial flow into a well, under both confined and unconfined aquifer conditions, is presented in the next section.

4.4.1.1 Confined Flow

As discussed in previous lectures, due to discharge water from a completely confined aquifer by water pumping, drawdowns of the potentiometric surface occur. Water is released from storage and the fluid pressure decreases by the expulsion of water due to aquifer compaction under increased effective stress. A confined aquifer with steady-state radial flow to the fully penetrating well being pumped is shown in Figure 4.9. The well discharge at any radial distance r from the pumped well is equal but by increasing the radial distance from the well, the area of discharging increases. Thus, the farther away from the well, the flow velocity decreases, the surface is flatter, and piezometric surface is almost static.

For a homogeneous, isotropic aquifer shown in Figure 4.10, the well discharge at any radial distance r from the pumped well is

$$\sum_{CS} \rho V dA = Q = 2\pi Krb \frac{dh}{dr} \tag{4.33}$$

Equation 4.33 can be presented for boundary conditions of $h = h_1$, at $r = r_1$, and $h = h_2$, at $r = r_2$:

$$\int_{h_1}^{h_2} dh = \frac{Q}{2\pi Kb} \int_{r_1}^{r_2} \frac{dr}{r} \tag{4.34}$$

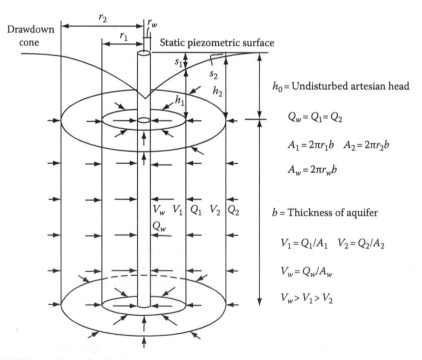

FIGURE 4.9 Flow distribution to a discharging well in a confined aquifer. (From U.S. Bureau of Reclamation, *Groundwater Manual*, U.S. Government Printing Office, Denver, CO, 1981.)

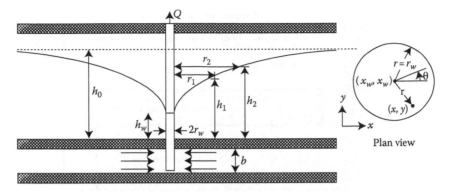

FIGURE 4.10 Well hydraulics for a confined aquifer.

$$h_2 - h_1 = \frac{Q}{2\pi Kb} \ln \frac{r_2}{r_1} \qquad (4.35)$$

Solving Equation 4.35 for Q gives

$$Q = 2\pi Kb \left[\frac{h_2 - h_1}{\ln(r_2/r_1)} \right] \qquad (4.36)$$

The general form of Equation 4.36 at any radial distance r from the pumped well is

$$Q = 2\pi Kb \left[\frac{h - h_1}{\ln(r/r_1)} \right] \qquad (4.37)$$

The equation is known as the equilibrium or Theim equation (Theim, 1906).

Example 4.8

An extensive *homogeneous* and isotropic leaky aquifer of thickness b is overlain by a bed of glacial till shown in Figure 4.11. The thickness of the semi-confining stratum is m_a and the hydraulic conductivity is K_a. The head in the aquifer overlying the aquitard is H_a and is assumed to be unaffected by pumping in the aquifer. Determine the steady-state response equation for a fully penetrating well in the aquifer system. As it is shown in the figure, the interior well area is sealed in the upper layer. (Why?)

Solution

Assuming that the pumping well is located at the center of the aquifer system since groundwater flow is an essential radial, the confined flow equation may be expressed in radial coordinates. The flow equation in radial coordinates and applying Darcy's law for estimating the vertical leakage from the upper semi-confining layer is

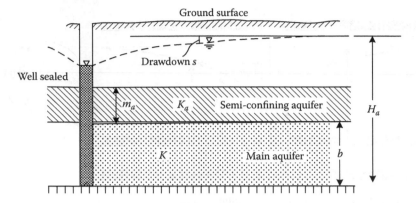

FIGURE 4.11   Leakage into aquifer for Example 4.8.

$$\frac{\partial^2 h}{\partial r^2} + \frac{1}{r}\frac{\partial h}{\partial r} = \frac{Q_w}{T}$$

or

$$\frac{T}{r}\frac{d}{dr}\left(r\frac{dh}{dr}\right) + \frac{K_a(H_a - h)}{m_a} = 0$$

Further assuming that the well diameter is infinitesimally small, the boundary condition at the well is given by Darcy's law, or

$$\lim_{r\to 0} r\frac{dh}{dr} = \frac{Q}{2\pi T}$$

Defining the new dependent variable, y as

$$y = H_a - h$$

the differential equation may be written as

$$\frac{d^2 y}{dr^2} + \frac{1}{r}\frac{dy}{dr} - \frac{K_a}{m_a}y = 0$$

Introducing the new independent variable, ρ,

$$\rho = r\left(\frac{K_a}{m_a T}\right)^{1/2}$$

the differential equation becomes

$$\frac{d^2 y}{d\rho^2} + \frac{1}{\rho}\frac{dy}{d\rho} - y = 0$$

which is a modified form of Bessel's equation. The general solution of the equation is (Willis and Yeh, 1987)

$$y = AI_0(\rho) + BK_0(\rho)$$

where
  A and B are constants of integration
  I_0 and K_0 are modified Bessel functions of the first and second kind of order zero

Requiring y to be finite as $\rho \to \infty$, A is 0 and the solution may be expressed as

$$y = BK_0(\rho)$$

Using Darcy's law to evaluate the constant of integration and recovering the original variables, the steady-state response equation is

$$h = \frac{Q}{2\pi T} K_0 \left[r \left(\frac{K_a}{m_a T} \right)^{1/2} \right]$$

For a system of pumping and injection wells, the solutions may be superimposed again to obtain the total hydraulic response of the aquifer system. The solutions are valid for finite diameter wells as shown by Marino and Luthin (1982).

4.4.1.2 Unconfined Flow

In unconfined groundwater systems, water of saturated zone is in open contact with atmospheric pressures. Groundwater release from storage occurs along permeable zones under the force of gravity as the aquifer responds to pumping. The streamline at the free surface of the aquifer is curved, and at the bottom of the aquifer it is horizontal lined, converging to the well.

Assuming that the aquifer is homogeneous and isotropic, develop the steady-state response equation relating the velocities (q_r, q_θ) and the groundwater extraction rate(s), Q as shown in Figure 4.12.

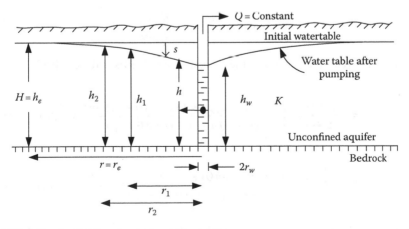

FIGURE 4.12 Radial flow in a water table aquifer.

In radial coordinates, the Boussinesq equation may be expressed as

$$\frac{1}{r}\frac{d}{dr}\left[rh\frac{dh}{dr}\right]=0 \tag{4.38}$$

with the Darcy boundary condition,

$$\lim_{r\to 0}\left[rh\frac{dh}{dr}\right]=\frac{Q}{2\pi K} \tag{4.39}$$

The solution of the differential equation is

$$h^2=\frac{Q}{\pi K}\ln r+C$$

$$h^2=\frac{Q}{2\pi K}\ln\left[(x-x_w)^2+(y-y_w)^2\right]+C \tag{4.40}$$

where
C is a constant of integration
(x_w, y_w) are the coordinates of the pumping well

For w pumping or injection ($-Q$) wells in the basin, the system response equation is found from superposition of the individual well responses, or

$$h^2=\sum_w\frac{Q_i}{2\pi K}\ln\left[(x-x_w)^2+(y-y_w)^2\right]+C \tag{4.41}$$

The velocity field is given by Darcy's law, or

$$q_r=-K\frac{\partial h}{\partial r},\quad q_\theta=\frac{K}{r}\frac{\partial h}{\partial\theta} \tag{4.42}$$

Evaluating the partial derivatives, the velocities are (Willis and Yeh, 1987)

$$q_r=-\frac{K}{\pi}\sum_w Q_i\left\{\frac{(r\cos\theta-x_w)\cos\theta+(r\sin\theta-y_w)\sin\theta}{(r\cos\theta-x_w)^2+(r\sin\theta-y_w)^2}\right\}\bigg/(2h) \tag{4.43}$$

$$q_\theta=\frac{K}{\pi}\sum_w Q_i\left\{\frac{(r\cos\theta-x_w)(-r\sin\theta)+(r\sin\theta-y_w)(r\cos\theta)}{(r\cos\theta-x_w)^2+(r\sin\theta-y_w)^2}\right\}\bigg/(2rh) \tag{4.44}$$

The velocities, in this case, are nonlinear functions of the pumping or injection rates.

By rearranging Equation 4.39,

$$hdh = \frac{Q}{2\pi K} \frac{r}{dr} \tag{4.45}$$

Integrating between lines r_1 and r_2 where the water table depths are h_1 and h_2, respectively, and on rearranging,

$$Q = \frac{\pi K (h_2^2 - h_1^2)}{\ln(r_2/r_1)} \tag{4.46}$$

There are large vertical-flow components near the well so that this equation fails to describe accurately the drawdown curve near the well, but can be defined for any two distances away from the pumped well. At the edge of the zone of influence of radius r_e, H_e = saturated thickness of the aquifer, Equation 4.46 can be written as

$$Q = \frac{\pi K (He^2 - h_w^2)}{\ln(r_e/r_w)} \tag{4.47}$$

Example 4.9

A well penetrates an unconfined aquifer. Before pumping, the water table is in $h_0 = 25$ m. After a long period of pumping at a constant rate of 0.5 m³/s, the drawdowns at distances of 50 m from the well were observed 3 m. If $K = 7.26 \times 10^{-2}$ m/s find the drawdowns at distances of 150 m from the well.

Solution
Applying Equation 4.46,

$$h_2^2 - h_1^2 = \frac{Q}{k\pi} \ln\left(\frac{r_2}{r_1}\right)$$

$$(25 - s)^2 - (25 - 3)^2 = \frac{0.5\ln(150/50)}{\pi \times 7.26 \times 10^{-2}} \Rightarrow s = 2.46 \text{ m}$$

4.4.2 Unsteady State in a Confined Aquifer

When a fully penetrated well is pumped in a confined aquifer at a constant rate Q, water is released from storage by the expansion of the water as pressure in the aquifer is reduced and by expulsion as the pore space is reduced as the aquifer compacts. During pumping of such a well, the cone of depression causes to progress outward from the well and pumping affects a relatively large area of the aquifer. Further, if no recharge occurs, the area of drawdown of the potentiometric surface will expand indefinitely and the rate of decline of the head continuously decreases as the cone of depression spreads.

The confined flow equation in radial coordinates can be expressed as

$$\frac{\partial^2 s}{\partial r^2} + \frac{1}{r}\frac{\partial s}{\partial r} = \frac{S}{T}\frac{\partial s}{\partial t} \qquad (4.48)$$

where
s is the drawdown ($s = h_0 - h$)
h_0 is the initial piezometric head

The basic assumption can be expressed mathematically as initial and boundary conditions. The initial condition of a horizontal potentiometric surface is

$$s(r,0) = 0 \quad \text{for all } r \qquad (4.49)$$

The boundary condition signifying an infinite horizontal extent with no drawdown at any time:

$$s(\infty,t) = 0 \quad \text{for all } r \qquad (4.50)$$

$$\lim_{r \to 0} r\frac{\partial s}{\partial r} = -\frac{Q}{2\pi T} \qquad (4.51)$$

The solution that Theis arrived for Equation 4.48 under the initial and boundary conditions is known as the Theis or nonequilibrium equation (Theis, 1935). In this method, the Boltzmann transformation is used for transforming the confined flow equation, a second-order linear diffusion equation, into an ordinary differential equation as follows:

$$u = \frac{r^2 S}{4Tt} \qquad (4.52)$$

Substituting this variable in Equation 4.48, the flow equation in radial coordinates can be expressed as

$$\frac{d^2 s}{du^2} + \left(1 + \frac{1}{u}\right)\frac{ds}{du} = 0 \qquad (4.53)$$

Then the initial and boundary conditions are transformed into

$$s(\infty) = 0 \qquad (4.54a)$$

$$\lim_{u \to 0} u\frac{ds}{du} = -\frac{Q}{4\pi T} \qquad (4.54b)$$

Using the substitution, ds/du, a first integration of the equation yields

$$u\frac{ds}{du} = C_1 e^{-u} \qquad (4.55)$$

where C_1 is a constant of integration. Using the second boundary condition, the equation simplifies to

$$\frac{ds}{du} = \frac{Q}{4\pi T} \times \frac{e^{-u}}{u} \tag{4.56}$$

Performing the final integration considering the first boundary condition, the drawdown equation is

$$s = \frac{Q}{4\pi T} \int_{u}^{\infty} \frac{e^{-a}}{a} \, da \tag{4.57}$$

The integral in Equation 4.57 is well known in mathematics. It is called the exponential integral. It can be approximately presented by an infinite series so that the Theis equation becomes

$$s = \frac{Q}{4\pi T} \left[-0.5772 - \ln u + u - \frac{u^2}{2 \times 2!} + \frac{u^3}{3 \times 3!} - \frac{u^4}{4 \times 4!} + \cdots \right] \tag{4.58}$$

where $W(u)$ is called the dimensionless well function for nonleaky, isotropic, artesian aquifers, fully penetrated by wells having constant discharge conditions. The tables of values are widely available and values of the well function are listed in Table 4.1. The well function is also expressed in the form of a type curve, as shown in Figure 4.13. Both u and $W(u)$ are dimensionless.

TABLE 4.1

Values of $W(u)$ for Various Values of u

	1.0	2.0	3.0	4.0	5.0	6.0	7.0	8.0	9.0
×1	0.219	0.049	0.013	0.0038	0.0011	0.00036	0.00012	0.000038	0.000012
1×10^{-1}	1.82	1.22	0.91	0.70	0.56	0.45	0.37	0.31	0.26
1×10^{-2}	4.04	3.35	2.96	2.68	2.47	2.30	2.15	2.03	1.92
1×10^{-3}	6.33	5.64	5.23	4.95	4.73	4.54	4.39	4.26	4.14
1×10^{-4}	8.63	7.94	7.53	7.25	7.02	6.84	6.69	6.55	6.44
1×10^{-5}	10.94	10.24	9.84	9.55	9.33	9.14	8.99	8.86	8.74
1×10^{-6}	13.24	12.55	12.14	11.85	11.63	11.45	11.29	11.16	11.04
1×10^{-7}	15.54	14.85	14.44	14.15	13.93	13.75	13.60	13.46	13.34
1×10^{-8}	17.84	17.15	16.74	16.46	16.2 3	16.05	15.90	15.76	15.65
1×10^{-9}	20.15	19.45	19.05	18.76	18.54	18.35	18.20	18.07	17.95
1×10^{-10}	22.45	21.76	21.35	21.06	20.84	20.66	20.50	20.37	20.25
1×10^{-11}	24.75	24.06	23.65	23.36	23.14	22.96	22.81	22.67	22.55
1×10^{-12}	27.05	26.36	25.96	25.67	25.44	25.26	25.11	24.97	24.86
1×10^{-13}	29.36	28.66	28.26	27.97	27.75	27.56	27.41	27.28	27.16
1×10^{-14}	31.66	30.97	30.56	30.27	30.05	29.87	29.71	29.58	29.46
1×10^{-15}	33.96	33.27	32.86	32.58	32.35	32.17	32.02	31.88	31.76

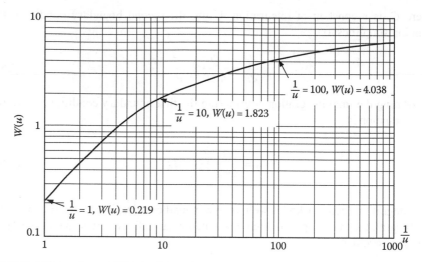

FIGURE 4.13   values of $W(u)$ (well function of u) corresponding to values of $1/u$.

This model can also be utilized to incorporate a variable groundwater pumping schedule. Assume the aquifer is pumped at a variable extraction rate for n planning periods. The drawdown for any period can be found from the superposition of the well responses because the response equations are linear functions of the extraction rates. The mathematical form of the equation for any period can be written as follows:

$$s = \frac{Q_1}{4\pi T} W\left(\frac{r^2 S}{4Tt}\right), \quad 0 \le t \le t_1 \tag{4.59a}$$

$$s = \frac{Q_1}{4\pi T} W\left(\frac{r^2 S}{4Tt}\right) + \left(\frac{Q_2 - Q_1}{4\pi T}\right) W\left(\frac{r^2 S}{4T(t - t_1)}\right), \quad t_1 \le t \le t_2 \tag{4.59b}$$

$$s = \frac{Q_1}{4\pi T} W\left(\frac{r^2 S}{4Tt}\right) + \left(\frac{Q_2 - Q_1}{4\pi T}\right) W\left(\frac{r^2 S}{4T(t - t_1)}\right) + \cdots$$

$$+ \left(\frac{Q_n - Q_{n-1}}{4\pi T}\right) W\left(\frac{r^2 S}{4T(t - t_{n-1})}\right), \quad t \ge t_{n-1} \tag{4.59c}$$

where Q_k is the extraction rate for planning period k.

The drawdown commonly referred to as the nonequilibrium well pumping equation is expressed as

$$s = \frac{Q}{4\pi T} W(u) \tag{4.60}$$

Equations 4.52 and 4.60 can be expressed in U.S. customary units (gallon-day-foot system) where s is in ft, Q is in gpm, T is in gpd/ft, r is in ft, and t is in days,

$$s = \frac{114.6Q}{T} W(u) \tag{4.61}$$

and

$$u = \frac{1.87r^2 S}{Tt} \tag{4.62}$$

or, for t in minutes,

$$u = \frac{2693r^2 S}{Tt} \tag{4.63}$$

Example 4.10

A well is located in a 25-m confined aquifer of hydraulic conductivity 30 m/day and storativity coefficient 0.005. If the well is being pumped at the rate of 1500 m³/day, calculate the drawdown at a distance of 100 m from the well after 20 h of pumping.

Solution

$$T = KB = \frac{30}{86,400} \times 25 = 8.68 \times 10^{-3} \text{ m}^2/\text{s}$$

$$u = \frac{r^2 S}{4Tt} = \frac{100^2 \times 0.005}{4 \times 8.68 \times 10^{-3} \times 20 \times 3600} = 0.08$$

Using Theis method and calculating $W(u)$ to four significant digits,

$$W(u) = -0.5772 - \ln(0.02) + (0.02) - \frac{(0.02)^2}{2 \times 2!} + \frac{(0.02)^3}{3 \times 3!} = 3.3547$$

$$s = \frac{Q}{4\pi T} W(u) = \frac{1500 \times 3.3547}{4 \times \pi \times 8.68 \times 10^{-3} \times 24 \times 3600} = 0.32 \text{ m}$$

4.4.2.1 Aquifer Test Application

An aquifer test (or a pumping test) is performed to evaluate an aquifer by observing the aquifer's drawdown in observation wells. One or more monitoring wells or piezometers are used for aquifer testing as the observation/monitoring wells. Water is not being pumped from the observation well just utilized to monitor the water drawdown. While water is being pumped from one well at a steady rate, the water tables are monitored in the observation wells. Typically, monitoring and pumping wells are screened across the same aquifers. Aquifer testing is based on the data processing to yield valuable qualitative and quantitative features about the subsurface geological composition of the aquifer domain. The test data processing is achieved by matching the data with a suitable type curve. An analytical or numerical model of aquifer flow is used to match the data observed in the real world, assuming that the parameters from the idealized model apply to the real-world aquifer.

Most aquifer tests evaluate the aquifer characteristics including

- Hydraulic conductivity or transmissivity, which shows how permeable the aquifer is and the ability of an aquifer to transmit water.
- Specific storage/specific yield or storativity, which is the measure of the amount of water an aquifer will give up for a certain change in head.
- Boundary determination, which determines the type of boundaries that give up water to the aquifer, providing additional water to reduce drawdown or no-flow boundaries.

Three major components for the aquifer tests to the ultimate objectives of quantifying the hydraulic properties, boundary determination, indication of the general type of aquifer, and hypothesis testing include design, field observations, and data analysis. In aquifer tests, in addition to knowing the geology of the site and construction of the wells (width, depth, materials used, development), measuring devices such as time, discharge, and water level should be calibrated and verified. Then the observed data during the test are plotted and the best curve is fitted on data. Finally, the aquifer characteristics are determined based on the fitted curve.

Graphical or computer estimation of parameter values using observed drawdown and well-produced rate records can be utilized as the method of analysis of aquifer test data. For confined aquifers, there are two graphical methods including the log-log plot of drawdown versus time for the Theis method and the semilog plot for the Cooper–Jacob method.

4.4.2.2 Theis Method of Solution

The following equation was adopted by Theis (1935) and is used to determining T and S by expressing Equation 4.60 and 4.52 as

$$\log s = \log\left(\frac{Q}{4\pi T}\right) + \log W(u) \tag{4.64}$$

$$\log\frac{t}{r^2} = \log\left(\frac{S}{4T}\right) + \log\frac{1}{u} \tag{4.65}$$

Because $Q/4\pi T$ and $S/4T$ are constants, the relation between $\log(s)$ and $\log(t/r^2)$ must be similar to the relation between $\log W(u)$ and $\log 1/u$. Therefore, drawdown measurements derived from pumping tests are plotted on double-logarithmic paper and $W(u)$ against $1/u$ on the same double-logarithmic paper; the two plotted curves have the same shape, but they are offsets of the constants $(Q/4\pi T)$ and $(S/4T)$, vertically

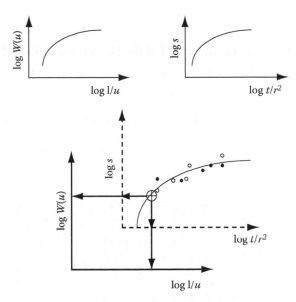

FIGURE 4.14 Graphical procedure to determine t and S from pump test data.

and horizontally, respectively, with keeping the coordinate axis parallel. Therefore, if plotting each curve on a separate sheet, match them by placing one graph on top of the other and moving it horizontally and vertically until the curves are matched. When superimposed over the type curve, values of $W(u)$, $1/u$, t/r^2 and s may be selected at any desirable point and substituted in the Theis equation. This is further illustrated in Figure 4.14.

Example 4.11

The drawdown time data recorded at two observation wells situated at a distance of 100 and 200 m from the pumping well are given in Table 4.2. If the well discharge is 1000 m³/day, calculate the transmissibility and storage coefficient of the aquifer.

Solution

The first step is to calculate r^2/t from the given data. The calculated values are presented in Table 4.3. Then the measured drawdown versus r^2/t is plotted on log paper as shown in Figure 4.15. The data plotted are superimposed on the $W(u)$ versus u plot (Figure 4.13) as presented in Figure 4.16. An arbitrary match point is selected as

TABLE 4.2
The Drawdown Time Data Recorded at the Distance of 100 and 200 m

t (Day)	0.001	0.005	0.01	0.05	0.1	0.5	1	5	10
s1 (m)	0.083	0.196	0.249	0.379	0.431	0.559	0.614	0.742	0.797
s2 (m)	0.017	0.097	0.145	0.267	0.322	0.449	0.504	0.632	0.687

TABLE 4.3

The Drawdown Time Data Recorded at the Distance of 100 and 200 m

t (Day)	0.001	0.005	0.01	0.05	0.1	0.5	1	5	10
t/r^2, $d = 100$ m (m³/day)	10^{-7}	5×10^{-7}	10^{-6}	5×10^{-6}	10^{-5}	5×10^{-5}	10^{-4}	5×10^{-5}	10^{-3}
t/r^2, $d = 200$ m (m³/day)	0.25×10^7	0.13×10^6	0.25×10^6	0.13×10^5	0.25×10^5	0.13×10^4	0.25×10^4	0.13×10^3	0.25×10^3

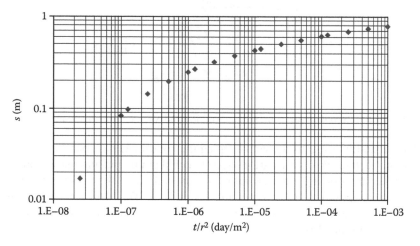

FIGURE 4.15   The drawdown versus t/r^2 for Example 4.11.

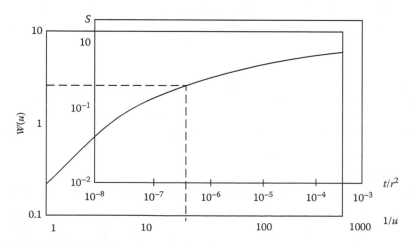

FIGURE 4.16   Matching the type curve with drawdown data.

$$s = 0.167 \text{ m}, \quad \frac{t}{r^2} = 3 \times 10^{-7} \text{ m}^2/\text{day}, \quad W(u) = 2.1, \quad \frac{1}{u} = 12.5,$$

From Equations 4.52 and 4.60:

$$T = \frac{1000}{4\pi \times 0.167} \times 2.1 = 1000.7 \text{ m}^2/\text{day}$$

$$S = \frac{4 \times 1000.7}{12.5 \times 3 \times 10^{-7}} = 0.000107 \text{ m}^2/\text{day}$$

4.4.2.3 Cooper–Jacob Method of Solution

Cooper and Jacob noted that for small values of r and large values of t, the parameter $u = r^2 S/4T$ becomes very small and the infinite series can be approximated by

$$W(u) = -0.5772 - \ln u \tag{4.66}$$

Thus,

$$s' = \frac{Q}{4\pi T}\left(-0.5772 - \ln\left(\frac{r^2 S}{4Tt}\right)\right) \tag{4.67}$$

Rearranging and conversing to decimal logs yield

$$s' = \frac{2.3Q}{4\pi T} \log\left(\frac{2.25Tt}{r^2 S}\right) \tag{4.68}$$

A plot of drawdown s via log of t forms a straight line. A projection of the line back to $s = 0$, where $t = t_0$ yields the following relation:

$$S = \left(\frac{2.25Tt_0}{r^2}\right) \tag{4.69}$$

$$T = \left(\frac{2.3Q}{4\pi s'}\right) \tag{4.70}$$

The Cooper–Jacob method first solves for T and then for S and is only applicable for small values of $u < 0.01$.

Example 4.12

A well penetrating a confined aquifer is pumped at a rate of a 2500 m³/s. The drawdown at an observation well at a radial distance of 30 m is presented in Table 4.4. Calculate the transmissibility and storage coefficient of the aquifer.

TABLE 4.4

The Drawdown Time Data Recorded at the Distance of 30 m

Time (min)	0	1	1.5	2	2.5	3	4	5	6	8	10	12	14
Drawdown (m)	0	0.2	0.27	0.3	0.34	0.37	0.41	0.45	0.48	0.53	0.57	0.6	0.63
Time (min)	18	24	30	40	50	60	80	100	120	150	180	210	240
Drawdown (m)	0.67	0.72	0.76	0.81	0.85	0.9	0.93	0.96	1	1.04	1.07	1.1	1.12

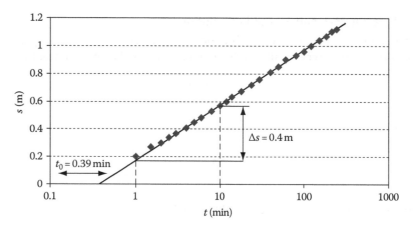

FIGURE 4.17 Time-drawdown plot for Example 4.12.

Solution

The drawdown is plotted against time on a semilog plot (Figure 4.17). From the fitted line,

$$t_0 = 0.39 \, \text{min} = 2.7 \times 10^{-4} \, \text{day}$$

$$s' = 0.4 \, \text{m}$$

$$T = \left(\frac{2.3Q}{4\pi s'} \right) = \frac{2.3 \times 2500}{4 \times \pi \times 0.4} = 1144.5 \, \text{m}^2/\text{day}$$

$$S = \left(\frac{2.25 T t_0}{r^2} \right) = \frac{2.25 \times 1144.5 \times 2.7 \times 10^{-4}}{(60)^2} = 0.000193$$

Example 4.13

A well was pumped at a rate of 200 gal/min. Drawdown as shown was measured in an observation well 250 ft away from the pumped well (Table 4.5).

(A) Plot the time drawdown data on four-cycle semilogarithmic paper. Use the Cooper–Jacob straight-line method to find the aquifer transmissivity and storativity.

TABLE 4.5

The Drawdown Time Data Recorded at the Distance of 250 ft

Time (min)	0	1	1.5	2	2.5	3	4	5	6	8	10	12	14
Drawdown (ft)	0	0.66	0.87	0.99	1.11	1.21	1.36	1.49	1.59	1.75	1.86	1.97	2.08
Time (min)		18	24	30	40	50	60	80	100	120	150	180	210	240
Drawdown (ft)		2.2	2.36	2.49	2.65	2.78	2.88	3.04	3.16	3.28	3.42	3.51	3.61	3.67

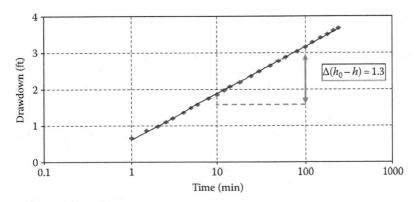

FIGURE 4.18   Time drawdown plot for Example 4.13.

Solution

The drawdown is plotted against time on a semilog plot (Figure 4.18). From the fitted line,

$$t_0 = 0.337 \text{ min} \times \frac{1}{1440} \text{day/min} = 0.23 \times 10^{-3} \text{day}$$

$$Q = 200 \text{ gal/min} \times \frac{1}{7.48} \text{ft}^3/\text{gal} \times 1440 \text{ min/day} = 38502.7 \text{ ft}^3/\text{day}$$

$$\Delta(h_0 - h) = 1.3 \text{ ft}$$

$$T = \frac{2.3Q}{4\pi\Delta(h_0 - h)} = \frac{2.3 \times 38502.7}{4 \times 3.14 \times 1.3} = 5423.6 \text{ ft}^3/\text{day}$$

$$r = 250 \text{ ft}$$

$$S = \frac{2.25Tt_0}{r^2} = \frac{2.25 \times 5423.6 \times 0.23 \times 10^{-3}}{(250)^2} = 0.044 \times 10^{-6}$$

4.4.3   UNSTEADY STATE FOR UNCONFINED AQUIFER

When water is being pumped from an unconfined aquifer, the water is derived from the storage by gravity drainage of tile interstices above the cone of depression. In the

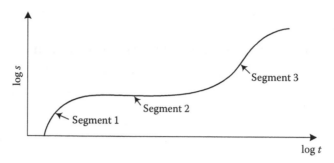

FIGURE 4.19   Three segments of time-drawdown curve for water table conditions.

nonequilibrium solution, the effects of gravity drainage are not considered. Gravity drainage is not immediate and, for unsteady flow of water toward a well for unconfined conditions, is characterized by the slow drainage of interstices. A graphical method for the analysis of an aquifer test in an unconfined aquifer is developed by Neuman (1975).

This method is based on the specific assumption. Figure 4.19 shows three distinct segments of the time-drawdown curve for water table conditions and it does not require the definition of any empirical constants. The first segment occurs for early drawdown data, when instantaneous release of water from storage occurs in the same manner as an artesian aquifer. It is possible to determine the coefficient of transmissibility under certain conditions by applying the nonequilibrium solution to the time-drawdown curve data. The gravity drainage is not immediate and the water is released instantaneously from the storage. As time elapses, the effects of gravity drainage and vertical flow cause deviation from the nonequilibrium-type curve. The coefficient of storage is computed by using the early time-drawdown and it is in the artesian range and cannot be used to predict long-term drawdown.

The second segment represents an intermediate stage of the decline of water level due to the slow expansion of the cone of depression and it is replenished by the gravity drainage. Recharge is reflected as the slope of the time-drawdown curve decreases. Pump test data deviate significantly from the nonequilibrium theory during the second segment.

The third segment is used for late drawdown data, when effects of gravity drainage become smaller. This segment occurs at later times. After the effects of delayed gravity drainage cease to influence the drawdown in observation wells, the time-drawdown field data conform closely to the nonequilibrium solution as illustrated above. By applying the nonequilibrium solution to the third segment of the time-drawdown data, the coefficient of transmissibility of an aquifer can be determined. The coefficient of storage is in the unconfined range and can be used to predict long-term effects.

The solution of Neuman (1975) also reproduced three segments of the time-drawdown curve and it does not require the definition of any empirical constants. The proposed equation for drawdown in an unconfined aquifer with fully penetrating wells and a constant discharge condition Q was presented:

$$s = h_0 - h = \frac{Q}{4\pi T} W(u_a, u_y, \eta) \tag{4.71}$$

where $W(u_a, u_y, \eta)$ is the well function for water table aquifer and

$$u_a = \frac{r^2 S}{4Tt} \quad \text{(for small values of } t) \tag{4.72}$$

$$u_y = \frac{r^2 S_y}{4Tt} \quad \text{(for large values of } t) \tag{4.73}$$

$$\eta = \frac{r^2 K_v}{b^2 K_h} \tag{4.74}$$

where
$s = h_0 - h$ is the drawdown (L)
Q is the pumping rate (L^3/T)
T is transmissivity
r is the radial distance from the pumping well
S is storativity
S_y is the specific yield
t is time
K_v is the vertical hydraulic conductivity
K_h is the vertical hydraulic conductivity
b is the initial saturated thickness of the aquifer

Figure 4.20 is a plot of this function for various values of η with two sets of type curves. Type-A curves that grow out of the left-hand Theis curve of Figure 4.20 are good for early drawdown data. Type-B curves that are asymptotic to the right-hand

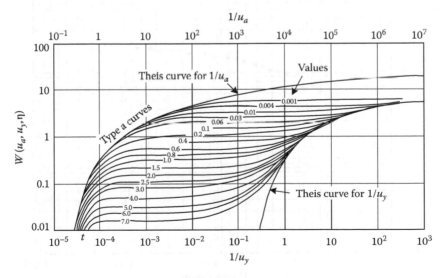

FIGURE 4.20 Theoretical curves of $W(u)$ versus $1/u$, for an unconfined aquifer. (From Neuman, S.P., *Water Resour. Res.*, 11, 329, 1975.)

Theis curve of Figure 4.20 are followed at a later time. The Type-B curves end on a
Theis curve and the distance-drawdown data can be analyzed using the nonequilibrium solution technique.

Example 4.13

A well in a water table aquifer was pumped at a rate of 873 m³/day. Drawdown
was measured in a fully penetrating observation well located 90 m away (Figure
4.21). The following data were obtained (Kruseman and de Ridder, 1991). Find the
transmissivity, storativity, and specific yield of the aquifer.

Solution

The best match for Type-A curve and the early drawdown data is with the $\eta =$
0.01 curve. At the match point $1/u_A = 10$, $W(u_a, \eta) = 1$, $h_0 - h = 4.8 \times 10^{-2}$ m and
$t = 10.5$ min (7.3×10^{-3} day). Substitute these values in Equations 4.71 and 4.72.

$$T = \frac{Q}{4\pi(h_0 - h)}W(u_a, \eta), \quad \text{given } Q = 873 \text{ m}^3/\text{day}$$

$$T = \frac{873 \text{ m}^2/\text{day} \times 1.0}{4\pi \times 4.8 \times 10^{-2}} = 1400 \text{ m}^2/\text{day}$$

$$S = \frac{4Tut}{r^2}, \quad \text{given } r = 90 \text{ m}$$

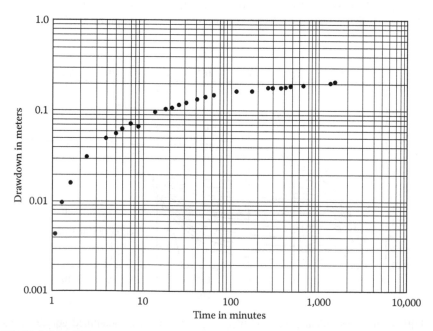

FIGURE 4.21 Drawdown in an unconfined aquifer as a function of time.

$$1/u_A = 10, \quad u_A = 0.1$$

$$S = \frac{4 \times 1400 \text{ m}^2/\text{day} \times 7.3 \times 10^{-3} \text{ day}}{90 \times 90 \text{ m}} \times 0.1 = 0.0005$$

The Type-B curve for $\eta = 0.01$ is matched to the late drawdown data. The match point values are $1/u_y = 10^2$, $W(u_y, \eta) = 1$, and $t = 880 \text{ min}$ (6.1×10^{-1} day).

$$T = \frac{873 \text{ m}^2/\text{day} \times 1.0}{4\pi \times 4.3 \times 10^{-2} \text{ m}} = 1600 \text{ m}^2/\text{day}$$

$$S_y = \frac{4 \times 1600 \text{ m}^2/\text{day} \times 6.1 \times 10^{-1} \text{ day} \times 0.01}{90 \times 90 \text{ m}} = 0.005$$

4.5 MULTIPLE-WELL SYSTEMS

This model can also be utilized to incorporate a variable groundwater pumping schedule. Assume the aquifer is pumped at a variable extraction rate for n planning periods. The drawdown for any period can be found from superposition of the well responses because the response equations are linear functions of the extraction rates. The mathematical form of the equation for any period can be written as follows:

The drawdown in hydraulic head at any point of the aquifer with more than one well is equal to the summation of the drawdowns that occur from each of the wells independently. The total drawdown of n wells pumping at rates $Q_1, Q_2, \ldots, Q_n$ can be found from the superposition of well response individually. Where pumping is from a confined aquifer,

$$H_e - h = \sum_{i=1}^{n} \frac{Q_i}{2\pi K b} \ln\left(\frac{r_{ei}}{r_i}\right) \tag{4.75}$$

where
  $H_e - h$ is the drawdown at any given point in the area of influence
  r_{ei} is the distance from the ith well to a point at which the drawdown becomes negligible
  r_i is the distance from the ith well to the given point

For an unconfined aquifer the total drawdown is calculated as

$$H_e^2 - h^2 = \sum_{i=1}^{n} \frac{Q}{\pi K} \ln\left(\frac{r_{ei}}{r_i}\right) \tag{4.76}$$

For example, for an unconfined aquifer with unsteady-state pumping, the drawdown at a point whose radial distance from each well is given by $r_1, r_2, \ldots, r_n$ is the arithmetic summation of the Theis solutions as follows:

$$h_0 - h = \frac{Q_1}{4\pi T} W(u_1) + \frac{Q_2}{4\pi T} W(u_2) + \cdots + \frac{Q_n}{4\pi T} W(u_n) \tag{4.77}$$

where

$$u_i = \frac{r_i^2 S}{4Tt_i}, \quad i = 1, 2, 3, \ldots, n$$

t_i is the time since pumping started at the well whose discharge is Q_i.

Example 4.14

A building is located on an island. It needs to decrease the water table under the building to the allowable depth. Therefore, four wells as shown in Figure 4.22 are utilized to withdraw the water. Determine the discharge from each well to decrease the water table in the building center to 20 m, assuming the water table before pumping is 50 m, the drawdown effect is predicted to 500 m (Point A) and $K = 10^{-3}$ m/s.

Solution

The drawdown in the building center is equal to the sum of the drawdowns that would arise from each of the wells independently. r_{ei} is the distance between the well i and the negligible drawdown point, and r_i is the distance between the well i and the central point.

$$H_e^2 - h^2 = \sum_{i=1}^{n} \frac{Q}{\pi K} \ln\left(\frac{r_{ei}}{r_i}\right)$$

$$50^2 - 20^2 = \frac{Q}{\pi \times 10^{-3}} \left[2\ln\left(\frac{\sqrt{530^2 + 30^2}}{\sqrt{30^2 + 30^2}}\right) + 2\ln\left(\frac{\sqrt{470^2 + 30^2}}{\sqrt{30^2 + 30^2}}\right) \right]$$

$$Q = 0.67 \text{ m}^3/\text{s}$$

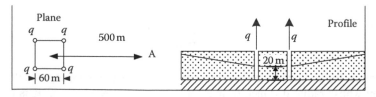

FIGURE 4.22 Plane and profile views of Example 4.14.

4.6 EFFECTIVE CONDITIONS ON TIME-DRAWDOWN DATA

In order to apply the Theis equation, basic assumptions are made in developing the equations for groundwater flow. However, geologic or hydrologic conditions do not conform fully in real aquifers. One of these assumptions is that aquifers have an infinite areal extent and clearly it is necessary in order to solve many groundwater problems. It is obvious that this type of aquifer cannot exist in reality and the aquifers have some boundaries. A hydrologic boundary could be the edge of the aquifer, a region of the recharge to a fully confined artesian aquifer, or the source of the recharge such as a stream or lake (Fetter, 2001). The boundary conditions can be divided into two categories: recharge ones and impermeable ones. Another assumption for applying the developed formula is the fully penetrated well that leads to change in the flow pattern.

4.6.1 RECHARGE BOUNDARY

Equilibrium conditions that stabilize the cone of depression around a pumping well may develop in several general situations. One of these is when an aquifer is recharged from a river or lake. A recharge boundary is a region in which the aquifer is replenished. Figure 4.23 shows a recharge boundary after equilibrium has been reached.

The drawdown data due to water pumping show that the cone of depression does not extend to the river and no recharge is evident. By keeping on pumping, the water table in the well decreases. A hydraulic gradient develops between the groundwater in the aquifer and the water in the river when the cone of depression intersects a river channel. If the stream bed is hydraulically connected with the aquifer, river water will percolate downward through the pervious stream bed under the influence of the hydraulic gradient. Thus, the river recharges the aquifer at an increasing rate as the cone of depression enlarges. When the rate of recharge to the aquifer equals the rate of discharge from the well, the cone of depression and the pumping level become stable (Driscoll, 1986).

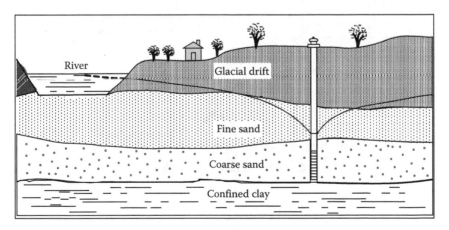

FIGURE 4.23 Cone of depression between the aquifer and river. (From Driscoll, F.G., *Groundwater and Wells*, 2nd edn., Johnson Division, St. Paul, MN, 1986.)

In order to predict aquifer behavior in the presence of boundaries, introduce imaginary wells such that the response at the boundary is made true. Image well method is widely used in heat-flow theory and has been adapted for application in the groundwater milieu, in order to predict the decreased drawdowns that occur in a confined aquifer in the vicinity of a constant-head boundary, such as would be produced by the slightly unrealistic case of a fully penetrating stream (Figure 4.24a). For this case, the imaginary infinite stream (Figure 4.24b) is introduced in order to set up a hydraulic flow system which will be equivalent to the effects of a known physical boundary. The recharge boundary can be simulated by a recharging image well duplicated in an equal and opposite side from the real well (Figure 4.24c).

The summation of the cone of depression from the pumping well and the cone of impression from the recharge well leads to an expression for the drawdown in an aquifer bounded by a constant-head boundary:

$$h_0 - h = \frac{Q}{4\pi T}\left[W(u_r) - W(u_i)\right] \tag{4.78}$$

$$u_r = \frac{r_r^2 S}{4Tt} \quad \text{and} \quad u_i = \frac{r_i^2 S}{4Tt} \tag{4.79}$$

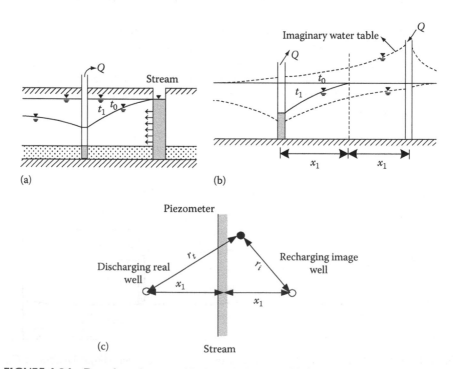

FIGURE 4.24 Drawdown in an aquifer bounded on one side by a stream.

4.6.2 IMPERMEABLE BOUNDARY

In many localities, definite geologic and hydraulic boundaries limit aquifers to areas. An impermeable boundary is an edge of the aquifer, where it terminates by thinning or abutting a low-permeability formation or has been eroded away (Fetter, 2001). Assuming that there is no groundwater flow across this layer, it is called a zero-flux boundary. The effects of an impermeable boundary on the time-drawdown data is opposite to the effects of recharge boundary. When an aquifer is bounded on one side by a straight-line impermeable boundary, drawdowns due to pumping will be greater near the boundary (Figure 4.25a). The effect of the impermeable boundary to flow in some region of the aquifer is to accelerate the drawdown (Figure 4.25b). In order to simulate an impermeable boundary, a discharging image well is duplicated in an equal and opposite side from the real well (Figure 4.24c). The summation of the cone of depression from the pumping well and the cone of impression from the discharge well leads to an expression of the drawdown in an aquifer bounded by a constant-head boundary:

$$h_0 - h = \frac{Q}{4\pi T}\left[W(u_r) + W(u_i)\right] \tag{4.80}$$

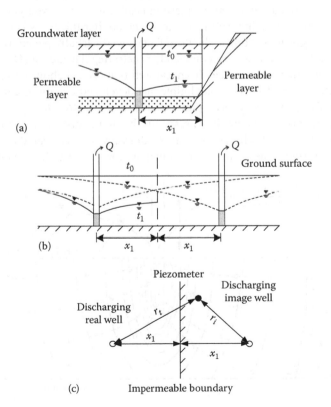

FIGURE 4.25   Drawdown in an aquifer bounded on one side by an impermeable boundary.

u_r and u_i are as defined in connection with Equation 4.79. It is possible to use the image well approach to provide predictions of drawdown in systems with more than one boundary.

4.6.3 PARTIALLY PENETRATING WELLS

In practice and well construction, it happens that water enters the well bore over a length which is less than the aquifer thickness. This is called a partially penetrating well. The flow pattern will differ from the radial flow which is thought to exist around wells which fully penetrate. Flow toward a partially penetrating well experiences convergence that is in addition to the convergence in flow toward a fully penetrating well. By changing the flow pattern, the presented equations can not be used to estimate the drawdowns because flow is not strictly radial.

Partial penetration causes the flow to have a vertical component in the vicinity of the well as shown in Figure 4.26. Some typical flow lines or paths of water particles are presented by the arrows as they move through the formation toward the intake portion of the well. Water in the lower part of the aquifer moves upward along the curved lines to reach the well screen. Therefore, the path that water must take is longer than radial flow lines. Also, the flow must converge through a smaller cross-sectional area while approaching the short screen. The result of the longer flow paths and the smaller cross-sectional area is an increase in head loss. For a given yield, therefore, the drawdown in a pumping well is greater if the aquifer thickness is only partially screened. For a given drawdown, the yield from a well partially penetrating the aquifer is less than the yield from one completely penetrating the aquifer (Driscoll, 1986).

It is assumed that for distances from the well that exceed about $1.5b\sqrt{K_h/K_v}$ the vertical components of flow are negligible and developed equations can be applied for the partially penetrating case, provided that $r > 1.5b\sqrt{K_h/K_v}$. $\sqrt{K_h/K_v}$ are the horizontal and vertical conductivity, respectively, in an anisotropic aquifer. In case $r < 1.5b\sqrt{K_h/K_v}$, the discharge per unit of drawdown is smaller than if the well is fully penetrating. One of the basic equations for the partially penetrating case

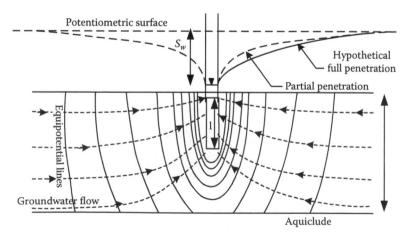

FIGURE 4.26   Flow lines toward a partially penetrating well.

utilizes a continuous superposition of sinks to represent the partially penetrating well to derive an approximate relationship between the discharge of a partially penetrating well and that for a fully penetrating well (Musket, 1946). The discharge per unit drawdown is called specific capacity and is widely used to characterize the discharge capacity of pumped wells.

$$\left(\frac{Q}{s_w}\right)_p = \left(\frac{Q}{s_w}\right)\left[\frac{l}{b}\left\{1+7\left(\frac{r_w}{2l}\right)^{1/2}\cos\frac{\pi l}{2b}\right\}\right] \tag{4.81}$$

where
  l is the length of well bore over which water enters the well
  Subscript p denotes the partially penetrating case

Example 4.15

A well with an effective well radius of 0.15 m exhibits a specific capacity of 0.032 $m^3/s \cdot m$ in an aquifer for which b is 35 m. The length of screened well bore is 10 m. If the length of well screen is increased to 20 m, estimate the increase in specific capacity.

Solution
By applying Equation 4.81 for partially penetrating case, the specific capacity for full penetration is

$$\left(\frac{Q}{s_w}\right) = \frac{(Q/s_w)_p}{[(l/b)\{1+7(r_w/2l)^{1/2}\cos\pi l/2b\}]}$$

$$\left(\frac{Q}{s_w}\right) = \frac{0.032}{[10/35\{1+7(0.15/20)^{1/2}\cos 10\pi/70\}]} = 0.072\,m^2/s$$

The specific capacity with $l = 20\ m$ is

$$\left(\frac{Q}{s_w}\right)_p = 0.072\left[20/35(1+7(0.15/40)^{1/2}\cos 20\pi/70)\right] = 0.052\,m^2/s$$

Increasing the length of well screen to 20 m will change the specific capacity from 0.032 to 0.052 m²/s, an increase of 63%.

4.7 DESIGN OF WELLS

A well is an intake structure dug in the ground to remove water from groundwater resources. Exploitation wells are drilled for the supply of municipal, industrial, and irrigation water demands, and for water table control for drainage purposes. The wells can be made as open, dug wells, or tube wells; tapping an unconfined aquifer; or penetrating the ground to tap a confined aquifer. The design of tube wells, a typical installation, is given in Figure 4.27.

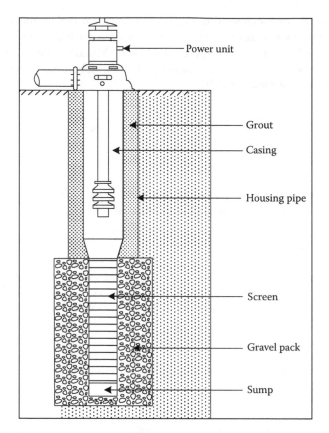

FIGURE 4.27   Typical installation of tube wells.

The design of a well is the process of selecting appropriate dimensions of various components and determining the physical materials for a well construction. Dimensional factors, strength requirements, and construction/maintenance cost play an important role in establishing the design parameters. A good design of a well minimizes economic costs including capital and maintenance costs and the possibility of collapse of service in large water-supply developments and also increases the life of the well.

A well is composed of two main elements, which include the casing and the screen.

Casing is used to maintain an open access in the earth and to avoid any entrance or leakage into the well. Housing for the pumping equipment and as a vertical conduit for water flowing upward from the aquifer to the pump intake is accommodated in the casing. The pump housing is the upper section of blind casing which supports the well against collapse. The pump housing diameter should be such that the pump accommodates with enough clearance for installation. It is recommended that the casing diameter be two pipe sizes larger than the nominal diameter of the pump. The length of the pump housing should be tall enough so that the pump remain below the water level in the well during the lifetime of the well. Thus, the maximum expected drawdown determines the length of the pump casing for the water-supply well and the water table should be used instead for drainage purposes. The casing material is chosen based on water quality, well depth, diameter, drilling procedure, construction cost, and well regulations. The most popular materials used for casing are black steel, galvanized steel, *PVC* pipe, and concrete pipe. The space around the outside of the well is filled by grout to prevent the intrusion of contaminants. A mix of cement, bentonite, or concrete can be used as grout.

The intake portion of wells is generally screened to prevent sand and fine material from entering the well during pumping and to minimize the head loss and entrance velocity. The screen keeps sand and gravel out of the well while allowing groundwater and water from formations to enter into the well. The wall of the well is supported from the loose formation material and the aquifer materials in many consolidated formations, especially sandstone and limestone, are established, and chemical and physical corrosion by the pumped water is resisted.

The optimum length of the well screen is based on the thickness of the aquifer and the total cost for water discharge or drainage. The total cost includes capital cost and pumping cost. In thick aquifers, while a fully penetrating well has high construction cost, it reduces the pumping cost since it reduces head loss. For confined aquifers, about 80%–90% of the depth at the center of the aquifer is advised to be screened. For unconfined aquifers, it is recommended to provide screen in the lower one-third thickness.

The screen length can be chosen to keep the actual screen entrance velocity in the allowable ranges. The minimum length of the well screen can be calculated using the relationship between different screen parameters (AWWA, 1998):

$$l_{min} = \frac{Q}{86400 A_e V_e} \qquad (4.82)$$

where
l_{min} is the minimum screen length (m)
Q is the discharge of the well (m³/day)
A_e is the effective open area per meter screen length (m²/m)
V_e is the screen entrance velocity (m/s)

The entrance velocity near the well screen does not exceed 0.03 m/s (Walton, 1962).

The size of screen slot openings depends on the grain-size-distribution curve for the aquifer materials. The size is selected to limit the moving of the finer formation materials near the borehole into the screen and pumped from the well during development. The typical approach to determine the slot openings size for nonhomogeneous sediments is to select a slot through which 60% of the material will pass and 40% will be retained. With corrosive water, the 50%-retained size should be chosen, because even a small enlargement of the slot openings due to corrosion could cause sand to be pumped (Driscoll, 1986).

The screen materials' strength should withstand the physical forces acting upon them during and following their installation, and during their use, including forces due to suspension in the borehole, grouting, development, purging, pumping, and sampling, and forces exerted on them by the surrounding geologic materials (Wilson and Strock, 1995). Screens are available in many materials, the most popular being stainless steel and slotted *PVC* pipe.

They should be resistant to incrustation and corrosion and should maintain their structural integrity and durability in the environment in which they are used over their operating life. Also the quality of groundwater, diameter, and depth of the well and type of strata encountered are important factors in selecting screen material.

In order to minimize the passage of formation materials into the well and to stabilize the well assembly, the annular space between the borehole wall and the screen or slotted casing should be filled. To make the zones around the well screen more permeable, some formation materials around the well are removed and replaced with graded material. A filter pack is selected to retain about 90% of the gravel pack after development. A filter pack is appropriate when the natural formation is a uniform fine sand, silt, or clay.

4.8   WELL CONSTRUCTION

Well construction usually includes drilling, installing the casing, placing a well screen and filter pack, grouting, and developing the well. There are various well drilling and installation methods depending on the size of the tube well, depth and formation to be drilled, and available facility and technical proficiency.

The cable-tool and rotary drilling methods are the most common methods for drilling deep wells.

4.8.1   CABLE-TOOL DRILLING

Cable tool drilling machines or percussion rigs operate by repeatedly lifting and dropping a heavy string of drilling tools into the borehole (Delluer, 1999). The consolidated rock is crushed into small fragments by impacts of the drill bit on the lower end with a relatively sharp tool. The cutting tool for breaking the rock is suspended from a cable and by up-and-down movement of the sharp tool, the drilling is accomplished. During the drilling, water is injected to the well and the slurry is made after mixing with crushed or loosened particles. Its use increases as drilling proceeds. Slurry is removed from the borehole by a sand pump when the penetration rate decreases.

A rig consists of a mast, lines of hoist for operating the drilling tool, and a sand pump as shown in Figure 4.28. A full string of drilling tool as shown in the figure consists of drill bit, drill stem, drilling jars, and rope socket. The drill stem is a long steel bar that adds weight and length to the drill so it will cut rapidly, and vertically fixed above the bit it provides additional tools in order to maintain a straight line. The drilling jars consist of a pair of linked steel bars and can be moved in vertical direction relative to each other. Its purpose is only to loosen the tools should they stick in the hole. The rope socket connects the string of tools to the cable (Todd and Mays, 2008).

4.8.2   ROTARY DRILLING METHOD

In this method, a rotating bit drills the hole and circulation of a drilling fluid removes the crushed or loosened particles. Drilling mud, which consists of a suspension of

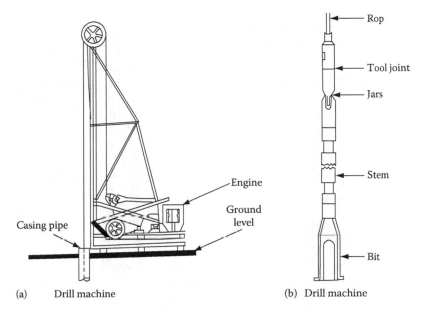

| (a) | Drill machine | (b) | Drill machine |

FIGURE 4.28 Cable-tool percussion drilling.

water, bentonite, clay, and various organic additives, is entered into the hollow to loosen and to lubricate the rotating bit hole. Then the mud is removed from the hole. In this method, the walls of the well are sealed and the holes of the walls are filled.

The method operates continuously with a hollow rotating bit through which a mixture of clay and water, or drilling mud, is forced. Material loosened by the bit is carried upward in the hole by the rising mud. No casing is ordinarily required during drilling because the mud forms a clay lining, or mud cake, on the wall of the well by filtration. This seals the walls, thereby preventing caving, entry of groundwater, and loss of drilling mud. For the rotary method of drilling without casing, screens are lowered into place as drilling mud is diluted and again are sealed by a lead packer to an upper permanent casing.

Direct rotary and reverse rotary methods are two primary types of rotary drilling methods. In the direct rotary method, the drilling mud is picked up from the annular space between the hole and drill pipe to ground surface by the pump. The fluid is cleaned in a settling pit and then a storage pit from where it is pumped back into the hole and it is recirculated. The schematic of the direct rotary is shown in Figure 4.29. The major difference between the two methods as shown in this figure is in the direction of the flowing fluid.

4.8.2.1 Well Development

During well construction, unconsolidated aquifer materials and packed artificial gravel are placed around the well screens and the permeability around the borehole is reduced. Well development is used to increase well-specific capacity, prevent sanding, and maximize economical well life. Well development also removes smaller grains initially present in the formation immediately surrounding the well screen to create a more permeable and stable zone adjacent to the well screen.

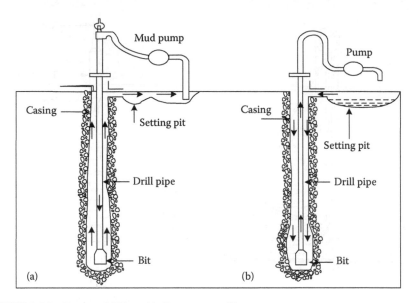

FIGURE 4.29   Rotary drilling: (a) direct rotary, (b) reverse rotary.

The development procedures are varied and include pumping, surging, use of compressed air, hydraulic jetting, addition of chemicals, and use of explosives. The simplest method of removing finer material from the borehole is pumping at a higher rate than the well will be pumped during exploitation or over pumping.

PROBLEMS

4.1 Consider two strata of the same soil material that lie between two channels. The first stratum is confined and the second one is unconfined, and the water surface elevations in the channels are 30 m and 15 m above the bottom of the unconfined aquifer. What is the thickness of the confined aquifer for which
a) The discharge through both strata are equal?
b) The discharge through the confined aquifer is half as compared to the unconfined aquifer?

4.2 Two parallel rivers A and B are separated by a land mass as shown in Figure 4.30. Estimate the seepage from river A to river B per unit length of the river.

4.3 Two parallel rivers A and B are separated by a land mass as shown in Figure 4.31. Estimate the seepage from river A to river B per unit length of the river.

4.4 For the unconfined aquifer shown in Example 4.6, determine the steady-state head distribution. Assume the aquifer is homogeneous and isotropic and has a uniform recharge rate, R with different water table in two sides of aquifer (h_1, h_2).

4.5 A well fully penetrates a 30-m thick confined aquifer. After a long period of pumping at a constant rate of 0.07 m³/s, the drawdowns at distances of 40 m and 120 m from the well were observed to be 3 in and 1.5 m, respectively. Determine the hydraulic conductivity and the transmissibility. What type of unconsolidated deposit would you expect this to be?

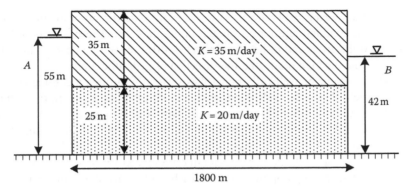

FIGURE 4.30 Cross section of parallel rivers as described in Problem 4.2.

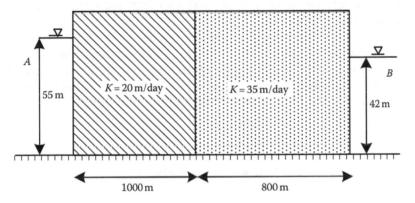

FIGURE 4.31 Cross section of parallel rivers as described in Problem 4.3.

4.6 A well penetrates an unconfined aquifer. Prior to pumping, the water level (head) is $h_o = 30$ m. After a long period of pumping at a constant rate of 0.07 (m³/s), the drawdowns at distances of 40 m and 120 m from the well were observed to be 3 m and 1.5 m, respectively. Determine the hydraulic conductivity. What type of deposit would the aquifer material probably be?

4.7 An unconfined aquifer ($K = 5$ m/day) situated on the top of a horizontal impervious layer connects two parallel water bodies M and N which are 1200 m apart. The water surface elevations of M and N, measured above the horizontal impervious bed, are 10.0 m and 8.00 m respectively. If a uniform recharge at the rate of 0.002 m³/day/m² of horizontal area occurs on the ground surface, estimate

(a) The water table profile

(b) The location and elevation of the water table divide

(c) The seepage discharges into the lakes and

4.8 Two wells, A and B are located 200 m apart and pump from an aquifer for which $T = 1.1$ (m²/min) and $S = 0.11$. The wells begin pumping simultaneously at rate $Q_A = 2$ (m³/min) and $Q_B = 3$ (m³/min). Compute the drawdown in an observation well located a distance of 50 m from well A and 200 m from well B after 7 days of continuous pumping from both wells.

4.9 Three monitoring wells are used to determine the direction of groundwater flow in a confined aquifer (as shown in Figure 4.32). The piezometric heads in the wells are found to be 45 m in well 1. 40 m in well 2 and 42 m in well 3. Determine the magnitude and the direction of hydraulic gradient.

4.10 A 40-cm diameter well fully penetrates vertically through a confined aquifer 15 m thick. When the well is pumped at 0.035 m³/s. the heads in the pumped well and the two other observation wells were found to be as shown in Figure 4.33. Does this test suggest that the aquifer material is fairly homogeneous in the directions of the observation wells?

4.11 Three pumping wells along a straight line are spaced 150 m apart. What should be the steady state pumping rate from each well so that the drawdown in each well will not exceed 1.5 m: The transmissibility of the confined aquifer at which all the wells penetrate fully is 2500 m²/day and all the wells are 30 cm in diameter. Take the thickness of the aquifer $b = 50$ m and the radius of influence of each well to be 1000 m.

4.12 The following data were collected during a test of a confined aquifer in a monitoring well located 305 m away from a well pumping. The pumped well fully penetrated the aquifer and the observation wells were constructed so that the observed drawdown are average values over the entire saturated thickness ($Q = 1200$ m³/day). Calculate the aquifer properties using Theis type curve (Table 4.6).

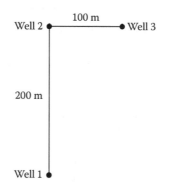

FIGURE 4.32 Location of the wells, Problem 4.9.

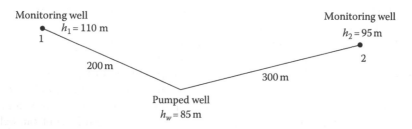

FIGURE 4.33 Location of the wells, Problem 4.10.

TABLE 4.6
Test of Confined Aquifer Data, Problem 4.12

Time (day)	1.30E-04	3.47E-04	6.94E-04	1.39E-03	3.47E-03	6.94E-03	1.39E-02
s (m)	0	0	0.006	0.043	0.168	0.302	0.445
Time (day)	3.47E-02	6.94E-02	1.39E-01	3.47E-01	6.94E-01	2.3	
s (m)	0.594	0.634	0.637	0.637	0.64	0.643	

4.13 Table 4.7 lists drawdown versus time data in an unconfined aquifer. Data are collected at an observation well 50m away from a well being pumped at 2 m³/min. Both the pumped and the observation well are fully-penetrating. The original saturated thickness of the aquifer is 50 meters. Calculate the hydraulic parameters of the aquifer using the type-curve method. Please use the type curves provided at the end of this assignment.

4.14 A well with an effective well radius of 0.3 m exhibits a specific capacity of 0.032 m³/sm in an aquifer with 28 m thickness. Estimate the specific capacity for full penetration when the length of screened well bore is 15 m.

4.15 Find the minimum screen length of a well with the designed discharge of 3.6 m³/min, a well screen with an open area of 30% and a diameter of 0.25 m.

TABLE 4.7
Drawdown Versus Time Data in an Unconfined Aquifer, Problem 4.13

Time (min)	S (m)	Time (min)	S (m)
0.1	0	19.36	1.06
0.13	0	25.19	1.06
0.17	0.01	32.78	1.06
0.22	0.02	42.65	1.06
0.29	0.06	55.49	1.07
0.37	0.11	72.21	1.07
0.49	0.18	93.96	1.08
0.63	0.28	122.26	1.08
0.82	0.39	159.08	1.09
1.07	0.51	206.99	1.1
1.39	0.64	269.34	1.12
1.81	0.75	350.47	1.14
2.36	0.85	456.03	1.16
3.07	0.93	593.38	1.19
3.99	0.99	772.1	1.24
5.19	1.04	1004.66	1.29
6.75	1.05	1307	1.36
8.79	1.05	1701.01	1.44
11.43	1.06	2213.34	1.55
14.88	1.06	2280	1.68

REFERENCES

AWWA (American Water Works Association). 1998. AWWA standard for water wells; American National Standard. ANSI/AWWA A100-97, AWWA, Denver, Co.

Bear, J. 1979. *Hydraulics of Groundwater*. McGraw-Hill, New York, 38 pp.

Delleur, J.W. 1999. *The Handbook of Groundwater Engineering*. CRC Press/Springer-Verlag, Boca Raton, FL.

Driscoll, F.G. 1986. *Groundwater and Wells*, 2nd edn. Johnson Division, St. Paul, MN.

Fetter, C.V. 2001. *Applied Hydrogeology*, 4th edn. Prentice Hall, Englewood Cliffs, NJ, ISBN 0-13-066239-9.

Hantosh, M.S and Jacob, C.E. 1955. Non-steady radial flow in an infinite leaky aquifer. Transactions, American Geophysical Union, 36(1), 95–100.

Hantush, M.S. and Jacob, C.E. 1955. Non-steady radial flow in an infinite leaky aquifer. Transactions, American Geophysical Union, 36(1), 95–100.

Karamouz, M., Szidarovszky, F., and Zahraie, B. 2003. *Water Resources Systems Analysis*. Lewis Publishers, Boca Raton, FL.

Kruseman, G.P. and de Ridder, N.A. 1991. *Analysis and Evaluation of Pumping Test Data*, 2nd edn. International Institute for Land Reclamation and Improvement, Wageningen, the Netherlands.

Marino, M.A. and Luthin, J.N. 1982. *Seepage and Groundwater: Developments in Water Science*. Elsevier Scientific Publishing Co, Amsterdam, the Netherlands, 489 p.

Musket, M. 1946. *The Flow of Homogeneous Fluids through Porous Media*. J.W. Edwards, Ann Arbor, MI.

Neuman, S.P. 1975. Analysis of pumping test data from anisotropic unconfined aquifers considering delayed gravity response, *Water Resources Research*, 11, 329–342.

Theim, G. 1906. *Hydrologische Methoden*. Gebhardt, Leipzig, Germany, 56 pp.

Theis, C.V. 1935. The lowering of the piezometer surface and the rate and discharge of a well using groundwater storage. *Transactions, American Geophysical Union*, 16, 519–524.

Todd, D.K. and Mays, L.W. 2008. *Groundwater Hydrology*, 3rd edn. John Wiley & Sons, Inc., New York.

U.S. Bureau of Reclamation, 1981. *Groundwater Manual*. U.S. Government Printing Office, Denver, CO.

Walton, W.C. 1962. Selected analytical methods for well and aquifer evaluation. *Illinois State Water Survey Bulletin*, 49, 81.

Willis, R. and Yeh, W.W.-G. 1987. *Groundwater Systems Planning and Management*. Prentice Hall, Englewood Cliffs, NJ.

Wilson, P. and Strock, J. 1995. *Monitoring Well Design and Construction for Hydrogeologic Characterization. Guidance Manual for Ground Water Investigations*. The California Environmental Protection Agency, CA.

5 Groundwater Quality

5.1 INTRODUCTION

Societal dependency on water and the shortage of adequate surface water have increased the use of groundwater. Since groundwater is used to supply the industrial, agricultural, and domestic needs, quality aspects are very important. A number of factors can contribute to the changes in the quality of groundwater, including changes in atmospheric composition, changes in soil chemistry, and human activities that directly alter the physical or chemical conditions of soil layers.

There are many parameters that affect groundwater solubility, such as pH, sorption to solids, ionic strength, temperature, and microbiological factors.

Many activities are responsible for the pollution of groundwater, including leaching from municipal and chemical landfills and abandoned dump sites, accidental spills of chemicals or waste materials, the improper underground injection of liquid wastes, and the placement of septic tank systems in hydrologically and geologically unsuitable locations.

These substances may be introduced to the aquifer system either through the hydraulic interaction with polluted surface waters or directly through injection wells and/or spreading basins or seepage from waste disposal sites. Hydraulic and mass transport equations are used to describe how the groundwater quality system is affected by waste contaminants.

In this chapter, groundwater constituents and contaminants are introduced. Then, a description of water quality standards and the factors affecting groundwater solubility is provided. Finally, the sources of groundwater contamination as well as contaminant transport in the saturated and unsaturated zones are evaluated.

5.2 GROUNDWATER CONSTITUENTS AND CONTAMINANTS

Groundwater contains various constituents and contaminants such as inorganics, organics, dissolved gasses, and particles, which are explained in the following subsections.

5.2.1 INORGANIC CONTAMINANTS

Nitrogen: Nitrogen is mostly from agricultural activities and the disposal of sewage on or beneath the land surface. The main form of nitrogen that exists in groundwater is NO_3. However, dissolved nitrogen can also be found in the form of ammonium (NH_4^+). Other forms of nitrogen that might be found in groundwater are ammonia (NH_3), nitrite (NO_2^-), nitrogen (N_2), nitrous oxide (N_2O), and organic nitrogen.

Metals: The mobility of metals in groundwater has received considerable attention due to increasing industrial activities. The metals for which maximum permissible or recommended limits have been set in drinking water standards include silver (Ag), cadmium (Cd), chromium (Cr), copper (Cu), mercury (Hg), iron (Fe), manganese (Mn), and zinc (Zn). Most of the above metals are known as heavy metals. The occurrence and mobility of the heavy metals in groundwater environments can be influenced by adsorption processes.

Nonmetals: The most common nonmetals that have been considered in groundwater studies include carbon, chlorine, sulfur, nitrogen, fluorine, arsenic, selenium, phosphorus, and boron. Carbon, chlorine, and sulfur may exist in high concentrations in dissolved forms in most natural and contaminated groundwater systems. Arsenic and its compounds have been used as insecticides and herbicides for a long time and can be found in groundwater systems influenced by agricultural activities. Selenium can exist in high concentrations in some rocks, such as shale, as well as in coal, uranium, and some soil particles. Fluoride is used as a municipal water-supply additive in many cities due to its beneficial effects on dental health. It is also a natural constituent of groundwater. Phosphorus causes accelerated growth of algae and aquatic vegetation, and in consequence eutrophication in ponds, lakes, reservoirs, and streams. However, it is not a harmful constituent in drinking water. The main sources of phosphorus are the widespread use of fertilizers and the disposal of sewage on land.

5.2.2 ORGANIC CONTAMINANTS

Carbon is the key element in organic compounds. Generally, organic compounds have carbon and usually hydrogen and oxygen as the main elements in their chemical structure. Dissolved organic carbon (DOC) is most commonly in concentrations in the range 0.1–10 mg/L in groundwater. However, in some areas, values might be as high as several tens of milligrams per liter. The species H_2CO_3, CO_2, HCO_3^-, and CO_3^{2-} are some exceptions that are not considered as organic components. These species are important constituents in all groundwater systems, though.

The prevalent organic constituents in uncontaminated groundwater systems are humic substances that are from decay in plant and animal matter. Fulvic acid, humic acid, and humin are three classes of humic substances. Among these classes, only fulvic acid is soluble in groundwater.

Many organic substances have extremely low solubility in water. Table 5.1 presents the maximum permissible concentrations and solubilities of some most commonly used pesticides.

5.2.3 DISSOLVED GASSES

The natural gases involved in the geochemical cycle of groundwater are carbon dioxide (CO_2), oxygen (O_2), and nitrogen (N_2). These gases mostly originate from the atmosphere. Besides, some biochemical processes produce dissolved gases in groundwater. These include the flammable gases methane (CH_4) and hydrogen sulfide (H_2S). The typical compositions of different gases in the atmosphere are presented in Table 5.2. Some of these constituents have significant roles in determining the chemistry of rain and snow and in consequence the chemistry of water resources influenced by the quality of rain and snow.

TABLE 5.1
Typical Limits on Drinking Water and the Solubilities of Four Pesticides

Pesticide	Maximum Permissible Concentration (mg/L)	Solubility in Water (mg/L)
Methoxychlor	0.1	0.1
Toxaphene	0.005	3
2,4-D	0.1	620
2,4,5-TP silvex	0.01	—

Source: Oregon State University, *Disposal of Environmentally Hazardous Wastes*, Oregon State University, Corvallis, OR, 210 p., 1974.

TABLE 5.2
Typical Gas-Phase Compositions in Atmosphere

Gas	Percentage by Volume
N_2	78.1
O_2	20.9
Ar	0.93
H_2O	0.1–2.8
CO_2	0.03
Ne	1.8×10^{-3}
He	5.2×10^{-4}
CH_4	1.5×10^{-1}
Kr	1.1×10^{-4}
CO	$(0.6–1) \times 10^{-4}$
SO_2	1×10^{-4}
N_2O	5×10^{-5}
H_2	5×10^{-5}
O_3	$(0.1–1.0) \times 10^{-5}$
Xe	8.7×10^{-6}
NO_2	$(0.05–2) \times 10^{-5}$

Source: Mirtov, B.A., *Gaseous Composition of the Atmosphere and its Analysis*, U.S. Department of Commerce, Office of Technical Services, Washington DC, 209 p., 1961.

Some important atmosphere gases, which affect the quality of groundwater are defined in the following paragraphs.

Carbon dioxide (CO_2) dissolves in water and forms carbonic acid. This acid can be dissociated from bicarbonate (HCO_3^-) and carbonate (CO_3^{2-}) ions. The carbonate system is a key component, which plays an important role in groundwater composition through precipitation reactions (explained in Chapter 9). Besides, the carbonic acid cycle influences the determination of the pH and buffering capacity of precipitation. Water at equilibrium with atmospheric CO_2 will have a pH of approximately 5.7, which can be used as a benchmark in the evaluation of the effects of atmospheric constituents.

Oxygen (O_2) plays an important role in microbiological activities depending on aerobic, anoxic, or anaerobic conditions. The common concentrations in the liquid phase are approximately 8–10 mg/L, depending on the altitude and temperature.

Nitrogen and sulfur oxides (NO_x and SO_x) may cause the production of strong acids, such as HNO_3 and H_2SO_4. Many natural processes, such as volcanic and microbiological activity release NO_x and SO_x into the atmosphere.

Hydrogen sulfide (H_2S) rarely accumulates to dangerous proportions and has a distinctive rotten-egg odor.

Methane (CH_4) is produced by the decomposition of buried plant and animal matter in unconsolidated and geologically young deposits. It is a colorless, tasteless, and odorless gas, which can cause serious problems. For example, if there is as little as 1–2 mg/L of methane in water, an explosion in a poorly ventilated air space can take place.

5.2.4 PARTICLES

Constituents that exist in the solid phase either within or apart from the groundwater are referred to as particulate matters. Generally, constituents with a size of 10 nm or greater are considered as particles in groundwater system. A graphical representation of the spectrum of particle sizes is depicted in Figure 5.1.

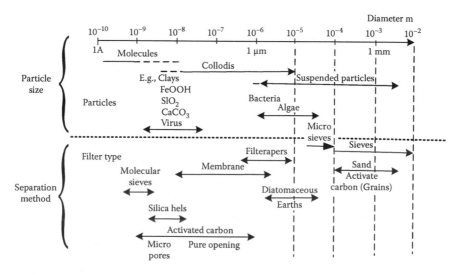

FIGURE 5.1   Spectrum of particle sizes to be seen in natural waters, including groundwater and physical processes for separation of particles. (From Stumm, W., Chemical interaction in particle separation, *Environmental Science and Technology*, 11, 1070 p., 1977.)

Microorganisms are also important particulate contaminants in groundwater systems. Microbial contaminants in groundwater can be categorized into viruses, bacteria, and protozoan cysts. All these microorganisms move in groundwater systems and the larger microorganisms generally move slower than viruses. Protozoan cysts are more difficult to inactivate than either bacteria or viruses. They might be found in higher concentrations in groundwater supplies that are under the direct influence of surface waters. Therefore, more rigorous disinfection requirements are recommended for potabilization in case protozoan cysts exist in the groundwater system.

5.3 WATER QUALITY STANDARDS

To evaluate the suitability of a water resource for specific uses, chemical properties of the water need to be compared with water quality standards assigned for that specific use. Therefore, standards are defined to serve as a basis for water quality evaluations. Drinking water standards (Table 5.3) are the most important of these standards.

In 2006, the World Health Organization (WHO) recommended values for chemicals that are of health significance in drinking water. These values are presented in Table 5.4.

In many regions, the most important uses of groundwater are for agriculture. In these situations, it is appropriate to appraise the quality of groundwater relative to criteria or guidelines established for livestock or irrigation. Recommended concentration limits for these uses are listed in Table 5.5. The list of constituents and the concentration limits are not as stringent as for drinking water. The serious degradation of groundwater quality could be due to man's activity that forces the concentration of a variety of constituents to increase.

In 2006, the state of Wisconsin established numerical groundwater quality standards for substances of public health or welfare concern. These standards are listed in Table 5.6.

It should be stated that *standards* may not have a scientific basis but must be regarded as a law or regulation. *Criteria* have a scientific basis but may not be regarded as a law or regulation. McKee (1960) illustrated the difference between standards and criteria as:

The term *standard* applies to any definite rule, principle, or measure established by authority. The fact that it has been established by an authority makes a standard somewhat rigid, official, or quasi-legal, but this fact does not necessarily mean that the standard is fair, equitable, or based on sound scientific knowledge, for it may have been established somewhat arbitrary on the basis of inadequate technical data tempered by a caution factor of safety. Where health is involved and where scientific sparse, such arbitrary standards may be justified. A *criterion* designates a mean by which anything is tried in forming a correct judgment concerning it. Unlike a standard, it carries no connotation of authority other than of fairness and equity nor does it imply an ideal condition. Where scientific data are being accumulated to serve as yardsticks of water quality, without regard for legal authority, the term *criterion* is most applicable.

TABLE 5.3
Summary of EPA's National Primary Drinking Water Standards (MCLG, MCL)

Contaminant	MCLG (mg/L)	MCL (mg/L)
Inorganics		
Antimony	0.006	0.006
Asbestos (>10 nm)	7 MFL[a]	7 MFL
Barium	2	2
Beryllium	0.004	0.004
Cadmium	0.005	0.005
Chromium (total)	0.1	0.1
Copper	1.3	TT[b]
Cyanide	0.2	0.2
Fluoride	4	4
Lead	0	TT
Mercury (inorganic)	0.002	0.002
Nitrate	10	10
Nitrite	1	1
Selenium	0.05	0.05
Thallium	0.0005	0.002
Coliform and surface water treatment		
Giardia lambia	0 detected	TT
Legionella	0 detected	TT
Standard plate count	N/A	TT
Total coliform	0 detected	c
Turbidity	N/A	TT
Viruses	0 detected	TT
Adionuclides		
Radium 226 + radium 228	0	5 pCi/L
Gross alpha particle activity	0	15 pCi/L
Beta particle + photon	0	4 mrem/year
Radioactivity		

Source: Delleur, J.W., *The Handbook of Groundwater Engineering*, CRC Press/Springer-Verlag, Boca Raton, FL, 992 p., 1999.

[a] MFL, million fibers per liter.
[b] TT, treatment techniques in lieu of a numerical standard.
[c] No more than 5% of samples can be total coliform positive.

TABLE 5.4
Some Selected Guideline Values for Chemicals That Are of Health Significant in Drinking Water

Chemical	Guideline Value (mg/L)
Arsenic	0.01 (P)
Barium	0.7
Benzene	0.1
Boron	0.5 (T)
Cadmium	0.003
Carbon tetrachloride	0.004
Chlorate	0.7 (D)
Chlordane	0.0002
Chlorine	5 (C)
Chlorite	0.7 (D)
Chloroform	0.3
Chlorotoluron	0.03
Chlorprifos	0.03
Chromium	0.05 (P)
Copper	2
Cyanazine	0.0006
Cyanide	0.07
Cyanogen choride	0.07
2,4-D(2,4-dichlorophenoxyaceticacid)	0.03
2,4-DB	0.09
DDT and metabolites	0.001
Dibromochloromethane	0.1
Dichlorobenzene, 1,2-	1 (C)
Dichlorobenzene, 1,4	0.3 (C)
Dichloroethene, 1,2	0.05
Dichloromethane	0.02
Fluoride	1.5
Nickel	0.07
Nitrate (as NO_3)	50
Nitrilotnacetic acid (NTA)	0.2
Nitrite (as NO_2)	0.2 (P)
Tetrachloroethene	0.04
Toluene	0.7 (C)
Trichloroacetate	0.2
Trichbrcethene	0.02 (P)
Trichlorophend,2,4,6-	0.2 (C)
Tdhalomethanet	
Uranium	0.015 (P) [only chemical aspects of uranium addressed]
Vinyl chloride	0.0003
Xylenes	0.5 (C)

(continued)

TABLE 5.4 (continued)
Some Selected Guideline Values for Chemicals That Are of Health Significant in Drinking Water

Source: WHO, *Guidelines for Drinking-Water Quality*, First Addendum to 3rd edn., Vol. 1-Recommendations, 2006.

Notes: P, provisional guideline value, as there is evidence of a hazard but the available information on health effects is limited; T, provisional guideline value because the calculated guideline value is below the level that can be achieved through practical treatment methods, source protection, etc.; D, provisional guideline value because disinfection is likely to result in the guideline value being exceeded; C, concentrations of the substance at or below the health-based guideline value may affect the appearance, taste and odor of the water leading to consumer complaints.

TABLE 5.5
Recommended Concentration Limits for Water Used for Livestock and Irrigation Crop Production

	Livestock: Recommended Limits (mg/L)	Irrigation Crops: Recommended Limits (mg/L)
Total dissolved solids		
Small animals	3000	700
Poultry	5000	
Other animals	7000	
Nitrate	45	—
Arsenic	0.2	0.1
Boron	5	0.75
Cadmium	0.05	0.01
Chromium	1	0.1
Fluoride	2	1
Lead	0.1	5
Mercury	0.01	—
Selenium	0.05	0.02

Source: U.S. Environmental Protection Agency, *Water Quality Criteria 1972.* EPA R373033, Government Printing Office, Washington, DC, 1973.

TABLE 5.6
Public Health Groundwater Quality Standards

Substance	Enforcement Standard (µg/L)	Preventive Action Limit (µg/L)
Acetone	1000	200
Alachor	2	0.2
Aldicarb	10	2
Antimony	6	1.2
Anthracene	3000	600
Arsenic	10	1
Atrazine, total chlorinated residues	3	0.3
Bacteria, total coliform	0	0
Barium	2 mg/L	0.4 mg/L
Bentazon	300	60
Benzene	5	0.5
Benzo(b)flouranthene	0.2	0.02
Benzo(a)pyrene	0.2	0.02
Beryllium	4	0.4
Boron	960	190
Bromodichloromethane	0.6	0.06
Bromoform	4.4	0.44
Bromomethane	10	1
Butylate	67	6.7
Cadmium	5	0.5
Carbaryl	960	192
Carbofuran	40	8
Carbon disulfide	1000	200
Carbon tetrachloride	5	0.5
Chloramben	150	30
Chlordane	2	0.2
Chloroethane	400	80
Chloroform	6	0.6
Chloromethane	3	0.3
Chromium	100	10
Chrysene	0.2	0.02
Cobalt	40	8
Capper	1300	130
Cyanazine	1	0.1
Cyanide	200	40
Dachtal	4 mg/L	0.8 mg/L
1,2-Dibromoethane (EDB)	0.05	0.005
Dibromochloromethane	60	6
1,2-Dibromo-3-chloropropane (BDCP)	0.2	0.02
Dybutyl phthalate	100	20
Dicamba	300	60

(*continued*)

TABLE 5.6 (continued)
Public Health Groundwater Quality Standards

Substance	Enforcement Standard (µg/L)	Preventive Action Limit (µg/L)
1,2-Dichlorobenzene	600	60
1,3-Dichlorobenzene	1250	125
1,4-Dichlorobenzene	75	15
Dichlorodifluoromethane	1000	200
1,1-Dichloroethane	850	85
1,2-Dichloroethane	5	0.5
1,1-Dichloroethylene	7	0.7
1,2-Dichloroethylene(*cis*)	70	7
1,2-Dichloroethylene(*trans*)	100	20
2,4-Dichlorophenoxy acetic acid (2,4-D)	70	7
1,2-Dichloropropane	5	0.5
1,3-Dichloropropane(*cis/trans*)	0.2	0.02
Di (2-ethylhexyl)phthalate	6	0.6
Dimethoate	2	0.4
2,4-Dinitrotoluene	0.05	0.005
2,6-Dinitrotoluene	0.05	0.005
Dinosed	7	1.4
Dioxin (2,3,7,8-TCDD)	0.00003	0.000003
Endrin	2	0.4
EPTC	250	50
Ethylbenzene	700	140
Ethylene glycol	7 mg/L	0.7 mg/L
Fluoranthene	400	80
Fluorene	400	80
Fluride	4 mg/L	0.8 mg/L
Fluorotrichloromethane	3490	698
Formaldehyde	1000	100
Heptachlor	0.4	0.04
Heptachlor epoxide	0.2	0.02
Hexachlorobenzene	1	0.1
N-hexane	600	120
Hydrogen sulfide	30	6
Lead	15	1.5
Lindane	0.2	0.02
Mercury	2	0.2
Methanol	5000	1000
Methoxychlor	40	4
Methylene chloride	5	0.5
Methyl ethyl ktone (MEK)	460	90
Methyl isobutyl ketone (MIBK)	500	50
Methyl *tert*-butyl ether (MTBE)	60	12

TABLE 5.6 (continued)
Public Health Groundwater Quality Standards

Substance	Enforcement Standard (μg/L)	Preventive Action Limit (μg/L)
Metolachlor	15	1.5
Metribuzin	250	50
Monochlorobenzene	100	20
Naphthalene	40	8
Nickel	100	20
Nitrate (as N)	10 mg/L	2 mg/L
Nitrate + nitrite (as N)	11 mg/L	2 mg/L
Nitrite (as N)	1 mg/L	0.2 mg/L
N-Nitrosodiphenylamine	7	0.7
Pentachlorophenol (PCP)	1	0.1
Phenol	6 mg/L	1.2 mg/L
Picloram	500	100
Toxaphene	3	0.3
1,2,4-Trichlorobenzene	70	14
1,1,1-Trichloroethane	200	40
1,1,2-Trichloroethane	5	0.5
Trichloroethylene (TCE)	5	0.5
2,4,5-Trichlorophenoxy-propoinic acid (2,4,5-TP)	50	5
1,2,3-Trichloropropane	60	12
Trifluralin	7.5	0.75
Trimethylbenzenes (1,2,4-and1,3,5-combincd)	480	96
Vanadium	30	6
Viny chloride	0.2	0.02
Xylene	10 mg/L	1 mg/L

5.4 GROUNDWATER SOLUBILITY

The solubility of a contaminant compound will determine the transport, fate, and toxicology of that compound in a groundwater system. The main characteristics of a system, which may affect solubility are pH, *sorption to solids*, and temperature.

For a chemical process, the concentrations of the reacting components and reaction products play the role of the driving force for the entire reaction process. Most chemical reactions are, to some extent, reversible, which means that they proceed in both directions at once. In the generalized reversible reaction consider the constituents A and B reacting to produce the products C and D,

$$aA + bB \Leftrightarrow cC + dD \tag{5.1}$$

where a, b, c, and d are coefficients corresponding to the number of moles of the chemical constituents A, B, C, D, respectively. The double arrow designation indicates that the reaction proceeds in both directions at the same time, i.e., it is a reversible reaction.

When the reaction reaches an equilibrium, the law of mass action expresses the relation between the reactants and the products as

$$K = \frac{[C]^c [D]^d}{[A]^a [B]^b} \tag{5.2}$$

where

K is the thermodynamic equilibrium constant or the stability constant.

The brackets represent the concentration of the substances at equilibrium, expressed in moles per liter.

Due to Equation 5.2, for any initial condition, the corresponding reaction (Equation 5.1) will proceed until both the reactants and products reach their equilibrium condition. Initial activities cause the reaction to proceed to the left or to the right until the equilibrium condition is reached. It should also be emphasized that Equation 5.2 is valid only when a chemical equilibrium is established, if ever. However, the law of mass action does not express the rate at which the reaction proceeds and does not indicate the kinetics of the chemical process.

A disturbance in the system, e.g., an addition of reactants, causes the system to continue to proceed toward a new equilibrium condition. Changes in the temperature or pressure can also result in a move toward a new equilibrium.

If auto-ionization occurs in water, H^+ and OH will be produced.

$$H_2O \Leftrightarrow H^+ + OH^- \tag{5.3}$$

The notation H^+ is a representation for the hydrated proton. It is used to determine the pH of water as

$$pH = -\log_{10}(H^+) \tag{5.4}$$

where (H^+) is the activity of the hydrogen ion.

The speciation of all acids and bases in solution can be specified by the pH of the solution. Table 5.7 lists some important acid-base pairs. These constants indicate the importance of pH in determining the characteristics of the acid-base behavior of groundwater.

Another term commonly used for describing the activity of a constituent is pC, which is defined as the negative $\log_{10}$ of the activity of the constituent. pC–pH diagrams are used to illustrate acid-base behavior over a broad range of conditions.

TABLE 5.7
Acid Dissociation Constants for Common Ionizable Groundwater Constituents at 25°C

Compound (Formula)	Conjugate Base (Formula)	pK_a
Hydrochloric acid (HCl)	Chloride (Cl$^-$)	−3
Sulfuric acid (H$_2$SO$_4$)	Bisulfate (HSO$_4^-$)	−3
Nitric acid (HNO$_3$)	Nitrate (NO$_3^-$)	−1
Bisulfate (HSO$_4^-$)	Sulfate (SO$_4^{2-}$)	1.9
Phosphoric acid (H$_3$PO$_4$)	Dihydrogen phosphate (H$_2$PO$_4^-$)	2.1
Acetic acid (CH$_3$COOH)	Acetate (CH$_3$COO$^-$)	4.7
Hydrogen sulfide (H$_2$S)	Bisulfide (HS$^-$)	7.1
Dihydrogen phosphate (H$_2$PO$_4^-$)	Hydrogen phosphate (HPO$_4^{2-}$)	7.2
Hydrogen cyanide (HCN)	Cyanide (CN$^-$)	9.2
Ammonium (NH$_4^+$)	Ammonia (NH$_3$)	9.3
Asilicic acid (Si(OH)$_4$)	Silicate (SiO(OH)$_3^-$)	9.5
Silicate (SiO(OH)$_3^-$)	(SiO$_2$(OH)$_2^{2-}$)	12.6
Bisulfide (HS$^-$)	Sulfide (S^{2-})	17

Source: Stumm, W. and Morgan, J.J., *Aquatic Chemistry*, 3rd edn., John Wiley & Sons, New York, 1022 p., 1996.

These diagrams also demonstrate the equilibrium relationships and the total concentration of the compounds.

Some important conditions in evaluating the rates and equilibrium of chemical reactions and processes are discussed in the following paragraphs.

Sorption to solids: There are two types of associations among aqueous and solid phases: (1) the accumulation of material at a water-solid interface, which is called adsorption; (2) the intermingling of solute molecules with the molecules of the solid phase, which is termed absorption. In other words, absorption is the dissolution of a liquid material in a solid solvent. It is usually hard, or sometimes even impossible, to distinguish between the two processes. The term sorption can be used for both these processes.

In general, there are three types of sorptions for groundwater systems. In all these three types, bonds are formed between the constituent and the solid phase. These types are

- *Physisorption*: Relatively weak physical bonds, such as van der Waal's forces, are formed in this type.
- *Chemisorption*: Under certain conditions, a chemical bond is formed between the solute and a solid surface, and the process is referred to as chemisorption.
- *Ion exchange*: Ion exchange occurs due to an electrostatic interaction between the solute and a solid surface.

In physisorption and ion exchange processes, the forces responsible for the attraction of a solute to the solid are relatively weak. Therefore, these processes tend to be readily reversible. The selectivity order in ion exchange for some common cations is

$$Ba^{2+} > Sr^{2+} > Ca^{2+} > Mg^{2+} > Cs^+ > K^+ > Na^+ > Li^+ \tag{5.5}$$

Ions will replace those ions on their right in the selectivity order. As an example, Mg^{2+} will replace Na^+. However, based on the circumstances, the specific selectivity order may vary from that defined in Equation 5.5.

Temperature: The rates and equilibria of chemical reactions are highly dependent on thermodynamic conditions. In higher temperatures, more thermal energy is available, and in consequence, reaction rates increase. If a given reaction consumes energy (an endothermic reaction), an increase in temperature promotes the forward reaction and the equilibrium will be shifted to the right (more product formation). For example, heat may affect groundwater contaminants as it (1) changes the equilibria and kinetics of precipitation and the dissolution of minerals and (2) changes the rates of biological transformation processes.

In groundwater systems, temperature is usually assumed to be at a value of 1°C–2°C greater than the annual average ambient air temperature for the region (Freeze and Cherry, 1979). In higher depths, greater than approximately 10 m, diurnal and seasonal variations of air temperature have minimal effects on groundwater; therefore, groundwater temperature at these depths is often assumed to be approximately 13°C.

Microbiological factors: In groundwater, bacteria are the most important microorganisms in redox processes. Algae, fungi, yeasts, and protozoans are important in other aqueous environments. The size of bacteria generally ranges from about 0.5 to 3 μm. The main effect of microorganisms on groundwater contaminants is that they catalyze the important redox reactions in groundwater.

5.5   DISEQUILIBRIUM AND SATURATION INDEX

In a condition of disequilibrium, if the reaction expressed in Equation 5.1 is considered, the relation between the reactants and the products can be shown as

$$Q = \frac{[A]^a [B]^b}{[C]^c [D]^d} \tag{5.6}$$

where Q is the reaction quotient and the other parameters are as expressed in Equation 5.2. In thermodynamic equilibrium condition, the following ratio can be used to compare the status of a mineral; the dissolution–precipitation reaction at a particular time or space:

$$S_t = \frac{Q}{K_{eq}} \tag{5.7}$$

TABLE 5.8

Equilibrium Constants for Calcite, Dolomite, and Major Aqueous Carbonate Species in Pure Water, 0°C–30°C, and 1 Bar Total Pressure

Temperature (°C)	$pK^a_{CO_2}$	$pK_{H_2CO_3}$	$pK_{HCO_3^-}$	pK_{cal}	pK_{dol}
0	1.12	6.58	10.62	8.34	16.56
5	1.2	6.52	10.56	8.345	16.63
10	1.27	6.47	10.49	8.355	16.71
15	1.34	6.42	10.43	8.37	16.79
20	1.41	6.38	10.38	8.385	16.89
25	1.47	6.35	10.33	8.4	17
30	1.67	6.33	10.29	8.51	17.9

Source: Freeze, R. and Cherry, J., *Groundwater*, Prentice-Hall, Englewood Cliffs, NJ, 1979, 604 p.

a $pK = -\log K$.

---

where S_t is the *saturation index*. For example, the saturation index for calcite is defined as

$$S_t = \frac{\left[Ca^{2+}\right]\left[CO_3^{2-}\right]}{K_{cal}} \tag{5.8}$$

The ion activities for Ca^{2+} and CO_3^{2-} are obtained by analyzing groundwater samples and the equilibrium constant, K_{cal}, is obtained from equilibrium constant tabulations. Some equilibrium constants in particular temperature and pressure conditions are presented in Table 5.8.

$S_t > 1$ shows that the ionic constituents exceed the normal conditions. Therefore, the reaction proceeds to the left, and in consequence, mineral precipitation occurs. If $S_t < 1$, the reaction proceeds to the right and results in the dissolve of the mineral. $S_t = 1$ shows that the reaction is at equilibrium and the water is saturated by the mineral. The saturation index helps to compare the status of water samples with the computed equilibrium conditions.

5.6 SOURCES OF GROUNDWATER CONTAMINATION

Water in the hydrologic cycle is subject to quality changes due to physical, chemical, and biological characteristics of the environment. The changes can be natural or manmade. For example, the quality of groundwater is widely affected by waste disposal, underground storage of waste materials, and seepage from agricultural fertilizers and pesticides as well as leakage from underground tanks.

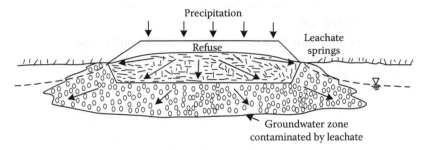

FIGURE 5.2   Water-table plume under a landfill, causing leachate springs and migration of contaminants deeper into the groundwater zone.

5.6.1 DISPOSAL OF SOLID WASTES

A considerable number of examinations on the infiltration of water through refuses in Europe have shown that this penetration is ascribed to the water table below the landfill. As a result of this event, leachate descends from the landfill into groundwater as shown in Figure 5.2. Under this condition, the seepage of the leachate source at the perimeter of the landfill into surface-water bodies will be probable. It should be mentioned that if the leachate does not pass through the aquifers including potable water, its downward movement will not endanger and pollute groundwater resources.

5.6.2 UNDERGROUND PETROLEUM TANK LEAKAGE

In industrial countries and urban areas, numerous storage tanks are buried at gas stations, and in residential, commercial, and industrial compounds to produce thermal and other kinds of energy. These petroleum products are carried by underground pipelines passing through continents. Also, a small amount of oil and gases are carried by tanker trucks that are continuously on the move. Therefore, groundwater quality will be increasingly menaced by seepages and penetrations from these sources. A strict testing of the buried storage tanks is required, but it is only gradually being performed in most countries. As a result, leakage problems due to rust on the tanks, especially those that are located in regions with high water tables and frequent infiltration, are common. It should be noted that as oils and gasoline are not mixable in water, they migrate into the unsaturated zone, form leakages, and stay on the water table. Some dense chemical products such as heavy metals and other materials move downward to deeper parts of the aquifer and threaten deep-water supply wells.

5.6.3 DISPOSAL OF LIQUID WASTES

In many areas, liquid waste injection has been widely used to practice waste disposal. Its main purpose is to isolate hazardous substances from the biosphere (Freeze and Cherry, 1979). Strict regulations have been implemented to prevent the discharge of contaminants to rivers and lakes and to protect surface water from pollution. Therefore, in waste management, using the deep permeable zones for liquid waste disposal is an interesting option for many industries.

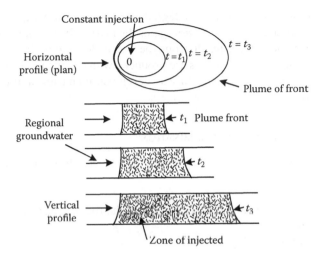

FIGURE 5.3   Potentiometric mound caused by waste disposal injection and the expansion of the affected zone occupied at times t_1, t_2, and t_3.

Figure 5.3 shows the impact of an injection well in a hypothetical horizontal aquifer on the hydrodynamic conditions in a regional flow. The injection well leads to an increase in the potentiometric surface. The mound enlarges in the flow direction asymmetrically. On continuing with the injection, the mound expands over an ever-increasing area.

In regions with injection wells close to each other, the potentiometric mounds merge similarly to the drawdown interface in fields of pumping wells. The front of the mound spreads very quickly in comparison with the spread of the zone of injected waste. The pressure translation leads to a spread in front of the potentiometric mound, which occurs as volume displacement. There is direct proportion between the affected area and the injected cumulative volume of waste. The dispersion also progressively decreases the interface between the water and the waste.

5.6.4   SEWAGE DISPOSAL ON LAND

Sewage is reaching subsurface flow in a variety of ways. Widespread use of septic tanks and drains in rural, recreational, urban, and suburban areas contributes filtered or unfiltered sewage effluent directly to the ground. Septic tanks and cesspools are the largest of all contributors of wastewater to the ground. Sewage disposal on land can cause the transportation of many kinds of contaminants to groundwater such as nitrogen and phosphorus, detergents, pathogenic bacteria and viruses, etc. There are absorption wells in many cities, even some megacities, such as Tehran/Iran with three million absorption wells.

5.6.5   AGRICULTURAL ACTIVITIES

Of all the man-made activities that influence the quality of groundwater, the use of chemicals in agriculture is probably the most important factor. Among the main agricultural activities that can cause a degradation of groundwater quality is the use of fertilizers and pesticides and the storage or disposal of livestock wastes on land. The most widespread effects result from the use of fertilizers. In developed and developing countries, most fertilizers are produced with chemicals. These fertilizers are known as inorganic fertilizers. In less developed countries, animal or human wastes are widely used as organic fertilizers.

5.6.6 OTHER SOURCES OF CONTAMINATION

There are many other sources that contribute to the contamination of groundwater, including

- Large quantities of salts that are spread on the roads to ease ice conditions during winter
- Activities of the mining industry

TABLE 5.9
Sources of Groundwater Quality Degradation

Groundwater Quality Problems that Originate on the Land Surface
Infiltration of polluted surface water
Land disposal of either solid or liquid wastes Dumps
Disposal of sewage and water-treatment plant sludge Deicing salt usage and storage
Animal feedlots
Fertilizers and pesticides
Accidental spills
Particulate matter from airborne sources

Groundwater Quality Problems that Originate in the Ground above the Water Table
Septic tanks, cesspools, and privies
Holding ponds and lagoons
Sanitary landfills
Waste disposal in excavations
Leakage from underground storage tanks
Leakage from underground pipelines
Artificial recharge

Groundwater Quality Problems that Originate in the Ground below the Water Table
Waste disposal in well excavations
Drainage wells and canals
Well disposal of wastes
Underground storage
Exploratory wells
Groundwater development

Source: U.S. Environmental Protection Agency, *Protection of Public Water Supplies from Ground-Water Contamination*, Center for Environmental Research Information, U.S. Environmental Protection Agency, Washington, DC, 2005.

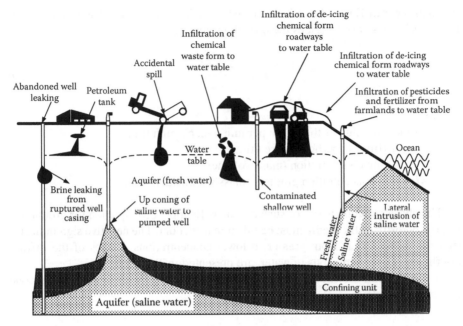

FIGURE 5.4 Sources of groundwater contamination. (From Delleur, J.W., *The Handbook of Groundwater Engineering*, CRC Press/Springer-Verlag, Boca Raton, FL, 992 p., 1999.)

- Seepage from industrial waste lagoons
- Urbanization and urban surface runoff spreading into the recharge areas of major aquifers

Table 5.9 and Figure 5.4 show groundwater contamination in different categories.

5.7 MASS TRANSPORT OF DISSOLVED CONTAMINANTS

In groundwater contamination studies, the basic theories and processes behind the movement of contaminants are required to be understood. In the subsurface area, complex processes occur that operate the movement of substances through porous media. These processes can be expressed mathematically. The main processes for the transportation of dissolved contaminants are advection, dispersion and diffusion. In the advection process, moving groundwater carries dissolved solutes. Along with advection, the dispersion process occurs that dilutes the solute. In the diffusion process, ionic and molecular species dissolved in water move from areas of higher concentration to areas of lower concentration. There are also chemical and physical processes that cause the retardation of the solute movement so that it may not move as fast as the advection rate would indicate.

5.7.1 DIFFUSION

Diffusion is a net transport of molecules from a region of higher concentration to one of lower concentration by random molecular motion. The diffusion of a contaminant through water in 1D condition is described by Fick's laws. Under steady-state conditions, Fick's first law describes the flux of a solute as

$$F_{dif} = \frac{-DdC}{dx}$$
(5.9)

where

F_{dif} is the mass flux of the solute per unit area per unit time
D is the diffusion coefficient (area/time)
C is the solute concentration (mass/volume)
dC/dx is the concentration gradient (mass/volume/distance)

The above equation is set for surface water. To apply this equation for groundwater, the porosity of the media must be taken into account. The negative sign indicates that the movement is from greater to lower concentrations. Values of the diffuse coefficient, D, for some ions in water, are presented in Table 5.10.

If the concentrations of contaminants in water change with time, Fick's second law may be applied:

$$\frac{\partial C}{\partial t} = \frac{D \partial^2 C}{\partial x^2}$$
(5.10)

where $\partial C / \partial t$ indicates the change in concentration with time.

TABLE 5.10
Diffusion Coefficients in Water
for Certain Ions at 25°C

Cation	D (10^{-6} cm²/s)	Anion	D (10^{-6} cm²/s)
H^+	93.1	OH^-	52.7
Na^+	13.3	F^-	14.6
K^+	19.6	Cl^-	20.3
Rb^+	20.6	Br^-	20.1
Cs^+	20.7	HS^-	17.3
Mg^{2+}	7.05	HCO_3^-	11.8
Ca^{2+}	7.93	CO_3^{2-}	9.55
Sr^{2+}	7.94	SO_4^{2-}	10.7
Ba^{2+}	8.48	PO_4^{3-}	6.12
Ra^{2+}	8.89		
Mn^{2+}	6.88		
Fe^{2+}	7.19		
Cr^{3+}	5.94		
Fe^{3+}	6.07		

Example 5.1

A site has been contaminated by Dieldrin. The porosity of the soil is 0.35 in this site. If the gradient of Dieldrin concentration is 0.0003 mg/L/m, calculate the diffusive flux. The diffusion coefficient of Dieldrin is 4.74×10^{-6} cm²/s.

Solution

$$F_{dif} = -nDdC/dx$$

$$= 0.35 \times 4.74 \times 10^{-6} \text{ cm}^2/\text{s} \times 10^{-6} \text{ m}^2/\text{cm}^2 \times 0.0003 \text{ mg/L/m}$$

$$= 5.00 \times 10^{-16} \text{ mg/m}^2/\text{s} = 4.32 \times 10^{-11} \text{ mg/m}^2/\text{day}$$

In porous media, the ions must follow longer pathways as they travel around mineral grains. Therefore, the diffusion process in groundwater is slower than in surface water. Therefore, an effective diffusion coefficient, D^*, is used to take this decrease in velocity into account. This coefficient is calculated as

$$D^* = \tau D \tag{5.11}$$

where τ is an empirical coefficient determined by laboratory experiments. It ranges from 0.5 to 0.01 for species that are not adsorbed onto the mineral surface (Freeze and Cherry, 1979).

Even in the absence of groundwater movement, contaminants existing in the aquifer tend to diffuse in all directions, which cause a blurring of the boundary between contaminated groundwater and the surrounding groundwater. So, if the groundwater is not flowing, contaminants may still move through the porous medium by diffusion, i.e., even in the condition of a zero hydraulic gradient, a contaminant could still move. Diffusion is dominant especially in rocks and soil with very low permeability that the water may be moving very slow.

If a solid waste is placed on a soil liner and both are saturated with water, contaminant ions from the solid waste (where the concentration is greater) will diffuse into the soil liner even if there is no water flow. Assume that the concentrations of the contaminant in the solid waste and soil liner are C_0 and C_i, respectively. C_i at distance d and time t is calculated as (Crank, 1956)

$$C_i(d,t) = C_0 erfc \frac{d}{2(D^*t)^{0.5}} \tag{5.12}$$

Example 5.2

A landfill holds a solid waste that has chloride ions in its leachate. This landfill has a clay liner. The concentration of chloride in the leachate is 150 mg/L. What is the concentration of chloride 7 m away from the interface of the landfill and the liner after 80 years? The diffuse coefficient is 1.8×10^{-8} m²/s and τ is 0.5.

Solution

$$D^* = \tau D = 0.5 \times 1.8 \times 10^{-8} = 9.0 \times 10^{-9} \text{ m}^2/\text{s}$$

$$t = 80 \text{ years} \times 365 \text{ day/year} \times 86400 \text{ s/day} = 2.52 \times 10^9 \text{ s}$$

$$C_i = 150 \text{ mg/L} \times erfc \frac{7}{2 \times (9 \times 10^{-9} \text{ m}^2/\text{s} \times 2.52 \times 10^9)^{0.5}}$$

$$C_i = 150 \text{ mg/L} \times erfc(0.735) = 44.8 \text{ mg/L}$$

5.7.2 Advection

As explained in Chapter 3, the rate of groundwater flow is determined from Darcy's law (Equation 3.1). Contaminant transport by the movement of a bulk of water is called advection. The rate of transport is then the same rate as the average linear velocity of the groundwater.

Example 5.3

Assume that a bulk of dissolved contaminants reaches an unconfined aquifer that has a hydraulic conductivity of 5×10^{-5} m/s and a porosity of 0.4. If the hydraulic gradient in this aquifer is 0.08, how far might this contaminant move in 3 years?

Solution

Since it is assumed that the contaminant is being transported only by complete mixing, it has the same velocity as the groundwater. Therefore,

$$v = \frac{Ki}{n} = \frac{5 \times 10^{-5} \times 0.08}{0.4} = 1.0 \times 10^{-5} \text{ m/s}$$

$$= 10^{-5} \text{ m/s} \times 86,400 \text{ s/day} \times 365 \text{ day/year} = 31.54 \text{ m/year}$$

So the distance that the contaminant moves in 3 years is

$$d = 31.54 \text{ m/year} \times 3 \text{ years} = 94.62 \text{ m}$$

The 1D mass flux, F_x, due to advection can be calculated as

$$F_x = v_x nC \tag{5.13}$$

where
v_x is the velocity of groundwater in x direction
n is the porosity
C is the concentration of dissolved contaminants

Example 5.4

A stream is being recharged by an unconfined aquifer which is 2.2 m thick and 98 m wide. The porosity of the aquifer is 0.23 and groundwater is flowing in this aquifer with the velocity of 0.3 m/day. Assuming that the concentration of nitrate is 22 mg/L in the groundwater, what is the mass flux of nitrate being discharged into the stream?

Solution

$$F_x = v_x nC$$

$$= 0.3 \text{ m/day} \times 0.23 \times 22 \text{ mg/L} \times 0.001 \text{ g/mg} \times 1000 \text{ L/m}^3$$

$$= 1.52 \text{ g/m}^2\text{-day}$$

Therefore, the total flux discharged to the stream is

$$F_T = 1.52 \text{ g/m}^2\text{-day} \times 2.2 \text{ m} \times 98 \text{ m}$$

$$= 327.7 \text{ g/day}$$

5.7.3 MECHANICAL DISPERSION

Another process that causes a contaminant plume to be spread out is dispersion. A contaminant plume follows irregular pathways as it moves. In other words, a portion of the contaminant plume may find large pore spaces in which it can move quickly, while other portions have to force their way through more confining voids. Therefore, there will be a difference in speed of an advancing plume that tends to cause the plume to be spread out. So, dispersion can be defined as the dilution of contaminants because of being mixed with noncontaminated water. If the mixing occurs along the streamline of fluid flow, it is called longitudinal dispersion. Transverse dispersion occurs perpendicular to the main direction of flow. Since contamination spreads out as it moves, it does not arrive all at once at a given location down-gradient. This effect is easily demonstrated in the laboratory by establishing a steady-state flow regime in a column packed with a homogeneous granular material and then introducing a continuous stream of a nonreactive tracer, as shown in Figure 5.5. If the tracer had no dispersion, a plot of concentration versus the time

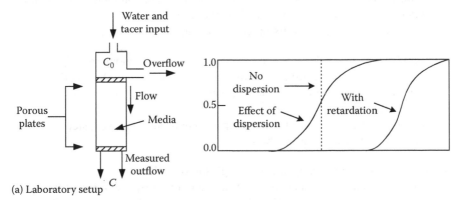

FIGURE 5.5 Dispersion and retardation as a continuous feed of tracer passes through a column. With no dispersion, the tracer emerges all at once. With retardation and dispersion, the tracer smears out and emerges with some delay. (From Masters, G.M. and Ela, W.P., *Introduction to Environmental Engineering and Science*, Prentice Hall Inc., Simon and Schuster, Englewood Cliffs, NJ, 576 p., 2008.)

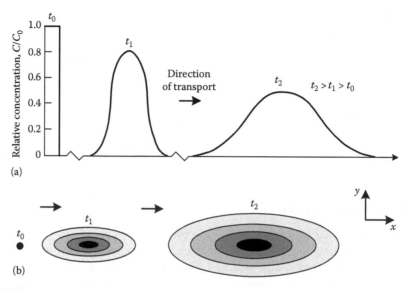

FIGURE 5.6 An instantaneous (pulse) source in a flowfield creates a plume that spreads as it moves down-gradient: (a) in 1D and (b) in 2Ds (darker colors mean higher concentrations). (From Bedient, P.B. et al., *Ground Water Contamination: Transport and Remediation*, Prentice-Hall Publishing Co., Englewood Cliffs, NJ, 560 p., 1994.)

that it takes to leave the column would show a sharp jump. Instead, the front arrives smeared out, as shown in Figure 5.6.

While the column experiment described in Figure 5.5 illustrates longitudinal dispersion, there is also some dispersion normal to the main flow. Figure 5.6 shows what might be expected if an instantaneous (pulse) source of contaminant is injected into a flowfield, such as what might occur with an accidental spill that contaminates groundwater. As the plume moves down-gradient, dispersion causes the plume to spread in the longitudinal as well as orthogonal directions.

Longitudinal dispersion occurs based on the following reasons:

1. Fluid moves faster through the center of the pore than along the edges.
2. Some portions of the fluid travel in longer pathways than other portions.
3. The movement of fluid through larger pores travels faster than that in smaller pores.
4. Heterogeneity of the aquifer causes the groundwater to move faster in some layers and slower in some others.

Figure 5.7 shows the fluid movement through soil particles.

The main reason for the occurrence of the transverse dispersion is that flow paths branch out to the sides while passing through a porous medium (Figure 5.8).

For uniform flow fields, the mechanical dispersion is equal to the product of the dynamic dispersivity (α_L) and the average linear velocity:

$$D_m = \alpha_L \times v_i \tag{5.14}$$

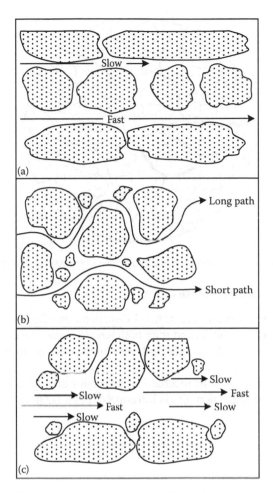

FIGURE 5.7   Main factors that cause longitudinal dispersion in subsurface area. (a) Larger pores let fluid move faster, (b) travel time through longer paths is more than shorter ones, and (c) fluid moves faster through the center of the pore.

where
  D_m is the mechanical dispersion
  v_i is the linear velocity in the i direction

Xu and Eckstein (1995) found a relationship between the apparent longitudinal dynamic dispersivity and flow length. This relationship is expressed as

$$\alpha_L = 0.83(\log L)^{2.414} \tag{5.15}$$

where
  α_L is the apparent longitudinal dynamic dispersivity
  L is the length of the flow path

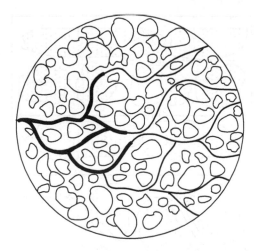

FIGURE 5.8   Flow paths in the porous medium causing transverse dispersion.

5.7.4   Hydrodynamic Dispersion

Since molecular diffusion and mechanical dispersion usually take place coincidentally in groundwater movement and they both tend to smear the edges of the plume, they are sometimes linked together and are simply referred to as *Hydrodynamic Dispersion*.

To take into account both the mechanical dispersion and diffusion, the coefficient of the hydrodynamic dispersion, D_L, is introduced. For a 1D flow, this coefficient can be represented as

$$D_L = \alpha_L v_i + D^* \tag{5.16}$$

where
    D_L is the longitudinal coefficient of the hydrodynamic dispersion
    α_L is the dynamic dispersivity
    v_i is the average linear groundwater velocity
    D^* is the molecular diffusion

For calculating the transverse coefficient of dispersion, Equation 5.16 is rewritten as

$$D_T = \alpha_T v_i + D^* \tag{5.17}$$

where α_T is the transverse dynamic dispersivity.

5.7.5   Derivation of the Advection–Dispersion Equation

Imagine a representative elementary volume of the porous media (Figure 5.9). To derive the advection–dispersion equation, the conservation of the mass of contaminant flux into and out of this volume is considered.

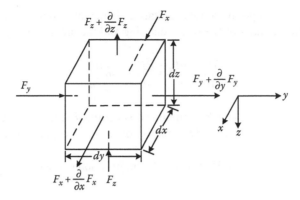

FIGURE 5.9 Representative elementary volume for groundwater flow.

Assuming a homogenious aquifer, the contaminant transport in the i direction based on advection and hydrodynamic dispersion can be calculated as

$$T_a = v_i n C dA \qquad (5.18)$$

$$T_d = n D_i \frac{\partial C}{\partial i} dA \qquad (5.19)$$

where
T_a and T_d are the advective and dispersive transport, respectively
dA is the cross-sectional area of the element

The total mass of contaminant per unit cross-sectional area transported in the i direction per unit time, F_i, can be computed as the sum of the advective and the discursive transport:

$$F_i = v_i n C - n D_i \frac{\partial C}{\partial i} \qquad (5.20)$$

Since the dispersive flux is from areas of greater to areas of lesser concentration, the second term has a negative sign.

The total amount of contaminant entering the representative elementary volume is $F_x d_z d_y + F_y d_z d_x + F_z d_x d_y$. The amount of contaminant leaving the representative elementary volume is

$$\left(F_x + \frac{\partial F_x}{\partial x} dx \right) dz\, dy + \left(F_y + \frac{\partial F_y}{\partial x} dy \right) dz\, dx + \left(F_z + \frac{\partial F_z}{\partial x} dz \right) dx\, dy \qquad (5.21)$$

On the other hand, the rate of mass change in the representative elementary volume is

$$n \frac{\partial C}{\partial t} dx\, dy\, dz \qquad (5.22)$$

This rate of mass change in the representative elementary volume must be equal to the difference in the mass of the solute entering and the mass leaving. Therefore,

$$\frac{\partial F_x}{\partial x} + \frac{\partial F_y}{\partial y} + \frac{\partial F_z}{\partial z} = -n\frac{\partial C}{\partial t} \tag{5.23}$$

Values of F_x, F_y, and F_z can be found and substituted using Equation 5.18. The resulting equation will be the 3D equation of mass transport for a conservative contaminant.

The general equation for 1D transportation in a homogeneous, isotropic porous media is

$$D_L\frac{\partial^2 C}{\partial x^2} - v_x\frac{\partial C}{\partial x} = \frac{\partial C}{\partial t} \tag{5.24}$$

The solution of this equation depends on boundary and initial conditions.

Contamination problems in aquifers can be because of the instantaneous injection or continuous release of contaminants. In both cases, the concentration of the contaminant well decreases with time and distance (Figure 5.10).

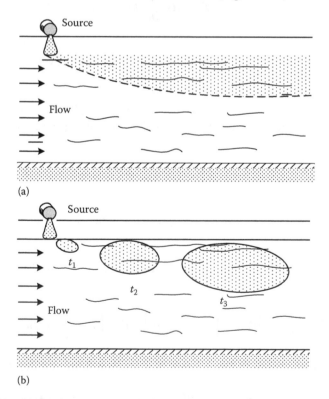

FIGURE 5.10   (a) Continuous release (leaching). (b) Instantaneous injection (spill) of contaminant from a point source into an aquifer with isotropic sand in a 2D uniform field.

If there is step change in the concentration of the contaminant (instantaneous injection), Equation 5.24 is solved as (Ogata and Banks, 1961):

$$C = \frac{C_0}{2}\left[erfc\left(\frac{L - v_x t}{2\sqrt{D_L t}} \right) + \exp\left(\frac{v_x L}{D_L} \right) erfc\left(\frac{L + v_x t}{2\sqrt{D_L t}} \right) \right] \qquad (5.25)$$

In Table 5.11 values of the error function, $erf(\psi)$, and the complementary error function, $erfc(\psi)$, for different numbers are presented.

$$erf(\psi) = \frac{2}{\pi}\int_0^\beta e^{-\varepsilon^2} d\varepsilon \qquad (5.26)$$

$$erf(-\psi) = -erf(\psi) \qquad (5.27)$$

$$erfc(\psi) = 1 - erf(\psi) \qquad (5.28)$$

If there is continuous change in the concentration of the contaminant (continuous release), the equation is solved as (Sauty, 1978):

$$C = \frac{C_0}{2}\left[erfc\left(\frac{L - v_x t}{2\sqrt{D_L t}} \right) - \exp\left(\frac{v_x L}{D_L} \right) erfc\left(\frac{L + v_x t}{2\sqrt{D_L t}} \right) \right] \qquad (5.29)$$

As can be seen in Equations 5.23 and 5.24, the only difference is that the second term is subtracted in the case of continuous injection rather than being added if the contaminant is applied to the field at once.

When the distance or time is large, the second term on the right-hand side of Equations 5.25 and 5.29 can be neglected. Therefore,

$$C = \frac{C_0}{2}\left[erfc\left(\frac{L - v_x t}{2\sqrt{D_L t}} \right) \right] \qquad (5.30)$$

Example 5.5

Nitrobenzene is being released from a point source to an aquifer. The hydraulic conductivity of this aquifer is 4.5×10^{-5} m/s and its porosity is 0.35. The hydraulic gradient in this aquifer is 0.0015 m/m. Assuming that the concentration of nitrobenzene is 670 mg/L at the source, compute the concentration after 2 years at 50 m away from the source. Molecular diffusion for nitrobenzene is assumed equal to 1.12×10^{-8} m²/s.

TABLE 5.11
Values of Complementary Error Function for Different Numbers

ψ	$erf(\psi)$	$erfc(\psi)$
0.00	0.00	1.00
0.05	0.056372	0.943628
0.10	0.112463	0.887537
0.15	0.167996	0.832004
0.20	0.222703	0.777297
0.25	0.276326	0.723674
0.30	0.328627	0.671373
0.35	0.379382	0.620618
0.40	0.428392	0.571608
0.45	0.475482	0.524518
0.50	0.5205	0.4795
0.55	0.563323	0.436677
0.60	0.603856	0.396144
0.65	0.642029	0.357971
0.70	0.677801	0.322199
0.75	0.711156	0.288844
0.80	0.742101	0.257899
0.85	0.770668	0.229332
0.90	0.796908	0.203092
0.95	0.820891	0.179109
1.00	0.842701	0.157299
1.10	0.880205	0.119795
1.20	0.910314	0.089686
1.30	0.934008	0.065992
1.40	0.952285	0.047715
1.50	0.966105	0.033895
1.60	0.976348	0.023652
1.70	0.98379	0.01621
1.80	0.989091	0.010909
1.90	0.99279	0.00721
2.00	0.995322	0.004678
2.10	0.997021	0.002979
2.20	0.998137	0.001863
2.30	0.998857	0.001143
2.40	0.999311	0.000689
2.50	0.999593	0.000407
2.60	0.999764	0.000236
2.70	0.999866	0.000134
2.80	0.999925	0.000075
2.90	0.999959	0.000041
3.00	0.999978	0.000022

Solution

To be able to calculate the value of the longitudinal dispersion coefficient, first the velocity of groundwater and the apparent longitudinal dynamic dispersivity must be computed. Therefore,

$$v_x = \frac{Ki}{n} = \frac{4.5 \times 10^{-5}\,\text{m/s} \times 0.0015}{0.35} = 1.9 \times 10^{-7}\,\text{m/s}$$

$$\alpha_L = 0.83(\log L)^{2.414} = 0.83(\log 50\,\text{m})^{2.414} = 2.98$$

D_L is calculated as

$$D_L = \alpha_L \times v_x + D^* = (2.98\,\text{m} \times 1.9 \times 10^{-7}\,\text{m/s}) + 1.12 \times 10^{-8}\,\text{m}^2/\text{s}$$
$$= 5.77 \times 10^{-7}\,\text{m}^2/\text{s}$$

$$t = 2\,\text{years} \times 86{,}400\,\text{s/day} \times 365\,\text{days/year} = 6.3 \times 10^7\,\text{s}$$

Now, the concentration of nitrobenzene after 2 years at 50 m away from the source is calculated as

$$C = \frac{C_0}{2}\left[erfc\left(\frac{L - v_x t}{2\sqrt{D_L t}}\right) + \exp\left(\frac{v_x L}{D_L}\right) erfc\left(\frac{L + v_x t}{2\sqrt{D_L t}}\right)\right]$$

$$= \frac{670\,\text{mg/L}}{2}\left\{ erfc\left(\frac{50 - (1.9 \times 10^{-7}\,\text{m/s} \times 6.3 \times 10^7\,\text{s})}{2\sqrt{5.77 \times 10^{-7}\,\text{m}^2/\text{s} \times 6.3 \times 10^7\,\text{s}}}\right)\right.$$

$$\left. + \exp\left(\frac{1.9 \times 10^{-7}\,\text{m/s} \times 50\,\text{m}}{5.77 \times 10^{-7}\,\text{m}^2/\text{s}}\right) erfc\left(\frac{50\,\text{m} + (1.9 \times 10^{-7}\,\text{m/s} \times 6.3 \times 10^7\,\text{s})}{2\sqrt{5.77 \times 10^{-7}\,\text{m}^2/\text{s} \times 6.3 \times 10^7\,\text{s}}}\right)\right\}$$

$$C = 335\,\text{mg/L}\left[erfc\left(\frac{50\,\text{m} - 11.97\,\text{m}}{12.06\,\text{m}}\right) + \exp(16.46) erfc\left(\frac{50\,\text{m} + 11.97\,\text{m}}{12.06\,\text{m}}\right)\right]$$

$$= 335\,\text{mg/L}\left[erfc(3.15) + \exp(16.46) \times erfc(5.14)\right]$$

$$= 335\,\text{mg/L} \times 1.35 \times 10^{-5}$$

$$= 4.5 \times 10^{-3}\,\text{mg/L} = 4.5\,\mu\text{g/L}$$

Example 5.6

An underground tank leaches benzene continuously into an aquifer. The hydraulic conductivity of the aquifer is 1.90 m/day, the effective porosity is 0.15 and the hydraulic gradient is 0.06 m/m. Assuming that the coefficient of the hydraulic depression is 0.55 m²/s, find the time taken for the contaminant concentration to reach 10% of the initial concentration at $L = 600\,m$ in a 1D movement. Neglect any other degradation processes.

Solution

Using Diary's Law, the seepage velocity v_x is first calculated:

$$v_x = \frac{Ki}{n} = \frac{1.9 \times 0.06}{0.15} = 0.76\,m/day$$

$$\frac{C}{C_0} = \frac{1}{2}erfc\left(\frac{L - v_x t}{2\sqrt{D_x t}}\right)$$

$$0.1 = \frac{1}{2}erfc\left(\frac{600 - 0.76t}{2\sqrt{0.55t}}\right)$$

By trial and error or using solver in Excel:

$$t = 741\ days$$

Therefore, it takes more than 2 years for the contaminant to reach 10% of its initial concentration.

5.7.6 Solute Transport Equation

Within a control volume in an aquifer, there are several processes that act as sources and sinks for a solute. These processes include sorption/desorption, chemical or biological reactions, and decay. In the sorption/desorption process, the net rate of reaction, r, can be expressed as

$$r = \theta \frac{\partial C}{\partial t} = -\rho_b \frac{\partial \overline{C}}{\partial t} \tag{5.31}$$

where
θ is the volumetric water content of the porous media
ρ_b is the bulk density of the porous media
C is the concentration of the dissolved contaminants
$\overline{C}$ is the concentration of the sorbed contaminants

C can be calculated as the mass of solute divided by the volume of groundwater and $\overline{C}$ can be obtained by dividing the mass of solute by the mass of dry porous media. If the reaction is in equilibrium, the concentration of the sorbed contaminants is

$$\overline{C} = K_d C \tag{5.32}$$

where K_d is the equilibrium distribution coefficient (L^3/M). Combining Equations 5.31 and 5.32:

$$\left.\frac{\partial(\theta C)}{\partial t}\right)_{sorption} = -\rho_b K_d \frac{\partial C}{\partial t} \qquad (5.33)$$

Equation 5.33 gives the net rate of solute production based on a sorption/desorption reaction between a solute and the porous media within a control volume.

In case biological degradation or decay is imposed on the solute, the net rate of solute production is expressed as

$$\left.\frac{\partial(\theta C)}{\partial t}\right)_{decay} = -\lambda(\theta C + \rho_b K_d C) \qquad (5.34)$$

where λ is the decay rate constant for the solute.

On the other hand, the net rate of solute inflow can be expressed as

$$\text{Net rate of solute inflow} = -\frac{\partial}{\partial x}(v_x C) + D_x \frac{\partial^2}{\partial x^2}(\theta C) + D_y \frac{\partial^2}{\partial y^2}(\theta C) + D_z \frac{\partial^2}{\partial z^2}(\theta C)$$

$$(5.35)$$

Based on the law of conservation of mass for solute transport, the rate of change of solute mass within a control volume is equal to the summation of the net rate of solute inflow and the net rate of solute production. Substituting Equations 5.33 through 5.35 into the law of conservation of mass for solute transport, the solute transport equation will be obtained as

$$\frac{\partial(\theta C)}{\partial t} = D_x \frac{\partial^2}{\partial x^2}(\theta C) + D_y \frac{\partial^2}{\partial y^2}(\theta C) + D_z \frac{\partial^2}{\partial z^2}(\theta C) - \frac{\partial}{\partial x}(v_x C)$$

$$-\frac{\partial}{\partial t}(\rho_b K_d C) - \lambda(\theta C + \rho_b K_d C) \qquad (5.36)$$

In this equation, the rate of change of solute transport in the control volume has been shown as $\partial(\theta C)/\partial t$ The solution of this equation will be explained in Chapter 6.

5.7.7 CAPTURE-ZONE CURVES

Installing extraction wells has become the most common way for beginning the cleanup of contaminated groundwater. In this process, the contaminated groundwater is pumped out through the extraction well. Then, it is cleaned in an aboveground treatment facility and may be returned to the aquifer. The process is referred to as *pump-and-treat* technology and will be discussed in Chapter 9.

In Figure 5.11, an extraction well is located in a region with a uniform and steady regional groundwater flow. This well is parallel to and in the reverse

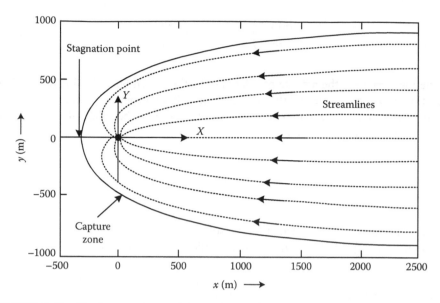

FIGURE 5.11   Single extraction well located at $x = 0$, $y = 0$, in an aquifer with A regional flow along the x-axis. The capture zone is the region in which all flow lines converge on the extraction well. Drawn for $(Q/Bv) = 2000$. (From Javandel, I. and Tsang, C.F., *Ground Water*, 24(5), 616, 1986.)

direction of the x-axis. As demonstrated in this figure, the natural streamlines are bent toward the well as water is extracted through the well. The outer envelope of the streamlines that converge on the well is called the *capture-zone* curve. Only that portion of groundwater, which is inside the capture zone, is extracted. There is no extraction of groundwater outside the capture zone. Although the flow lines outside of the capture zone curve toward the well, there is no extraction though those flow lines. The reason is that the regional flow associated with the natural hydraulic gradient is so strong that it carries groundwater past the well (Masters and Ela, 2008).

Javandel and Tsang (1986) assumed an ideal aquifer that is homogeneous, isotropic, uniform in cross section, and infinite in width and developed the use of capture-zone type curves. These curves can be used for the design of extraction well fields for aquifer cleanup. These curves are appropriate for confined or unconfined aquifers with an insignificant drawdown relative to the total thickness of the aquifer. In their analysis, they considered that the extraction wells extend downward through the entire thickness of the aquifer and they assumed that the wells are screened to extract uniformly from every level. Although these assumptions are so strict and never happen in actual conditions, acceptable insight into the main factors for even more complex models can be obtained by this analysis.

Consider the extraction well located at the origin of the coordinate system shown in Figure 5.11. The following relationship between the x and y coordinates of the envelope surrounding the capture zone were derived by Javandel and Tsang (1986):

$$y = \pm \frac{Q}{2Bv} - \frac{Q}{2\pi Bv} \tan^{-1} \frac{y}{x} \qquad (5.37)$$

where
   B is the aquifer thickness (m)
   v is the Darcy velocity (conductivity × gradient (m/day))
   Q is the pumping rate from the well (m³/day)

Equation 5.37 is then rewritten to consider an angle ϕ (in radians), which is drawn from the origin of the coordinate system to the coordinate of interest on the capture-zone curve (Figure 5.11):

$$\tan \phi = \frac{y}{x} \qquad (5.38)$$

For $0 \leq \phi \leq 2\pi$,

$$y = \frac{Q}{2Bv} \left(1 - \frac{\phi}{\pi} \right) \qquad (5.39)$$

In Equation 5.39, if x approaches infinity, $\phi = 0$ and $y = Q/(2Bv)$. Therefore, the maximum total width of the capture zone will be $2[Q/(2Bv)] = Q/Bv$. If $\phi = \pi/2$, then $x = 0$ and y will be equal to $Q/(4Bv)$. Thus, the width of the capture zone along the y-axis is $Q/(2Bv)$, which is half of its width at an infinite distance. Figure 5.12 illustrates these relationships.

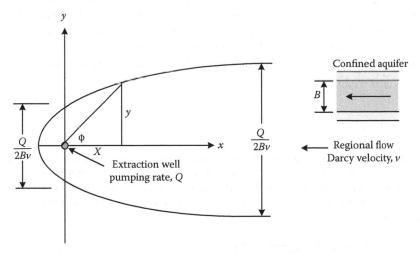

FIGURE 5.12 Capture-zone curve for a single extraction well located at the origin in an aquifer with regional flow velocity v, thickness B, and pumping rate Q. (From Masters, G.M. and Ela, W.P., *Introduction to Environmental Engineering and Science*, Prentice Hall Inc., Simon and Schuster, Englewood Cliffs, NJ, 576 p., 2008.)

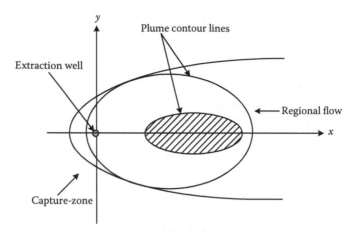

FIGURE 5.13 Superimposing the plume onto a capture-zone type curve for a single extraction well. (From Masters, G.M. and Ela, W.P., *Introduction to Environmental Engineering and Science*, Prentice Hall Inc., Simon and Schuster, Englewood Cliffs, NJ, 576 p., 2008.)

Pumping rates, Q, have a direct impact on the width of the capture zone, while the product of the Darcy flow velocity, v, and the aquifer thickness, B, inversely affects the width of this zone. Therefore, to capture the same area at higher regional flow velocities, higher pumping rates are required. However, there are some limitations for maximum pumping rates that restrict the size of the capture zone (Masters and Ela, 2008).

In capture-zone type curve analysis, the curve corresponding to the maximum acceptable pumping rate is first drawn. Then, the plume is superimposed onto the capture-zone curve. By evaluating this superposition, it can be determined whether or not the well of interest is sufficient to extract the entire plume. It can also be used to specify the location of the well. It should be noted that both the capture-zone type curve and the plume need to be drawn to the same scale. Figure 5.13 illustrates this analysis (Masters and Ela, 2008).

Example 5.7

A confined aquifer has been contaminated and a rectangular (for simplicity) plume has been created in this aquifer. To clean up this aquifer, it is decided to pump out the groundwater and treat it at the aboveground facilities. Determine the location of a single well that can totally extract the plume. The characteristics of the aquifer are as follows:

- Thickness of the aquifer = 25 m
- Hydraulic conductivity = 1.5×10^{-3} m/s
- Regional hydraulic gradient = 0.001
- Maximum pumping rate = 0.003 m³/s
- Width of the plume = 60 m

Solution

First, the regional Darcy velocity is calculated as

$$v = K\frac{dh}{dx} = 1.5 \times 10^{-3} \text{m/s} \times 0.001 = 1.5 \times 10^{-6} \text{m/s}$$

The width of the capture zone along the y-axis is

$$\frac{Q}{2Bv} = \frac{0.003 \text{m}^3/\text{s}}{2 \times 25\text{m} \times 1.5 \times 10^{-6} \text{m/s}} = 40\text{m}$$

As discussed earlier in this section, at an infinite distance up-gradient, this width becomes twice

$$\frac{Q}{Bv} = 80\text{m}$$

Therefore, if the well is located some distance downgrading from the front edge, the capture zone encompasses the 60 m wide plume (Figure 5.14). Considering y = 30 m,

$$y = \frac{Q}{2Bv}\left(1 - \frac{\varphi}{\pi}\right) = 30 = 40\left(1 - \frac{\varphi}{\pi}\right)$$

From Figure 5.14,

$$x = \frac{y}{\tan\varphi} = \frac{30}{\tan(0.25\pi)} = 30\text{m}$$

Therefore, the extraction well should be placed in line with the oncoming plume and 30 m ahead of it.

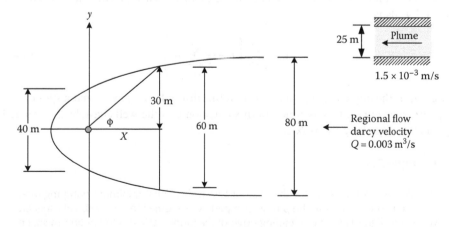

FIGURE 5.14 Problem with extraction well in Example 5.7.

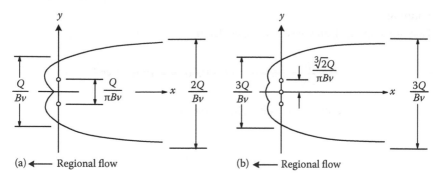

FIGURE 5.15 Capture-zone type curves for optimally spaced wells along the y-axis, each pumping at the rate Q: (a) two wells and (b) three wells. (From Masters, G.M. and Ela, W.P., *Introduction to Environmental Engineering and Science*, Prentice Hall Inc., Simon and Schuster, Englewood Cliffs, NJ, 576 p., 2008.)

The main issue in Example 5.7 is that the extraction well will be located far down-gradient from the plume, which results in the extraction of a large volume of clean groundwater before the contaminated plume reaches the well. This situation imposes significant pumping costs. To solve this situation is to install more extraction wells placed closer to the head of the plume.

Capture-zone type curves for a series of n optimally placed wells were also derived by Javandel and Tsang (1986). They assumed each well is pumping at the same rate, Q, and the wells are lined up along the y-axis. In this analysis, the maximum spacing between wells is optimized while the wells prevent any flow from passing between them. Therefore, the optimized distance between two wells is $Q/(\pi Bv)$. Locating wells with this distance apart from each other provides the possibility of capturing a plume as wide as $Q/(Bv)$ along the y-axis and as wide as $2Q/(Bv)$ along the x-axis, i.e., far up-gradient from the wells, (Figure 5.15a). Figure 5.15b also depicts correspondent parameters for the case of three optimally spaced wells.

A general form for the positive half of the capture-zone type curve for n optimally spaced wells is

$$y = \frac{Q}{2Bv}\left(n - \frac{1}{\pi}\sum_{i=1}^{n}\varphi_i\right) \tag{5.40}$$

where φ_i is the angle between a horizontal line through the ith well and a spot on the capture-zone curve. It is assumed in this equation that the wells have been arranged symmetrically along the y-axis.

Example 5.8

Recalculate Example 5.7 for the case of two wells. What minimum pumping rate, Q, is required to capture the plume completely? Assume the two optimally spaced wells are aligned along the leading edge of the plume. What is the optimum space between the wells?

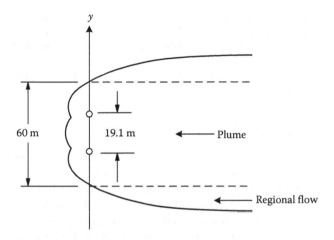

FIGURE 5.16 Problem with two extraction wells in Example 5.8.

Imagine the length of the plume is 750 m, how long would it take to pump out all of the contaminated groundwater? The aquifer porosity is 0.45.

Solution

The plume width along the y-axis (also the leading edge of the plume) is 60 m, so from Figure 5.16,

$$\frac{Q}{Bv} = \frac{Q}{25\,m \times 1.5 \times 10^{-6}\,m/s} = 60\,m$$

$$Q = 0.00225\,m^3/s$$

Therefore, each well has the pumping rate of 0.00225 m³/s. From Figure 5.16, optimal spacing between the wells $= \dfrac{Q}{\pi Bv}$

$$= \frac{0.00225\,m^3/s}{\pi \times 25 \times 1.5 \times 10^{-6}\,m/s} = 19.1\,m$$

The porosity times the plume volume gives the volume of contaminated water in the plume:

$$V = 0.45 \times 60 \times 25 \times 750\,m = 506,250\,m^3$$

The total pumping rate is 2 × 0.00225 m³/s = 0.0045 m³/s, therefore, the time to pump out the total contaminated groundwater is

$$t = \frac{506,250\,m^3}{0.0045\,m^3/s \times 3,600\,s/h \times 24\,h/day \times 365\,day/year} = 3.57\,year$$

However, it should be noted that it would take much longer to pump out the whole plume due to retardation (which will be discussed in Chapter 9) and there will also be some uncontaminated groundwater removed with the plume.

Details about designing more complicated well fields can be found in Gupta (1989) and Bedient et al. (1994).

5.8 MODELING CONTAMINANT RELEASE

Analytical solutions of the advection–dispersion equation can be used for many applications, such as assessing the potential impacts of releases of contaminants to groundwater, estimating the potential exposure concentrations at different locations, and providing tools for parameter estimation. Analytical models can also be used for establishing the soil cleanup level. In this regard, critical exposure locations are identified and exposure concentrations are determined for these locations. Then, the models can be used "backwards" to calculate the concentration at source that would result in this critical exposure. This procedure can be performed to establish the minimum cleanup standard levels.

5.8.1 MODELING INSTANTANEOUS RELEASE OF CONTAMINANTS

A few models of chemical spills are briefly discussed in this section. Generally in these models, the initial distribution of chemical concentrations is given and the models determine how the changes of concentration are through time and space.

To simplify calculations, flow is usually considered 1D, despite the fact that dispersion transport can occur in 3Ds. Depending on the number of dimensions considered and the different initial distributions of contaminants, the models may vary.

5.8.1.1 Fourier Analysis in Solute Transport

The problem of transport of a solute in a uniform flow field in the x-direction, with mixing in all three directions for an arbitrary initial distribution of contaminants can be expressed as

$$\frac{\partial c}{\partial t} + v'\frac{\partial c}{\partial x} + \lambda c = D'_{xx}\frac{\partial^2 c}{\partial x^2} + D'_{yy}\frac{\partial^2 c}{\partial y^2} + D'_{zz}\frac{\partial^2 c}{\partial z^2} \quad (5.41)$$

where
$v' = v/R$ is the retarded velocity
$D'_{xx} = D_{xx}/R$ is the retarded x-dispersion coefficient, etc.

Fourier methods can be used to solve this equation. Based on these methods, the solution to Equation 5.41 for a 1D case is

$$c(x,t) = \int_{-\infty}^{\infty} c(\xi,0)G(x,t|\xi,0)d\xi e^{-\lambda t} \quad (5.42)$$

where

$$G(x,t|\xi,0) \equiv \frac{1}{\sqrt{4\pi D't}} \exp\left(-\frac{(x-\xi-v't)^2}{4D't}\right) \qquad (5.43)$$

The function from Equation 5.43 plays the part of an impulse response function for the partial differential Equation 5.41. It is called Green's function. This function gives the contribution to the solution at point x and time t due to the initial unit mass at point $x = \xi$. The contaminant decays with the rate λ. The solution with decay is equal to the solution without decay times the exponential decay term.

5.8.1.2 Point Spill Model

Imagine a contaminant with concentration c_0 is being released with the volume V_0 at point $x = 0$. If the time period of release is so small that we can consider the release instantaneous, the mass, M, released from the source is equal to $c_0 V_0$. Then the concentration of the contaminant at point x and time t is

$$c(\tilde{x},t) = \frac{V_0 c_0 \exp\left(-\dfrac{(x-v't)^2}{4D'_{xx}t} - \dfrac{y^2}{4D'_{yy}t} - \dfrac{z^2}{4D'_{zz}t}\right)}{nR\sqrt{64\pi^3 D'_{xx}D'_{yy}D'_{zz}t^3}} e^{-\lambda t} \qquad (5.44)$$

where R is the retardation factor. Also, the maximum concentration at any time is

$$c_{max}(t) = \frac{V_0 c_0}{nR\sqrt{64\pi^3 D'_{xx}D'_{yy}D'_{zz}t^3}} e^{-\lambda t}; \quad \tilde{x} = (x,y,z) = (v't,0,0) \qquad (5.45)$$

5.8.1.3 Vertically Mixed Spill Model

If the release is mixed over the thickness of the aquifer, i.e., if the release occurs through a fully penetrating and screened well, the transport equation is the same as Equation 5.44, except that there are no vertical gradients and the term involving z does not appear.

$$c(\tilde{x},t) = \frac{V_0 c_0 \exp\left(-\dfrac{(x-v't)^2}{4D'_{xx}t} - \dfrac{y^2}{4D'_{yy}t}\right)}{4\pi nbt\sqrt{D_{xx}D_{yy}}} e^{-\lambda t} \qquad (5.46)$$

where b is the aquifer thickness and the retarded dispersion coefficient, R, is no longer being used.

5.8.1.4 Vertical Mixing Region

If the point release of a contaminant is at the top of the aquifer, near the source, the spreading is in 3Ds. In this case, Equation 5.44 is not applicable anymore and the correct solution near the source is twice that of Equation 5.44. This is because it has half of the mass spreading above the water table. The solution is then

$$c(\tilde{x},t) = \frac{V_0 c_0 \exp\left(-\dfrac{(x-v't)^2}{4D'_{xx}t} - \dfrac{y^2}{4D'_{yy}t} - \dfrac{z^2}{4D'_{zz}t}\right)}{4nR\sqrt{\pi^3 D'_{xx}D'_{yy}D'_{zz}t^3}} e^{-\lambda t} \qquad (5.47)$$

Far from the source, the contaminant is spread over the thickness of the aquifer and is only spread in the two lateral directions. That is because the base of the aquifer prevents further vertical migration (Figure 5.17). Therefore, Equation 5.47 cannot be applied anymore and Equation 5.46 is used. Now, at what horizontal distance the 3D solution ceased to be valid must be determined.

Equation 5.48 is appropriate both for the near-field and far-field solutions. However, it is inconvenient to use the infinite system of images. Therefore, when the simple 3D and 2D solutions can be applied in place of Equation 5.48 must be identified.

$$c(x,y,z,t) = \frac{V_0 c_0 \exp\left(-\dfrac{(x-v't)^2}{4D'_{xx}t} - \dfrac{y^2}{4D'_{yy}t}\right)}{4nR\sqrt{\pi^3 D'_{xx}D'_{yy}D'_{zz}t^3}} e^{-\lambda t} \sum_{j=-\infty}^{\infty} \exp\left(-\frac{(z+2jb)^2}{4D'_{zz}t}\right) \qquad (5.48)$$

Charbeneau (2000) shows that the 2D solution can be used when

$$t > \frac{Rb^2}{D_{zz}} \qquad (5.49)$$

By this time, the center of the spill would have moved a distance:

$$x = \frac{vt}{R} > \frac{vb^2}{D_{zz}} \qquad (5.50)$$

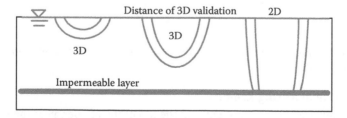

FIGURE 5.17 Schematic spreading of a point release.

If mechanical dispersion dominates the mixing process, the 2D solution is valid after

$$x > \frac{b^2}{a_v} \tag{5.51}$$

where a_v is the vertical dispersivity. Likewise, the 3D solution is valid up to the distance:

$$x < \frac{0.2\,v\,b^2}{D_{zz}} \tag{5.52}$$

Therefore, the general solution of Equation 5.48 is used when

$$\frac{0.2\,v\,b^2}{D_{zz}} < x < \frac{v\,b^2}{D_{zz}} \tag{5.53}$$

For shorter and longer distances, the single source 3D and 2D solutions are used, respectively.

Example 5.9

An industrial facility releases its wastewater on the adjacent ground surface at the top of an aquifer. The wastewater infiltrates into the underlying soil and reaches the aquifer. If the velocity of groundwater in the aquifer is 1.5 m/day and the thickness of the aquifer is 5 m, calculate the extent to which a 3D solution of the contaminant distribution is applicable. Also, determine from what distance the 2D solution can be used. Assume that dynamic dispersivity is 6 m.

Solution

$$D_{zz} = a_l v = 6\,\text{m} \times 1.5\,\text{m/day} - 9\,\text{m}^2/\text{day}$$

The 3D solution can be applied up to

$$x < \frac{0.2\,v\,b^2}{D_{zz}} = \frac{0.2 \times 1.5 \times 5^2}{9} = 0.83\,\text{m}$$

And the 2D solution can be used from

$$x > \frac{v\,b^2}{D_{zz}} = \frac{1.5 \times 5^2}{9} = 4.17\,\text{m}$$

5.8.2 MODELING CONTINUOUS RELEASE OF CONTAMINANTS

If the source release continues at a constant rate, the contaminant distribution appears as a plume that initially grows in length and width and eventually reaches a steady state distribution. The general mathematical framework from which plume models can be developed as well as the steady state concentration distribution near a source of releasing mass at a constant rate for a long time period are discussed in this section.

5.8.2.1 Development of Contaminant Plume Models

The simplest way to develop a contaminant plume model is to consider the plume as being derived from a continuing series of spills.

Assume that $c_s^*(x, y, z, t)$ represents the mathematical model for a chemical spill of unit mass, $M = 1$. Then, the solution for a plume that develops from a source with a time-variable mass release rate $\dot{m}(t)$ is given by

$$c_p(x, y, z, t) = \int \dot{m}(\tau) c_s^*(x, y, z, t - \tau) d\tau \qquad (5.54)$$

This equation gives the plume concentration distribution at time t. If the mass release rate is constant $\dot{m}$, the concentration distribution is obtained by

$$c_p(x, y, z, t) = \dot{m} \int_0^1 c_s^*(x, y, z, \tau) d\tau \qquad (5.55)$$

Assume that $\chi(\xi)$ represents an indicator variable for the source region. $\chi(\xi)$ is equal to one, $\chi(\xi) = 1$, if the point spill is within the initial source zone of the spill. Otherwise, $\chi(\xi) = 0$. Also, if the volume of the source region is represented by $V(\Omega)$, the formal solution for a contaminant plume model, when the source strength is variable through time, can be obtained by

$$c_p(\tilde{x}, t) = \frac{1}{V(\Omega)} \int_0^1 \dot{m}(\tau) \int_\Omega \chi(\tilde{\xi}) G(\tilde{x}, t - \tau | \tilde{\xi}, 0) d\tilde{\xi} \cdot e^{-\lambda(t-\tau)} d\tau \qquad (5.56)$$

5.8.2.2 Simple Plume Model

Imagine that a constant mass release rate $\dot{m}$ occurs at the point $(x, y) = (0, 0)$. Considering only addiction and transverse dispersion and assuming that longitudinal dispersion is negligible, the steady-state transport equation is expressed as

$$v \frac{\partial c}{\partial x} - D_{yy} \frac{\partial^2 c}{\partial y^2} + \lambda R c = 0 \qquad (5.57)$$

and the solution of this equation is

$$c(x, y) = \frac{\dot{m}}{q b} \frac{1}{\sqrt{4 \pi \dfrac{D_{yy}}{v} x}} \exp\left(-\frac{y^2}{4 \dfrac{D_{yy}}{v} x}\right) \exp\left(-\frac{\lambda R x}{v}\right) \qquad (5.58)$$

5.8.2.3 Point-Source Plume Model

If the longitudinal dispersion is also supposed to be considered in a point-source model, the transient period of development of the plume as well as the steady-state plume solution is expressed as

$$c_p(x, y, t \to \infty) = \frac{\dot{m}\exp(vx/2D_{xx})}{2\pi nb\sqrt{D_{xx}D_{yy}}} K_0\left(\frac{r}{B}\right) \tag{5.59}$$

where $K_0()$ is the modified Vessel function of the second kind of order zero. Also, the steady state (maximum) concentration along the x-axis is obtained by

$$c_p(x, y = 0, t \to \infty) = \frac{\dot{m}\exp\left(\dfrac{v \cdot x}{2D_{xx}}\left(1 - \sqrt{1 + \dfrac{4\lambda RD_{xx}}{v^2}}\right)\right)}{nb\sqrt{4\pi D_{yy}\,vx\sqrt{1 + \dfrac{4\lambda RD_{xx}}{v^2}}}} \tag{5.60}$$

If $\lambda = 0$, Equations 5.58 and 5.60 become identical for the solution along the x-axis. Figure 5.18 depicts the configuration for the point-source plume model.

5.8.2.4 Gaussian-Source Plume Model

The point-source plume model explained previously predicts infinite concentrations at the source, which is required to introduce a finite mass flux to the aquifer through a single point. If it is supposed to predict concentrations near the source as well as in the far field, then a source of finite size must be considered. The Gaussian-source plume model can be considered for this purpose.

   In the Gaussian-source plume model, it is assumed that the leachate from a surface source migrates through the unsaturated zone and mixes with groundwater flowing underneath the source (Figure 5.19). Assume that the source has the length L and width W with respect to the mean flow direction, and the total penetration depth of leachate at the down-gradient end of the source is H. Both the vertical dispersion and the vertical advection of water as it moves beneath the source cause penetration. The depth of penetration of the contaminant into the aquifer beneath the source is calculated as

$$H = H_{adv} + H_{dis} = b\left(1 - \exp\left(-\frac{i_f L}{q_x b}\right)\right) + \sqrt{2a_y L} \tag{5.61}$$

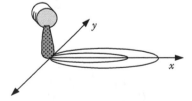

FIGURE 5.18   Configuration of the point source plume model.

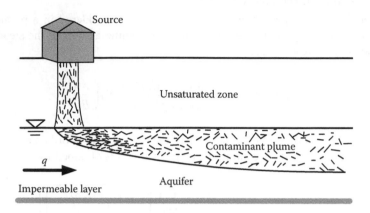

FIGURE 5.19 Set up of Gaussian source plume model.

where

i_f is the infiltration through the contaminant source
a_v is the vertical dispersivity

If the value of H calculated with Equation 5.61 exceeds b, $H > b$, H is considered equal to b, $H = b$.

If $H < b$, recharge forces the plume deeper into the aquifer and H remains constant. In the case $H = b$, the recharge serves to dilute the plume and acts as an equivalent decay term. In this latter case, the transport equation is

$$\frac{\partial c}{\partial t} + v' \frac{\partial c}{\partial x} + \left(\lambda + \frac{i_r}{nRH} \right) c = D'_{xx} \frac{\partial^2 c}{\partial x^2} + D'_{yy} \frac{\partial^2 c}{\partial y^2} \qquad (5.62)$$

where i_r is the diffuse recharge rate outside of the source. This equation assumes that the flow is steady and the velocity remains uniform in the x-direction. The solution for this steady state problem can be developed using Fourier transforms. Smith and Charbeneau (1990) proposed

$$c(x, y, t \rightarrow \infty) = \frac{c_0 \exp\left(A - \dfrac{y^2}{B} \right)}{C} \qquad (5.63)$$

In this equation,

$$A = \frac{vx}{2D_{xx}} \left(1 - \sqrt{1 + \frac{4\lambda^* R D_{xx}}{v^2}} \right)$$

$$B = 2\sigma^2 \left(1 + \frac{2xD_{yy}}{\sigma^2 v \sqrt{1 + \dfrac{4\lambda^* RD_{xx}}{v^2}}} \right)$$

$$C = \sqrt{1 + \frac{2xD_{yy}}{\sigma^2 v \sqrt{1 + \dfrac{4\lambda^* RD_{xx}}{v^2}}}}$$

where σ is a measure of the width of the source and $\lambda^* = \lambda + \dfrac{i_r}{nRH}$.

More details about these models can be found in Charbeneau (2000). He compared the results obtained from the simple plume model, the point source model, and the Gaussian-source plume model for a hypothetical aquifer in steady-state conditions. Figure 5.20 shows the concentration contours for these models. The most significant feature is the similarity between the contours suggesting equivalence in solution. In this figure, the smaller contours represent the higher concentrations. As demonstrated in this figure, there is a significant similarity between the contours that

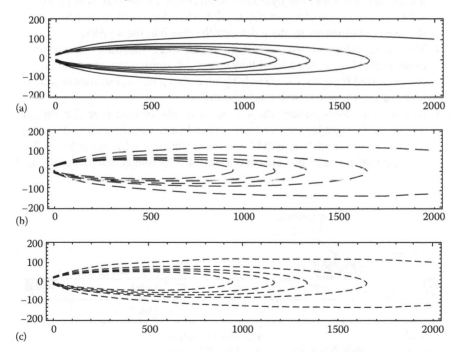

FIGURE 5.20 Steady-state plume model concentration contours for (a) simple plume model, (b) point-source plume model, and (c) Gaussian source plume model. (From Charbeneau, R.J., *Groundwater Hydraulics and Pollutant Transport*, Waveland Press Inc., Long Grove, IL, 593 p., 2000.)

show the equivalence in solutions. Longitudinal dispersion has minor importance in determining steady-state plume concentration distributions.

Charbeneau concluded that based on this comparison, if the source is small and maximum concentrations from a continual source are supposed to be determined, then the simple plume model is adequate. For small sources, if transient effects are important, then the point source plume model is preferred because the calculations are simple. The Gaussian-source model is used when the size of the source is important.

PROBLEMS

5.1 Groundwater deep in a sedimentary basin has electrical conductance of 300 micro-siemens. Make a rough estimate of the total dissolved solids of this water (in mg/L).

5.2 A dissolved contaminant is being leached from a point source with the concentration of 114 mg/L. There is a stream 7.5 m away from the point source. What concentration of the contaminant reached the stream after 5 years? Diffuse coefficient and τ are 2.15×10^{-8} and 0.48, respectively.

5.3 A landfill is located on 6 m thick dense clay overlying an aquifer, which is the primary source of drinking water to a small town. Leachate-contaminated groundwater has accumulated at the base of the landfill. Hydraulic gradient in the aquifer is 0.9 and the hydraulic conductivity of the clay is approximately 1.8×10^{-10} m/s, and the porosity is 22%. How long will it take for the non-reactive contaminants to move through the clay into the aquifer.

5.4 The hydrogeologic cross section in Figure 5.21 illustrates the pattern of groundwater flow along a local flow system. Assuming that the contamination is transported from the source only by advection process, estimate at what time in the future the plume will reach the stream.

5.5 Groundwater flows through the left face of a cube of sandstone (2 m on a side) and out of the right face with a linear groundwater velocity of 3×10^{-5} m/s. The porosity of the sandstone is 0.12 and the diffusion coefficient is 9×10^{-9} m²/s. Assume that a tracer has concentrations of 1.25 and 0.75×10^4 mg/m³ at the

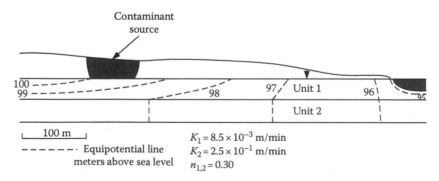

FIGURE 5.21   The hydrogeologic cross section of the area in Problem 5.4.

inflow and the outflow faces, respectively. Calculate the mass flux at the outflow due to advection and diffusion.

5.6 An unconfined aquifer has the hydraulic conductivity of 1.5×10^{-4} m/s and porosity of 0.35. If a dissolved contaminant with the initial contamination of 470 mg/L reaches this aquifer, how far may this contaminant move in 1 year? Assume that the hydraulic gradient in this aquifer is 0.05 and molecular diffusion of the contaminant is 2.34×10^{-8} m²/s.

5.7 A contaminated site has the porosity of 0.28. If the gradient of the contaminant concentration is 0.0018 mg/L/m, calculate the diffusive flux. Diffusion coefficient of the contaminant is 5.86×10^{-6} cm²/s.

5.8 A contaminant is being released from a point source to an aquifer. This aquifer is being discharged to a stream 40 m away from the contaminant source. Compute the concentration of the contaminant that reaches the stream through the aquifer after 3 years.

Hydraulic conductivity $= 2.5 \times 10^{-6}$ m/s

Porosity $= 0.25$

Hydraulic gradient $= 0.035$ m/m

Concentration of nitrobenzene at the source $= 520$ mg/L

Molecular diffusion for nitrobenzene $= 1.93 \times 10^{-8}$ m²/s.

5.9 Nitrate is being leaked from an industrial facility with the concentration of 22 mg/L. If the diffuse coefficient is 1.95×10^{-7} m²/s and τ is 0.35, what is the concentration of nitrate 500 m away from the facility in 20 years? The velocity of groundwater is 4.74×10^{-7} m/s.

5.10 An unconfined aquifer with porosity of 0.3 is being discharged to a stream flow. The groundwater flow velocity is 0.25 m/day and the thickness and width of the aquifer are 1.5 m and 75 m, respectively. What mass flux of bicarbonate may be discharged into the stream by advection if the concentration of bicarbonate in groundwater is 18 mg/L in the groundwater.

5.11 How long does it take for a contaminant to reach 20 percent of its initial concentration, 500 m away from the source. Assume one-dimensional movement. The hydraulic conductivity of the aquifer is 2.50 m/day, effective porosity is 0.35 and hydraulic gradient is 0.08 m/m. The longitudinal coefficient of hydraulic depression is 0.62 m²/s.

REFERENCES

Bedient, P.B., Rifai, H.S., and Newell, C.J. 1994. *Ground Water Contamination: Transport and Remediation*. Prentice-Hall Publishing Co., Englewood Cliffs, NJ, 560 p.

Charbeneau, R.J. 2000. *Groundwater Hydraulics and Pollutant Transport*. Waveland Press Inc., Long Grove, IL, 593 p.

Crank, J. 1956. *The Mathematics of Diffusion*. Oxford University Press, New York, 424 p.

Delleur, J.W. 1999. *The Handbook of Groundwater Engineering*. CRC Press/Springer-Verlag, Boca Raton, FL, 992 p.

Gupta, R.S. *Hydrology and Hydraulic Systems*. Prentice-Hall, Engleweood Cliffs, NJ.

Javandel, I. and Tsang, C.F. 1986. Capture-zone type curves: A tool for aquifer cleanup. *Ground Water*, 24(5), September–October, 616–625.

Masters, G.M. and Ela, W.P. 2008. *Introduction to Environmental Engineering and Science*. Prentice Hall Inc., Simon and Schuster, Englewood Cliffs, NJ, 576 p.

McKee, J.E. 1960. The need for water quality criteria, *Conference of the Physiological Aspects of Water Quality*, September, 1960, Division of Water Supply and Pollution Control, U.S. Public Health Service, Washington, DC.

Mirtov, B.A. 1961. *Gaseous Composition of the Atmosphere and Its Analysis*. Akad. Nauk SSSR, Inst. Prikl. Geofiz Moskva (translated by the Israel Program for Scientific Translations). U.S. Department of Commerce, Office of Technical Services, Washington, DC, 209 p.

Ogata, A. and Banks, R.B. 1961. *A Solution of the Differential Equation of Longitudinal Dispersion in Porous Media*. U.S. Geological Survey Professional Paper 411-A.

Oregon State University. 1974. *Disposal of Environmentally Hazardous Wastes*. Task Force Report, Environment Health Science Center, Oregon State University, Corvallis, OR, 210 p.

Sauty, J.P. 1978. Identification des parameters du transport hydrodispersif das les aquiferes par interpretation de tracages en ecoulement cyclndrique ou divergent. *Journal of Hydrology*, 39(1), 145–158.

Smith, V.J. and Charbaneau, R.J. 1990. Probabilistic soil contamination exposure assessment procedures. *Journal of Environmental Engineering*, 116(6), 1143–1163.

Stumm, W. 1977. Chemical interaction in particle separation, *Environmental Science and Technology*, 11, 1070.

Stumm, W. and Morgan, J.J. 1996. *Aquatic Chemistry*, 3rd edn. John Wiley & Sons, New York, 1022 p.

U.S. Environmental Protection Agency. 1975. Water programs: National interim primary drinking water regulations. *Federal Register*, 40(248), 59566.

U.S. Environmental Protection Agency. 1973. *Water Quality Criteria 1972*. EPA R3 73033. Government Printing Office, Washington, DC.

WHO 2006, *Guidelines for Drinking-Water Quality*. First Addendum to 3rd edn. Vol. 1-Recommendations.

Xu, M. and Eckstein, Y. 1995. Use of weighted least squares method in evaluation of the relationship between dispersivity and field scale. *Ground Water*, 33(6), 905–908.

6 Groundwater Modeling

6.1 INTRODUCTION

A mathematical groundwater model is used to simulate and describe real-world groundwater flow. The mathematical model is developed by translating a conceptual model in the form of governing equations, with associated boundary and initial conditions. This model can then be solved using a numerical model, which is developed through the implementation of computer programs (codes). A groundwater simulation model is a nonunique model due to different sets of assumptions used for simplifying the mathematical description of groundwater flow. It can be less burdened with approximating the investigated groundwater system by the simplifying assumptions such as homogeneity, isotropy, direction of flow, geometry of the aquifer, mechanisms of contaminant transport and its reaction. Models can also simulate more complicated problems with higher accuracy, utilizing more inputs, system parameters, and boundary conditions. A successful model can result from a complete site investigation and filed data. Thus, the model selection is a trade-off between the computational burdens including boundary conditions, grid discretization, time steps, the model accuracy, and ways to avoid truncation errors and numerical oscillations.

The performance and efficiency of a model depends upon how accurate the mathematical equations approximate the physical system being modeled. However, it should be noted that the developed model is an approximation and not an exact simulation of real-world groundwater flow.

The first step in groundwater system modeling is the conceptualization of the model, which includes a set of assumptions for the system's components, the media properties, and the transport processes in the system. The mathematical model contains the same information as the conceptual one and it is transformed to a numerical model by the fundamental laws of hydraulics.

Numerous numerical methods such as finite difference method (FDM), finite element method (FEM), and finite volume method (FVM) have been developed to solve and simulate the numerical models of groundwater flow system in porous media. These models and their functions are explained in this chapter.

6.2 PROCESS OF MODELING

The process of aquifer modeling consists of the following activities (Thangarajan, 2007):

- The parameters characterizing the physical framework of the aquifer and the system condition are identified.
- The hydrogeological parameters are estimated using field data in specific points.

- The spatial distribution of parameters is estimated utilizing interpolation/extrapolation methods.
- The entire estimated parameters and field data are utilized to make the conceptual model structure.
- A mathematical model is developed to describe the conceptual model by expressing the system condition using the groundwater flow equations.
- The mathematical model is transformed to a numerical model to find the aquifer response including hydraulic head or pollutants concentrations.
- The generated model is solved by numerical methods of solution.
- The model is calibrated for predicting the behavior of a considered system by simulating the available field data.
- The model is verified to eliminate errors resulting from the numerical approximations.
- The sensitivity analysis is done to select the estimates of model coefficients, which need to be estimated more accurately, and also to decipher the error bounds.
- The management strategies are suggested for aquifer restoration and optimal utilization of the groundwater resources.

The model accuracy depends upon the level of conceptualization and understanding of the groundwater system as well as the assumptions embedded in the derivation of the mathematical equations.

6.3   MATHEMATICAL MODELING

Groundwater simulation models have been widely used in groundwater system analysis and management. These models generally require the solution of partial differential equation. Prediction of subsurface flow, water table level, solute transport, and simulation of natural or human-induced stresses are necessary for groundwater management (Karamouz et al., 2003).

Mathematical models may be deterministic, stochastic (statistical), or a combination of both. In stochastic models, a range of predictions, based on probabilities of occurrence, is provided.

Such predictions can be used in planning and decision-making processes for the groundwater resource. Stochastic models can also help to evaluate the uncertainties of a system.

Deterministic models widely used for solving regional groundwater problems are based on cause-and-effect relationship of known systems and processes. Deterministic models can be further classified as analytical and numerical.

Another class of mathematical models in solving groundwater flow is analytical modeling, which is an easy method to evaluate the physical characteristics of an aquifer. This method of solution provides a rapid preliminary analysis of groundwater system utilizing a number of simplifying assumptions. These models cannot be used for solving the problems with the irregularity of the domain's shape, the heterogeneity of the domain, and complex boundary conditions. The numerical models' implementation is then carried out using computer programs for addressing more complicated problems.

The exact solutions for some simple or idealized problems can be found by numerical models. These models can yield approximate solutions by discretization of time and space. Numerical models can be further classified as FDM, FEM, and FVM.

In FDM, the first derivative in partial differential equations is approximated by the difference between values of independent variables at adjacent nodes considering the distance between the nodes, and considering the duration of time step increment at two successive time levels. In FEM, functions of dependent variables and parameters are used to evaluate equivalent integral formulation of partial differential equations (PDEs) (Delleur, 1999). Although each approach has some advantages and disadvantages, the FDMs are generally easier to program because of their conceptual and mathematical simplicity.

There are two ways in solving the PDEs in the mentioned methods to obtain the grid-point values of the dependent variable. One way, used in the FEM, is that the head variable consists of the grid point values and the shape function is employed to interpolate between the grid points. In the other way, used in FDM, without any consideration of the head variation between the grids, the equations include the grid-point values of the head. These values are obtained at some discrete locations without any statement about the variation between these locations similar to a laboratory experiment.

The flexibility of FEM in close spatial approximation of irregular boundaries of the aquifer and parameters of zones within the aquifer is a major advantage of this method. However, mesh generation and specification, and construction of input data sets for an irregular finite element grid is much more difficult in comparison with a regular rectangular finite difference grid.

The FVM is a method for representing and evaluating PDEs as algebraic equations. Similar to the FDM, values are calculated at discrete places on a meshed geometry. Finite volume refers to the small volume surrounding each node point on a mesh.

One advantage of the FVM over FDM is that it does not require a structured mesh. Furthermore, the FVM is preferable to other methods as a result of the fact that boundary conditions can be applied noninvasively. This is true, because the values of the conserved variables are located within the volume element, and not at nodes or surface. FVMs are especially powerful on coarse nonuniform grids and in calculations where the mesh moves to track interfaces or shocks.

6.3.1 ANALYTICAL MODELING

An analytical model provides a solution for a mathematical description of a physical process. In this order, groundwater flow equations require several simplifying assumptions including the domain's shape, the boundary and initial conditions.

Because of the simplifications inherent with analytical models, it is not possible to account for field conditions that change with time or space (Mandle, 2002). Thus, the system under study may vary from actual conditions. Analytical models have mostly been in use for particular sets of conditions and involve manually solving equations, such as Darcy's law, the Theis equation, generating solutions utilizing curve-matching techniques, inverse solutions for interpretation of flow tests and verification of numerical models.

The earliest work on analytical solutions for groundwater flow can be found in Carslaw and Jaeger (1959) and the latest one on the above is from Bear (1979), van Genuchten and Alves (1982), Walton (1989), Strack (1989). In most of the analytical methods for solution of two-dimensional (2D) groundwater problems, a suitable function is first determined to transform the problem from a geometrical domain into a domain with a more straightforward solution algorithm (Karamouz et al., 2003). A transformation that possesses the property of preserving angles of intersections and the approximate image of small shapes is conformal mapping. Harr (1962) presented various functions and discussed a manner in which these functions transform geometric figures from one complex plan to another. Toth (1962, 1963) developed two analytical solutions for the boundary value problem with steady-state flow in 2D, saturated, homogeneous, isotropic flow field bounded on top by a water table and on the other three sides by impermeable boundaries.

6.3.2 NUMERICAL MODELING

Numerical modeling is used to solve the PDEs that represent the groundwater flow. It gives an approximate solution, which may vary with respect to the assumptions. There are several types of numerical models depending on the considered numerical techniques. These models include finite difference, finite element, integrated finite difference, boundary element, and particle tracking. The main features of the various numerical models are (Bear and Verruijt, 1992)

1. The models are solved only at specified points in the space and time domains defined for the problem (discrete values). These points are considered as discontinuous state variables for the entire domain.
2. The PDEs that describe the groundwater flow are transformed by a set of mathematical equations in certain points as discrete values of the state variables.
3. The solution is for a specified set of numerical values of the various model coefficients rather than a general relationship in terms of these coefficients.
4. A computer program is employed to solve the large number of equations that must be solved simultaneously.

6.4 FINITE DIFFERENCE METHOD

The FDM consists of transforming the partial derivatives in difference equations over a small interval by algebraic expressions written in terms of aquifer characteristics and conditions. The problem domain is divided into connected series of discrete points called nodes. This method replaces a continuous media by a discrete set of points and assigns various hydrogeological parameters to each node (Thangarajan, 2007). FDM can be used to discretize both time and space. The partial derivatives are replaced utilizing the difference operators that define the spatial–temporal relationships between some parameters. The developed model is solved in each node by obtaining the solution for a set of algebraic equations in that node. There are various methods to solve the simplified equations. Some of the prevalent iterative numerical methods are (1) successive over relaxation (SOR) method (Aziz and Settari, 1972; Watts, 1971, 1973); (2) alternating direction implicit (ADI) procedure (Bredehoeft and Pinder, 1970; Peaceman and Rachford, 1955; Remson et al., 1971; Rushton, 1974); (3) modified iterative ADI method (Prickett and Lonnquist, 1971); (4) strongly implicit procedure (Stone, 1968; Trescott et al., 1976; Weinstein et al., 1970, etc.).

The advantage of FDM is the ease of following a complex system including complicated loading path and highly nonlinear behavior, since it is easy to understand and to prepare the program. Thus, it is an economic method for solving large nonlinear groundwater flow problems. The use of FDM is difficult for irregular geometrical domain. Generally, the conventional FDM with regular grid systems suffers from the shortcomings in irregular shape domain, complex boundary conditions, and material heterogeneity.

The continuous media is expressed in Equation 6.1. It is replaced by a finite set of discrete points in space, and also time is discretized:

$$\frac{\partial}{\partial x}\left(K_x \frac{\partial h}{\partial x}\right) + \frac{\partial}{\partial y}\left(K_y \frac{\partial h}{\partial y}\right) + \frac{\partial}{\partial z}\left(K_z \frac{\partial h}{\partial z}\right) = S_s \frac{\partial h}{\partial t} \tag{6.1}$$

The partial derivatives are replaced by terms calculated from the differences in head values at each node. Simultaneous linear algebraic difference equations are achieved through the derivatives of variables. Their solutions lead to the head values at nodes and time by solving the linear algebraic equations. The solutions obtained by this method are just approximations. Figure 6.1 shows a three-dimensional (3D) discretized aquifer. In this aquifer, terms of (i, j, k) address the location of each node in the mesh, where i, j, and k represent rows, columns, and layers, respectively.

The head in each node is a function of both space and time and it is required to discretize the continuous media into discrete nodes and time. The calculation is solved for the nodes that are located within each cell. There are many schemes for locating nodes in cells such as a mesh-centered node or block-centered node as shown in Figure 6.2.

Finite-difference approximations for the first- and second-order derivatives of the groundwater flow and mass transport equations can be developed directly from the Taylor series approximations of the temporal and spatial derivatives. In the discrete

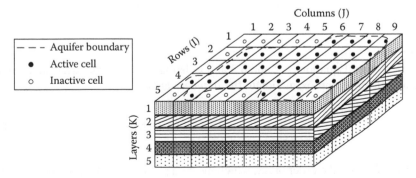

FIGURE 6.1 Illustration of discretization of continuous media into finite difference cells.

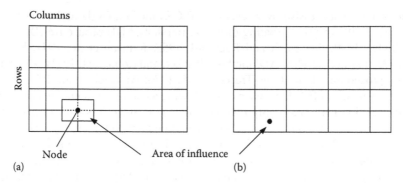

FIGURE 6.2 Illustration of 2D space discretization methods: (a) mesh-centered and (b) block-centered.

groundwater model, the head or mass concentration at the nodal points $x + \Delta x$ or $x - \Delta x$ is expressed in terms of the state variables and all the derivatives are evaluated at an adjacent nodal point, x,

$$h(x - \Delta x) = h(x) - \frac{\partial h}{\partial x}\bigg|_x \Delta x + \frac{\partial^2 h}{\partial x^2}\bigg|_x \frac{\Delta x^2}{2!} - \frac{\partial^3 h}{\partial x^3}\bigg|_x \frac{\Delta x^3}{3!} + \cdots \tag{6.2}$$

By rearranging Equation 6.2,

$$\frac{\partial h}{\partial x}\bigg|_x = \frac{h(x + \Delta x) - h(x)}{\Delta x} - \frac{\Delta x}{2!} \frac{\partial^2 h}{\partial x^2}\bigg|_x - \frac{\Delta x^2}{3!} \frac{\partial^3 h}{\partial x^3}\bigg|_x - \cdots \tag{6.3}$$

The forward and backward first differences are obtained as

$$\frac{\partial h}{\partial x}\bigg|_x = \frac{h(x + \Delta x) - h(x)}{\Delta x} + O(\Delta x) \tag{6.4}$$

$$\frac{\partial h}{\partial x}\bigg|_x = \frac{h(x) - h(x - \Delta x)}{\Delta x} + O(\Delta x) \tag{6.5}$$

respectively.

Differencing Equations 6.2 and 6.3 and isolating the first-order derivative gives a central difference approximation to the first derivative:

$$\frac{\partial h}{\partial x}\bigg|_x = \frac{h(x + \Delta x) - h(x)}{\Delta x} + O(\Delta x^2) \tag{6.6}$$

The error, $\bar{E}$, is on the order of Δx, which means a positive constant δ exists, which is independent of Δx, and $|\bar{E}| \leq \delta |\Delta x|$ for all sufficiently small Δxs. The errors of these approximations are called truncation errors. The truncation error of the central difference approximation is

$$\bar{E} = -\frac{\Delta x^2}{3!} \frac{\partial^3 h}{\partial x^3}\bigg| = O(\Delta x^2) \tag{6.7}$$

It is decreased by decreasing the values of Δx and Δt. An approximation to the second derivative can be obtained by Equation 6.2:

$$\frac{\partial^2 h}{\partial x^2}\bigg|_x \cong \frac{h(x+\Delta x) - 2h(x) + h(x - \Delta x)}{\Delta x^2} \tag{6.8}$$

The truncation error of the second-order approximation is

$$\bar{E} = \frac{\Delta x^2}{12} \frac{\partial^4 h}{\partial x^4}\bigg|_\xi = O(\Delta x^2) \tag{6.9}$$

Higher-order derivative approximations can be developed using linear difference operators.

Similar discretization could be used for time intervals. Therefore, the approximations of $\partial h/\partial t$ using forward and backward differences are as follows:

$$\frac{\partial h}{\partial t}\bigg|_t = \frac{h^{t+1} - h^t}{\Delta t} \tag{6.10}$$

$$\frac{\partial h}{\partial t}\bigg|_t = \frac{h^t - h^{t-1}}{\Delta t} \tag{6.11}$$

respectively.

Figure 6.3 shows time and space discretization at node (i, j) in a 2D finite-difference grid, and illustrates the application of some simple finite-difference methods for solving groundwater flow equation. These methods are explained later in this section.

6.4.1 FORWARD DIFFERENCE EQUATION

The head at point (i, j) at time step $(n + 1)$ can be obtained utilizing the value of h at time step n and forward difference time derivative. Therefore, there is a finite-difference equation for each node at time step $n + 1$ with only one unknown variable. As shown in Figure 6.3a, all values of head h are known at all spatial nodes at time n. This method is called forward difference or explicit method.

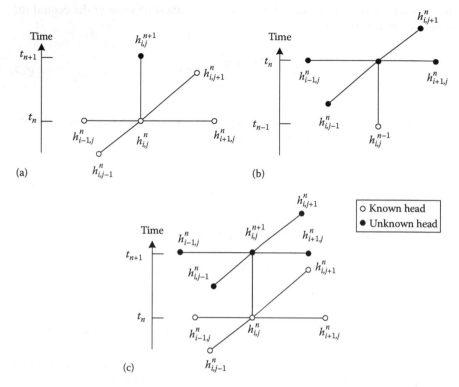

FIGURE 6.3   (a) The forward difference, (b) backward difference, and (c) Crank–Nicholson method in a 2D finite-difference grid.

In a 2D groundwater flow equation for a heterogeneous, anisotropic aquifer, Equation 6.1 can be written as

$$S_s\left(\frac{h_{i,j}^{n+1}-h_{i,j}^n}{\Delta t}\right) = K_{x(i-1/2,j)}\left(\frac{h_{i-1,j}^n-h_{i,j}^n}{(\Delta x)^2}\right) + K_{x(i+1/2,j)}\left(\frac{h_{i+1,j}^n-h_{i,j}^n}{(\Delta x)^2}\right)$$

$$+ K_{y(i,j-1/2)}\left(\frac{h_{i,j-1}^n-h_{i,j}^n}{(\Delta y)^2}\right) + K_{y(i,j+1/2)}\left(\frac{h_{i,j+1}^n-h_{i,j}^n}{(\Delta y)^2}\right) \quad (6.12)$$

In this equation, only h_{ij}^{n+1} is unknown. Equation 6.12 can be solved explicitly at each grid for the head at the new time level. Since the solution is dependent only upon known values of heads in the adjacent grids at the beginning of the time periods, the computation for h_{ij}^{n+1} in any grid can be made in any order without regard to values of h_{ij}^{n+1} for any other grid (Karamouz et al., 2003).

For example, in a one-dimensional (1D) groundwater flow equation for a heterogeneous, isotropic, and confined aquifer, Equation 6.12 can be written as

$$S\left(\frac{h_i^{n+1} - h_i^n}{\Delta t}\right) = Kb\left(\frac{h_{i-1}^n - h_i^n}{(\Delta x)^2}\right) + Kb\left(\frac{h_{i+1}^n - h_i^n}{(\Delta x)^2}\right) \tag{6.13}$$

or

$$h_i^{n+1} = \frac{T\Delta t}{S(\Delta x)^2}(h_{i-1}^n + h_{i+1}^n) + h_i^n\left(1 - \frac{2T\Delta t}{S(\Delta x)^2}\right) \tag{6.14}$$

Explicit finite-difference equations are simple to solve but when time increments are too large, small numerical errors can propagate into larger errors in the next computational stages. A stable solution is ensured in 1D heterogeneous case if

$$\frac{T\Delta t}{S(\Delta x)^2} < \frac{1}{2} \tag{6.15}$$

Consequently, the time increment cannot be selected independently of the space increment.

Example 6.1

Consider a nonsteady, 1D flow in a confined aquifer shown in Figure 6.4. Let $\Delta x = 3\,m$, $b = 3\,m$, $h_1 = 5\,m$, $h_5 = 1\,m$ for $t > 0$, $K = 0.5\,m/day$, $S = 0.03$. The initial conditions are $h_1 = h_2 = h_3 = h_4 = h_5 = 5\,m$. Determine the spatial variation of piezometric head.

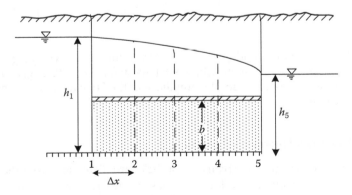

FIGURE 6.4 The confined aquifer in Example 6.1.

Solution

To satisfy the stability requirement of Equation 6.15, the maximum time step Δt is computed as

$$h_i^{n+1} = \frac{T\Delta t}{S(\Delta x)^2}(h_{i-1}^n + h_{i+1}^n) + h_i^n\left(1 - \frac{2T\Delta t}{S(\Delta x)^2}\right)$$

$$\Delta t < \frac{S(\Delta x)^2}{2T} = \frac{1}{2}\frac{(0.03)(3)^2}{0.5 \times 3} = 0.09\,\text{day}$$

Therefore, the time increment is selected as 0.08 days. With assumption of $h_1 = h_2 = h_3 = h_4 = h_5 = 5\,\text{m}$, $h_1^{0.08} = h_2^{0.08} = h_3^{0.08} = h_4^{0.08} = 5\,\text{m}$, $h_5^{0.08} = 1\,\text{m}$. For the first time, step grid (4) is affected and Equation 6.14 becomes

$$h_4^{2\times0.08} = \frac{T\Delta t}{S(\Delta x)^2}(h_3^{0.08} + h_5^{0.08}) + h_4^{0.08}\left(1 - \frac{2T\Delta t}{S(\Delta x)^2}\right)$$

$$= \frac{1.5 \times 0.08}{0.03(3)^2}(5+1) + 5 \times \left(1 - \frac{2 \times 1.5 \times 0.08}{0.03(3)^2}\right) = 3.22\,\text{m}$$

$$h_2^{2\times0.08} = h_3^{2\times0.08} = 5\,\text{m}$$

For the second time step $t = 3 \times 0.08$ for grid (4):

$$h_2^{3\times0.08} = 5\,\text{m}, \quad h_3^{3\times0.08} = 4.21\,\text{m}, \quad h_4^{3\times0.08} = 3.02\,\text{m}$$

The above process is repeated until the head at each grid is calculated at the desired time. To illustrate the stability problem, a set of calculations was made in which Δt was selected to be 0.12 days so that the expression for stability results in

$$\frac{T\Delta t}{S(\Delta x)^2} = \frac{1.5 \times 0.12}{0.03(3)^2} > \frac{1}{2}$$

The calculated head in grid (4) as a function of time is shown in Figure 6.5. The computed values fluctuate with each time step for $\Delta t = 0.12$, giving completely erroneous results. Also, the amplitude of the fluctuation increases with increasing time.

6.4.2 Backward Difference Equation

Figure 6.3b shows the time derivative as a backward difference from the heads at time level, $n - 1$, which are the known heads. Therefore, the difference equation of each node will have five unknown variables. For a grid, which has N nodes, there is a system of N equations containing N unknown variables. This system

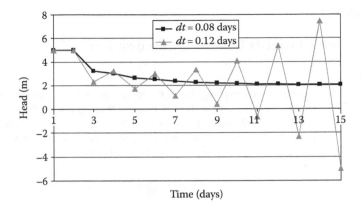

FIGURE 6.5   Calculated piezometric head in grid (4), using forward difference equation.

of equations can be solved simultaneously considering the boundary conditions. This method is called forward difference or implicit method. The implicit finite differential form of 2D groundwater equation, (Equation 6.1) can be expressed as follows:

$$S_s \left(\frac{h_{i,j}^n - h_{i,j}^{n-1}}{\Delta t} \right) = k_{x(i-1/2,j)} \left(\frac{h_{i-1,j}^n - h_{i,j}^n}{(\Delta x)^2} \right) + k_{x(i+1/2,j)} \left(\frac{h_{i+1,j}^n - h_{i,j}^n}{(\Delta x)^2} \right)$$

$$+ k_{y(i,j-1/2)} \left(\frac{h_{i,j-1}^n - h_{i,j}^n}{(\Delta y)^2} \right) + k_{y(i,j+1/2)} \left(\frac{h_{i,j+1}^n - h_{i,j}^n}{(\Delta y)^2} \right) \quad (6.16)$$

For example, in a 1D groundwater flow equation for a heterogeneous, isotropic, and confined aquifer, Equation 6.16 can be written as

$$S \left(\frac{h_i^n - h_i^{n-1}}{\Delta t} \right) = Kb \left(\frac{h_{i-1}^n - h_i^n}{(\Delta x)^2} \right) + Kb \left(\frac{h_{i+1}^n - h_i^n}{(\Delta x)^2} \right) \quad (6.17)$$

Rearranging Equation 6.17 so that all of the known values are on the right-hand side of the equal sign results in

$$h_{i-1}^n - h_i^n \left(2 + \frac{S(\Delta x)^2}{T \Delta t} \right) + h_{i+1}^n = - \frac{S(\Delta x)^2}{T \Delta t} h_i^{n-1} \quad (6.18)$$

The head in grid (*i*) depends upon the value of head at time *n* in the adjacent grids, (*i* + 1) and (*i* − 1). Thus, Equation 6.18 represents a set of algebraic equations that must be solved simultaneously.

Example 6.2

Solve Example 6.1 using the backward difference equation.

Solution

Equation 6.18 is used for determining the three interior grids (2), (3), and (4). Grids (1) and (5) are boundary grids and values of head at these grids are specified as 5 and 1 m, respectively. With assumption of $h_1 = h_2 = h_3 = h_4 = h_5 = 5\,m$, and $h_1^{0.08} = h_2^{0.08} = h_3^{0.08} = h_4^{0.08} = 5\,m$ and $h_5^{0.08} = 1\,m$, for the second time step $\Delta t = 0.08$ days, the following equations for grids (2), (3), (4) are obtained:

$$5 - 4.25h_2^{0.16} + h_3^{0.16} = -11.25$$

$$h_2^{0.16} - 4.25h_3^{0.16} + h_4^{0.16} = -11.25$$

$$h_3^{0.16} - 4.25h_4^{0.16} + 1 = -11.25$$

Rearranging the above equations so that all known values are placed on the right-hand side and summing them up, we get

$$h_2^{0.16} = 4.9451, \quad h_3^{0.16} = 4.7665, \quad h_4^{0.16} = 4.0627\,m$$

Example 6.3

Develop the finite-difference approximation (FDA) of flow equation for a 2D semi-confined, inhomogeneous, and anisotropic aquifer.

Solution

The FDA of flow equation in semi-confined aquifer equation in an inhomogeneous and isotropic aquifer is as follows:

$$\frac{1}{\Delta x^2}\left\{T_x^{i+1/2,j}(h_{i+1,j} - h_{i,j}) - T_x^{i-1/2,j}(h_{i,j} - h_{i-1,j})\right\} + \frac{1}{\Delta y^2}\left\{T_y^{i,j+1/2}(h_{i,j+1} - h_{i,j})\right.$$

$$\left. -T_y^{i,j-1/2}(h_{i,j} - h_{i,j-1})\right\} \pm \frac{Q_{w,i,j}}{\Delta x\,\Delta y} + \frac{K_{a,i,j}(H_{a,i,j} - h_{i,j})}{b_a} = S_{i,j}h_{i,j} \quad (6.19)$$

By rewriting Equation 6.19 for all internal nodes $(i, j;\ i = 1, \ldots, m$ and $j = 1, \ldots, m)$ of the domain, a set of linear equations can be expressed in the following form:

$$A h^{\bullet} + B h + g = 0 \quad (6.20)$$

In these equations, which are called *dynamic response equations*, h is a column vector of unknown heads, $h = (h_{11}, \ldots, h_{1m}, \ldots, h_{n,1}, \ldots, h_{nm})^T$. Vector g contains the rate of pumping and injections that can be the decision variables in the groundwater planning models. The coefficient matrices A and B depend on the flow and hydraulic properties of porous media. Dynamic response equations can easily provide the hydraulic head at the nodes as a linear function of initial condition and the pumping/injection rate of well sites.

6.4.3 ALTERNATING DIRECTION IMPLICIT METHOD

Considering that the first space derivative of head is calculated at time $(n + 1)$ and the second at the current time (n), the groundwater flow equation can be written as follows:

$$\frac{h_{i-1,j}^{n+1} - 2h_{i,j}^{n+1} + h_{i+1,j}^{n+1}}{\Delta x^2} + \frac{h_{i,j-1}^{n} - 2h_{i,j}^{n} + h_{i,j+1}^{n}}{\Delta y^2} = \frac{S}{T}\frac{h_{i,j}^{n+1} - h_{i,j}^{n}}{\Delta t} \tag{6.21}$$

There are three unknown values at time $(n + 1)$. Other values at time (n) are known. The Equation 6.21 can be rewritten as

$$h_{i-1,j}^{n+1} + B_i h_{i,j}^{n+1} + h_{i+1,j}^{n+1} = D_i \tag{6.22}$$

where B_i and D_i are constants and can be calculated as

$$B_i = -\left(2 + \frac{S}{T}\frac{\Delta x^2}{\Delta t}\right) \tag{6.23}$$

$$D_i = -h_{i,j-1}^{n}\frac{\Delta x^2}{\Delta y^2} - h_{i,j}^{n}\left(\frac{S}{T}\frac{\Delta x^2}{\Delta t} - \frac{2\Delta x^2}{\Delta y^2}\right) - h_{i,j+1}^{n}\frac{\Delta x^2}{\Delta y^2} \tag{6.24}$$

If Equation 6.22 is applied to all nodes along the x-axis (j rows), there will be $(k - 2)$ simultaneous equations since the value of head at two outer nodes is known due to the boundary conditions (k is the number of nodes along the x-axis). Therefore, by solving these equations, it is possible to calculate the heads on a line parallel to x-axis. The coefficient matrix of this set of equations is tridiagonal.

In the next time step, $\partial^2 h/\partial y^2$ is approximated at time $(k + 2)$ while $\partial^2 h/\partial x^2$ retained at current time $(n + 1)$; thus (Gupta, 1989),

$$\frac{h_{i-1,j}^{n+1} - 2h_{i,j}^{n+1} + h_{i+1,j}^{n+1}}{\Delta x^2} + \frac{h_{i,j-1}^{n} - 2h_{i,j}^{n} + h_{i,j+1}^{n}}{\Delta y^2} = \frac{S}{T}\frac{h_{i,j}^{n+1} - h_{i,j}^{n}}{\Delta t} \tag{6.25}$$

There are three unknown values at time $(n + 2)$ and other values at time $(n + 1)$ are known. Equation 6.25 can be rewritten as follows:

$$h_{i,j-1}^{n+2} + B_j h_{i,j}^{n+2} + h_{i,j+1}^{n+2} = D_j \tag{6.26}$$

where

$$B_i = -\left(2 + \frac{S}{T}\frac{\Delta y^2}{\Delta t}\right) \tag{6.27}$$

$$D_i = -h_{i-1,j}^{n+1}\frac{\Delta y^2}{\Delta x^2} - h_{i,j}^{n+1}\left(\frac{S}{T}\frac{\Delta y^2}{\Delta t} - \frac{2\Delta y^2}{\Delta x^2}\right) - h_{i+1,j}^{n+1}\frac{\Delta y^2}{\Delta x^2} \tag{6.28}$$

Applying Equation 6.26 to all nodes along the y-axis (i columns), there will be (m − 2) simultaneous equations since the value of head at two outer nodes is known due to the boundary conditions (m is the number of nodes along the y-axis). Therefore by solving (m − 2) equations, it is possible to calculate the heads of a line parallel to y-axis.

Example 6.4

Two fully penetrating wells in a confined aquifer system are shown in Figure 6.6, Wells 1 and 2 pump at rates of 90 and 60 L/s, respectively. The aquifer is bounded by a stream on one side and impermeable boundaries on two sides. It is extended semi-infinitely on the remaining side, but a width of 100 m is represented on the flow system under consideration. The initial peizometric head is horizontal, in level with the stream. Determine the spatial variation of piezometric head after one day of pumping (this example is adapted from Karamouz et al., 2003).

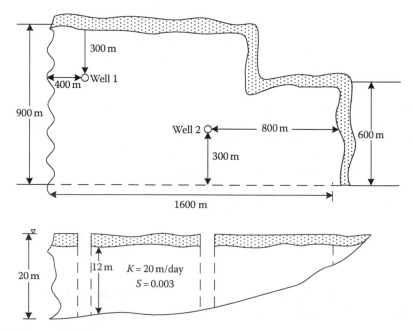

FIGURE 6.6 Plan and cross section of confined aquifer. (From Gupta, R.S., *Hydrology and Hydraulic Systems*, Prentice-Hall, Englewood Cliffs, NJ, 1989.)

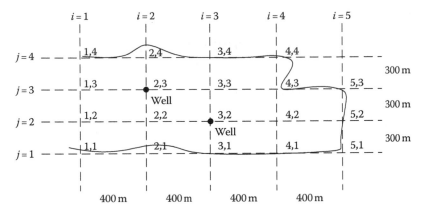

FIGURE 6.7 Finite-difference grid and its coordinates.

Solution

The grid layout is shown in Figure 6.7. As the initial condition, the heads of all nodes are 20 m at $t = 0$ and boundary conditions are as follows:

Recharge boundary

$$h_{1,1} = h_{1,2} = h_{1,3} = h_{1,4} = 20\,m$$

Impermeable boundaries

$$h_{5,1} = h_{4,1} \quad h_{2,4} = h_{2,3}$$

$$h_{5,2} = h_{4,1} \quad h_{3,4} = h_{3,3}$$

$$h_{4,3} = h_{3,3} \quad h_{4,4} = h_{4,3}$$

$$h_{4,4} = h_{3,4}$$

$$T = h \times b = 20 \times 12 = 240\,m^2/day$$

First time step: A time step Δt of 0.5 day is selected as the time interval. Since there is no pumping on the first row, piezometric heads in this row will not change in the first time step. For row 2, $j = 2$, we have

$$B_2 = -\left(2 + \frac{S}{T} \times \frac{(\Delta x)^2}{\Delta t}\right) = -\left(2 + \frac{0.003}{240} \times \frac{(400)^2}{0.5}\right) = -6.0$$

$$B_2 = B_3 = B_4 = -6.0$$

and

$$D_2 = -20 \times \frac{400^2}{300^2} - 20\left(\frac{0.003}{240} \times \frac{400^2}{0.5} - 2 \times \frac{400^2}{300^2}\right) - 20 \times \frac{400^2}{300^2} = -80$$

Since

$$q = \frac{0.06}{300 \times 400} = 0.5 \times 10^{-6}\,m/s = 0.0432\,m/day$$

hence,

$$D_3 = -80 + \frac{0.0432(400)^2}{240} = -51.2$$

$$D_4 = D_2 = -80$$

For node 2,

$$h_{1,2}^{0.5} + B_2 h_{2,2}^{0.5} + h_{3,2}^{0.5} = D_2$$

For node 3,

$$h_{2,2}^{0.5} + B_3 h_{3,2}^{0.5} + h_{4,2}^{0.5} = D_3$$

For node 4,

$$h_{3,2}^{0.5} + B_4 h_{4,2}^{0.5} + h_{5,2}^{0.5} = D_4$$

Substituting the B, D, and initial and boundary conditions, we have

$$20 - 6 h_{2,2}^{0.5} + h_{3,2}^{0.5} = -80$$

$$h_{2,2}^{0.5} - 6 h_{3,2}^{0.5} + h_{4,2}^{0.5} = -51.2$$

$$h_{3,2}^{0.5} - 6 h_{4,2}^{0.5} + h_{5,2}^{0.5} = -80$$

or

$$\begin{bmatrix} -6 & 1 & 0 \\ 1 & -6 & 1 \\ 0 & 1 & -5 \end{bmatrix} \begin{bmatrix} h_{2,2}^{0.5} \\ h_{3,2}^{0.5} \\ h_{4,2}^{0.5} \end{bmatrix} = \begin{bmatrix} -100 \\ -51.2 \\ -80 \end{bmatrix}$$

Solving gives

$$h_{2,2}^{0.5} = 19.15\,\text{m}$$

$$h_{3,2}^{0.5} = 14.89\,\text{m}$$

$$h_{4,2}^{0.5} = 18.97\,\text{m}$$

The values of heads at nodes of row 3 are obtained by a similar procedure. Since row 4 is an impermeable boundary, it will have the same heads as row 3. The values of nodal heads after the first time step are shown in Figure 6.8.

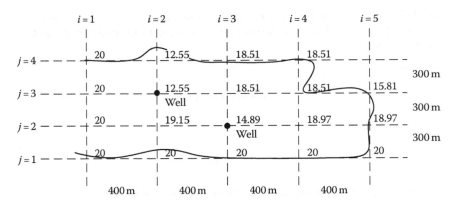

FIGURE 6.8   Heads at end of first time step ($\Delta t = 0.5$ day).

Second time step: The results of the first time step are the initial heads of second time step. In this time step, computations are made by column. Therefore, for column 2 ($i = 2$), we have

$$B_1 = -\left(2 + \frac{0.003}{240} \times \frac{400^2}{0.5}\right) = -4.25$$

$$B_1 = B_2 = B_3 = -4.25$$

and

$$D_1 = -20 \times \frac{300^2}{400^2} - 20\left(\frac{0.003}{240} \times \frac{300^2}{0.5} - 2 \times \frac{300^2}{400^2}\right) - 20 \times \frac{300^2}{400^2} = -45$$

$$D_2 = -20 \times \frac{300^2}{400^2} - 19.15\left(\frac{0.003}{240} \times \frac{300^2}{0.5} - 2 \times \frac{300^2}{400^2}\right) - 14.89 \times \frac{300^2}{400^2} = -41.17$$

Since

$$q = \frac{0.09}{300 \times 400} = 0.75 \times 10^{-6} \text{ m/s} = 0.0648 \text{ m/day}$$

hence,

$$D_3 = -20 \times \frac{300^2}{400^2} - 12.55\left(\frac{0.003}{240} \times \frac{300^2}{0.5} - 2 \times \frac{300^2}{400^2}\right) - 18.51 \times \frac{300^2}{400^2}$$

$$+ \frac{0.0648(300)^2}{240} = -11.48$$

For node 1,

$$h_{2,0}^1 + B_1 h_{2,1}^1 + h_{2,2}^1 = D_1$$

As the aquifer in this direction extends beyond the grid boundary, the head outside of the boundary is included and is assumed to be equal to the initial head of 20 m.

For node 2,

$$h_{2,1}^1 + B_2 h_{2,2}^2 + h_{2,3}^1 = D_2$$

For node 3,

$$h_{2,2}^1 + B_3 h_{2,3}^1 + h_{2,4}^1 = D_3$$

Substituting B, D, and initial and boundary conditions, we have

$$20 - 4.25 h_{2,1}^1 + h_{2,2}^1 = -45$$

$$h_{2,1}^1 - 4.25 h_{2,2}^1 + h_{2,3}^1 = -41.17$$

$$h_{2,2}^1 - 4.25 h_{2,3}^1 + h_{2,3}^1 = -11.48$$

or

$$\begin{bmatrix} -4.25 & 1 & 0 \\ 1 & -4.25 & 1 \\ 0 & 1 & -4.25 \end{bmatrix} \begin{bmatrix} h_{2,1}^1 \\ h_{2,2}^1 \\ h_{2,3}^1 \end{bmatrix} = \begin{bmatrix} -45 \\ -41.17 \\ -11.48 \end{bmatrix}$$

Solving gives

$$h_{2,1}^1 = 19.10\,\mathrm{m}$$

$$h_{2,2}^1 = 16.18\,\mathrm{m}$$

$$h_{2,3}^1 = 8.51\,\mathrm{m}$$

Again, the heads at all nodes in columns are obtained by a similar procedure. The values of nodal heads after the second time step (after $t = 1$ day) are shown in Figure 6.9.

6.4.4 CRANK–NICOLSON DIFFERENCE EQUATION

The Crank–Nicolson method is based on central difference in space. It is the average of forward Euler and backward Euler in time. For example, in one dimension, if the partial differential equation is

$$\frac{\partial u}{\partial t} = F\left(u, x, t, \frac{\partial h}{\partial x}, \frac{\partial^2 h}{\partial x^2}\right) \tag{6.29}$$

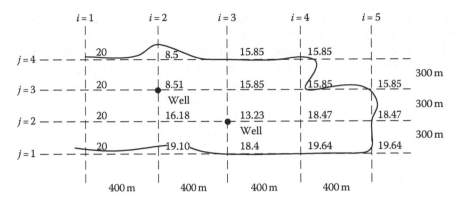

FIGURE 6.9 Heads at the end of second time step ($\Delta t = 1$ day).

then, letting $h(i\Delta x, n\Delta t) = h_i^n$, the Crank–Nicolson method is the average of the forward Euler method at n and the backward Euler method at $n + 1$.

Forward difference equation can be expressed as

$$\frac{u_i^{n+1} - u_i^n}{\Delta t} = F_i^n \left(u, x, t, \frac{\partial h}{\partial x}, \frac{\partial^2 h}{\partial x^2} \right) \tag{6.30}$$

Backward difference equation can be shown as

$$\frac{u_i^{n+1} - u_i^n}{\Delta t} = F_i^{n+1} \left(u, x, t, \frac{\partial h}{\partial x}, \frac{\partial^2 h}{\partial x^2} \right) \tag{6.31}$$

Crank–Nicolson difference equation is formulated as

$$\frac{u_i^{n+1} - u_i^n}{\Delta t} = \frac{1}{2} \left(F_i^n \left(u, x, t, \frac{\partial h}{\partial x}, \frac{\partial^2 h}{\partial x^2} \right) + F_i^{n+1} \left(u, x, t, \frac{\partial h}{\partial x}, \frac{\partial^2 h}{\partial x^2} \right) \right) \tag{6.32}$$

The function F must be discretized spatially with a central difference. The discretization will also be nonlinear if the PDE is nonlinear. Thus, advancing in time will involve the solution of a system of nonlinear algebraic equations, although linearization is possible.

In groundwater modeling, the heads at all nodes are changed between time steps (n) and ($n + 1$). Using head values at time (n) to approximate the space derivatives is valid only if relatively small time steps are considered. The approximation is improved by evaluating the space derivatives between $t = n\Delta t$ and $t = (n + 1)\Delta t$. It can be improved using a weighted average of the approximations at (n) and ($n + 1$) as shown in Figure 6.3c. The weighting parameter is represented by α, and it lies between 0 and 1. If time step ($n + 1$) is weighted by α and time step (n) is weighted by ($1 - \alpha$), then

$$\frac{\partial^2 h}{\partial x^2} \cong \alpha \frac{h_{i-1,j}^{n+1} - 2h_{i,j}^{n+1} + h_{i+1,j}^{n+1}}{\Delta x^2} + (1 - \alpha) \frac{h_{i-1,j}^n - 2h_{i,j}^n + h_{i+1,j}^n}{\Delta x^2} \tag{6.33}$$

The parameter α is selected by the model. For $\alpha = 1$, the space derivatives are approximated solely at the advanced time level $(n + 1)$, and the finite-difference scheme is said to be fully implicit. Use of the fully implicit scheme implies that the value of the space derivatives at the future time is the best approximation. Use of the explicit scheme $\alpha = 0$ implies that the value of the space derivatives at the old time level is the best approximation. When $\alpha = 1/2$, the best value lies halfway between time levels (n) and $(n + 1)$. The FDA associated with a choice of $\alpha = 1/2$ is called the Crank–Nicolson method (Wang and Anderson, 1995).

Example 6.5

A plan view of an unconfined aquifer is shown in Figure 6.10. A mesh-centered grid system has been overlain such that $\Delta x = \Delta y = 100\,\text{m}$. The shaded area in the figure represents a gravel pit including 12 wells along its boundary for water pumping. Each well is to be pumped at a rate of $3500\,\text{m}^3/\text{day}$. The hydraulic conductivity of the aquifer is $50\,\text{m/day}$ and $S = 0.2$. The boundary conditions are also given. The heads are relative to a datum at the base of the aquifer. Compute the drawdowns in the gravel pit after 200 days of pumping. For initial conditions, use the steady-state head configuration without pumping. Write mass–balance calculation and incorporate it into your program (Wang and Anderson, 1995).

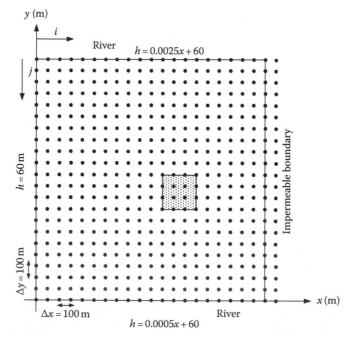

FIGURE 6.10　Discretization and boundary conditions of unconfined aquifer in Example 6.5.

Solution

The groundwater flow equation can be written as follows:

$$\frac{\partial}{\partial x}\left(K_x h \frac{\partial h}{\partial x}\right) + \frac{\partial}{\partial y}\left(K_y h \frac{\partial h}{\partial y}\right) = S \frac{\partial h}{\partial t} - R(x,y,t)$$

If $K_x = K_y = K$, then

$$\frac{\partial}{\partial x}\left(\frac{K}{2}\frac{\partial h^2}{\partial x}\right) + \frac{\partial}{\partial y}\left(\frac{K}{2}\frac{\partial h^2}{\partial y}\right) = S \frac{\partial h}{\partial t} - R(x,y,t)$$

or

$$\frac{K}{2}\left[\frac{\partial^2 h^2}{\partial x^2} + \frac{\partial^2 h^2}{\partial y^2}\right] = S \frac{\partial h}{\partial t} - R(x,y,t)$$

With assumption $V = h^2$, we have

$$\frac{K}{2}\left[\frac{\partial^2 V}{\partial x^2} + \frac{\partial^2 V}{\partial y^2}\right] = S \frac{\partial h}{\partial t} - R(x,y,t)$$

Substituting Equation 6.33 using Crank–Nicolson method, the transition flow equation can be approximated by

$$\alpha(\tilde{V}_{i,j}^{n+1} - V_{i,j}^{n+1}) + (1-\alpha)(\tilde{V}_{i,j}^{n} - V_{i,j}^{n}) = \frac{a^2 S}{2}\frac{\sqrt{V_{i,j}^{n+1}} - \sqrt{V_{i,j}^{n}}}{K\Delta t} - \frac{a^2 R_{i,j}^{n}}{2K}$$

$$\tilde{V}_{i,j}^{n} = \frac{V_{i-1,j}^{n} + V_{i+1,j}^{n} + V_{i,j-1}^{n} + V_{i,j+1}^{n}}{4}$$

where $\Delta x = \Delta y = a$. All the unknowns are put on the left-hand side and all the known parameters on tile right-hand side and multiply each side by −1:

$$\left(\frac{a^2 S}{4K\Delta t \sqrt{V_{i,j}^{n+1}}} + \alpha\right)V_{i,j}^{n+1} - \alpha\tilde{V}_{i,j}^{n+1} = \frac{a^2 S}{4K\Delta t\sqrt{V_{i,j}^{n+1}}}V_{i,j}^{n} + (1-\alpha)(\tilde{V}_{i,j}^{n} - V_{i,j}^{n}) + \frac{a^2 R_{i,j}^{n}}{2K}$$

The above equation represents a system of linear equations for $V_{i,j}$ at time level $(n + 1)$. As expressed in Figure 6.10, in the mesh-centered grid, $\Delta x = \Delta y = 100\,\text{m}$. The right boundary is approximated as symmetry boundaries. In the computer model, the lower, upper, and left boundaries are also considered to be known boundaries.

The groundwater flow equation for initial conditions, the steady-state flow without pumping, can be written as follows:

$$\frac{\partial}{\partial x}\left(K_x h \frac{\partial h}{\partial x}\right) + \frac{\partial}{\partial y}\left(K_y h \frac{\partial h}{\partial y}\right) = 0$$

If $K_x = K_y = K$, then

$$\frac{K}{2}\left[\frac{\partial^2 h^2}{\partial x^2} + \frac{\partial^2 h^2}{\partial y^2}\right] = 0$$

With assumption $V = h^2$, we have

$$\frac{K}{2}\left[\frac{\partial^2 V}{\partial x^2} + \frac{\partial^2 V}{\partial y^2}\right] = 0$$

Substituting Equation 6.33 using Crank–Nicolson method, the transition flow equation can be approximated by

$$\alpha(\tilde{V}_{i,j}^{n+1} - V_{i,j}^{n+1}) + (1-\alpha)(\tilde{V}_{i,j}^{n} - V_{i,j}^{n}) = 0$$

$$\tilde{V}_{i,j}^{n} = \frac{V_{i-1,j}^{n} + V_{i+1,j}^{n} + V_{i,j-1}^{n} + V_{i,j+1}^{n}}{4}$$

This system can be solved directly using matrix methods or by iteration with the same way as what is used to solve the system of linear equations for the steady-state Laplace's or Poisson's equation. The difference is that, for the transient flow problem, a new system of equations must be solved at each time step. A FORTRAN code is developed to solve this problem, which is shown in Figure 6.11. It calculates the volume of water removed by pumping in the 12 nodes, the volume of water removed from storage over the entire aquifer and the volume of water flowing across the boundaries. Figures 6.12 and 6.13 show the computed water head at initial condition and 200 days after pumping, respectively.

6.5 FINITE ELEMENT METHOD

Another numerical technique for the solution of groundwater flow and contaminant mass transport problems is the FEM. For many groundwater problems, the FEM is superior to classical finite-difference models. Groundwater problems with irregular shape domain, complex boundary conditions, and material heterogeneity can be modeled using FEM, while the FDM implies the complicated interpolation schemes to approximate complex boundary conditions.

In FEM, the irregular shape domain can be divided into a set of elements with different sizes or shapes. In order to reflect variation of state variables or parameter

```
drawdown example-unconfined aquifer- transient conditions
      DIMENSION HNEW(22,22),HOLD(22,22),R(22,22),DD(22,22),
     &H0(22,22),VNEW(22,22),VOLD(22,22)
      !DEFINE INPUT PARAMETERS
      S=0.2
      K=50.
      !H0 IS INITIAL HEAD
      P=60.
      DX=100.
      DT=.01
      SUML=0.
      SUMB=0.
      SUMT=0.
      Drawd1=0.
      Drawd2=0.
      SUMR=0.
      !USE CRANK-NICOLSON APPOXIMATION
      ALPHA=0.5
      !SET RELAXATION FACTOR
      OMEGA=1.8
      !INITIALIZE ARRAYS
      !HOLD IS HEAD AT THE TIME STEP N
      !HNEW IS HEAD AT THE TIME STEP N+1
      DO 3 I=1,21
      DO 4 J=1,22
      HNEW(I,J)=P
      HOLD(I,J)=P
      R(I,J)=0.
4     CONTINUE
3     CONTINUE
      DO 33 J=1,22
      HNEW(1,J)=60.
      HOLD(1,J)=60.
33    CONTINUE
      DO 30 I=2,21
      HNEW(I,1)=.0025*(I-1)*100.+60.
      HOLD(I,1)=.0025*(I-1)*100.+60.
      HNEW(I,22)=.0005*(I-1)*100.+60.
      HOLD(I,22)=.0005*(I-1)*100.+60.
30    CONTINUE
      DO 36 J=1,22
      HNEW(22,J)=HNEW(20,J)
      HOLD(22,J)=HOLD(20,J)
36    CONTINUE
      !DEFINE PUMPING RATE AS RECHARGE TO CELL AT ORIGIN
      R0=-3500./DX/DX
      DO 13 I=1,22
      DO 14 J=1,22
      VNEW(I,J)=HNEW(I,J)*HNEW(I,J)
      VOLD(I,J)=HOLD(I,J)*HOLD(I,J)
      R(I,J)=0.
```

FIGURE 6.11 Fortran program for solving Example 6.5.

(*continued*)

```
14        CONTINUE
13        CONTINUE
          TIME=-0.01
          !START TIME STEPS
          !AT EACH TIME STEP SOLVE SYSTEM OF EQUATIONS BY ITERATION
          NEND=53
          DO 5 N=1,NEND
          TIME=TIME+DT
          NUMIT=0
10        AMAX=0.
          NUMIT=NUMIT+1
          IF (NUMIT.GT.500) GO TO 50
          IF (TIME.EQ.0.0) THEN
          DO 45 J=2,22
          VNEW(22,J)=VNEW(20,J)
45        CONTINUE
          DO 15 I=2,21
          DO 16 J=2,21
          ! FOR T=0 :R=0,H=H0,INDEPENDENT OF TIME
          OLDVAL=VNEW(I,J)
          VNEW(I,J)=(VNEW(I-1,J)+VNEW(I+1,J)+VNEW(I,J-1)+
          VNEW(I,J+1)+2.*R(I,J)*DX* DX/K)/4.
          VNEW(I,J)=OMEGA*VNEW(I,J)+(1.-OMEGA)*OLDVAL
          ERR=ABS(VNEW(I,J)-OLDVAL)
          IF (ERR.GT.AMAX) AMAX=ERR
16        CONTINUE
15        CONTINUE
          ELSE
          DO 87 I=2,21
          DO 88 J=2,21
          OLDVAL=VNEW(I,J)
          V1=(VOLD(I,J+1)+VOLD(I,J-1)+VOLD(I+1,J)+VOLD(I-1,J))/4.
          V2=(VNEW(I,J+1)+VNEW(I,J-1)+VNEW(I+1,J)+VNEW(I-1,J))/4.
          F1=(DX*DX*S/(4*K*SQRT(VOLD(I,J))*DT)+ALPHA)
          VNEW(I,J)=(ALPHA*V2+(F1-ALPHA)*VOLD(I,J)+
          &(1-ALPHA)*(V1-VOLD(I,J))+DX*DX*R(I,J)/2/K)/F1
          ERR=ABS(VNEW(I,J)-OLDVAL)
          IF (ERR.GT.AMAX) AMAX=ERR
88        CONTINUE
87        CONTINUE
          !BOUNDARY CONDITION ON RIGHT
          DO 17 J=2,22
          VNEW(22,J)=VNEW(20,J)
17        CONTINUE
          END IF
          IF (AMAX.GT.0.001) GO TO 10
50        DO 47 I=1,22
          DO 48 J=1,22
          IF (VNEW(I,J).LT.0)VNEW(I,J)=0.
```

FIGURE 6.11 (continued)

```
         HNEW(I,J)=SQRT(VNEW(I,J))
         VOLD(I,J)=VNEW(I,J)
         HOLD(I,J)=HNEW(I,J)
48       CONTINUE
47       CONTINUE
         ! PREPARE FOR NEXT TIME STEP
         ! PUT HNEW VALUES INTO HOLD ARRAY
         IF (TIME.EQ.0.) THEN
         DO 27 I=1,22
         DO 28 J=1,22
         H0(I,J)=HNEW(I,J)
28       CONTINUE
27       CONTINUE
         WRITE (50,74)
74       FORMAT(10X,"HEAD AT TIME=0 ",/7X,"**********************")
         WRITE (50,31) ((HNEW(I,J),I=1,21),J=1,22)
         WRITE(50,75)
75       FORMAT(//10X,"HEAD IN STEADY STATE SOLUTION",/)
         ELSE
         DT=DT*1.5
         IF(DT.GT.5.0) DT=5.0
         END IF
         R(12,11)=R0
         R(13,11)=R0
         R(14,11)=R0
         R(15,11)=R0
         R(12,14)=R0
         R(13,14)=R0
         R(14,14)=R0
         R(15,14)=R0
         R(12,12)=R0
         R(12,13)=R0
         R(15,12)=R0
         R(15,13)=R0
         ! COMPUTE DRAWDOWN
         DO 25 I=1,22
         DO 26 J=1,22
         DD(I,J)=H0(I,J)-HNEW(I,J)
26       CONTINUE
25       CONTINUE
         DO 64 J=1,22
         SUML=SUML+K/2*(HNEW(1,J)**2-HNEW(2,J)**2)*DT
64       CONTINUE
         DO 65 I=1,22
         SUMT=SUMT+K/2*(HNEW(I,1)**2-HNEW(I,2)**2)*DT
         SUMB=SUMB+K/2*(HNEW(I,22)**2-HNEW(I,21)**2)*DT
65       CONTINUE
         ! PRINT RESULTS
         WRITE(50,29) TIME,DT,NUMIT
```

FIGURE 6.11 (continued)

(*continued*)

```
29      FORMAT(1X,"TIME=",F9.2,5X,"DT=",F9.2,5X,"NUMIT=",I8,/)
        WRITE (50,32)((DD(I,J),I=1,21),J=1,22)
31      FORMAT(1X,21F6.2)
32      FORMAT(21F4.2)
5       CONTINUE
        DO 81 I=1,20
        DO 82 J=1,22
        Drawd1=Drawd1+DD(I,J)
82      CONTINUE
81      CONTINUE
        DO 83 J=1,22
        Drawd2=Drawd2+DD(21,J)
83      CONTINUE
        Drawd1=.2*DX*DX*Drawd1
        Drawd2=Drawd1+.1*DX*DX*Drawd2
        SUMTOTAL=SUML+SUMB+SUMT
        ERROR=100*(SUMTOTAL+Drawd2-3500*12*200)/3500/200/12
        WRITE(50,66) SUML,SUMT,SUMB,SUMTOTAL,Drawd2,ERROR
66      FORMAT(//5X,"SUML=",F10.1," SUMT=",F10.1,"
        &SUMB=",F10.1," SUMTOTAL=",F11.1,/10X, "Drawdown=",
        &F14.3,/5X,"ERROR=",F7.3,"%")
        WRITE(50,85)
85      FORMAT(//10X,"FINAL HEAD"/,10X,"****************************")
        WRITE (50,31) ((HNEW(I,J),I=1,21),J=1,22)
        STOP
        END
```

FIGURE 6.11 (continued)

values, the element sizes can be changed. The PDEs for the elements are approximated utilizing several approaches such as direct approach, the weighted residual approach, and the variational approach to obtain a set of algebraic equations. The piecewise continuous representation of the dependent variables and, possibly, the parameters of the groundwater system can increase the accuracy of the numerical approximations (Karamouz et al., 2003).

6.5.1 DISCRETIZE THE PROBLEM DOMAIN

The problem domain is divided into some nonoverlapping elements as the first step in solving a groundwater flow or solute transport problem by the FEM. In this regard, the problem domain is replaced with a series of nodes and discrete elements or finite element mesh. The elements are generated by connecting two or more nodes together utilizing the lines as shown in Figure 6.14. Then, a node number is assigned to each node and an element number is assigned to each element. However, there is much computer software that creates this mesh. Elements can have one, two, or thee dimensions with any size. Figure 6.15 shows different element types. In each element, the groundwater flow characteristics should be specified.

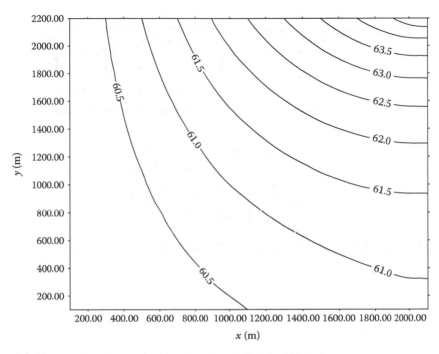

FIGURE 6.12 Water heads at initial condition (Example 6.5).

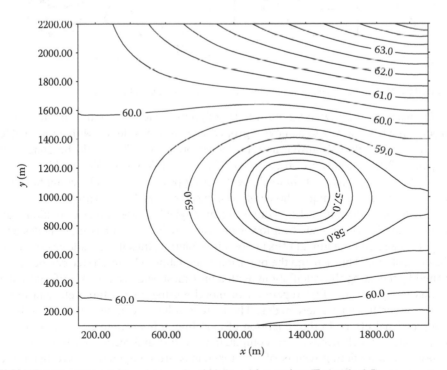

FIGURE 6.13 Water table contour after 200 days of pumping (Example 6.5).

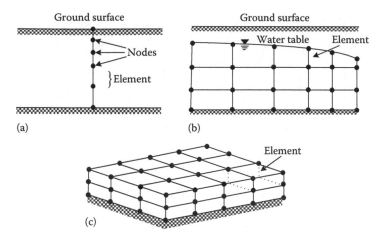

FIGURE 6.14 Discretization of (a) 1D, (b) 2D, and (c) 3D problem domains.

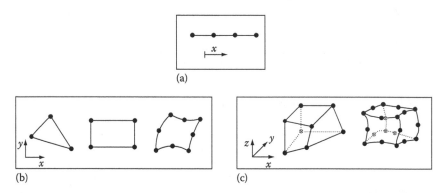

FIGURE 6.15 Some types of finite elements: (a) 1D, (b) 2D, and (c) 3D.

Drawing the mesh with the knowledge of groundwater flow and solute transport process is a helpful way to achieve a groundwater system solution with a reasonable cost for computational burden and acceptable precision. This could be done with respect to flow or transport process visualization and flow nets. The problem could be modeled using different finite element mesh types and the results could be similar. Thus, there is no unique choice of mesh type and size for modeling.

The finite element modeling with a fine mesh yields a more precise solution than a coarse mesh, which leads to a lower precision. A fine mesh has more nodes and requires more computational effort to obtain a solution, thus it is a trade-off between the computational burden and the precision of modeling. The mesh size can be determined by repeating the calculations with a finer mesh and see how significantly the results change. In the first repetition of modeling, the coarse finite element mesh can be generated with a few nodes. Therefore, a solution can be derived with little computational effort. In the repetition of modeling, a finer finite element mesh can be prepared, which needs more computational effort and leads to a more precise solution. The modeling results of two repetitions are compared to investigate the improving results. If the change in the results is more than the tolerance modeling parameter, which is a predetermined parameter and it is defined based on the acceptable level of precision, more mesh refinement is needed. Otherwise, the modeling process is stopped because no significant change in the computed values is achieved (Istok, 1989).

6.5.2 DERIVE THE APPROXIMATING EQUATIONS

After preparing the finite element mesh, the next step of groundwater modeling is to evaluate the integral formulation for the governing groundwater flow or solute transport equation. A system of algebraic equations is generated by implementation of the integral formulation over the domain. By solving the algebraic equations, the groundwater flow is simulated and the values of the field variable such as hydraulic head, pressure head, or solute concentration can be determined at each node in the mesh. Numerical integration methods are extensively employed to derive the integral formulation for a PDE that presents the groundwater flow and solute transport. The method of weighted residuals is a most common approach that is widely used in groundwater modeling.

In the method of weighted residuals, an approximate solution is developed and by replacing the approximate solution into the governing equation, the residual at each node is determined. The weighted average of the residuals for each point is supposed to be equal to zero. In this method, the trial solution to the system's equations is expressed in terms of the differential operator L, as

$$L(h - F) = 0 \qquad (6.34)$$

where

h is the field variable
F is a known function

The approximate solution can be written as follows:

$$h \approx \hat{h}(x,t) = \sum_{i=1}^{n} N_i(x)\, \tilde{h}_i(t) \qquad (6.35)$$

where

$\hat{h}$ is the approximate solution
$N_i(x)$ are the basis or shape functions of node i, defined over the entire domain
n is the number of linearly independent shape functions
$\tilde{h}_i(t)$ are the unknown variables (heads) that are determined for each node of the finite element grid

Regarding spatial variation of the state variables in the domain, shape function for each node can be defined, which are interpolation functions (Karamouz et al., 2003).

For example, the approximate solution for a 1D element with two nodes i and j, as shown in Figure 6.16, can be formulated as follows:

$$\hat{h}^{(e)}(x) = N_i^{(e)}(x)\, h_i + N_j^{(e)}(x)\, h_j \qquad (6.36)$$

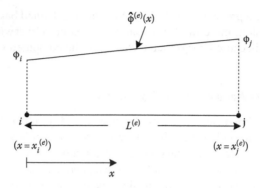

FIGURE 6.16   Approximate solution for an element.

or in matrix form

$$\hat{h}^{(e)}(x) = \left[N^{(e)} \right]\{h\}$$

(6.37)

where

$$[N^{(e)}] = [N_i^{(e)}(x)N_j^{(e)}(x)]$$

(6.38)

$$\{h\} = \begin{Bmatrix} h_i(t) \\ h_j(t) \end{Bmatrix}$$

(6.39)

The interpolation functions are equal to one at each node and decrease linearly to zero at the node in the other side. The interpolation functions can be written as

$$N_i^{(e)}(x) = \frac{x_j^{(e)} - x}{x_j^{(e)} - x_i^{(e)}} = \frac{x_j^{(e)} - x}{L^{(e)}}$$

(6.40)

$$N_j^{(e)}(x) = \frac{x - x_i^{(e)}}{x_j^{(e)} - x_i^{(e)}} = \frac{x - x_i^{(e)}}{L^{(e)}}$$

(6.41)

where $L^{(e)}$ is the element length. These interpolation functions are shown in Figure 6.17. The value of $\hat{h}(x, t)$ is h_i at node i and decreases linearly to h_j at node j and at the midpoint of the element $x = (x_j + x_i)/2$ and $\hat{h}(x, t)$ is $h = (h_j + h_i)/2$.

Although many types of interpolation functions have been defined to develop an approximate solution for $\hat{h}(x, t)$, the linear interpolation functions are the most commonly used. Because the approximate solution does not satisfy Equation 6.34, the residual or error R can be estimated by substituting the approximate solution $\hat{h}$ into Equation 6.34 as follows:

$$R = L(\hat{h} - F) \neq 0$$

(6.42)

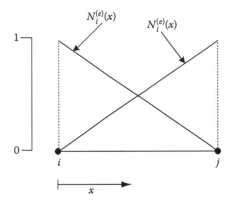

FIGURE 6.17 Linear interpolation functions for an element.

In order to determine the unknown coefficients $\tilde{h}_i$, the residual of approximate solution is minimized. The estimation error may be different in each point in the calculation domain, hence the residual cannot be zero at certain specified points. The residual might have low value in some points while it has large value in the other points. Thus, minimizing the errors is achieved by integrating the weighted average of the error over the domain (D) and setting it equal to zero as follows:

$$\int_{D} W_k R \, dD = \int_{D} W_k L(\hat{h}) dD = 0, \quad k = 1, 2, \dots, n \tag{6.43}$$

where
  W_k is weighting function in node i
  D is a length in 1D mesh or an area in 2D mesh or a volume in 3D mesh

To determine the n state variables at nodal points, the weighting function should be specified and the integral equation can be calculated by breaking it to n simple equations.

The form of the weighting function W in Equation 6.43 must also be specified. The type of weighted residual method is selected by choosing the n weighting functions. Although several weighted residual methods (WRMs) such as subdomain method, collocation method, and Galerkin's method have been applied in water resource engineering, the Galerkin method is commonly used to represent the hydraulic and mass transport response equations of the aquifer system.

In Galerkin's method, the weighting function for a node is considered similar to the interpolation function used in definition of the approximate solution for $\hat{h}(x, t)$. The modified integral equation (Equation 6.43) is then given by

$$\int_{D} N_k L(\hat{h}) \, dD = 0, \quad k = 1, 2, \dots, n \tag{6.44}$$

The modified integral equation is straightforward to solve and can be converted to n simultaneous algebraic equations of the form

$$[K]\{h\} = \{F\} \tag{6.45}$$

where $[K]$ is the global conductance matrix. In order to develop this matrix, conductance matrix is computed for all nodes and then the global conductance matrix is obtained for the finite element mesh. In Equation 6.45, $\{F\}$ addresses specified flow rates at boundary conditions. The system of algebraic equations is revised based on the known variables on the boundary condition, and unknown values are obtained by solving the equations.

Example 6.6

Solve the differential equation $d^2h/dx^2 - h = 0$ using the Galerkin method. Consider the following conditions:

$$0 \leq x \leq 5$$

$$h = 0\,\text{m} \quad \text{where } x = 0$$

$$h = 8\,\text{m} \quad \text{where } x = 5$$

Solution
As shown in Figure 6.18, the problem domain is divided into four nodes/three elements and for each node, a partially linear shape function $N_i(x)$ is considered. Therefore, the approximate function $\hat{h}(x, t)$, at node i, is defined as follows:

$$h \approx \hat{h}(x,t) = \sum_{i=1}^{w} N_i(x)\hat{h}_i(t), \quad 0 \leq x \leq 5$$

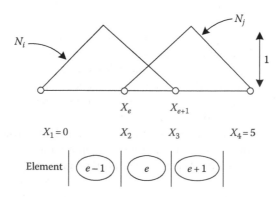

FIGURE 6.18 General numbering of nodes and elements of Example 6.6.

Considering the unknown variables $h_1, h_2, ..., h_4$, the Equation 6.43 can be rewritten as follows:

$$\int_0^5 W_i \left(\frac{d^2\hat{h}}{dx^2} - \hat{h} \right) dx = 0, \quad i = 1, 2, ..., 4$$

or

$$\int_0^5 \left(\frac{dW_i}{dx} \frac{d\hat{h}}{dx} + W_i\hat{h} \right) dx + \left[W_i \frac{d\hat{h}}{dx} \right] = 0, \quad i = 1, 2, ..., 4$$

This equation can be rewritten as follows using the Galerkin method ($W_i = N_i$):

$$Kh = F$$

$$K_{ij} = \int_0^5 \left(\frac{dN_i}{dx} \frac{dN_j}{dx} + N_iN_j \right) dx, \quad 1 \le i, j \le 4$$

$$F_i = \left[N_i \frac{dh}{dx} \right]_0^5, \quad 1 \le i \le 4$$

If the above equation is used for each element (e) between nodes m and n, the coefficients of equation can be calculated using Equations 6.40 and 6.41. Assuming $X - x - x_m$, the shape functions are as follows:

$$N_m = N_m^e = \frac{X}{L^e}$$

$$N_n = N_n^e = \frac{L^e - X}{L^e}$$

where $L^e = x_n - x_m$. Therefore, the components of matrix A are calculated as follows:

$$K_{mn}^e = K_{nm}^e = \int_0^{L^e} \left[\frac{dN_m^e}{dx} \frac{dN_n^e}{dx} + N_m^e N_n^e \right] dx = \frac{-1}{L^e} + \frac{L^e}{6}$$

$$K_{mm}^e = K_{nn}^e = \int_0^{L^e} \left[\left(\frac{dN_m^e}{dx} \right)^2 + (N_n)^2 \right] dx = \frac{1}{L^e} + \frac{L^e}{3}$$

The matrix K is calculated from submatrices K^e, considering the position of the element in Figure 6.18. Assuming $L_1 = 2\,\text{m}$, $L_2 = 1\,\text{m}$, $L_3 = 2\,\text{m}$, the matrices K^e and matrix K can be calculated as

$$K^1 = \begin{bmatrix} \dfrac{1}{L_1} + \dfrac{L_1}{3} & \dfrac{-1}{L_1} + \dfrac{L_1}{6} & 0 & 0 \\[2ex] \dfrac{-1}{L_1} + \dfrac{L_1}{6} & \dfrac{1}{L_1} + \dfrac{L_1}{3} & 0 & 0 \\[2ex] 0 & 0 & 0 & 0 \\[1ex] 0 & 0 & 0 & 0 \end{bmatrix}$$

$$K^2 = \begin{bmatrix} 0 & 0 & 0 & 0 \\[1ex] 0 & \dfrac{1}{L_2} + \dfrac{L_2}{3} & \dfrac{-1}{L_2} + \dfrac{L_2}{6} & 0 \\[2ex] 0 & \dfrac{-1}{L_2} + \dfrac{L_2}{6} & \dfrac{1}{L_2} + \dfrac{L_2}{3} & 0 \\[2ex] 0 & 0 & 0 & 0 \end{bmatrix}$$

$$K^3 = \begin{bmatrix} 0 & 0 & 0 & 0 \\[1ex] 0 & 0 & 0 & 0 \\[1ex] 0 & 0 & \dfrac{1}{L_3} + \dfrac{L_3}{3} & \dfrac{-1}{L_3} + \dfrac{L_3}{6} \\[2ex] 0 & 0 & \dfrac{-1}{L_3} + \dfrac{L_3}{6} & \dfrac{1}{L_3} + \dfrac{L_3}{3} \end{bmatrix}$$

$$K = \begin{bmatrix} \dfrac{1}{L_1} + \dfrac{L_1}{3} & \dfrac{-1}{L_1} + \dfrac{L_1}{6} & 0 & 0 \\[2ex] \dfrac{-1}{L_1} + \dfrac{L_1}{6} & \left(\dfrac{1}{L_1} + \dfrac{L_1}{3}\right) + \left(\dfrac{1}{L_2} + \dfrac{L_2}{3}\right) & \dfrac{-1}{L_2} + \dfrac{L_2}{6} & 0 \\[2ex] 0 & \dfrac{-1}{L_2} + \dfrac{L_2}{6} & \left(\dfrac{1}{L_2} + \dfrac{L_2}{3}\right) + \left(\dfrac{1}{L_3} + \dfrac{L_3}{3}\right) & \dfrac{-1}{L_3} + \dfrac{L_3}{6} \\[2ex] 0 & 0 & \dfrac{-1}{L_3} + \dfrac{L_3}{6} & \dfrac{1}{L_3} + \dfrac{L_3}{3} \end{bmatrix}$$

Considering $Kh = F$, we have

$$K \begin{bmatrix} h_1 \\ h_2 \\ h_3 \\ h_4 \end{bmatrix} = \begin{bmatrix} -\dfrac{d\hat{h}}{dx}\Big|_{x=0} \\[2ex] 0 \\[1ex] 0 \\[1ex] -\dfrac{d\hat{h}}{dx}\Big|_{x=5} \end{bmatrix}$$

h_1 and h_4 are known, therefore from the above equation set, we have

$$\left(\frac{1}{L_1} + \frac{L_1}{3} + \frac{1}{L_2} + \frac{L_2}{3}\right)h_2 + \left(-\frac{1}{L_2} + \frac{L_2}{6}\right)h_3 = 0$$

$$\left(-\frac{1}{L_2} + \frac{L_2}{6}\right)h_2 + \left(\frac{1}{L_2} + \frac{L_2}{3} + \frac{1}{L_3} + \frac{L_3}{3}\right)h_3 = -8\left(-\frac{1}{L_3} + \frac{L_3}{6}\right)$$

if $L_1 = 2\,\text{m}$, $L_2 = 1\,\text{m}$, $L_3 = 2\,\text{m}$, the results are as follows:

$$h_2 = 0.2\,\text{m} \quad \text{and} \quad h_3 = 0.6\,\text{m}$$

Example 6.7

Two parallel rivers A and B are separated by a confined aquifer as shown in Figure 6.19. Consider the aquifer is multilayered and the lengths of layers are $L_1 = 2\,\text{m}$, $L_2 = 3\,\text{m}$, $L_3 = 4\,\text{m}$ with $K_1 = 2$, $K_2 = 4$, and $K_3 = 1$. Determine the spatial variation of piezometric head using FEM.

Solution

In order to solve the problem, the problem domain is divided into a mesh with five nodes and four different elements. The governing differential equation is the 1D form of the steady-state, saturated groundwater flow equation:

$$\frac{\partial}{\partial x}\left(K_x \frac{\partial h}{\partial x}\right) = 0$$

where
K_x is the saturated hydraulic conductivity in the x direction
h is hydraulic head

Using the method of weighted residuals, an approximate solution of h is defined. If this approximate solution is substituted into the above equation, the differential equation is not satisfied exactly:

$$\frac{\partial}{\partial x}\left(K_x \frac{\partial h}{\partial x}\right) = R(x) \neq 0$$

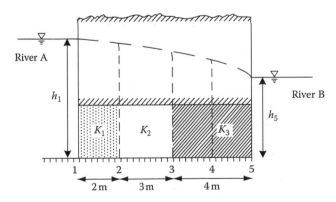

FIGURE 6.19 Two parallel rivers and a confined aquifer of Example 6.7.

where R is the error of the estimation of the residual in each node that varies from point-to-point within the problem domain. For the residual at any node i, R represents the error between the exact value of hydraulic head and the approximate solution of h at that node:

$$\{R\} = \begin{Bmatrix} R(x = 0) \\ R(x = 2) \\ R(x = 5) \\ R(x = 7) \\ R(x = 9) \end{Bmatrix} = \begin{Bmatrix} R_1 \\ R_2 \\ R_3 \\ R_4 \\ R_5 \end{Bmatrix}$$

where R_i is the value of the residual at node i ($i = 1, ..., 5$). The approximate solution at a node is estimated utilizing the values of hydraulic head at the nodes in all elements and assembling the contributed element to node i. As shown in Figure 6.20, elements 1 and 2 are joined to node 2. Thus, the values of hydraulic head for the other nodes in these elements contribute to the residual at node 2. This can be written as

$$R_2 = R_2^{(1)} + R_2^{(2)}$$

where the first term in the left-hand side of the equation is the contribution of element 1 to the residual at node 2 and the second term is the contribution of element 2 to the residual at node 2. Thus, the residual at each node is (Istok, 1989)

$$R_i = \sum_{e=1} R_i^{(e)}$$

where e is the number of elements that are joined to node i. The contribution of element e to the residual at node i can be obtained from the integral formulation for that node.

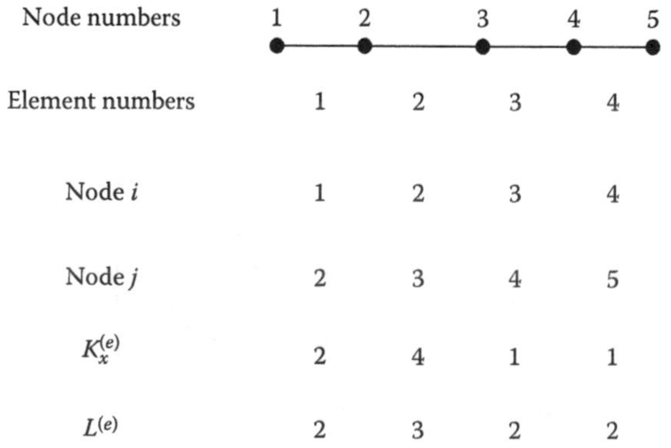

Node numbers	1	2	3	4	5
Element numbers	1	2	3	4	
Node i	1	2	3	4	
Node j	2	3	4	5	
$K_x^{(e)}$	2	4	1	1	
$L^{(e)}$	2	3	2	2	

FIGURE 6.20   Assigning node numbers and element numbers j.

For the 1D elements in this example, we have

$$R_i^{(e)} = -\int_{x_i^{(e)}}^{x_j^{(e)}} N_i^{(e)} \left[K_x^{(e)} \frac{\partial^2 \hat{h}^{(e)}}{\partial x^2} \right] dx$$

A similar equation can be written for the contribution of element e to the residual at any other node j joined to the element:

$$R_j^{(e)} = -\int_{x_i^{(e)}}^{x_j^{(e)}} N_j^{(e)} \left[K_x^{(e)} \frac{\partial^2 \hat{h}^{(e)}}{\partial x^2} \right] dx$$

where

$x_i^{(e)}$ and $x_j^{(e)}$ are the coordinates of the nodes at each end of the element

$N_i^{(e)}$ is the weighting function for node i in element e (which is identical to the interpolation function for node i in element e because Galerkin's method is used)

$K_x^{(e)}$ is the saturated hydraulic conductivity for element e

$K_x^{(e)}$ is assumed to be constant within an element but can vary from one element to another. The equation was multiplied by –1 for later convenience.

Using Equations 6.36, 6.40, and 6.41, the approximate solution $\hat{h}$ in each element can be written as follows based on the interpolation functions and the unknown element:

$$\hat{h}^{(e)}(x) = N_i^{(e)} h_i + N_j^{(e)} h_j = \left(\frac{x_j^{(e)} - x}{L^{(e)}} \right) h_i + \left(\frac{x - x_i^{(e)}}{L^{(e)}} \right) h_j$$

Because the approximate solution is a linear function of x, $\partial^2 h / \partial x^2$ is not defined. The approximate solution does have a continuous first derivative, however, we can evaluate $R_i^{(e)}$ if $\partial^2 h / \partial x^2$ is rewritten in the other form:

$$\int_{x_i^{(e)}}^{x_j^{(e)}} N_i^{(e)} \left[K_x^{(e)} \frac{\partial^2 \hat{h}^{(e)}}{\partial x^2} \right] dx = -\int_{x_i^{(e)}}^{x_j^{(e)}} K_x^{(e)} \frac{\partial N_i^{(e)}}{\partial x} \frac{\partial \hat{h}^{(e)}}{\partial x} dx + \left[N_i^{(e)} K_x^{(e)} \frac{\partial \hat{h}^{(e)}}{\partial x} \right]_{x_i^{(e)}}^{x_j^{(e)}}$$

$$F_i^{(e)} = \left[N_i^{(e)} K_x^{(e)} \frac{\partial \hat{h}^{(e)}}{\partial x} \right]_{x_i^{(e)}}^{x_j^{(e)}}$$

$F_i^{(e)}$ shows the inflow/outflow from the element. It is positive for inflow to the element. If the flow is not specified or at impermeable aquifer boundaries, $F_i^{(e)}$ will be zero. For elements on the interior of the mesh, the term $F_i^{(e)}$ for adjacent elements with opposite signs will be zero and their contribution in $F_i^{(e)}$ is not considered:

$$R_i^{(e)} = -\int_{x_i^{(e)}}^{x_j^{(e)}} N_i^{(e)} \left[K_x^{(e)} \frac{\partial^2 \hat{h}^{(e)}}{\partial x^2} \right] dx = \int_{x_i^{(e)}}^{x_j^{(e)}} K_x^{(e)} \frac{\partial N_i^{(e)}}{\partial x} \frac{\partial \hat{h}^{(e)}}{\partial x} dx - \left[N_i^{(e)} K_x^{(e)} \frac{\partial \hat{h}^{(e)}}{\partial x} \right]_{x_i^{(e)}}^{x_j^{(e)}}$$

$$= \int_{x_i^{(e)}}^{x_j^{(e)}} K_x^{(e)} \frac{\partial N_i^{(e)}}{\partial x} \frac{\partial \hat{h}^{(e)}}{\partial x} dx$$

From the definition for h in Equation 6.36 and definitions of the interpolation functions, we have

$$\frac{\partial N_i^{(e)}}{\partial x} = \frac{\partial}{\partial x}\left(\frac{x_j^{(e)} - x}{L^{(e)}}\right) = -\frac{1}{L^{(e)}}$$

$$\frac{\partial N_j^{(e)}}{\partial x} = \frac{\partial}{\partial x}\left(\frac{x - x_i^{(e)}}{L^{(e)}}\right) = \frac{1}{L^{(e)}}$$

So

$$\frac{\partial \hat{h}^{(e)}}{\partial x} = -\frac{1}{L^{(e)}}h_i + \frac{1}{L^{(e)}}h_j = \frac{1}{L^{(e)}}(-h_i + h_j)$$

$$R_i^{(e)} = \int_{x_i^{(e)}}^{x_j^{(e)}} K_x^{(e)}\left(-\frac{1}{L^{(e)}}\right)\left(\frac{1}{L^{(e)}}\right)(-h_i + h_j)dx = -\frac{K_x^{(e)}}{L^{(e)2}}(x_j^{(e)} - x_i^{(e)})(-h_i + h_j)$$

But

$$x_j^{(e)} - x_i^{(e)} = L^{(e)}$$

$$R_i^{(e)} = \frac{K_x^{(e)}}{L^{(e)}}(h_i - h_j)$$

Similarly, for the contribution of element e to the residual at node j, we have

$$R_j^{(e)} = \frac{K_x^{(e)}}{L^{(e)}}(-h_i + h_j)$$

Last two equations can be combined and written in matrix form as follows:

$$\left\{\begin{matrix} R_i^{(e)} \\ R_j^{(e)} \end{matrix}\right\} = \frac{K_x^{(e)}}{L^{(e)}}\begin{bmatrix} 1 & -1 \\ -1 & 1 \end{bmatrix}\left\{\begin{matrix} h_i \\ h_j \end{matrix}\right\} = \left[K_x^{(e)}\right]\left\{\begin{matrix} h_i \\ h_j \end{matrix}\right\}$$

where

$$\left[K_x^{(e)}\right] = \frac{K_x^{(e)}}{L^{(e)}}\begin{bmatrix} 1 & -1 \\ -1 & 1 \end{bmatrix}$$

$K_x^{(e)}$ is the element conductance matrix. This matrix is determined based on the hydraulic conductivity of the aquifer material within the element $K_x^{(e)}$, the size, $L^{(e)}$, and shape (through the functions for the element) of the element. $[K_x^{(e)}]$ is a square, symmetric matrix with size $n \times n$ where n is the number of nodes in the element. Thus, for a 1D element with two nodes, the size of $[K_x^{(e)}]$ is 2×2, and for a 2D element with three nodes, the size of $[K_x^{(e)}]$ is 3×3.

The characteristics of each mesh element are illustrated in Figure 6.20. Node numbers i and j for each element are shown in this figure. Therefore, the element conductance matrices can be derived as follows:

$$\left[K_x^{(1)}\right] = \frac{2}{2}\begin{bmatrix} 1 & -1 \\ -1 & 1 \end{bmatrix} = \begin{bmatrix} 1 & -1 \\ -1 & 1 \end{bmatrix}$$

$$\left[K_x^{(2)}\right] = \frac{4}{3}\begin{bmatrix} 1 & -1 \\ -1 & 1 \end{bmatrix} = \begin{bmatrix} 4/3 & -4/3 \\ -4/3 & 4/3 \end{bmatrix}$$

$$\left[K_x^{(3)}\right] = \left[K_x^{(4)}\right] = \frac{1}{2}\begin{bmatrix} 1 & -1 \\ -1 & 1 \end{bmatrix} = \begin{bmatrix} 1/2 & -1/2 \\ -1/2 & 1/2 \end{bmatrix}$$

Consider the aquifer is multilayered $L_1 = 2\,\text{m}$, $l_2 = 3\,\text{m}$, $l_3 = 4\,\text{m}$ with $K_1 = 2$, $K_2 = 4$, and $K_3 = 1$. Determine the spatial variation of piezometric head using FEM.

By assembling the element conductance matrices for all the elements in the mesh, the global conductance matrix can be obtained as follows:

$$[K]_{global} = \sum_{e=1}^{m} \left[K^{(e)}\right]_{expanded}$$

where m is the number of elements in the mesh. The element conductance matrices give a system of linear equations of the following form:

$$\{R\} = [K]\{h\} - \{F\} = \{0\}$$

where
  $\{R\}$ is the global residual matrix
  $[K]$ is the global conductance matrix
  $\{h\}$ is the vector of unknown hydraulic heads
  $\{F\}$ is a vector containing the specified fluxes at Neumann nodes

$\{F\} = \{0\}$, because no fluxes were specified

$$\{R\}_{5\times1} = \begin{Bmatrix} R_1 \\ R_2 \\ R_3 \\ R_4 \\ R_5 \end{Bmatrix} \quad \{h\}_{5\times1} = \begin{Bmatrix} h_1 \\ h_2 \\ h_3 \\ h_4 \\ h_5 \end{Bmatrix} \quad \{F\}_{5\times1} = \begin{Bmatrix} 0 \\ 0 \\ 0 \\ 0 \\ 0 \end{Bmatrix}$$

The expanded form of the element conductance matrices is as follows:

$$\left[K_x^{(1)}\right] = \frac{2}{2}\begin{bmatrix} 1 & -1 \\ -1 & 1 \end{bmatrix} = \begin{bmatrix} 1 & -1 \\ -1 & 1 \end{bmatrix}$$

$$\left[K_x^{(2)}\right] = \frac{4}{3}\begin{bmatrix} 1 & -1 \\ -1 & 1 \end{bmatrix} = \begin{bmatrix} 4/3 & -4/3 \\ -4/3 & 4/3 \end{bmatrix}$$

$$\left[K_x^{(3)}\right] = \left[K_x^{(4)}\right] = \frac{1}{2}\begin{bmatrix} 1 & -1 \\ -1 & 1 \end{bmatrix} = \begin{bmatrix} 1/2 & -1/2 \\ -1/2 & 1/2 \end{bmatrix}$$

$$\left[K_x^{(1)}\right] = \begin{bmatrix} 1 & -1 & 0 & 0 & 0 \\ -1 & 1 & 0 & 0 & 0 \\ 0 & 0 & 0 & 0 & 0 \\ 0 & 0 & 0 & 0 & 0 \\ 0 & 0 & 0 & 0 & 0 \end{bmatrix} \quad \left[K_x^{(2)}\right] = \begin{bmatrix} 0 & 0 & 0 & 0 & 0 \\ 0 & 4/3 & -4/3 & 0 & 0 \\ 0 & -4/3 & 4/3 & 0 & 0 \\ 0 & 0 & 0 & 0 & 0 \\ 0 & 0 & 0 & 0 & 0 \end{bmatrix}$$

$$\left[K_x^{(3)}\right] = \begin{bmatrix} 0 & 0 & 0 & 0 & 0 \\ 0 & 0 & 0 & 0 & 0 \\ 0 & 0 & 1/2 & -1/2 & 0 \\ 0 & 0 & -1/2 & 1/2 & 0 \\ 0 & 0 & 0 & 0 & 0 \end{bmatrix} \quad \left[K_x^{(4)}\right] = \begin{bmatrix} 0 & 0 & 0 & 0 & 0 \\ 0 & 0 & 0 & 0 & 0 \\ 0 & 0 & 0 & 0 & 0 \\ 0 & 0 & 0 & 1/2 & -1/2 \\ 0 & 0 & 0 & -1/2 & 1/2 \end{bmatrix}$$

and the global conductance matrix is

$$\left[K\right]_{global} = \left[K^{(1)}\right] + \left[K^{(2)}\right] + \left[K^{(3)}\right] + \left[K^{(4)}\right]$$

$$\left[K_x\right]_{global} = \begin{bmatrix} 1 & -1 & 0 & 0 & 0 \\ -1 & 7/3 & -4/3 & 0 & 0 \\ 0 & -4/3 & 11/6 & -1/2 & 0 \\ 0 & 0 & -1/2 & 1 & -1/2 \\ 0 & 0 & 0 & -1/2 & 1/2 \end{bmatrix}$$

The resulting system of equations, when this global conductance matrix is substituted into the system of linear equations, is

$$\begin{bmatrix} 1 & -1 & 0 & 0 & 0 \\ -1 & 7/3 & -4/3 & 0 & 0 \\ 0 & -4/3 & 11/6 & -1/2 & 0 \\ 0 & 0 & -1/2 & 1 & -1/2 \\ 0 & 0 & 0 & -1/2 & 1/2 \end{bmatrix}\begin{Bmatrix} h_1 \\ h_2 \\ h_3 \\ h_4 \\ h_5 \end{Bmatrix} = \begin{Bmatrix} 0 \\ 0 \\ 0 \\ 0 \\ 0 \end{Bmatrix}$$

Considering $h_1 = 12$ and $h_5 = 6$ (nodes 1 and 5 are called Dirichlet nodes) from the boundary conditions, the system of equations can be modified as

$$
\begin{bmatrix}
1 & -1 & 0 & 0 & 0 \\
-1 & 7/3 & -4/3 & 0 & 0 \\
0 & -4/3 & 11/6 & -1/2 & 0 \\
0 & 0 & -1/2 & 1 & -1/2 \\
0 & 0 & 0 & -1/2 & 1/2
\end{bmatrix}
\begin{Bmatrix}
h_1 \\ h_2 \\ h_3 \\ h_4 \\ h_5
\end{Bmatrix}
=
\begin{Bmatrix}
0 \\ 0 \\ 0 \\ 0 \\ 0
\end{Bmatrix}
$$

$$
\begin{bmatrix}
7/3 & -4/3 & 0 \\
-4/3 & 11/6 & -1/2 \\
0 & -1/2 & 1
\end{bmatrix}
\begin{Bmatrix}
h_2 \\ h_3 \\ h_4
\end{Bmatrix}
=
\begin{Bmatrix}
12 \\ 0 \\ 3
\end{Bmatrix}
$$

The results are then given as $h_2 = 10.96\,\text{m}$, $h_3 = 10.17\,\text{m}$, and $h_4 = 8.09\,\text{m}$.

6.5.3 TRANSIENT SATURATED FLOW EQUATION

The 3D form of the equation for transient groundwater flow through saturated porous media is as follows:

$$
\frac{\partial}{\partial x}\left(K_X \frac{\partial h}{\partial x}\right) + \frac{\partial}{\partial y}\left(K_Y \frac{\partial h}{\partial y}\right) + \frac{\partial}{\partial z}\left(K_Z \frac{\partial h}{\partial z}\right) = S_s \frac{\partial h}{\partial t} \tag{6.46}
$$

where
S_s is the specific storage of the porous media
t is the time

In transient groundwater flow equations, the term $S_s\, \partial h/\partial t$ implies the steady-state condition. The residual at node i is calculated by replacing the approximate solution for hydraulic head, $\hat{h}$ in Equation 6.46. It can be formulated as

$$
R_i^{(e)} = -\iiint\limits_{V^{(e)}} W_i^{(e)}\left[\frac{\partial}{\partial x}\left(K_x^{(e)} \frac{\partial \hat{h}^{(e)}}{\partial x}\right) + \frac{\partial}{\partial y}\left(K_y^{(e)} \frac{\partial \hat{h}^{(e)}}{\partial y}\right) + \frac{\partial}{\partial x}\left(K_z^{(e)} \frac{\partial \hat{h}^{(e)}}{\partial z}\right) - S_s^{(e)} \frac{\partial \hat{h}^{(e)}}{\partial t}\right] dxdydz
$$

$$
\tag{6.47}
$$

where
$W_i^{(e)}$ is the weighting function for node i of element e, which is similar to $N_i^{(e)}$ in Galerkin's method
$S_s^{(e)}$ is the specific storage for element e

Assuming that the values of $K_x^{(e)}$, $K_y^{(e)}$, $K_z^{(e)}$, and $S_s^{(e)}$ are constant within an element, Equation 6.47 can be rearranged as

$$
R_i^{(e)} = -\iiint_{V^{(e)}} N_i^{(e)} \left[K_i^{(e)} \frac{\partial^2 h^{(e)}}{\partial x^2} + K_y^{(e)} \frac{\partial^2 h^{(e)}}{\partial y^2} + K_z^{(e)} \frac{\partial^2 h^{(e)}}{\partial z^2} - S_s^{(e)} \frac{\partial h^{(e)}}{\partial t} \right] dxdydz
$$

$$
= \iiint_{V^{(e)}} N_i^{(e)} \left[K_x^{(e)} \frac{\partial^2 h^{(e)}}{\partial x^2} + K_y^{(e)} \frac{\partial^2 \hat{h}^{(e)}}{\partial y^2} + K_z^{(e)} \frac{\partial \hat{h}^{(e)}}{\partial z^2} \right] dxdydz
$$

$$
+ \iiint_{V^{(e)}} N_i^{(e)} S_s^{(e)} \frac{\partial \hat{h}^{(e)}}{\partial t} dxdydz \tag{6.48}
$$

Based on the approximated solution as a function of interpolation function and unknown hydraulic head at the element's nodes, the above integral can be summarized in the matrix form as

$$
\left\{ \begin{array}{c} R_1^{(e)} \\ \vdots \\ R_n^{(e)} \end{array} \right\}_K = \left[K^{(e)} \right] \left\{ \begin{array}{c} h_1 \\ \vdots \\ h_n \end{array} \right\} + \left[C^{(e)} \right] \left\{ \begin{array}{c} \dfrac{h_1}{\partial t} \\ \vdots \\ \dfrac{h_n}{\partial t} \end{array} \right\} \tag{6.49}
$$

where $[K^{(e)}]$ is the element conductance matrix and $[C^{(e)}]$ is the element capacitance matrix (Istok, 1989):

$$
\left[K^{(e)} \right] = -\iiint_{V^{(e)}} \begin{bmatrix} \dfrac{\partial N_1^{(e)}}{\partial x} & \dfrac{\partial N_1^{(e)}}{\partial y} & \dfrac{\partial N_1^{(e)}}{\partial z} \\ \vdots & \vdots & \vdots \\ \dfrac{\partial N_n^{(e)}}{\partial x} & \dfrac{\partial N_n^{(e)}}{\partial x} & \dfrac{\partial N_n^{(e)}}{\partial x} \end{bmatrix} \begin{bmatrix} K_x^{(e)} & 0 & 0 \\ 0 & K_y^{(e)} & 0 \\ 0 & 0 & K_z^{(e)} \end{bmatrix} \begin{bmatrix} \dfrac{\partial N_1^{(e)}}{\partial x} & \cdots & \dfrac{\partial N_n^{(e)}}{\partial x} \\ \dfrac{\partial N_1^{(e)}}{\partial y} & \cdots & \dfrac{\partial N_n^{(e)}}{\partial y} \\ \dfrac{\partial N_1^{(e)}}{\partial z} & \cdots & \dfrac{\partial N_n^{(e)}}{\partial z} \end{bmatrix} dxdydz
$$

$$
\tag{6.50}
$$

$$
\left[C^{(e)} \right] = -\iiint_{V^{(e)}} \begin{bmatrix} N_1^{(e)} \\ \vdots \\ N_n^{(e)} \end{bmatrix} \left[S_s^{(e)} \right] \left[N_1^{(e)} \cdots N_n^{(e)} \right] dxdydz \tag{6.51}
$$

or

$$
\left[C^{(e)} \right] = S_s^{(e)} \begin{bmatrix} N_1^{(e)} N_1^{(e)} & \cdots & N_1^{(e)} N_n^{(e)} \\ \vdots & \cdots & \vdots \\ N_n^{(e)} N_1^{(e)} & \cdots & N_n^{(e)} N_n^{(e)} \end{bmatrix} -\iiint_{V^{(e)}} dxdydz \tag{6.52}
$$

If the lumped element formulation is used, the element capacitance matrix is

$$\left[C^{(e)} \right] = S_s^{(e)} \frac{V^{(e)}}{n} \begin{bmatrix} 1 & & 0 \\ & \ddots & \\ 0 & & 1 \end{bmatrix} \tag{6.53}$$

Thus, the time derivative of the approximate solution should be defined over the element domain for estimating the errors. For this purpose, the interpolation functions and the values of the time derivative at the nodes are required.

The global conductance matrix and global capacitance matrix are obtained by assembling the element matrices for all the elements in the finite mesh. These are square and symmetric matrices with size $p \times p$, where p is the number of nodes in the mesh. Thus, Equation 6.49 can be written as

$$[C] \left\{ \begin{array}{c} \frac{\partial h_1}{\partial t} \\ \vdots \\ \frac{\partial h_p}{\partial t} \end{array} \right\} + [K] \left\{ \begin{array}{c} h_1 \\ \vdots \\ h_p \end{array} \right\} = [F] \tag{6.54}$$

or

$$\underset{global}{[C]} \{\overset{\circ}{h}\} + \underset{global}{[K]} \{h\} = \underset{global}{\{F\}} \tag{6.55}$$

where

$$\{\overset{\circ}{h}\} = \left\{ \begin{array}{c} \frac{\partial h_1}{\partial t} \\ \vdots \\ \frac{\partial h_p}{\partial t} \end{array} \right\} \qquad \{h\} = \left\{ \begin{array}{c} h_1 \\ \vdots \\ h_p \end{array} \right\} \tag{6.56}$$

The solution of the system of algebraic equation determines the head and the head time derivative values at each node in the mesh. Among various methods for modeling the time derivative value, the FDM is commonly used. In this method, the head time derivative value is computed based on the difference between the head values of the two end points of the interval. FDA to the time derivative for hydraulic head is shown in Figure 6.21. The variation of hydraulic head is considered linear between given time intervals.

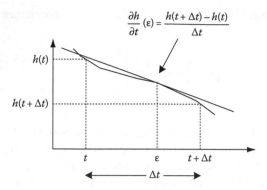

FIGURE 6.21 The time derivative approximation using FDM.

For the point with the position of ε in the interval t to $t + \Delta t$, the unknown hydraulic head can be calculated based on the FDM. Thus, the time derivative for hydraulic head is expressed as follows:

$$\frac{\partial h}{\partial t}(\varepsilon) = \frac{h(t + \Delta t) - h(t)}{\Delta t} \tag{6.57}$$

or

$$h(\varepsilon) = h(t) + (\varepsilon - t)\frac{\partial h}{\partial t}(\varepsilon) \tag{6.58}$$

A variable α is defined as

$$\alpha = \frac{\varepsilon - t}{\Delta t} \tag{6.59}$$

Thus, Equation 6.58 can be written as

$$h(\varepsilon) = (1 - \alpha)\, h(t) + \alpha h(t + \Delta t) \tag{6.60}$$

Therefore, based on the Equation 6.60, unknown hydraulic heads $\{h\}$ vector and $\{F\}$ vector can be written as

$$\{h\} = (1 - \alpha)\{h\}_t + \alpha\{h\}_{t+\Delta t} \tag{6.61}$$

$$\{F\} = (1 - \omega)\alpha\{F\}_t + \alpha\{F\}_{t+\Delta t} \tag{6.62}$$

By substituting Equations 6.61, 6.62, and 6.57 into Equation 6.55, the finite-difference formulation for the transient, saturated flow equation is derived as follows:

$$[C]\{\overset{\circ}{h}\} + [K]\{h\} = \{F\} \quad (6.63)$$
$$\underset{global}{\qquad} \underset{global}{\qquad} \underset{global}{\qquad}$$

$$([C] + \alpha \Delta t[K])\{h\}_{t+\Delta t} = ([C] - (1-\alpha)\Delta t[K])\{h\}_t + \Delta t((1-\alpha)\{F\}_t + \alpha\{F\}_{t+\Delta t}) \quad (6.64)$$

For solving the developed set of equations, the initial value of $\{h\}_{t_0}$ is required. This value is specified at time $t = t_0 = 0$ as the initial condition. Then, the system of linear equations is solved to obtain values of $\{h\}$ at the end of the first time step, $\{h\}_{t_0+\Delta t}$. This process is repeated for each time step (Istok, 1989). Similar to FDM, there are several ways to formulate the time derivative equations by selecting the value of α. If α is considered equal to zero, the method is called forward difference method and is expressed as

$$[C]\{h\}_{t+\Delta t} = ([C] - \Delta t[K])\{h\}_t + \Delta t\{F\}_t \quad (6.65)$$

If α is considered equal to 0.5, the method is called Crank–Nicolson difference method and is shown as

$$\left([C]\frac{\Delta t}{2}[K]\right)\{h\}_{t+\Delta t} = \left([C] - \frac{\Delta t}{2}[K]\right)\{h\}_t + \frac{\Delta t}{2}(\{F\}_t + \{F\}_{t+\Delta t}) \quad (6.66)$$

when α is equal to one, it is called the backward difference method, given as

$$([C] + \Delta t[K])\{h\}_{t+\Delta t} = [C]\{h\}_t + \Delta t\{F\}_{t+\Delta t} \quad (6.67)$$

Example 6.8

Consider the confined aquifer shown in Example 6.7. Initially, the aquifer is in steady state, saturated condition with a distribution of hydraulic head computed from the previous example (Figure 6.22). At time $t = 0$, the water level of river A increases (at the upper boundary, node 1) from 12 to 18 m. Assume $S_s^{(1)} = 0.02$, $S_s^{(2)} = 0.04$, and $S_s^{(3)} = S_s^{(4)} = 0.02$ for the mesh elements. Find the value of hydraulic head at each node at time $t = 1$ s.

Solution

The governing differential equation is a 1D form of groundwater flow:

$$\frac{\partial}{\partial z}\left(K_x \frac{\partial h}{\partial x}\right) = S_s \frac{\partial h}{\partial t}$$

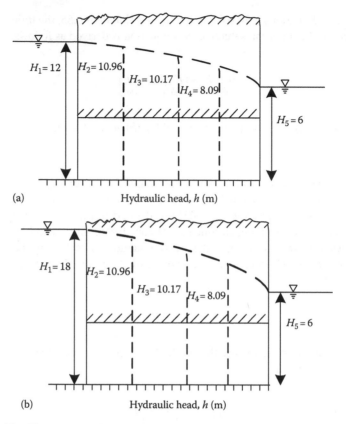

FIGURE 6.22   Flow problem in Example 6.8: (a) steady-state solution; (b) initial conditions.

where K_x is the saturated hydraulic conductivity in the direction of flow (the x-axis is directed vertically downward in this case). For 1D element with two nodes, the element capacitance matrices are given as

$$\left[C^{(1)}\right] = \frac{S_s^{(1)} L^{(1)}}{2}\begin{bmatrix} 1 & 0 \\ 0 & 1 \end{bmatrix} = \frac{(0.02)(2)}{2}\begin{bmatrix} 1 & 0 \\ 0 & 1 \end{bmatrix} = \begin{bmatrix} 0.02 & 0 \\ 0 & 0.02 \end{bmatrix}$$

$$\left[C^{(2)}\right] = \frac{S_s^{(2)} L^{(2)}}{2}\begin{bmatrix} 1 & 0 \\ 0 & 1 \end{bmatrix} = \frac{(0.04)(3)}{2}\begin{bmatrix} 1 & 0 \\ 0 & 1 \end{bmatrix} = \begin{bmatrix} 0.06 & 0 \\ 0 & 0.06 \end{bmatrix}$$

$$\left[C^{(3)}\right] = \left[C^{(4)}\right] = \frac{S_s^{(3)} L^{(3)}}{2}\begin{bmatrix} 1 & 0 \\ 0 & 1 \end{bmatrix} = \frac{(0.02)(2)}{2}\begin{bmatrix} 1 & 0 \\ 0 & 1 \end{bmatrix} = \begin{bmatrix} 0.02 & 0 \\ 0 & 0.02 \end{bmatrix}$$

The global capacitance matrix is obtained by adding the expanded form of the element capacitance matrices:

$$[C] = \begin{bmatrix} 0.02 & 0 & 0 & 0 & 0 \\ 0 & 0.02+0.06 & 0 & 0 & 0 \\ 0 & 0 & 0.06+0.02 & 0 & 0 \\ 0 & 0 & 0 & 0.02+0.02 & 0 \\ 0 & 0 & 0 & 0 & 0.02 \end{bmatrix}$$

$$= \begin{bmatrix} 0.02 & 0 & 0 & 0 & 0 \\ 0 & 0.08 & 0 & 0 & 0 \\ 0 & 0 & 0.08 & 0 & 0 \\ 0 & 0 & 0 & 0.04 & 0 \\ 0 & 0 & 0 & 0 & 0.02 \end{bmatrix}$$

From the previous example, the global conductance matrix is given as

$$[K_x]_{global} = \begin{bmatrix} 1 & -1 & 0 & 0 & 0 \\ -1 & 7/3 & -4/3 & 0 & 0 \\ 0 & -4/3 & 11/6 & -1/2 & 0 \\ 0 & 0 & -1/2 & 1 & -1/2 \\ 0 & 0 & 0 & -1/2 & 1/2 \end{bmatrix}$$

The initial values of hydraulic head at the nodes are

$$\{h\}_{t=0} = \begin{Bmatrix} h_1 \\ h_2 \\ h_3 \\ h_4 \\ h_5 \end{Bmatrix}_{t=0} = \begin{Bmatrix} 18 \\ 10.96 \\ 10.17 \\ 8.09 \\ 6 \end{Bmatrix}$$

The backward difference formulation is used with a time step $\Delta t = 1$ s. By setting $\{F\} = 0$ (no specified flow rates), the system of equations at the end of the first time step becomes as follows:

$$([C] + \Delta t[K])\{h\}_{t=1} = [C]\{h\}_{t=0} + \Delta t\{F\}_{T=1}^{0}$$

Substituting equations in the above equation gives

$$[K_x]_{global} = \begin{bmatrix} 1 & -1 & 0 & 0 & 0 \\ -1 & 7/3 & -4/3 & 0 & 0 \\ 0 & -4/3 & 11/6 & -1/2 & 0 \\ 0 & 0 & -1/2 & 1 & -1/2 \\ 0 & 0 & 0 & -1/2 & 1/2 \end{bmatrix}$$

$$\left(\begin{bmatrix} 0.02 & 0 & 0 & 0 & 0 \\ 0 & 0.08 & 0 & 0 & 0 \\ 0 & 0 & 0.08 & 0 & 0 \\ 0 & 0 & 0 & 0.04 & 0 \\ 0 & 0 & 0 & 0 & 0.02 \end{bmatrix} + (1) \begin{bmatrix} 1 & -1 & 0 & 0 & 0 \\ -1 & 7/3 & -4/3 & 0 & 0 \\ 0 & -4/3 & 11/6 & -1/2 & 0 \\ 0 & 0 & -1/2 & 1 & -1/2 \\ 0 & 0 & 0 & -1/2 & 1/2 \end{bmatrix} \right) \begin{Bmatrix} h_1 \\ h_2 \\ h_3 \\ h_4 \\ h_5 \end{Bmatrix}_{t=1}$$

$$= \begin{bmatrix} 0.02 & 0 & 0 & 0 & 0 \\ 0 & 0.08 & 0 & 0 & 0 \\ 0 & 0 & 0.08 & 0 & 0 \\ 0 & 0 & 0 & 0.04 & 0 \\ 0 & 0 & 0 & 0 & 0.02 \end{bmatrix} \begin{Bmatrix} 18 \\ 10.96 \\ 10.17 \\ 8.09 \\ 6 \end{Bmatrix}$$

which can be simplified to

$$\begin{bmatrix} 1.02 & -1 & 0 & 0 & 0 \\ -1 & 2.41 & -1.33 & 0 & 0 \\ 0 & -1.33 & 1.91 & -0.5 & 0 \\ 0 & 0 & -0.5 & 1.04 & -0.5 \\ 0 & 0 & 0 & -0.5 & 0.52 \end{bmatrix} \begin{Bmatrix} h_1 \\ h_2 \\ h_3 \\ h_4 \\ h_5 \end{Bmatrix}_{t=1} = \begin{Bmatrix} 0.36 \\ 0.88 \\ 0.81 \\ 0.32 \\ 0.12 \end{Bmatrix}$$

The hydraulic heads at the upper and lower boundary conditions are $h_1 = 18$ and $h_5 = 6$, respectively, for all values of t. Modifying the last equation for these known values, we have

$$\begin{bmatrix} 2.41 & -1.33 & 0 \\ -1.33 & 1.91 & -0.5 \\ 0 & -0.5 & 1.04 \end{bmatrix} \begin{Bmatrix} h_2 \\ h_3 \\ h_4 \end{Bmatrix}_{t=1} = \begin{Bmatrix} 18.88 \\ 0.81 \\ 3.32 \end{Bmatrix}$$

This equation can be solved to obtain the values of hydraulic head at the end of the time step as follows:

$$\begin{Bmatrix} h_1 \\ h_2 \\ h_3 \\ h_4 \\ h_5 \end{Bmatrix}_{t=1} = \begin{Bmatrix} 18.00 \\ 15.40 \\ 13.71 \\ 9.78 \\ 6.00 \end{Bmatrix}$$

6.5.4 Solute Transport Equation

The 3D form of the solute transport equation for uniform groundwater flow in the x-direction is

$$\frac{\partial(\theta C)}{\partial t} = D_x \frac{\partial^2}{\partial x^2}(\theta C) + D_y \frac{\partial^2}{\partial y^2}(\theta C) + D_z \frac{\partial^2}{\partial z^2}(\theta C) - \frac{\partial}{\partial x}(v_x C)$$

$$-\frac{\partial}{\partial t}(\rho_b K_d C) - \lambda(\theta C + \rho_b K_d C) \tag{6.68}$$

where
C is the solute concentration
θ is the volumetric water content of the porous media
D_x, D_y, D_z are the dispersion coefficients of the porous media in the x, y, and z directions
v_x is the apparent groundwater velocity in the x-direction
ρ_b is the density of the porous media
K_d is the distribution coefficient
λ is the solute decay constant.

The unknown concentration at each node is determined by solving the solute transport equation. Similar to applying FEM for solving groundwater flow, an approximate solution for unknown concentration, C, within element e is considered as $\hat{C}$:

$$\hat{C}^{(e)}(x, y, z) = \sum_{i=1}^{n} N_i^{(e)} C_i \tag{6.69}$$

where
$N_i^{(e)}$ is the interpolation function for node i within element e
C_i is the unknown solute concentration in this node

By substituting the approximate solution into Equation 6.68, the error or residual due to difference of the exact solution and the approximate solution at each node occurs. The residual can be expressed as

$$R_i^{(e)} = -\iiint_{V^{(e)}} W_i^{(e)}(x, y, z) \left[D_x \frac{\partial^2}{\partial x^2}(\theta \hat{C}_i^{(e)}) + D_y \frac{\partial^2}{\partial y^2}(\theta \hat{C}_i^{(e)}) + D_z \frac{\partial^2}{\partial z^2}(\theta \hat{C}_i^{(e)}) \right.$$

$$-\frac{\partial}{\partial x}(v_x \hat{C}_i^{(e)}) - \frac{\partial}{\partial t}(\rho_b K_d \hat{C}_i^{(e)}) - \lambda(\theta \hat{C}_i^{(e)}$$

$$\left. + \rho_b K_d \hat{C}_i^{(e)}) - \frac{\partial}{\partial t}(\theta \hat{C}_i^{(e)}) \right] \tag{6.70}$$

where $W_i^{(e)}$ is the element's weighting function for node i. The weighting function for each node in the element is considered to be equal to the element's interpolation function for that node in Galerkin's method ($W_i^{(e)} = N_i^{(e)}$):

$$R_i^{(e)} = -\iiint\limits_{V^{(e)}} N_i^{(e)}(x,y,z) \left[D_x^{(e)}\theta^{(e)} \frac{\partial^2 \hat{C}^{(e)}}{\partial x^2} + D_y^{(e)}\theta^{(e)} \frac{\partial^2 \hat{C}^{(e)}}{\partial y^2} \right.$$
$$+ D_z^{(e)}\theta^{(e)} \frac{\partial^2 \hat{C}^{(e)}}{\partial z^2} - v_x^{(e)} \frac{\partial \hat{C}^{(e)}}{\partial x} - \rho_b^{(e)} K_d^{(e)} \frac{\partial \hat{C}^{(e)}}{\partial t}$$
$$\left. - \lambda(\theta^{(e)}\hat{C}^{(e)} + \rho_b^{(e)} K_d^{(e)}\hat{C}^{(e)}) - \theta^{(e)} \frac{\partial \hat{C}^{(e)}}{\partial t} \right] dx\,dy\,dz \qquad (6.71)$$

Extending Equation 6.71, we have

$$R_i^{(e)} = -\iiint\limits_{V^{(e)}} N_i^{(e)} \left[D_x^{(e)}\theta^{(e)} \frac{\partial^2 \hat{C}^{(e)}}{\partial x^2} + D_y^{(e)}\theta^{(e)} \frac{\partial^2 \hat{C}^{(e)}}{\partial y^2} + D_z^{(e)}\theta^{(e)} \frac{\partial^2 \hat{C}^{(e)}}{\partial z^2} \right] dx\,dy\,dz$$
$$+ \iiint\limits_{V^{(e)}} N_i^{(e)} \left[v_x^{(e)} \frac{\partial \hat{C}^{(e)}}{\partial x} \right] dx\,dy\,dz + \iiint\limits_{V^{(e)}} N_i^{(e)} \left[\lambda(\theta^{(e)}\hat{C}^{(e)} + \rho_b^{(e)} K_d^{(e)}\hat{C}^{(e)}) \right] dx\,dy\,dz$$
$$+ \iiint\limits_{V^{(e)}} N_i^{(e)} \left[\rho_b^{(e)} K_d^{(e)} \frac{\partial \hat{C}^{(e)}}{\partial t} \right] dx\,dy\,dz + \iiint\limits_{V^{(e)}} N_i^{(e)} \left[\theta^{(e)} \frac{\partial \hat{C}^{(e)}}{\partial t} \right] dx\,dy\,dz$$
$$(6.72)$$

Equation 6.72 can be presented in matrix form as

$$\begin{Bmatrix} R_1^{(e)} \\ \vdots \\ R_n^{(e)} \end{Bmatrix} = [D^{(e)}] \begin{Bmatrix} C_1 \\ \vdots \\ C_n \end{Bmatrix} + [A^{(e)}] \begin{Bmatrix} \dfrac{\partial C_1}{\partial t} \\ \vdots \\ \dfrac{\partial C_n}{\partial t} \end{Bmatrix} \qquad (6.73)$$

where
   $[D^{(e)}]$ is the advection–dispersion matrix of the element
   $[A^{(e)}]$ is the sorption matrix of the element

The advection–dispersion matrix of the element is defined as

$$
\left[D^{(e)}\right] = \iiint\limits_{V^{(e)}}
\begin{bmatrix}
\dfrac{\partial N_1^{(e)}}{\partial x} & \dfrac{\partial N_1^{(e)}}{\partial y} & \dfrac{\partial N_1^{(e)}}{\partial z} \\
\vdots & \vdots & \vdots \\
\dfrac{\partial N_n^{(e)}}{\partial x} & \dfrac{\partial N_n^{(e)}}{\partial y} & \dfrac{\partial N_n^{(e)}}{\partial z}
\end{bmatrix}
\begin{bmatrix}
D_x^{(e)}\theta^{(e)} & 0 & 0 \\
0 & D_y^{(e)}\theta^{(e)} & 0 \\
0 & 0 & D_z^{(e)}\theta^{(e)}
\end{bmatrix}
\begin{bmatrix}
\dfrac{\partial N_1^{(e)}}{\partial x} & \cdots & \dfrac{\partial N_n^{(e)}}{\partial x} \\
\dfrac{\partial N_1^{(e)}}{\partial y} & \cdots & \dfrac{\partial N_n^{(e)}}{\partial y} \\
\dfrac{\partial N_1^{(e)}}{\partial z} & \cdots & \dfrac{\partial N_n^{(e)}}{\partial z}
\end{bmatrix} dx\,dy\,dz
$$

$$
\quad n\times n \qquad\qquad n\times 3 \qquad\qquad\quad 3\times 3 \qquad\qquad\quad 3\times n
$$

$$
+ \iiint\limits_{V^{(e)}}
\begin{bmatrix}
N_1^{(e)} \\
\vdots \\
N_n^{(e)}
\end{bmatrix}
v_x^{(e)}
\begin{bmatrix}
\dfrac{\partial N_1^{(e)}}{\partial x} & \cdots & \dfrac{\partial N_n^{(e)}}{\partial x}
\end{bmatrix} dx\,dy\,dz
$$

$$
\quad\quad n\times 1 \qquad 1\times 1 \qquad\quad 1\times n
$$

$$
+ \iiint\limits_{V^{(e)}}
\begin{bmatrix}
N_1^{(e)} \\
\vdots \\
N_n^{(e)}
\end{bmatrix}
\left[\lambda(\theta^{(e)} + \rho_b^{(e)} K_d^{(e)})\right]
\begin{bmatrix}
N_1^{(e)} & \cdots & N_n^{(e)}
\end{bmatrix} dx\,dy\,dz \tag{6.74}
$$

$$
\quad\quad n\times 1 \qquad\quad 1\times 1 \qquad\quad 1\times n
$$

where $V^{(e)}$ is the volume of element e. The sorption matrix of the element is defined as

$$
\left[A^{(e)}\right] = \iiint\limits_{V^{(e)}}
\begin{bmatrix}
N_1^{(e)} \\
\vdots \\
N_n^{(e)}
\end{bmatrix}
\left[\rho_b^{(e)} K_d^{(e)} + \theta^{(e)}\right]
\begin{bmatrix}
N_1^{(e)} & \cdots & N_n^{(e)}
\end{bmatrix} dx\,dy\,dz \tag{6.75}
$$

$$
\quad\quad n\times 1 \qquad\quad 1\times 1 \qquad\quad 1\times n
$$

or

$$
\left[A^{(e)}\right] = \left(\rho_b^{(e)} K_d^{(e)} + \theta^{(e)}\right)\left(\frac{V^{(e)}}{n}\right)
\begin{bmatrix}
1 & & 0 \\
& \ddots & \\
0 & & 1
\end{bmatrix} \tag{6.76}
$$

$$
\qquad\qquad\qquad\qquad n\times n
$$

The finite difference is used to estimate the time derivative of the approximate solution, $\partial \hat{C}/\partial t$. By summing up the advection–dispersion matrix $[D]$ of all elements in the mesh and their sorption matrix $[A]$, the global matrices can be obtained as follows:

$$\underset{\substack{global \\ p \times p}}{[D]} = \sum_{e=1}^{m} \underset{n \times n}{\left[D^{(e)} \right]} \tag{6.77}$$

$$\underset{\substack{global \\ p \times p}}{[A]} = \sum_{e=1}^{m} \underset{n \times n}{\left[A^{(e)} \right]} \tag{6.78}$$

where
   m is the number of elements
   p is the number of nodes in the mesh

The weighted residual formulation for the solute transport equation becomes

$$\underset{global}{[D]} \begin{Bmatrix} C_1 \\ \vdots \\ C_p \end{Bmatrix} + \underset{global}{[A]} \begin{Bmatrix} \dfrac{\partial C_1}{\partial t} \\ \vdots \\ \dfrac{\partial C_p}{\partial t} \end{Bmatrix} = \underset{global}{\{F\}} \tag{6.79}$$

Two vectors $\{C\}$ and $\{\dot{C}\}$ are defined as

$$\{C\} = \begin{Bmatrix} C_1 \\ \vdots \\ C_p \end{Bmatrix} \quad \{\dot{C}\} = \begin{Bmatrix} \dfrac{\partial C_1}{\partial t} \\ \vdots \\ \dfrac{\partial C_p}{\partial t} \end{Bmatrix} \tag{6.80}$$

Equation 6.79 can be written as

$$\underset{global}{[A]}\{\dot{C}\} + \underset{global}{[D]}\{C\} = \underset{global}{\{F\}} \tag{6.81}$$

The solution of the system of algebraic equations determines the concentration and the concentration time derivative values at each node in the mesh. The finite-difference formulation for Equation 6.81 can be expressed as

$$([A] + \alpha \Delta t[D])\{C\}_{t+\Delta t} = ([A] - (1-\alpha)\Delta t[D])\{C\}_t + \Delta t((1-\alpha)\{F\}_t + \alpha\{F\}_{t+\Delta t}) \tag{6.82}$$

Example 6.9

Consider the confined aquifer shown in Figure 6.23. Steady-state, saturated ground-water flow is occurring in this aquifer. Initially, the solute concentration is constant in the aquifer at 5 mg/L. At time $t = 0$, the solute concentration in River A increases to 25 mg/L and remains constant thereafter and the solute concentration in the River B remains constant as 5 mg/L. Find the solute concentration in the aquifer after 5 days.

$$V_x^{(e)} = 0.12\,\text{m/day}, \quad D_x^{(e)} = 1.0\,\text{m}^2/\text{day}, \quad \theta^{(e)} = n^{(e)} = 0.3$$

$$L_1^{(e)} = 2\,\text{m}, \quad L_2^{(e)} = 3\,\text{m}, \quad L_3^{(e)} = 2\,\text{m}, \quad L_4^{(e)} = 2\,\text{m}$$

Solution

The problem domain is discretized into a mesh with four elements and five nodes. The dispersion–advection matrix for each element is considered 1D, i.e., each element has two nodes:

$$\left[D^{(e)}\right] = \int_{x_i^{(e)}}^{x_j^{(e)}} \begin{bmatrix} \dfrac{\partial N_1^{(e)}}{\partial x} \\[2mm] \dfrac{\partial N_2^{(e)}}{\partial x} \end{bmatrix} \left[D_x^{(e)}\theta^{(e)}\right]\left[\dfrac{\partial N_1^{(e)}}{\partial x}\quad \dfrac{\partial N_2^{(e)}}{\partial x}\right] dx + \int_{x_i^{(e)}}^{x_j^{(e)}} \begin{bmatrix} N_1^{(e)} \\[2mm] N_2^{(e)} \end{bmatrix}\left[V_x^{(e)}\right]\left[\dfrac{\partial N_1^{(e)}}{\partial x}\quad \dfrac{\partial N_2^{(e)}}{\partial x}\right] dx$$

$$\qquad 2\times 2 \qquad\qquad 2\times 1 \qquad 1\times 1 \qquad 1\times 2 \qquad\qquad\qquad 2\times 1 \quad 1\times 1 \qquad 1\times 2$$

$$+ \int_{x_i^{(e)}}^{x_j^{(e)}} \begin{bmatrix} N_1^{(e)} \\[2mm] N_2^{(e)} \end{bmatrix}\left[\lambda(\theta^{(e)} + \rho_b^{(e)}K_d^{(e)})\right]\left[N_1^{(e)}\quad N_2^{(e)}\right] dx \qquad\qquad (6.83)$$

$$\qquad 2\times 1 \qquad\qquad 1\times 1 \qquad\qquad 1\times 2$$

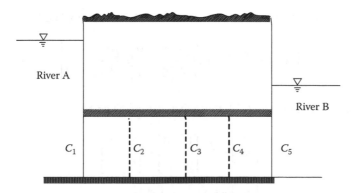

FIGURE 6.23 Confined aquifer with solute transportation for Example 6.9.

Based on Galerkin's method, the interpolation functions for all of the elements in the mesh are

$$N_1^e = \frac{x_j^{(e)} - x}{L^{(e)}}, \quad \frac{\partial N_1^e}{\partial x} = \frac{-1}{L^{(e)}}$$

$$N_2^e = \frac{x - x_i^{(e)}}{L^{(e)}}, \quad \frac{\partial N_2^e}{\partial x} = \frac{1}{L^{(e)}}$$

Since the solute does not react with the porous media and does not decay, $K_d^{(e)} = 0$ and $\lambda = 0$. In addition, because the porous media is saturated, $\theta^{(e)} = n^{(e)}$. Therefore,

$$\left[D^{(e)} \right] = \int_{x_i^{(e)}}^{x_j^{(e)}} \begin{bmatrix} \dfrac{-1}{L^{(e)}} \\ \dfrac{1}{L^{(e)}} \end{bmatrix} \left[D_x^{(e)} \right] \begin{bmatrix} \dfrac{-1}{L} & \dfrac{1}{L} \end{bmatrix} dx + \int_{x_i^{(e)}}^{x_j^{(e)}} \begin{bmatrix} \dfrac{x_j^{(e)} - x}{L^{(e)}} \\ \dfrac{x - x_i^{(e)}}{L^{(e)}} \end{bmatrix} \begin{bmatrix} \dfrac{V_x^{(e)}}{\theta^{(e)}} \end{bmatrix} \begin{bmatrix} \dfrac{-1}{L^{(e)}} & \dfrac{1}{L^{(e)}} \end{bmatrix} dx$$

2×2

$$= \frac{D_x^{(e)}}{L^{(e)}} \begin{bmatrix} 1 & -1 \\ -1 & 1 \end{bmatrix} + \frac{V_x^{(e)}}{2\theta^{(e)}} \begin{bmatrix} 1 & -1 \\ -1 & 1 \end{bmatrix}$$

For the elements in Figure 6.23, these matrices are given as

$$\left[D^{(1)} \right] = \frac{1}{2} \begin{bmatrix} 1 & -1 \\ -1 & 1 \end{bmatrix} + \frac{0.12}{2(0.3)} \begin{bmatrix} -1 & 1 \\ -1 & 1 \end{bmatrix} = \begin{bmatrix} 0.3 & -0.3 \\ -0.7 & 0.7 \end{bmatrix}$$

$$\left[D^{(2)} \right] = \frac{1}{3} \begin{bmatrix} 1 & -1 \\ -1 & 1 \end{bmatrix} + \frac{0.12}{2(0.3)} \begin{bmatrix} -1 & 1 \\ -1 & 1 \end{bmatrix} = \begin{bmatrix} 0.13 & -0.13 \\ -0.53 & 0.53 \end{bmatrix}$$

$$\left[D^{(3)} \right] = \left[D^{(4)} \right] = \left[D^{(1)} \right]$$

Using Equation 6.76, the sorption matrix of the element is simplified as

$$\left[A^{(e)} \right] = \left(\frac{\rho_b^{(e)} K_d^{(e)}}{n^{(e)}} + 1 \right) \frac{L^{(e)}}{2} \begin{bmatrix} 1 & 0 \\ 0 & 1 \end{bmatrix} = \frac{L^{(e)}}{2} \begin{bmatrix} 1 & 0 \\ 0 & 1 \end{bmatrix}$$

For the elements in Figure 6.23, these matrices are given as

$$\left[A^{(1)} \right] = \frac{2}{2} \begin{bmatrix} 1 & 0 \\ 0 & 1 \end{bmatrix} = \begin{bmatrix} 1 & 0 \\ 0 & 1 \end{bmatrix}$$

$$\left[A^{(1)} \right] = \left[A^{(3)} \right] = \left[A^{(4)} \right]$$

$$[A^{(2)}] = \frac{3}{2}\begin{bmatrix} 1 & 0 \\ 0 & 1 \end{bmatrix}$$

The global matrices $[D]$ and $[A]$ are obtained by combining the matrices $[D]$ and $[A]$ of all elements:

$$[D] = \begin{bmatrix} 0.3 & -0.3 & 0 & 0 & 0 \\ -0.7 & 0.7+0.13 & -0.13 & 0 & 0 \\ 0 & -0.53 & 0.53+0.3 & -0.3 & 0 \\ 0 & 0 & -0.7 & 0.7+0.3 & -0.3 \\ 0 & 0 & 0 & -0.7 & 0.7 \end{bmatrix}$$

$$= \begin{bmatrix} 0.3 & -0.3 & 0 & 0 & 0 \\ -0.7 & 0.83 & -0.13 & 0 & 0 \\ 0 & -0.53 & 0.83 & -0.3 & 0 \\ 0 & 0 & -0.7 & 1 & -0.3 \\ 0 & 0 & 0 & -0.7 & 0.7 \end{bmatrix}$$

$$[A] = \begin{bmatrix} 1 & 0 & 0 & 0 & 0 \\ 0 & 1+1.5 & 0 & 0 & 0 \\ 0 & 0 & 1.5+1 & 0 & 0 \\ 0 & 0 & 0 & 1+1 & 0 \\ 0 & 0 & 0 & 0 & 1 \end{bmatrix} = \begin{bmatrix} 1 & 0 & 0 & 0 & 0 \\ 0 & 2.5 & 0 & 0 & 0 \\ 0 & 0 & 2.5 & 0 & 0 \\ 0 & 0 & 0 & 2 & 0 \\ 0 & 0 & 0 & 0 & 1 \end{bmatrix}$$

The backward difference method is used, thus,

$$([A] + \Delta t[D])\{C\}_{t+\Delta t} = [A]\{C\}_t + \Delta t\{F\}_{t+\Delta t}$$

The solute concentrations at the nodes at time $t = 0$ are given as

$$\{C\}_{t=0} = \begin{Bmatrix} C_1 \\ C_2 \\ C_3 \\ C_4 \\ C_5 \end{Bmatrix}_{t=0} = \begin{Bmatrix} 5 \\ 5 \\ 5 \\ 5 \\ 5 \end{Bmatrix}$$

The equation can be solved to obtain the concentrations at the end of the time step ($\Delta t = 5$):

$$([A] + 5[D])\{C\}_{t=5} = [A]\{C\}_{t=0}$$

$$\begin{bmatrix} 2.5 & -1.5 & 0 & 0 & 0 \\ -3.5 & 6.65 & -0.65 & 0 & 0 \\ 0 & -2.65 & 6.65 & -1.5 & 0 \\ 0 & 0 & -3.5 & 7 & -1.5 \\ 0 & 0 & 0 & -3.5 & 4.5 \end{bmatrix} \begin{Bmatrix} C_1 \\ C_2 \\ C_3 \\ C_4 \\ C_5 \end{Bmatrix}_{t=5} = \begin{bmatrix} 1 & 0 & 0 & 0 & 0 \\ 0 & 2.5 & 0 & 0 & 0 \\ 0 & 0 & 2.5 & 0 & 0 \\ 0 & 0 & 0 & 2 & 0 \\ 0 & 0 & 0 & 0 & 1 \end{bmatrix} \begin{Bmatrix} 5 \\ 5 \\ 5 \\ 5 \\ 5 \end{Bmatrix}$$

$$= \begin{Bmatrix} 5 \\ 12.5 \\ 12.5 \\ 10 \\ 5 \end{Bmatrix}$$

This system of equations must be modified because of the boundary conditions, $C_{A,t=5} = 25$ and $C_{B,t=5} = 5$. Modifying this system of equations gives

$$\begin{bmatrix} 6.65 & -0.65 & 0 \\ -2.65 & 6.65 & -1.5 \\ 0 & -3.5 & 7 \end{bmatrix} \begin{Bmatrix} C_2 \\ C_3 \\ C_4 \end{Bmatrix}_{t=5} = \begin{Bmatrix} 100 \\ 12.5 \\ 17.5 \end{Bmatrix}$$

This equation is solved to obtain the values of C_2–C_4 at the end of the time step. The solution is

$$\begin{Bmatrix} C_1 \\ C_2 \\ C_3 \\ C_4 \\ C_5 \end{Bmatrix}_{t=5} = \begin{Bmatrix} 25 \\ 16.01 \\ 9.95 \\ 7.47 \\ 5 \end{Bmatrix}$$

This solution is then substituted into the right-hand side of Equation 6.97 and the procedure is repeated for the next time step if more time steps are supposed to be considered.

6.6 FINITE VOLUME METHOD

In FVM, a number of weighted residual equations are generated by dividing the calculation function to be control volumes, and setting the weighting function to be union over the control volume. In a control volume, the integral of the residual must become equal to zero (Patankar, 1980).

The basic idea of the control volume formulation is easy to understand and helps to direct physical interpretation. The problem domain is divided into a number of control volumes with no overlap. The differential equation is integrated over one control volume that surrounds each grid point. The piecewise continuous represents the variation of h between the grid points and is used to evaluate the integrals. The result is the discretization equation containing the values of h for a group of grid points. The conservation principle for h for the finite control volume is obtained in the discretization equation. The integral conservation of quantities such as mass, momentum, and energy is exactly satisfied over any group of control volumes and over the entire problem domain. This is the most attractive feature of the control volume formulation for any number of grid points; even the coarse grid solution shows the exact integral balances.

In FVM, a structured mesh is not needed and it is an advantage of the FVM over FDMs. The values of the conserved variables are located within the volume element, and not at nodes; hence, FVM is preferable to other methods as a result of the fact that boundary conditions can be applied noninvasively.

6.6.1 STEADY-STATE ONE-DIMENSIONAL FLOW EQUATION

Consider steady-state, 1D heat conduction equation as follows:

$$\frac{d}{dx}\left(K\frac{dh}{dx}\right) + R = 0 \tag{6.84}$$

To derive the discretization equation, the grid-point cluster is employed as shown in Figure 6.24. Consider the grid point P that has the grid points E and W as its neighbors. (E denotes the east side, i.e., the positive x-direction, while W stands for west or the negative x-direction.) The dashed lines show the faces of the control volume; their exact locations are unimportant at first. The letters e and w denote these faces. For a 1D problem, a unit thickness in the y and z directions is assumed. Thus, the volume of the control volume is $\Delta x \times 1 \times 1$. If Equation 6.84 is integrated over the control volume:

$$\left(K\frac{dh}{dx}\right)_e - \left(K\frac{dh}{dx}\right)_w + \int_w^e R\,dx = 0 \tag{6.85}$$

To make further progress, a profile assumption or an interpolation formula is required. Two simple profile assumptions are shown in Figure 6.25. The simplest possibility is to assume that the value of h at a grid point is the same as the value of h over the entire control volume. This gives the stepwise profile sketched in Figure 6.25a.

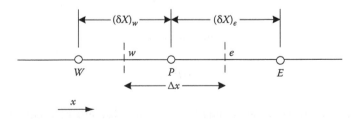

FIGURE 6.24 Grid-point cluster for a 1D problem.

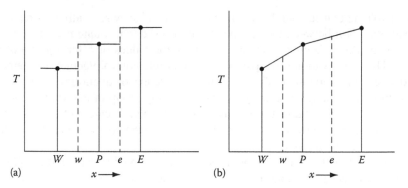

(a)

(b)

FIGURE 6.25   Two simple profile assumptions: (a) stepwise profile; (b) piecewise linear profile.

For this profile, the slope dh/dx is not defined at the control volume faces (i.e., at w or e). As shown in Figure 6.25b, the linear interpolation functions are used between the grid points. Therefore, in this profile, the slope dh/dx is defined for the control volume faces.

Evaluating the derivatives dh/dx in Equation 6.85 for the piecewise-linear profile, the resulting equation will be as follows:

$$\frac{K_e(h_E - h_P)}{(\delta x)_e} - \frac{K_w(h_P - h_W)}{(\delta x)_w} + \bar{R}\,\Delta x = 0 \tag{6.86}$$

where $\bar{q}_e$ is the average value of q_e over the control volume. It is assumed that the simplest profile is used to evaluate dh/dx. Several interpolation functions could be used for other variables. For example, $\bar{R}$ needs not to be calculated by a linear variation of R between the grid points. It is useful to modify the discretization equation (6.86) to the following form:

$$a_P h_P = a_E h_E + a_w h_W + b \tag{6.87}$$

$$a_e = \frac{k_e}{(\delta x)_e} \tag{6.88a}$$

$$a_w = \frac{k_w}{(\delta x)_w} \tag{6.88b}$$

$$a_P = a_E + a_w \tag{6.88c}$$

$$b = \bar{R}\,\Delta x \tag{6.88d}$$

Equation 6.87 represents the standard form of the discretization equations. h_p in the left side of this equation shows the hydraulic head at the central grid point.

By increasing the dimension of the problem, the number of neighbors increases. Generally, Equation 6.87 can be written as follows for more than 1D problems:

$$a_p h_p = \sum a_{nb} h_{nb} + b \qquad (6.89)$$

where the subscript nb denotes a neighbor, and the summation is to be taken over all the neighbors.

Example 6.10

Consider a steady-state 1D flow in a confined aquifer between two rivers shown in Figure 6.26. Assume $L = 50\,\text{m}$, $h_A = 10\,\text{m}$, $h_B = 50\,\text{m}$. Determine the spatial variation of piezometric head.

Solution

The groundwater flow equation for 1D flow in a confined aquifer can be written as follows:

$$\frac{d}{dx}\left(K\frac{dh}{dx}\right) = 0$$

As shown in Figure 6.26, for the grid point P, points E and W are its x-direction neighbors. For the control volume around P (when P is set at nodes 2, 3, 4), we have

$$\int_{C.V} \frac{d}{dx}\left(K\frac{dh}{dx}\right) A dx = 0$$

or

$$AK\frac{dh}{dx}\bigg|_w^e = 0$$

$$AK\frac{dh}{dx}\bigg|_e - AK\frac{dh}{dx}\bigg|_w = 0$$

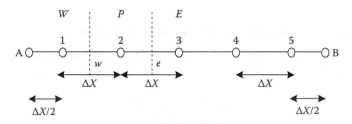

FIGURE 6.26   The elements of confined aquifer between Rivers A and B in Example 6.10.

$$A_e K_e \frac{h_E - h_P}{(\delta x)_e} - A_w K_w \frac{h_P - h_W}{(\delta x)_w} = 0$$

$$\left(A_e \frac{K_e}{(\delta x)_e} + A_w \frac{K_w}{(\delta x)_w}\right) h_P = \left(A_w \frac{K_w}{(\delta x)_w}\right) h_W + \left(A_e \frac{K_e}{(\delta x)_e}\right) h_E$$

If $A_e = A_w$, $K_e = K_w$ and $(\delta x)_e = (\delta x)_w = \Delta x$ then, $h_P = (h_W + h_E)/2$. For the control volume around P, when it is set at node 1, $(\delta x)_e = \Delta x$ and $(\delta x)_w = \Delta x/2$. Thus,

$$A_e K_e \frac{h_E - h_P}{\Delta x} - A_w K_A \frac{h_P - h_A}{\Delta x/2} = 0$$

$$\frac{h_E - h_P}{\Delta x} - \frac{h_P - h_A}{\Delta x/2} = 0$$

$$3h_P = h_E + 2h_A$$

When P is set at node 5, $(\delta x)_e = \Delta x/2$ and $(\delta x)_w = \Delta x$. So,

$$A_e K_e \frac{h_B - h_P}{\Delta x/2} - A_w K_A \frac{h_P - h_W}{\Delta x} = 0$$

$$3h_P = h_W + 2h_B$$

Therefore, the following matrix for grids 1 through 5 is obtained:

$$\begin{bmatrix} 3 & -1 & 0 & 0 & 0 \\ -1 & 2 & -1 & 0 & 0 \\ 0 & -1 & 2 & -1 & 0 \\ 0 & 0 & -1 & 2 & -1 \\ 0 & 0 & 0 & -1 & 3 \end{bmatrix} \begin{bmatrix} h_1 \\ h_2 \\ h_3 \\ h_4 \\ h_5 \end{bmatrix} = \begin{bmatrix} 2h_A \\ 0 \\ 0 \\ 0 \\ 2h_B \end{bmatrix}$$

Solving this matrix, the values of heads are obtained as follows:

$$h_1 = 14, \quad h_2 = 22, \quad h_3 = 30, \quad h_4 = 38, \quad h_5 = 46.$$

6.6.2 Unsteady-State One-Dimensional Flow Equation

In this section, the solution for the unsteady-state, 1D groundwater equation is sought. The general form of this equation can be expressed as follows:

$$\frac{S}{b} \frac{\partial h}{\partial t} = \frac{d}{dx}\left(K \frac{dh}{dx}\right) \tag{6.90}$$

In this equation, it is assumed that S/b is constant. Since time is a one-way coordinate, the solution is obtained by marching in time from a given initial distribution of temperature. In a typical time step, given the grid-point values of T at time t, the values of T at time $t + \Delta t$ are determined. The given values of T at the grid points will be specified by h_P^0, h_E^0, h_W^0, and the unknown values at time $t + \Delta t$ by h_P^1, h_E^1, h_W^1.

The discretization equation can now be derived by integrating Equation 6.90 over the control volume shown in Figure 6.24 and over the time interval from t to $t + \Delta t$:

$$\frac{S}{b} \int_w^e \int_t^{t+\Delta t} \frac{\partial h}{\partial t} \, dt \, dx = \int_w^e \int_t^{t+\Delta t} \frac{\partial}{\partial x}\left(K \frac{\partial h}{\partial x} \right) dt \, dx \tag{6.91}$$

where the order of the integration is chosen according to the nature of the term. For the representation of the term $\partial h/\partial t$, it is assumed that the grid-point value of h prevails throughout the control volume. Then,

$$\frac{S}{b} \int_w^e \int_t^{t+\Delta t} \frac{\partial h}{\partial t} \, dt \, dx = \frac{S}{b} \Delta x (h_P^1 - h_P^0) \tag{6.92}$$

Following the steady-state practice for $\partial h/\partial t$, we have

$$\frac{S}{b} \Delta x (h_P^1 - h_P^0) = \int_t^{t+\Delta t} \left[\frac{K_e(h_E - h_P)}{(\delta x)_e} - \frac{K_w(h_P - h_W)}{(\delta x)_w} \right] dt \tag{6.93}$$

It is at this point, an assumption about how h_P, h_E, h_W vary with time from t to $t + \Delta t$ is made. Some assumptions can be generalized by proposing the following:

$$\int_t^{t+\Delta t} h_P dt = \left[\alpha h_P^1 + (1-\alpha)h_P^0 \right] \Delta t \tag{6.94}$$

where α is a weighting factor, which ranges from 0 to 1. Using similar formulas for the integrals of h_E, h_W, Equation 6.93 can be rewritten as

$$\frac{S}{b} \Delta x (h_P^1 - h_P^0) = \int_t^{t+\Delta t} \alpha \left[\frac{K_e(h_E^1 - h_P^1)}{(\delta x)_e} - \frac{K_w(h_P^1 - h_W^1)}{(\delta x)_w} \right]$$

$$+ (1-\alpha)\left[\frac{K_e(h_E^0 - h_P^0)}{(\delta x)_e} - \frac{K_w(h_P^0 - h_W^0)}{(\delta x)_w} \right] \tag{6.95}$$

By rearranging the equation and considering that h_P, h_E, h_W henceforth stand for the new values of h at time $t + \Delta t$, we have (Patankar, 1980)

$$a_P h_P = a_E \left[\alpha\, h_E + (1-\alpha) h_E^0 \right] + a_w \left[\alpha\, h_W + (1-\alpha) h_W^0 \right]$$
$$+ \left[a_P^0 - (1-\alpha) a_E + (1-\alpha) a_w \right] h_P^0 \qquad (6.96)$$

where

$$a_E = \frac{K_e}{(\delta x)_e} \qquad (6.97a)$$

$$a_W = \frac{k_w}{(\delta x)_w} \qquad (6.97b)$$

$$a_P^0 = \frac{S\,\Delta x}{b\Delta t} \qquad (6.97c)$$

$$a_P = \alpha a_E + \alpha a_W + a_P^0 \qquad (6.97d)$$

6.6.3   Two-Dimensional Flow Equation

A portion of a 2D mesh is shown in Figure 6.27. In this figure, grid points E and W are neighbors of point P in x-direction and points N and S (denoting north and south) are the neighbors in y-direction. The hatched area around P shows the control volume. Its thickness in the z-direction is assumed to be unity. It is usually preferred

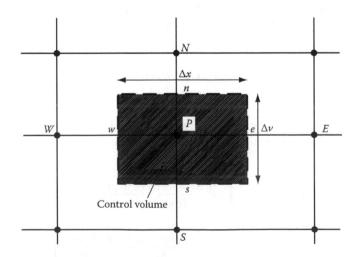

FIGURE 6.27   A control volume for a 2D mesh.

to set the control volume faces exactly *midway* between the neighboring grid points However, other practices can also be employed (Patankar, 1980). The general form of the 2D flow equation can be written as

$$\frac{S}{b}\frac{\partial h}{\partial t} = \frac{d}{dx}\left(K\frac{dh}{dx}\right) + \frac{d}{dy}\left(K\frac{dh}{dy}\right) + R \tag{6.98}$$

Similar to the 1D form, the differential equation can be discretized as

$$a_p h_p = a_E h_E + a_W h_W + a_N h_N + a_S h_S + b \tag{6.99}$$

where

$$a_E = \frac{K_e \, \Delta y}{(\delta x)_e} \tag{6.100a}$$

$$a_W = \frac{K_w \, \Delta y}{(\delta x)_w} \tag{6.100b}$$

$$a_N = \frac{K_n \, \Delta x}{(\delta x)_n} \tag{6.100c}$$

$$a_S = \frac{K_s \, \Delta x}{(\delta x)_s} \tag{6.100d}$$

$$a_p^0 = \frac{S \Delta x \Delta y}{b \Delta t} \tag{6.100e}$$

$$b = S_c \Delta x \, \Delta y + a_p^0 \, h_p^0 \tag{6.100f}$$

$$a_p = a_E + a_W + a_N + a_S + a_p^0 - S_p \Delta x \, \Delta y \tag{6.100g}$$

Since the thickness in the z-direction is unity, the product of $\Delta x \Delta y$ gives the volume of the control volume.

Example 6.11

Estimate the net flow for seepage under the sheet piling shown in Figure 6.28 for $K = 10^{-4}$ m/s, $L = 300$ m, and $\Delta x = \Delta y = 10$ m with steady-state condition.

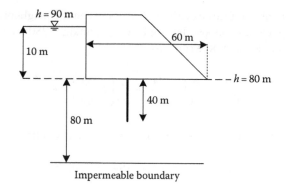

FIGURE 6.28   A dam with the sheet piling for Example 6.11.

Solution

The governing differential equation is a 2D equation for a confined aquifer:

$$\frac{\partial^2 h}{\partial x^2} + \frac{\partial^2 h}{\partial y^2} = 0$$

$$\int_s^n \int_w^e \frac{\partial}{\partial x}\left(\frac{\partial h}{\partial x}\right) dx\, dy + \int_w^e \int_s^n \frac{\partial}{\partial y}\left(\frac{\partial h}{\partial y}\right) dx\, dy = 0$$

$$\int_s^n \left.\frac{\partial h}{\partial x}\right|_w^e dy + \int_w^e \left.\frac{\partial h}{\partial y}\right|_s^n dx = 0$$

$$\int_s^n \left[\left(\frac{\partial h}{\partial x}\right)_e - \left(\frac{\partial h}{\partial x}\right)_w\right] dy + \int_w^e \left[\left(\frac{\partial h}{\partial y}\right)_n - \left(\frac{\partial h}{\partial y}\right)_s\right] dx = 0$$

$$\left[\left(\frac{\partial h}{\partial x}\right)_e - \left(\frac{\partial h}{\partial x}\right)_w\right]\Delta y + \left[\left(\frac{\partial h}{\partial y}\right)_n - \left(\frac{\partial h}{\partial y}\right)_s\right]\Delta x = 0$$

$$\left(\frac{h_E - h_P}{(\delta x)_e}\right)\Delta y - \left(\frac{h_P - h_W}{(\delta x)_w}\right)\Delta y + \left(\frac{h_N - h_P}{(\delta y)_n}\right)\Delta x - \left(\frac{h_P - h_S}{(\delta y)_s}\right)\Delta x = 0$$

$$h_P\left(\frac{\Delta y}{(\delta x)_e} + \frac{\Delta y}{(\delta x)_w} + \frac{\Delta x}{(\delta y)_n} + \frac{\Delta x}{(\delta y)_s}\right) = h_E\frac{\Delta y}{(\delta x)_e} + h_W\frac{\Delta y}{(\delta x)_w} + h_N\frac{\Delta x}{(\delta y)_n} + h_S\frac{\Delta x}{(\delta y)_s}$$

If $\delta x = \Delta x$, $\delta y = \Delta y$, then,

$$h_P = \frac{(h_E + h_W)(\Delta y/\Delta x) + (h_N + h_S)(\Delta x/\Delta y)}{2(\Delta y/\Delta x + \Delta x/\Delta y)}$$

If $\Delta x = \Delta y$ then

$$h_P = \frac{h_E + h_W + h_N + h_S}{4}$$

The boundary condition at the bottom of the aquifer is an impervious layer that has no inflow (Figure 6.29a):

$$\left(\frac{h_E - h_P}{(\delta x)_e}\right)\frac{\Delta y}{2} - \left(\frac{h_P - h_W}{(\delta x)_w}\right)\frac{\Delta y}{2} + \left(\frac{h_N - h_P}{(\delta y)_n}\right)\Delta x = 0$$

If $\delta x = \Delta x$ and $\delta y = \Delta y$, then

$$h_P = \frac{h_E(\Delta y 2\Delta x) + h_W(\Delta y 2\Delta x) + h_N(\Delta x \Delta y)}{(\Delta y \Delta x + \Delta x/\Delta y)}$$

In the left and right boundaries (Condition 2), it is assumed that $\partial h/\partial x = 0$. The finite volume problem domain is expanded by two additional columns to the left and right, which are called imaginary or fictitious nodes. The value of head along the fictitious column must reflect across the boundary.

For the base of the dam (Condition 3), the control volume can be considered as shown in Figure 6.29b:

$$\left(\frac{h_E - h_P}{(\delta x)_e}\right)\frac{\Delta y}{2} - \left(\frac{h_P - h_W}{(\delta x)_w}\right)\frac{\Delta y}{2} - \left(\frac{h_P - h_S}{(\delta y)_s}\right)\Delta x = 0$$

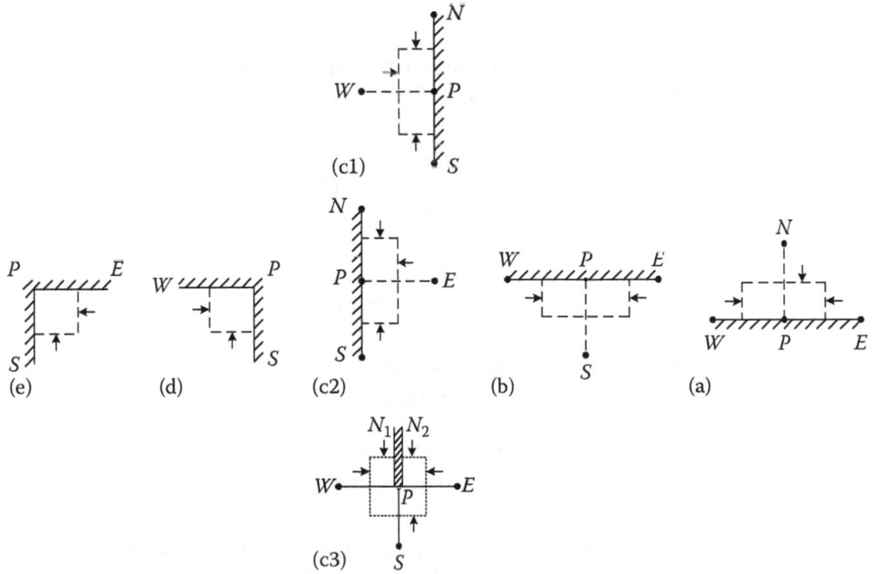

FIGURE 6.29 The boundary conditions for Example 6.11.

If $\delta x = \Delta x$ and $\delta y = \Delta y$, then

$$h_P = \frac{h_E(\Delta y/2\Delta x) + h_W(\Delta y/2\Delta x) + h_S(\Delta x/\Delta y)}{(\Delta y/\Delta x + \Delta x/\Delta y)}$$

At the sheet piling (Condition 4)

1. For the right side of the sheet piling, the control volume has been shown in Figure 6.29c1:

$$h_P = \frac{h_W(\Delta y/\Delta x) + h_N(\Delta x/2\Delta y) + h_S(\Delta x/2\Delta y)}{(\Delta y/\Delta x + \Delta x/\Delta y)}$$

2. For the left side of the sheet piling, the control volume has been shown in Figure 6.29c2:

$$h_P = \frac{h_E(\Delta y/\Delta x) + h_N(\Delta x/2\Delta y) + h_S(\Delta x/2\Delta y)}{(\Delta y/\Delta x + \Delta x/\Delta y)}$$

3. For the bottom edge of the sheet piling, the control volume has been shown in Figure 6.29c3:

$$\left(\frac{h_E - h_P}{(\delta x)_e}\right)\Delta y - \left(\frac{h_P - h_W}{(\delta x)_w}\right)\Delta y - \left(\frac{h_P - h_S}{(\delta y)_s}\right)\Delta x + \left(\frac{h_{N1} - h_P}{(\delta y)_s}\right)\frac{\Delta x}{2} + \left(\frac{h_{N2} - h_P}{(\delta y)_s}\right)\frac{\Delta x}{2} = 0$$

$$h_P = \frac{h_E(\Delta y/\Delta x) + h_W(\Delta y/\Delta x) + h_S(\Delta x/\Delta y) + h_{N1}(\Delta x/2\Delta y) + h_{N2}(\Delta x/2\Delta y)}{2(\Delta y/\Delta x + \Delta x/\Delta y)}$$

4. For the left edge of the sheet piling, where it is connected to the dam, the control volume has been shown in Figure 6.29d:

$$-\left(\frac{h_P - h_W}{(\delta x)_w}\right)\frac{\Delta y}{2} - \left(\frac{h_P - h_S}{(\delta y)_s}\right)\Delta x = 0$$

$$h_P = \frac{h_W(\Delta y/2\Delta x) + h_S(\Delta x/2\Delta y)}{1/2(\Delta y/\Delta x + \Delta x/\Delta y)}$$

5. For the right edge of the sheet piling, where it is connected to the dam, the control volume has been shown in Figure 6.29e:

$$h_P = \frac{h_E(\Delta y/2\Delta x) + h_S(\Delta x/2\Delta y)}{1/2(\Delta y/\Delta x + \Delta x/\Delta y)}$$

A FORTRAN code has been developed to solve this problem. This code is presented in Figure 6.30. The results of this code determine the equipotential lines.

```
DIMENSION F(1500,1500),F1(1500,1500)
      OPEN(1,FILE='A.TXT')
      READ (1,*) IMAX,JMAX,DX,DY,FO
      DO  I=2,IMAX-1
      DO  J=2,JMAX-1
      F(I,J)=0.0
      F1(I,J)=0.0
      ENDDO
      ENDDO
      DO I=1,27
      F(I,JMAX)=FO
      F1(I,JMAX)=FO
      ENDDO
      DO I=36,62
      F(I,JMAX)=0.0
      F1(I,JMAX)=0.0
      ENDDO
      DO J=1,JMAX-1
      F(IMAX,J)=0.0
      F(1,J)=FO
      F1(IMAX,J)=0.0
      F1(1,J)=FO
      ENDDO
      A=(DX/DY+DY/DX)
      ITER=0
5     DIFMX=0.0
      ITER=ITER|1
      IF(ITER.GT.10000) STOP
      DO I=2,30
      DO J=2,JMAX-1
      F1(I,J)=(DY/DX*(F(I+1,J)+F(I-1,J))+DX/
     &DY*(F(I,J+1)+F(I,J-1)))/(2*A)
      END DO
      END DO
      DO I=33,61
      DO J=2,JMAX-1
      F1(I,J)=(DY/DX*(F(I+1,J)+F(I-1,J))+DX/
     &DY*(F(I,J+1)+F(I,J-1)))/(2*A)
      END DO
      END DO
****** Condition 1
      DO I=2,IMAX-1
      F1(I,1)=((F(I+1,1)+F(I-1,1))*DY/(2*DX)+F(I,2)*DX/DY)/A
      END DO
      I=31
      DO J=2,4
      F1(I,J)=(DY/DX*(F(I+2,J)+F(I-1,J))+DX/DY*
      (F(I,J+1)+F(I,J-1)))/(2*A)
      END DO
```

FIGURE 6.30 Fortran program for solving Example 6.11.

(continued)

```
****** Condition 2
      DO J=1,5
      F1(32,J)=F1(31,J)
      END DO
****** Condition 3
      J=9
      DO I=28,30
      F1(I,J)=((F(I+1,J)+F(I-1,J))*DY/(2*DX)+F(I,J-1)*DX/DY)/A
      END DO
      DO I=33,35
      F1(I,J)=((F(I+1,J)+F(I-1,J))*DY/(2*DX)+F(I,J-1)*DX/DY)/A
      F1(I,J)=F1(I,J-1)
      END DO
****** Condition  4
*     Condition 4-1
      DO J=6,JMAX-1
      F1(31,J)=(F(30,J)*DY/DX+(F(31,J+1)+F(31,J-1))*DX/(2*DY))/A
      END DO
*     Condition 4-2
      DO J=6,JMAX-1
      F1(32,J)=(F(33,J)*DY/DX+(F(32,J+1)+F(32,J-1))*DX/(2*DY))/A
      END DO
*     Condition 4-3
      I=31
      J=5
      F1(I,J)=((F(I+2,J)+F(I-1,J))*DY/DX+(F(I,J-1)+
     &F(I,J+1)/2+F(I+1,J+1)/2)*DX/DY)/2/A
*     Condition 4-4
      J=9
      I=32
      F1(I,J)=(F(I+1,J)*DY/DX+F(I,J-1)*DX/DY)/A
*     Condition 4-5
      I=31
      F1(I,J)=(F(I-1,J)*DY/DX+F(I,J-1)*DX/DY)/A

      DO I=1,62
      DO J=1,JMAX
      DIF=ABS(F(I,J)-F1(I,J))
      F(I,J)=F1(I,J)
      IF(DIF.GT.DIFMX) DIFMX=DIF
      END DO
      END DO
      WRITE(*,*) "ITER=",ITER
      WRITE(1,11)(I,I=1,IMAX)
11    FORMAT(62I10)
      DO  J=JMAX,1,-1
      WRITE(1,10)(F(I,J)+80,I=1,IMAX)
10    FORMAT(62F10.4)
      ENDDO
      IF(DIFMX.GT.0.00001) GOTO 5
      END
```

FIGURE 6.30 (continued) Fortran program for solving Example 6.11.

Example 6.12

A building is supposed to be built on an island. For construction purposes, the water table needs to be lowered to a certain depth. For this purpose, the groundwater is pumped out from four wells as shown in Figure 6.31. Determine the discharge from each well in steady-state condition.

Solution

The governing differential equation is a 2D equation for an unconfined aquifer:

$$-q + \frac{\partial}{\partial x}\left(K_x h \frac{\partial h}{\partial x}\right) + \frac{\partial}{\partial y}\left(K_y h \frac{\partial h}{\partial y}\right) = 0$$

assuming $K_x = K_y$, we have

$$-q + \frac{\partial}{\partial x}\left(\frac{K}{2}\frac{\partial h^2}{\partial x}\right) + \frac{\partial}{\partial y}\left(\frac{K}{2}\frac{\partial h^2}{\partial y}\right) = 0$$

$$-q + \frac{K}{2}\left(\frac{\partial^2 h^2}{\partial x^2} + \frac{\partial^2 h^2}{\partial y^2}\right) = 0$$

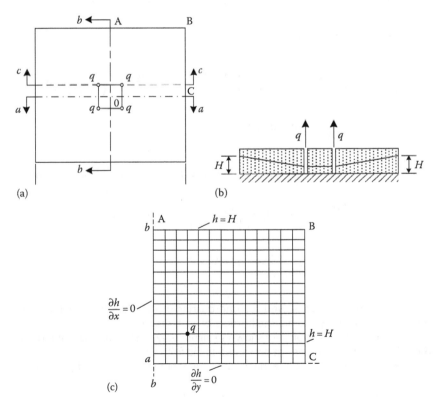

FIGURE 6.31 The definition sketch for Example 6.12: (a) plan view, (b) section view, and (c) one quarter of the plan.

$$\int_s^n \int_w^e \frac{K}{2} \frac{\partial^2 h^2}{\partial x^2} dx\, dy + \int_w^e \int_s^n \frac{K}{2} \frac{\partial^2 h^2}{\partial y^2} dx\, dy - \int_s^n \int_w^e q dx\, dy = 0$$

$$\int_s^n \frac{K}{2} \frac{\partial h^2}{\partial x}\Big|_w^e dy + \int_w^e \frac{K}{2} \frac{\partial h^2}{\partial y}\Big|_s^n dx - q\Delta x\, \Delta y = 0$$

$$\int_s^n \left[\frac{K_e}{2}\left(\frac{\partial h^2}{\partial x}\right)_e - \frac{K_w}{2}\left(\frac{\partial h^2}{\partial x}\right)_w \right] dy + \int_w^e \left[\frac{K_n}{2}\left(\frac{\partial h^2}{\partial y}\right)_n - \frac{K_s}{2}\left(\frac{\partial h^2}{\partial y}\right)_s \right] dx - q\Delta x\, \Delta y = 0$$

$$\left[\frac{K_e}{2}\left(\frac{\partial h^2}{\partial x}\right)_e - \frac{K_w}{2}\left(\frac{\partial h^2}{\partial x}\right)_w \right] \Delta y + \left[\frac{K_n}{2}\left(\frac{\partial h^2}{\partial y}\right)_n - \frac{K_s}{2}\left(\frac{\partial h^2}{\partial y}\right)_s \right] \Delta x - q\Delta x\, \Delta y = 0$$

$$\frac{K_e}{2}\left(\frac{h_E^2 - h_P^2}{(\delta x)_e} \right)\Delta y - \frac{K_w}{2}\left(\frac{h_P^2 - h_W^2}{(\delta x)_w} \right)\Delta y + \frac{K_n}{2}\left(\frac{h_N^2 - h_P^2}{(\delta y)_n} \right)\Delta x - \frac{K_s}{2}\left(\frac{h_P^2 - h_S^2}{(\delta y)_s} \right)\Delta x - q\Delta x\, \Delta y = 0$$

assuming $H = h^2$, we have

$$K_e \frac{h_E + h_P}{2}\left(\frac{h_E - h_P}{(\delta x)_e} \right)\Delta y + K_w \frac{h_P + h_W}{2}\left(\frac{h_W - h_P}{(\delta x)_w} \right)\Delta y + K_n \frac{h_N + h_P}{2}\left(\frac{h_N - h_P}{(\delta y)_n} \right)\Delta x$$

$$- K_s \frac{h_P + h_S}{2}\left(\frac{h_S - h_P}{(\delta y)_s} \right)\Delta x - q\Delta x\, \Delta y = 0$$

We have $h_e = (h_E + h_P)/2$, $h_w = (h_W + h_P)/2$, $h_n = (h_N + h_P)/2$, $h_s = (h_S + h_P)/2$, thus,

$$K_e h_e\left(\frac{h_E - h_P}{(\delta x)_e} \right)\Delta y + K_w h_w\left(\frac{h_W - h_P}{(\delta x)_w} \right)\Delta y + K_n h_n\left(\frac{h_N - h_P}{(\delta y)_n} \right)\Delta x$$

$$- K_s h_s\left(\frac{h_S - h_P}{(\delta y)_s} \right)\Delta x - q\Delta x\, \Delta y = 0$$

These equations can be solved for one quarter of the plan. It is assumed that $\partial h^2/\partial x = 0$ and $\partial h^2/\partial y = 0$ are the borders of each quarter.

The boundary conditions are shown in Figure 6.32. The mathematical forms of these boundary conditions are presented in the following.

For the left boundary, the equation for the control volume shown in Figure 6.32a is

$$\frac{K_e}{2}\left(\frac{h_E^2 - h_P^2}{(\delta x)_e} \right)\Delta y + \frac{K_n}{2}\left(\frac{h_N^2 - h_P^2}{(\delta x)_n} \right)\frac{\Delta x}{2} - \frac{K_s}{2}\left(\frac{h_P^2 - h_S^2}{(\delta x)_s} \right)\frac{\Delta x}{2} - q\Delta x\, \Delta y = 0$$

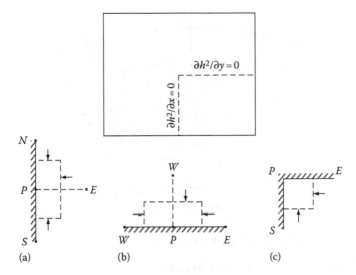

FIGURE 6.32 Boundary conditions of Example 6.12.

If $\Delta x = \Delta y$ and $K_x = K_y = K$, then

$$h_P = \left(\frac{h_E^2}{2} + \frac{h_N^2}{4} + \frac{h_S^2}{4} - \frac{q \Delta x \, \Delta y}{K} \right)^{1/2}$$

For the bottom boundary, the equation for the control volume shown in Figure 6.32b is

$$\frac{K_e}{2} \left(\frac{h_E^2 - h_P^2}{(\delta x)_e} \right) \frac{\Delta y}{2} - \frac{K_w}{2} \left(\frac{h_P^2 - h_W^2}{(\delta x)_w} \right) \frac{\Delta y}{2} - \frac{K_n}{2} \left(\frac{h_N^2 - h_P^2}{(\delta x)_n} \right) \Delta x - q \Delta x \, \Delta y = 0$$

If $\Delta x = \Delta y$ and $K_x = K_y = K$, then

$$h_P = \left(\frac{h_E^2}{4} + \frac{h_W^2}{4} + \frac{h_N^2}{2} - \frac{q \Delta x \, \Delta y}{K} \right)^{1/2}$$

For the corner boundary, the equation for the control volume shown in Figure 6.32c is

$$\frac{K_e}{2} \left(\frac{h_E^2 - h_P^2}{(\delta x)_e} \right) \frac{\Delta y}{2} + \frac{K_s}{2} \left(\frac{h_S^2 - h_P^2}{(\delta x)_s} \right) \frac{\Delta x}{2} - q \Delta x \, \Delta y = 0$$

If $\Delta x = \Delta y$ and $K_x = K_y = K$, then

$$h_P = \left(\frac{h_E^2}{2} + \frac{h_S^2}{2} + \frac{q \Delta x \, \Delta y}{K} \right)^{1/2}$$

The discharge from each well is determined with trial and error. The piezometric heads are calculated with some discharge values from wells to set the water table at the given depth.

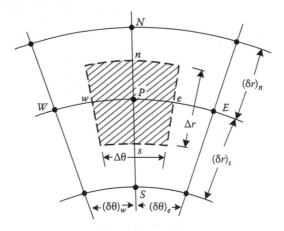

FIGURE 6.33　Control volume in polar coordinates.

6.6.4 ORTHOGONAL COORDINATE SYSTEM

The discretization equations in orthogonal coordinate system are formulated by using a grid in the Cartesian coordinate system. However, the presented method is not limited to Cartesian grids but they can be developed with a grid in any orthogonal coordinate system. In order to illustrate the derivation of the discretization equation in Cartesian coordinate systems, a 2D situation is considered in polar coordinates, r and θ. The $r\theta$ counterpart is

$$\frac{1}{r}\frac{d}{dr}\left(rK\frac{dh}{dr}\right)+\frac{1}{r}\frac{\partial}{\partial\theta}\left(\frac{K}{r}\frac{\partial h}{\partial\theta}\right)=\frac{S}{b}\frac{\partial h}{\partial t} \tag{6.101}$$

The grid and the control volume in $r\theta$ coordinates are shown in Figure 6.33. The z-direction, thickness of the control volume, is assumed to be unity.

The discretization equation is obtained by multiplying Equation 6.101 by r and integrating with respect to r and θ over the control volume. (This operation gives the volume integral, since $rdrd\theta$ represents a volume element of unit thickness). Discretization equations in orthogonal coordinate system are then derived similar to the Cartesian coordinate.

Example 6.13

Develop the finite volume approximation of flow equation for a 2D confined aquifer in the orthogonal coordinate system.

Solution

Since flow is symmetric, the governing equation can be written as

$$\frac{1}{r}\frac{d}{dr}\left(r\frac{dh}{dr}\right)=\frac{S}{T}\frac{\partial h}{\partial t}$$

For the grid and the control volume in $r\theta$ coordinates, as shown in Figure 6.33, we have

$$\int_t^{t+\Delta t}\int_w^e\int_s^n \frac{1}{r}\frac{\partial}{\partial r}\left(r\frac{\partial h}{\partial r}\right)dr\,(rd\theta)\,dt = \int_s^n\int_w^e\int_t^{t+\Delta t}\frac{S}{T}\frac{\partial h}{\partial t}\,dt\,(rd\theta)\,dr$$

The left side of the equation is

$$\int_t^{t+\Delta t}\int_w^e r\left.\frac{\partial h}{\partial r}\right|_s^n d\theta\,dt = \int_t^{t+\Delta t}\int_w^e\left[r_n\left.\frac{\partial h}{\partial r}\right|_n - r_s\left.\frac{\partial h}{\partial r}\right|_s\right]d\theta\,dt$$

$$\int_t^{t+\Delta t}\int_w^e r\left.\frac{\partial h}{\partial r}\right|_s^n d\theta\,dt = \int_t^{t+\Delta t}\left[r_n\frac{h_N - h_P}{(\delta r)_n} - r_s\frac{h_P - h_S}{(\delta r)_s}\right]\Delta\theta\,dt$$

$$\int_t^{t+\Delta t}\int_w^e r\left.\frac{\partial h}{\partial r}\right|_s^n d\theta\,dt = \left[r_n\frac{h_N^{m+1} - h_P^{m+1}}{(\delta r)_n} - r_s\frac{h_P^{m+1} - h_S^{m+1}}{(\delta r)_s}\right]\Delta\theta\,\Delta t$$

And the right side of the equation is

$$\int_s^n\int_w^e\int_t^{t+\Delta t}\frac{S}{T}\frac{\partial h}{\partial t}\,dt\,(rd\theta)\,dr = \int_s^n\int_w^e\frac{S}{T}[h_P]_t^{t+\Delta t}\,(rd\theta)\,dr$$

$$\int_s^n\int_w^e\int_t^{t+\Delta t}\frac{S}{T}\frac{\partial h}{\partial t}\,dt\,(rd\theta)\,dr = \int_s^n\int_w^e\frac{S}{T}(h_P^{m+1} - h_P^m)(rd\theta)\,dr$$

$$\int_s^n\int_w^e\int_t^{t+\Delta t}\frac{S}{T}\frac{\partial h}{\partial t}\,dt\,(rd\theta)\,dr = \int_s^n\int_w^e\frac{S}{T}(h_P^{m+1} - h_P^m)(rd\theta)\,dr$$

$$\int_s^n\int_w^e\int_t^{t+\Delta t}\frac{S}{T}\frac{\partial h}{\partial t}\,dt\,(rd\theta)\,dr = \int_s^n\frac{S}{T}(h_P^{m+1} - h_P^m)r\Delta\theta\,dr$$

$$\int_s^n\int_w^e\int_t^{t+\Delta t}\frac{S}{T}\frac{\partial h}{\partial t}\,dt\,(rd\theta)\,dr = \frac{S}{T}(h_P^{m+1} - h_P^m)r\Delta\theta\int_s^n r\,dr$$

$$\int_s^n\int_w^e\int_t^{t+\Delta t}\frac{S}{T}\frac{\partial h}{\partial t}\,dt\,(rd\theta)\,dr = \frac{S}{T}(h_P^{m+1} - h_P^m)r\Delta\theta\left(\frac{r_n^2 - r_s^2}{2}\right)$$

Therefore,

$$\left[r_n \frac{h_N^{m+1} - h_P^{m+1}}{(\delta r)_n} - r_s \frac{h_P^{m+1} - h_S^{m+1}}{(\delta r)_s} \right] \Delta\theta\Delta t = \frac{S}{T}(h_P^{m+1} - h_P^m) r \Delta\theta \left(\frac{r_n^2 - r_s^2}{2} \right)$$

The continuity equation is obtained by rearranging the equation:

$$K \left[\frac{h_N^{m+1} - h_P^{m+1}}{(\delta r)_n} (b \, r_n \Delta\theta) + \frac{h_S^{m+1} - h_P^{m+1}}{(\delta r)_s} (b \, r_s \Delta\theta) \right] = \frac{S}{T} \frac{(h_P^{m+1} - h_P^m)}{\Delta t} \Delta r \left(\Delta\theta \frac{r_n + r_s}{2} \right)$$

This equation shows that the volume of water flowing across the boundaries (the left side of the equation) is equal to the volume of water removed from storage over the entire control volume (the right side of the equation).

6.7 MODFLOW

MODFLOW is a 3D, finite-difference groundwater model that was first released in 1984. MODFLOW is a modular 3D, finite-difference groundwater flow model, developed by the U.S. Geological Survey (McDonald and Harbaugh, 1988). This program is designed to simplify model development and data input for groundwater modeling to develop maps, diagrams, and text files. It is one of the most widely used groundwater simulation models. This model can be found and downloaded from USGS Web site. Many new capabilities have been added to the original model in recent versions.

MODFLOW-2000 simulates steady and unsteady flow in an irregularly shaped flow system in which aquifer layers can be confined, unconfined, or a combination of confined and unconfined. Flow from external stresses, such as flow to wells, surface recharge, evapotranspiration, flow to drains, and flow through river beds, can be simulated. Hydraulic conductivities or transmissivities for any layer may differ spatially and be anisotropic (restricted to having the principal directions aligned with the grid axes), and the storage coefficient may be heterogeneous. Specified head and flux boundaries can be simulated by this model. It is also capable of modeling head-dependent flux across the outer boundary of the model, which allows water to be supplied to a boundary block in the modeled area at a rate proportional to the current head difference between a source of water outside the modeled area and the boundary block.

In addition to simulating groundwater flow, the scope of MODFLOW-2000 has been expanded to incorporate related capabilities such as solute transport and parameter estimation. In this model, the groundwater flow equation is solved using the FDA. The flow region is subdivided into blocks in which the medium properties are assumed to be uniform. In plan view, the blocks are made from a grid of mutually perpendicular lines that may be variably spaced. Model layers can have varying thickness. A flow equation is written for each block, or cell. Several methods are provided for solving the resulting matrix problem; the user can choose the best one for a particular problem. Flow rate and cumulative volume, which are balanced from each type of inflow and outflow, are computed for each time step.

MODFLOW-2005 is similar in design to MODFLOW-2000. The primary change in MODFLOW-2005 is the incorporation of a different approach for managing internal data. Fortran modules are used to declare data that can be shared among subroutines. This allows data to be shared without using subroutine arguments. As a result of using Fortran modules, a change in terminology for MODFLOW has been made.

Processing MODFLOW for Windows (PMWIN) is a simulation system for modeling groundwater flow and transport processes with the modular 3D, finite-difference groundwater model MODFLOW. PMWIN also supports the calculation of elastic and inelastic compaction of an aquifer due to changes of hydraulic heads (Chiang and Kinzelbach, 2005).

Groundwater modeling system (GMS) is a powerful graphical tool for model creation and visualization of results. Models can be built using digital maps and elevation models for reference and source data. One of GMS's greatest strengths traditionally has been the conceptual model approach. This approach makes it possible to build a conceptual model in the GMS map module using GIS feature objects. The conceptual model defines the boundary conditions, sources/sinks, and material property zones for a model. The model data can then be automatically discretized to the model grid or mesh. The conceptual model approach makes it possible to deal with large complex models in a simple and efficient manner. An example solved with MODFLOW is presented in the following section.

The first step toward creating a model is the development of a grid system, then layers are added and parameters of hydraulic conductivity, initial head, storage coefficient, top elevation, and bottom elevation are specified for each cell. The number of stress periods over which the model will run and the time of their operation should be set.

Results could be analyzed through contour plots, if desired. Superimposing a grid over a graphic image or digitized line drawing is one of the most powerful features of graphic groundwater. These images are generally 2D plan views representing the area that is going to be modeled.

Example 6.14

This tutorial presents a model to simulate the groundwater of Lenjanat aquifer and Zayandeh-Rud River in central Iran.

In this area, 750 and 220 MCM/year water is provided for agricultural and industrial consumption, respectively. Meanwhile, more than 800 million cubic meters is supplied from Zayandeh-Rud River. This river has an average of 1300 MCM discharge per year. The return flows from agricultural and industrial sectors, to groundwater are, respectively, 15% and 10% of the allocated water. The transmissivity in Lenjanat unconfined aquifer varies from 300 to 2400 m²/day. The average coefficient of hydraulic conductivity is about 7 m/day.

There exist about 40 piezometric wells to record the water table of this aquifer. Lenjanat aquifer is mainly recharged by precipitation, infiltration from the bed of river, inflow from neighbor aquifers, and return flows from industrial and agricultural sectors. The discharge from the aquifer occurs through water extraction from wells, springs, and qanats, as well as groundwater outflow and evapotranspiration.

The industrial and agricultural return flows are the most important sources of water pollution in the region. The recharged water from the aquifer to the river is

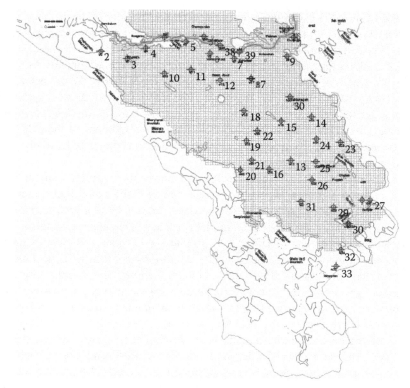

FIGURE 6.34 Study area, showing the location of the river and piezometric wells in the region.

also the cause of increasing water salinity. The TDS variation could be a good index for assessing these pollutants and, therefore, it was used in this study (Figure 6.34).

The problem consists of creating a mathematical model to simulate Lenjanat aquifer by means of PMWIN software. The goal is to define the monthly average groundwater table variations in the agriculture zones, which are functions of discharge, recharge, water head in the river, inflow, and outflow at the boundaries as well as physical characteristics of the aquifer.

Solution

Step 1: Grid creation

The model is created from the file menu. Next, the mesh size including the number of layers, columns, rows are entered in the model dimension box according to Figure 6.35. Then, the grid is generated as shown in Figure 6.36.

Step 2: Layer properties

Using the layer options from the grid menu, the type of layers is assigned. In this case, the unconfined type is chosen for the aquifer (Figure 6.37).

Step 3: Using background maps

In this step, using the maps options, up to five background DXF maps are displayed. As shown in Figure 6.38, the study area map is imported.

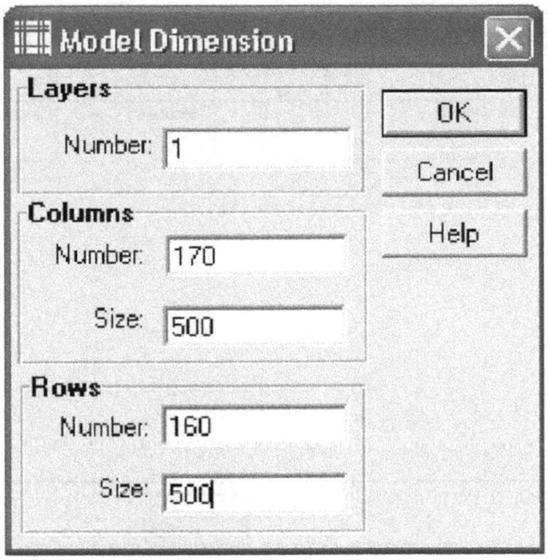

FIGURE 6.35   Model dimension input box.

FIGURE 6.36   Generated model grid.

Step 4: Coordinates properties

The grid coordinates and worksheet size are determined by entering the sizes in the environment option. The grid coordinates and worksheet size are determined by entering the sizes in the "environment option" as shown in Figure 6.39.

Step 5: Boundary conditions

In order to define boundary conditions, the boundary condition is selected from menu and the value 1 is entered as shown in Figure 6.40. The number of active gird cells is 6210.

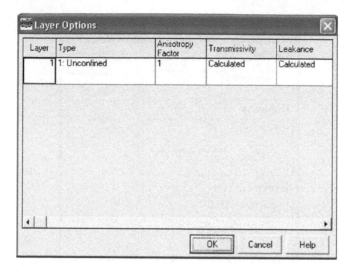

FIGURE 6.37   The layer options dialog box and the layer type drop-down list.

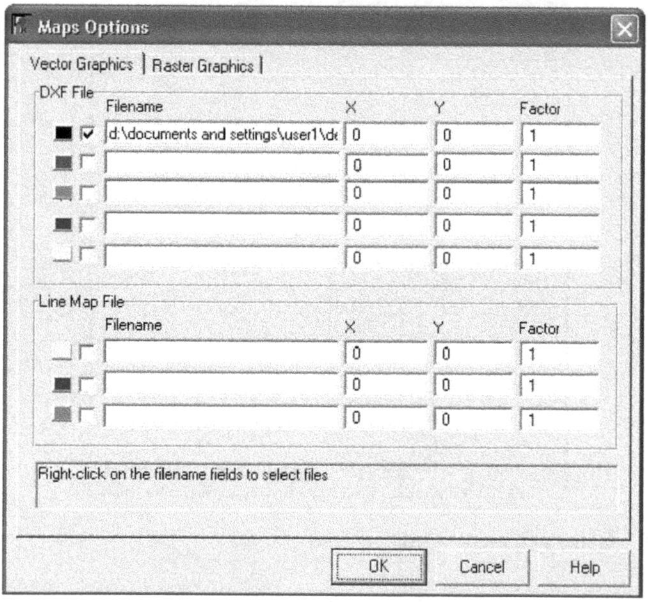

FIGURE 6.38   Map options window.

Step 6: Layer elevations

The top elevation of each aquifer is specified in the Top of Layer and Bottom of Layer menu. The values are entered based on the topographic map and geophysical studies in the study area. The top and the bottom elevations are 1865.01 and 1515.18 (meters above sea level), respectively.

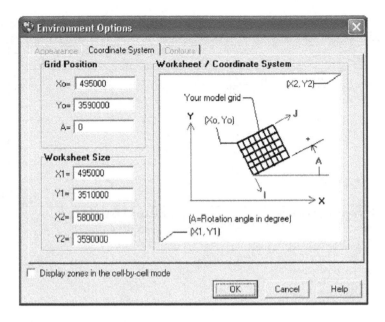

FIGURE 6.39 Environment option window.

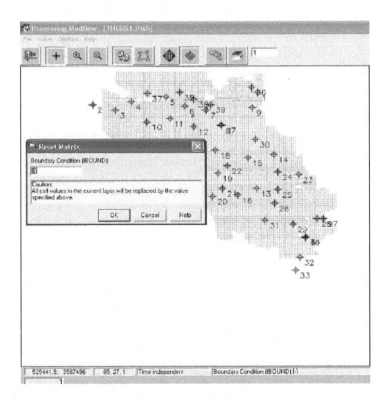

FIGURE 6.40 Boundary condition window.

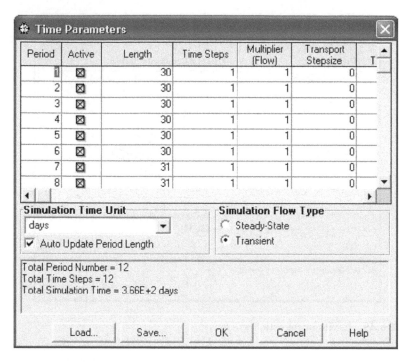

FIGURE 6.41   Time parameter window.

Step 7: Simulation time properties

After determining the aquifer elevation, the time parameters are indicated by choosing the time menu. The simulation time unit changes to day and the flow type is transient in this example as shown in Figure 6.41.

Step 8: Initial conditions

The initial hydraulic heads are loaded in browse matrix box. Its value in this study is considered 1874/484 m (Figure 6.42).

Also, the horizontal hydraulic conductivity, effective porosity, and specific yield are chosen from the parameter menu and their values are entered in the reset matrix box from the value menu. These parameter values in the study area are 3 m/day, 25%, and 0.58, respectively. The related boxes are shown in Figures 6.43 through 6.45.

In addition, the recharge rate with the value of 1.9778×10^{-4} m/day as a necessary input model is entered. This value is determined in recharge box from model menu (Figure 6.46). Also, in order to edit the model data for a specific stress period, a period from the table of temporal data is selected and then the edit data is pressed (Figure 6.47).

In the next step, the river data for each cell is entered. To do this, the river menu from MODFLOW model is chosen. Then, the river hydraulic conductivity, head and elevation of the riverbed bottom saved in three ASCII matrix files are imported. These files are loaded in reset matrix box from value menu as shown in Figure 6.48.

Then, the well data should be specified to the model. To enter the recharge rate of the well, which is $-38/77$ m³/day in this example, the well item is selected from

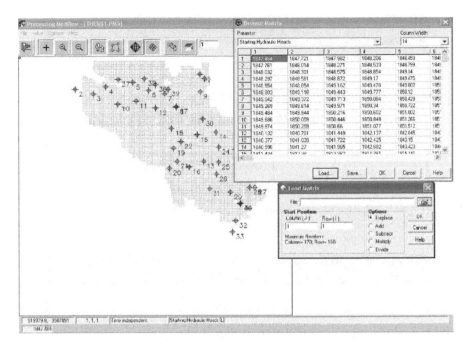

FIGURE 6.42   Browse matrix window.

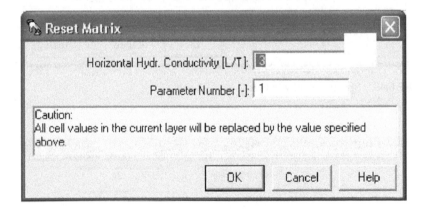

FIGURE 6.43   Horizontal hydraulic conductivity window.

the MODFLOW model. Then, by choosing the reset matrix menu, the cell value box, as shown in Figure 6.49, is appeared. The value is entered into the box and this step is repeated with wells from 2 to 39 in this example.

Step 9: Output control

After entering all parameters of initial conditions, the user should control the output of these unformatted files. This can be done by choosing the output control from the MODFLOW menu. The related box is shown in Figure 6.50. The output terms can be seen in this figure. The predefined values of non-flow cells and dry cells are −999.99 and 1E + 30, respectively.

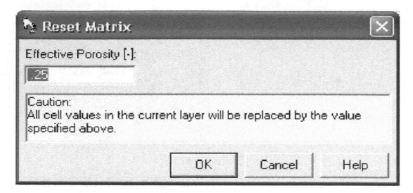

FIGURE 6.44   Effective porosity window.

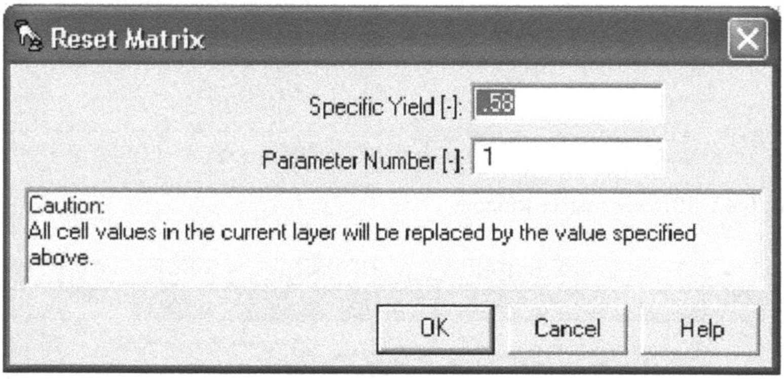

FIGURE 6.45   Specific yield window.

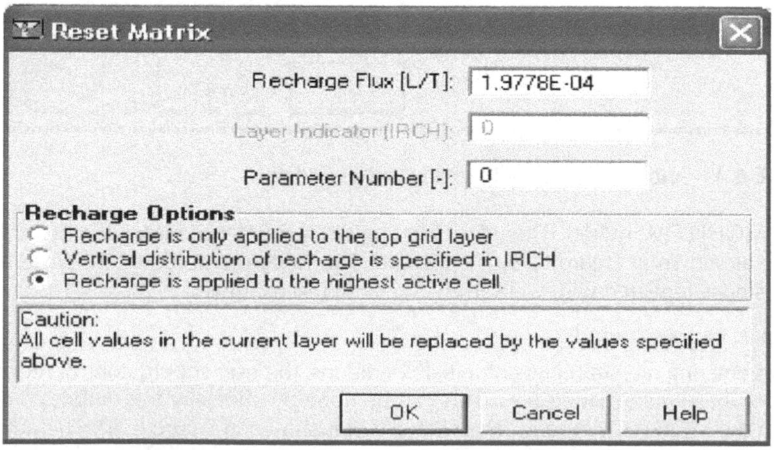

FIGURE 6.46   Recharge rate window.

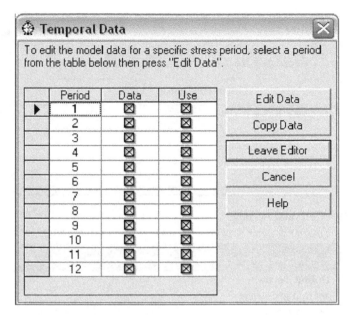

FIGURE 6.47 Temporal data window.

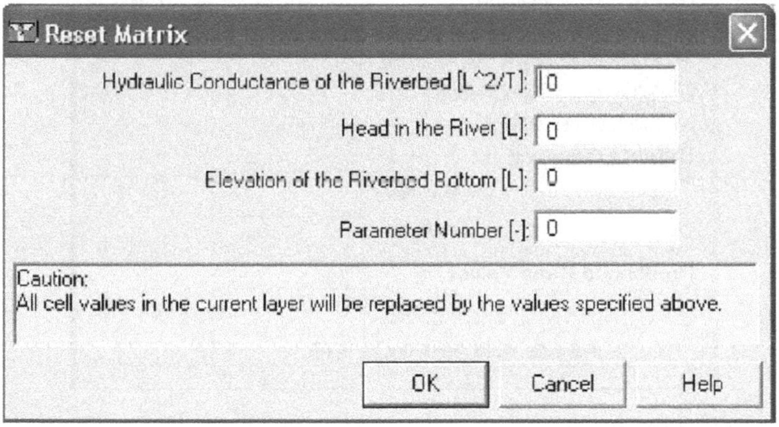

FIGURE 6.48 Reset matrix window for the river package.

Step 10: Run MODFLOW

In this step, the run item from the MODFLOW menu is selected to generate the required data files and to run the model. As shown in Figure 6.51, the (c:\program files\pmwin\modflw96\lkmt2\modflow2v.exe) file from the MODFLOW program is chosen and the model is run (Figure 6.52).

Step 11: MT3D data preparation

To evaluate the aquifer qualification, the salinity amount is needed. Then, by selecting the matrix item from the value menu, the subdivision MT3D is chosen and the initial concentration values are entered and loaded (Figure 6.53).

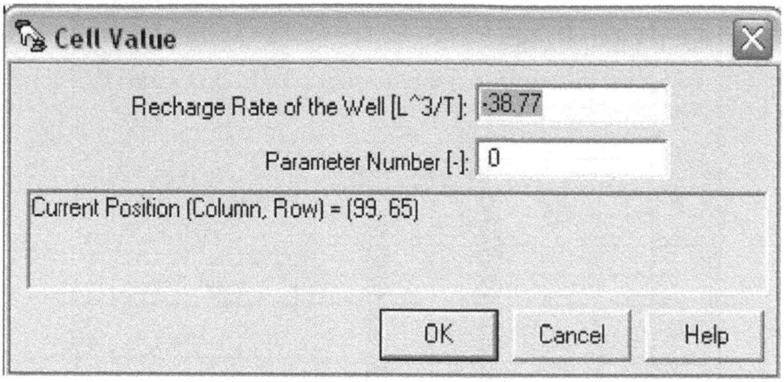

FIGURE 6.49   Cell value window.

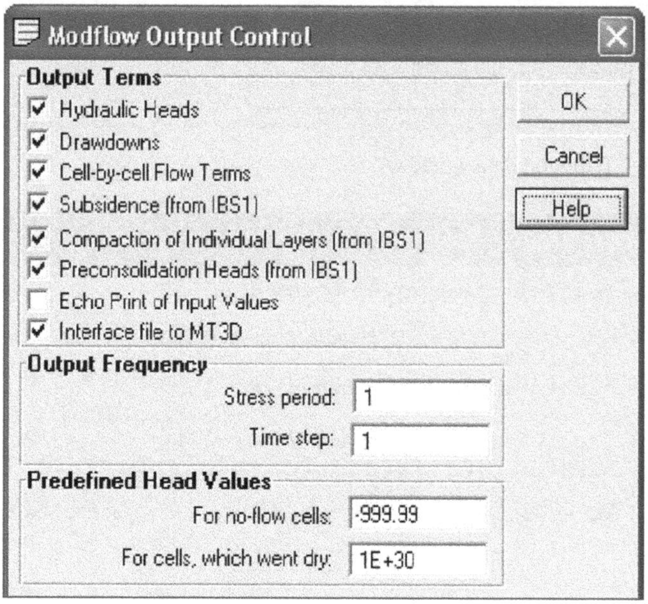

FIGURE 6.50   MODFLOW output control window.

In addition, the transport parameters to the advection package are assigned by choosing the MT3D item from the model menu. In the advection package box, as shown in Figure 6.54, the methods of characteristics for the solution scheme and first-order Euler for the particle tracking algorithm are selected.

In the next step, to specify the dispersion package of aquifer salinity, the dispersion item from the MT3D menu is selected; then, the ratio of the transverse dispersivity to longitudinal dispersivity is entered as shown in Figure 6.55. The DMCOEF value is considered $0.000004\,\text{m}^2/\text{day}$ for this example. This parameter is the effective molecular diffusion coefficient.

Moreover, to specify the output times, the parameters, which should be controlled, are selected. To do this, the output control from the model menu is chosen

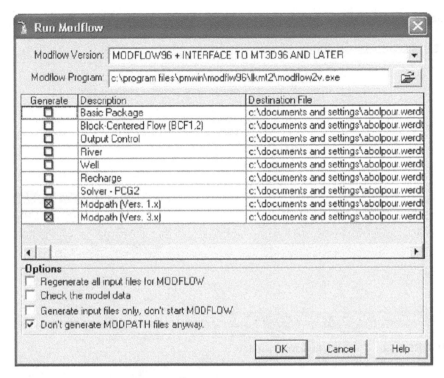

FIGURE 6.51 Configuration to run MODFLOW.

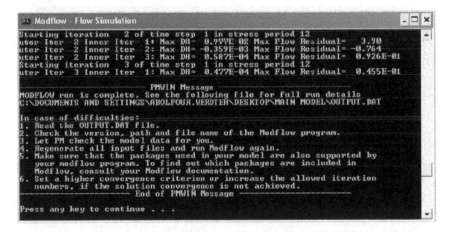

FIGURE 6.52 Completion of the flow simulation in MS-DOS environment.

and the related box is appeared. As shown in Figure 6.56, the options in this box are grouped in three tabs. For this example, the value of 30 for interval is entered.

To perform the transport simulation, the subdivision Run from MT3D program is selected. Figure 6.57 is opened and by choosing the (c:\program files\pmwin\mt3d\mt3dv.exe) file, the model is run. Prior to running MT3D, PMWIN will use user-specified data to generate input files for MT3D (Figure 6.58).

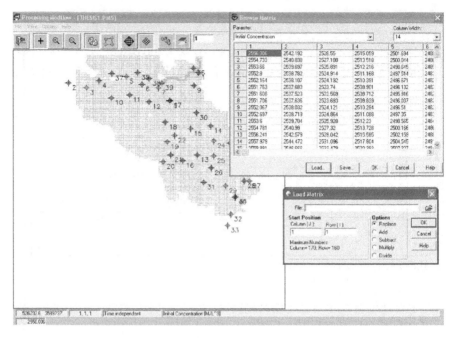

FIGURE 6.53   The browse matrix window.

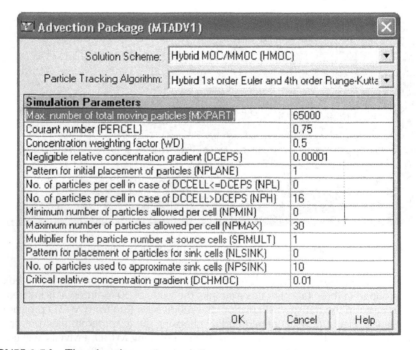

FIGURE 6.54   The advection package window.

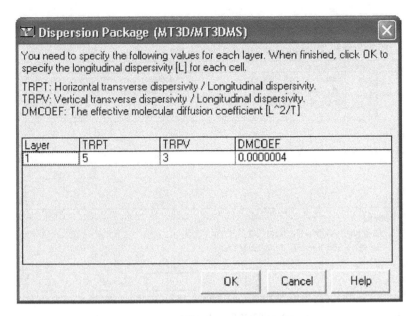

FIGURE 6.55 The advection package (MTADV1) dialog box.

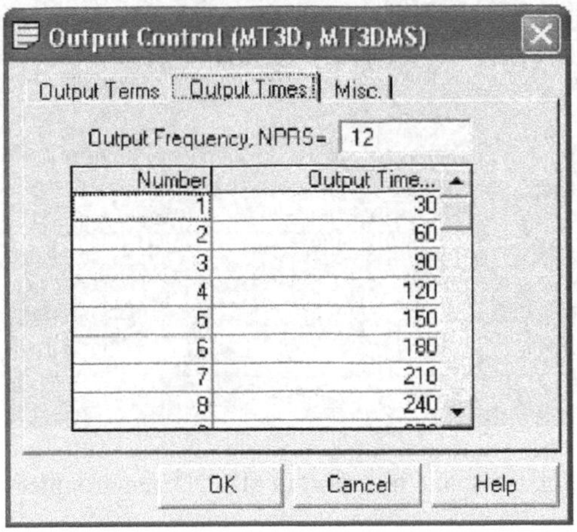

FIGURE 6.56 The output control dialog box.

Step 12: The output evaluation

The graphs of observed and calculated head can be compared using tools menu. As shown in Figure 6.59, the head-time curves are opened. Their scatter diagram can be seen in Figure 6.60. These two figures show the results in all existing wells.

As a result of comparison, the calculated head value is fitted on observed value.

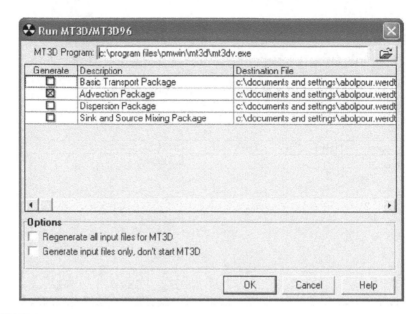

FIGURE 6.57   The run MT3D/MT3D96 dialog box.

```
MT3D - Solute Transport                                            - □ x
STRESS PERIOD NO.   11

TIME STEP NO.    1
FROM TIME =   304.00      TO    335.00
Transport Step:    2    Step Size:   5.000      Total Elapsed Time:    335.00

STRESS PERIOD NO.   12

TIME STEP NO.    1
FROM TIME =   335.00      TO    366.00
Transport Step:    1    Step Size:   31.00      Total Elapsed Time:    366.00

------------ PMWIN Message ------------
MT3D run is complete. See the following file for full run details
C:\DOCUMENTS AND SETTINGS\ABOLPOUR.WERDIEH\DESKTOP\MAIN MODEL\OUTPUT.MT3

In case of difficulties:
1. Check the version, path and file name of the MT3D program.
2. Regenerate all input files and run MT3D again.
3. Make sure that the packages used in your model are also supported by your
   MT3D and modflow program. To find out which packages are included in
   MT3D or Modflow, consult your program documentations.
------------ End of PMWIN Message ------------
Press any key to continue . . .
```

FIGURE 6.58   Running the solute transport in MS-DOS environment.

The model is used to stimulate one year data. The time step is monthly and the measurement unit is day. Also, the flow type is unsteady.

In Figures 6.61 and 6.62, the drawdown–time curve for the well 34 and concentration–time curve for the well 39 are shown.

Also, in this example, the TDS is selected as a quality's variable index for the study aquifer. This index, based on the availability of salinity data, is considered. The contour maps of concentration and head variation are shown in Figures 6.63 and 6.64.

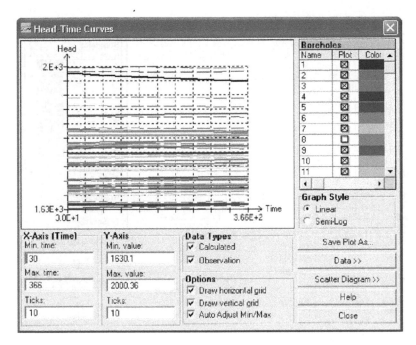

FIGURE 6.59   Head-time curves window.

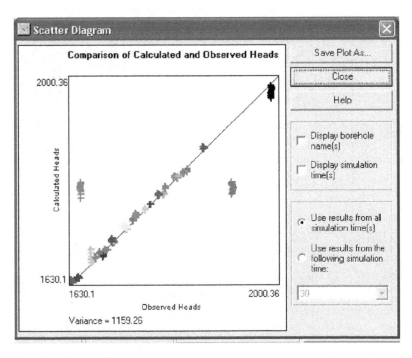

FIGURE 6.60   Scatter diagram window.

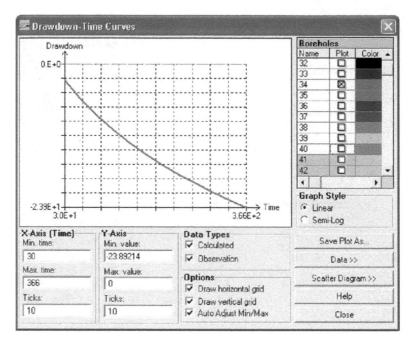

FIGURE 6.61 Drawdown–time curves window.

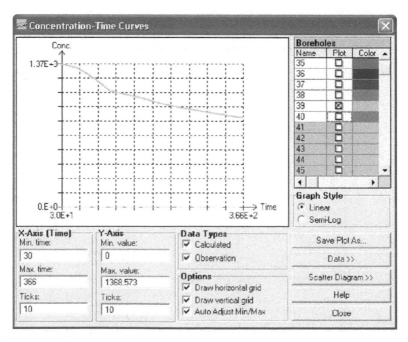

FIGURE 6.62 Concentration–time curves window.

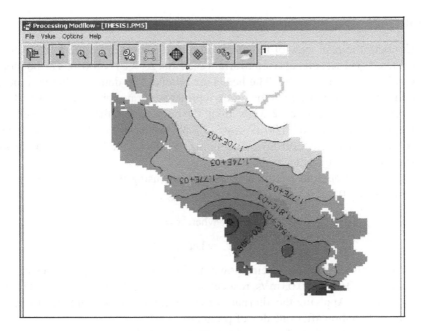

FIGURE 6.63 Water table elevation in the study area.

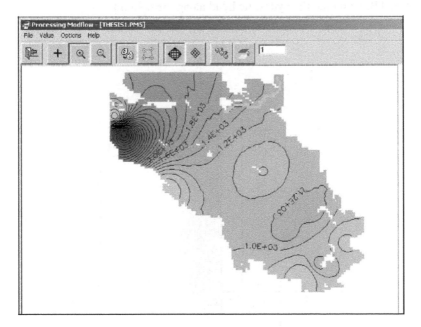

FIGURE 6.64 TDS concentration contours in the study area.

PROBLEMS

6.1 Consider the plan view of the confined aquifer shown in Figure 6.10. A mesh-centered grid system has been overlain such that $\Delta x = \Delta y = 100\,$m. The boundary conditions are also given. The heads are relative to a datum at the base of the aquifer. The shaded area in the figure represents a gravel pit which is to be dewatered by pumping from 12 wells along the boundary of the pit. Each well is to be pumped at a rate of $1200\,$m^3 day^{-1}, $T = 300\,$m^2 day^{-1} and $S = 0.002$. Compute the drawdowns after 200 days of pumping. For initial conditions, use the steady-state head configuration without pumping.

6.2 Solve the differential equation of $d^2h/dx^2 = 0$ using the Galerkin method and considering: $0 \leq x \leq 3$

$$h = 0\,\text{m} \quad \text{where } x = 0$$

$$h = 10\,\text{m} \quad \text{where } x = 3$$

6.3 In the confined aquifer system shown in Figure 6.6, the pumping rates in wells 1 and 2 are 0.02 and 0.03 m³/s, respectively. The initial piezometric head is 20 m everywhere. Applying the alternate-direction implicit procedure, determine the head distribution after one day of pumping.

6.4 The column of soil in Figure 6.65 is saturated and water is flowing vertically downwards at a constant rate Q. Hydraulic head is held constant at the upper and lower ends of the column and we wish to calculate the values of head at points A and B. Compute the hydraulic head along the column.

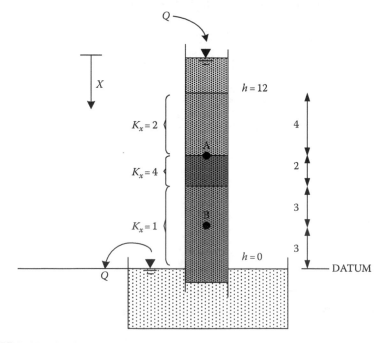

FIGURE 6.65   A soil column of Problem 6.4.

6.5 Consider the column of soil shown in the Figure 6.65. Initially the column is in steady-state saturated flow with a distribution of hydraulic head computed from the previous example. Then at time $t = 0$ the value of hydraulic head increases at the upper boundary (node 1) from 12 to 25 cm. Find the value of hydraulic head at each node at time $t = 1, 2, \ldots$ seconds. $S_S^{(1)} = 0.02$, $S_S^{(2)} = 0.04$, $S_S^{(3)} = S_S^{(4)} = 0.02$.

6.6 Consider a steady-state, saturated groundwater flow is occurring in a confined aquifer. Initially no solute is present. At time zero, the solute concentration along the left boundary of the aquifer is increased to 10 mg/1 and remains constant thereafter. Assuming $V_x^{(e)} = 0.03$ m/d, $D_x^{(e)} = 1$ m²/d, $L^{(e)} = 10$ m, $\theta^{(e)} = n^{(e)} = 0.3$ m, compute the hydraulic head in five elements with equal condition of the aquifer.

6.7 Consider one dimensional flow in a confined aquifer shown in Figure 6.66. Let $\Delta x = 3$ m, $b = 6$ m, $h_1 = 15$ m, $h_5 = 8$ m, $K = 0.5$ m/day. Determine the spatial variation of piezometric head using finite volume method.

6.8 Estimate the net flow for seepage under the sheet piling shown in Figure 6.67 for $K_1 = 10^{-4}$ m/s, $K_2 = 2 \times 10^{-4}$ m/s, $L = 400$ m and $\Delta x = \Delta y = 10$ m with steady state condition using finite volume method.

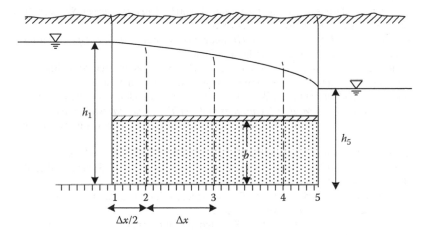

FIGURE 6.66 A confined aquifer between Rivers A and B.

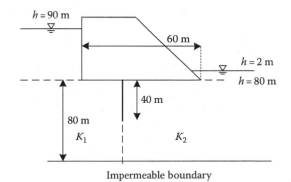

FIGURE 6.67 A dam with a sheet pile in nonhomogenous conductivity soil.

REFERENCES

Aziz, K. and Settari, A. 1972. A new interactive method for solving reservoir simulation equations. *Jour. Canadian Technology*, 11, 62–68.

Bear, J. 1979. *Hydraulics of Groundwater*. McGraw-Hill, New York, 38 p.

Bear, J. and Verruijt, A. 1992. *Modeling Groundwater Flow and Pollution with Computer Programs for Sample Cases*. Redial Publishing Company, Dordrecht, the Netherlands, 414 p.

Bredehoeft, J.D. and Pinder, G.F. 1970. Digital analysis of areal flow in multiaquifer groundwater systems-a quasi three-dimensional model. *Water Resources Research*, 6(3), 883–888.

Carslaw, H.S. and Jaeger, J.C. 1959. *Conduction of Heat in Solids*. Oxford University Press, New York, 510 p.

Chiang, W.H. and Kinzelbach, W. 2005. *3D-Groundwater Modeling with PMWIN: A Stimulation System for Modeling Groundwater Flow and Transport Processes*. Springer-Verlag, Berlin, Germany.

Delleur, J.W., ed. 1999. *The Handbook of Groundwater Engineering*. CRC Press/Springer-Verlag, Boca Raton, FL, 518 p.

Gupta, R.S. 1989. *Hydrology and Hydraulic Systems*. Prentice-Hall, Englewood Cliffs, NJ.

Harr, M.E. 1962. *Groundwater and Seepage*. McGraw-Hill, New York, 309 p.

Istok, J.D. 1989. *Groundwater Modeling by the Finite Element Method*. American Geophysical Union, Washington, DC, 415 p.

Karamouz, M., Szidarovszky, F., and Zahraie, B. 2003. *Water Resources Systems Analysis*. Lewis Publishers, Boca Raton, FL, 608 p.

Llamas, M.R. 2003. *Intensive Use of Groundwater: Challenges and Opportunities*. A.A. Balkema Press, Lisse, the Netherlands, 478 p.

Mandle, R.J. 2002. *Groundwater Modeling Guidance*. Groundwater Modeling Program, Michigan Department of Environmental Quality, Michigan 55 p.

McDonald, M.G. and Harbaugh, A.W. 1988. *A Modular Three-Dimensional Finite-Difference Ground-Water Flow Model: Techniques of Water-Resources Investigations of the United States Geological Survey*, Book 6, Chap. A1, U.S. Geological Survey, Washington, 586 p.

Patankar, S.V. 1980. Numerical heat transfer and fluid flow. Series in *Computational Methods in Mechanics and Thermal Sciences*, W.J. Minkowycz and E.M. Sparrow, eds., McGraw Hill, New York, 214 p.

Peaceman, D. and Rachford, M. 1955. The numerical solution of parabolic and elliptic differential equations. *Journal of SIAM*, 3, 28–41.

Prickett, T.A. and Lonnquist, C.G. 1971. Selected digital computer for groundwater resource evaluation. III. *State Water Survey Bulletin*, 55, 62.

Remson, I., Hormberger, G.M., and Molz, F.J. 1971. *Numerical Methods in Subsurface Hydrology With an Introduction to the Finite Element*. John Wiley & Sons, New York, 416 p.

Rushton, K.R. 1974. Aquifer analysis using backward difference methods. *Journal of Hydrology*, 22, 253–369.

Stone, H.L. 1968. Iterative solution of implicit approximations of multidimensional partial differential equations. *SIAM Journal of Numerical Analysis*, 5, 530–558.

Strack, O.D.L. 1989. *Groundwater Mechanics*. Prentice Hall, Englewood Cliffs, NJ, 732 p.

Thangarajan, M. 2007. *Groundwater: Resource Evaluation, Augmentation, Contamination, Restoration, Modeling and Management*. Springer, the Netherlands, 362 p.

Toth, J. 1962. A theory of groundwater motion in small drainage basins in Central Alberta, Canada. *Journal of Geophysical Research*, 67(11), 4375–4387.

Toth, J., 1963. A Theoretical analysis of groundwater flow in small drainage basins. *Journal of Geophysical Research*, 68, 4795–4812.

Trescott, P.C., Pinder, G.F., and Larson, S.P. 1976. *Finite-Difference Model for Aquifer Simulation in Two- Dimensions with Results of Numerical Experiments. Techniques of water-resources investigations of the United States Geological Survey*, U.S. Geological Survey, Washington, 116 p.

Van Genuchten, M.Th. and Alves, W.J. 1982. *Analytical Solutions of the One-Dimensional Convective-Dispersive Solute Transport Equation*. U.S. Department of Agriculture, Washington, DC, 151 p.

Walton, W. 1989. *Analytical Ground Water Modeling*. Lewis Publishers, Chelsea, MI, 173 p.

Wang, H.F. and Anderson, M.P. 1995. *Introduction to Groundwater Modeling: Finite Difference and Finite Element Methods*. Academic Press, New York, 237 p.

Watts, J.W. 1971. An interactive matrix solution method suitable for anisotropic problems. *Society of Petroleum Engineers Journal*, 11, 47–51.

Watts, J.W. 1973. A method for improving line successive over-relaxation in anisotropic problems: A theoretical analysis. *Society of Petroleum Engineers Journal*, 13, 105–118.

Weinstein, H.G., Stone, H.L., and Kwan, T.V. 1970. Simultaneous solution of multiphase reservoir flow equations. *Society of Petroleum Engineers Journal*, 10(2), 99–110.

7 Groundwater Planning and Management

7.1 INTRODUCTION

The development of groundwater resources involves planning in terms of an entire aquifer. The increasing demand for water throughout the world makes aquifers vast underground reservoirs and their proper management has become a matter of considerable interest. Groundwater development is characterized by unrestricted land use, growth of agriculture, and use of shallow aquifers, which allows human settlement in large areas including semiarid regions with scarce surface water resources (Kresic, 2008). A program for subsurface water development considering objectives of social, economic, and environmental issues is developed through the management of groundwater. The management objectives include the providing of an economic and continuous water supply to meet a continuous growing demand from an underground water resource.

The management objectives for developing and operating the aquifer consider geologic and hydrologic, economic, legal, political, and financial aspects. The integrated approach that coordinates the use of both surface water and groundwater resources should be considered in selecting the objective functions. The total water resources and water demands are evaluated in the region and then the optimal water development strategies in an area can be obtained considering alternative management plans. One scenario could obtain the quantity of water to meet the water demands with acceptable water quality considering the objective functions. Finally, action plans can be taken by appropriate public bodies or agencies.

The general perception is that groundwater is an unlimited, reliable, and safe drinking water source. This could only occur if there is a balance between water recharged to the aquifer and water removed from wells. Rapid groundwater development causes overexploitation, which leads to lowering of the water table and widespread shallow groundwater contamination. Continued development without a management plan could have more obvious consequences such as water table depletion and land subsidence.

Several technical tools and techniques that are classified as simulation, optimization, and economic analysis are utilized for groundwater planning and management. Engineering applications primarily use analysis and design techniques. Simulation models have been used to test the performance of groundwater systems and answer development questions under different design and operation alternatives in a physical-mathematical fashion. Simulation models are the overlay between engineering and management tools and they may require substantial data and detailed modeling. In the analysis of groundwater systems, three classes of management problems can be considered: the estimation of model parameters, the prediction of system responses, and knowing the set of inputs to the system. The simulation models need to be calibrated and verified utilizing collected and recorded data. One of the simulation techniques that have received a great deal of attention is system dynamics utilizing object-oriented programming. In this chapter, some simulation models are presented.

Considering economic, social, and environmental issues and conflicting objectives in groundwater management leads to a multi-objective planning problem. Therefore, the analysis of the system is a trade-off between different objective functions. Optimal pumping schedules (well locations and pumping rates) to satisfy given water demand, the optimal injection schedules (well locations and injection rates) necessary to satisfy a waste load demand, and optimal allocation schedule from groundwater and surface water resources in a region to meet different water demands are some examples in groundwater management. In this chapter, different optimization models and methods to solve them are introduced. The conventional and evolutionary optimization techniques are then presented. Economic analyses in groundwater systems are described. There are certain inserts from the Karamouz et al. (2003, 2010) books that are adapters for groundwater management. Some case studies on groundwater planning and management using simulation and optimization techniques are presented.

7.2 DATA COLLECTION

Groundwater simulation models require considerable data to define all nodal parameters. The accuracy of simulation depends on the reliability of the estimated parameters as well as accuracy of model and boundary conditions (Karamouz et al., 2003). Recently, computer-based models are used to expedite the modeling. In model calibration and verification stages, extensive filed data are needed to set variable parameters. Considerable technical skill is required as well. Computer models can extend the data collected and enhance findings. They can be utilized to generate management scenarios and run them to get the response of the system into different plans and find the best management practices (BMP) scenarios.

The following data should be collected for groundwater management according to Todd (1980):

- *Topographic data*: They are used for locating and identifying wells, measuring groundwater tables, and plotting areal data.
- *Geologic data*: They are utilized for modeling surface and subsurface water interactions based on the geologic maps.
- *Hydrologic data*: They are used to model the hydrologic cycle elements. They can be precipitation, soil moisture, evaporation and transpiration, temperature, land use, water demand, surface storage, groundwater storage, surface and subsurface inflow and outflow data. The primary purpose of hydrologic data collection is evaluation of the hydrologic conservation equation in an aquifer.

Nowadays, many simulation models utilize the geographic information system (GIS) for data input and presentation of results. Also GIS can be used to evaluate and find patterns of recharge within a region utilizing available information of observed groundwater table at specified locations. In groundwater studies, in order to study the distributions of nonpoint source (NPS) contamination of groundwater, spatial statistics can be used. It could be utilized to determine the spatial correlation of NPS pollution and to eliminate detections from point sources of pollution in a water quality database. Spatial variation of hydrogeologic factors affecting the solute transport including hydraulic conductivity of the aquifer, slope, and groundwater properties can then be assessed.

DRASTIC and SEEPAGE are simple techniques that can be used for developing regional-scale groundwater vulnerability maps. The capabilities of GIS in spatial statistics and grid design can improve the modeling effort. Linking GIS with MODFLOW is another example of a groundwater numerical model interfaced with a GIS.

7.3 SIMULATION TECHNIQUES

A groundwater simulation model is used for mathematical formulations. Generally, the simulation model structure addresses a detailed and realistic representation of the complex physical, economic, and social characteristics of a water resources system utilizing the mathematical formulations.

The simulation model represents a process or concept by means of a number of variables which are defined to represent the inputs, outputs, and internal states of the process, and it utilizes a set of equations and inequalities describing the interaction of these variables. The simulation techniques are used for analyzing the groundwater systems and evaluating the systems performance under a given set of inputs and operating policies. The simulation models may deal with steady-state or transient conditions. The pumping test in the aquifers could be done in both states.

One of the differences between simulation and optimization models is that the simulation model is not used for optimal decision determination in comparison with optimization model. Since groundwater planning models with highly nonlinear relationships and constraints cannot be handled by constrained optimization procedures, the simulation method is used in the optimization process. The optimal scenarios are usually selected based on simulation results of the system and evaluation of the system performance under different groundwater development projects. Therefore, linking the simulation and optimization models for estimation of groundwater system performance for different scenarios is required.

Thus, a simulation model is an input–output relation in a certain sense, when the input is the set of model parameters and our actions, and the output is the consequence of our actions. In Chapter 6, some numerical models for simulation of the groundwater systems are presented. Sometimes, running the numerical simulation models is time consuming and it has a high computational burden. Also in analyzing complex systems, the physical model and the mathematical description is so complicated that no solution algorithm can be developed. In many of these cases, no mathematical model is even available. The only choice we have then is conceptual simulation of the model. In this chapter, some conceptual models recently used for groundwater simulation are introduced.

- **Model calibration**

As the parameters of groundwater models cannot be measured directly, they can be estimated from historical data using an inverse parameter estimation procedure (ReVelle and McGarity, 1997).

The simulation models are categorized in two groups, namely deterministic and stochastic. If the system is subject to random input events, or generates them internally, the model is called stochastic. The model is deterministic if no random components are involved.

In deterministic groundwater simulation models, the difference between the observed data of aquifer responses (such as water table variation) and corresponding values calculated by the model considered as the objective function are minimized in the calibration procedure. The comparison between observed and calculated values is usually subjective and it should be considered that a good match does not necessarily show the validity of the model (Karamouz et al., 2003).

The model calibration is done through a trial-and-error procedure to modify the model parameters by evaluating the model output, using changing of aquifer properties. It could consist of model sensitivity analysis by changing the aquifer recharges and discharges, aquifer properties, and initial and boundary conditions and considering the uncertainties in sources, sinks, and initial and boundary conditions, as well as the uncertainties in aquifer properties. The model calibration procedure is usually time-consuming. Therefore, the experience and engineering judgment of the modeler and considerable technical skill and commitment from personnel are important factors in efficient calibration of the model.

The automated parameter estimation techniques, extended as the simulation model software, based on the least square deviation as a criterion to obtain the estimate of system parameters can be used for expediting the model calibration.

- **Model verification**

The calibrated model should be verified before using it for groundwater simulation. For this purpose, the calibrated model is used to simulate the groundwater system under the observed input data that have not been used in the model calibration process. Then the simulated results are compared with the observed aquifer responses and the reliability and accuracy of model simulation is evaluated. If the model has a good reliability in the verification procedure, it is selected as the simulation model.

One of the limitations in groundwater model verification is the data deficiency. In this case, the unverified model should be applied only if the sensitivity analyses for both model calibration and prediction are performed (ReVelle and McGarity, 1997, editors).

There are several performance functions that may be used to evaluate the simulation model performance, such as root mean squared error (RMSE), mean bias error (MBE), and mean absolute error (MAE). The RMSE, MBE, and MAE performance functions are formulated in the following equations, respectively:

$$\text{RMSE} = \sqrt{\frac{\sum_{t=1}^{n} (y_t - \hat{y}_t)^2}{n}} \tag{7.1}$$

$$\text{MBE} = \frac{\sum_{t=1}^{n} (\hat{y}_t - y_t)}{n} \tag{7.2}$$

$$\text{MAE} = \frac{\sum_{t=1}^{n} |\hat{y}_t - y_t|}{n} \tag{7.3}$$

where
y_t is observed data
$\hat{y}_t$ is simulated data
n is the total number of data

- **Model predictions**

A calibrated model should have a good performance in model predictions. The uncertainties of model parameters and variables as well as boundary conditions are the main sources of prediction errors. For example, if the aquifer boundary conditions such as imposed stresses have been changed during the prediction horizon, the model cannot simulate the future condition of a groundwater system with the calibrated model. Therefore, if a model is continuously used for the prediction of future conditions of a groundwater system, the model should be periodically postaudited, or calibrated using the field monitoring data (Delleur, 1999).

7.3.1 ARTIFICIAL NEURAL NETWORKS

Artificial neural networks (ANN) model is significantly used to simulate a groundwater system with a large number of observed data, but no known mathematical model exists. These models are used in different groundwater systems such as simulation of water quality in monitoring wells or simulation of water table fluctuation. This model is utilized to simplify an excessively complex model by a typical ANN formulation. ANNs are global nonlinear function approximates, and are powerful and easy to use (StatSoft, 2002).

An ANN is a computational model that tries to emulate the structure or functional aspects of biological neural networks. It can be trained to recognize patterns and can learn from their interaction with the environment. The objective of the neural network is to transform the inputs into meaningful outputs. The ANN structure is based on a biological neural system that a network of cells, known as neurons, is proceeded using the powerful functions. An ANN imitates this structure by distributing the computation to small and simple processing units, called artificial neurons or nodes.

Thus, an ANN is composed of many artificial neurons that are linked together according to the specific network architecture. Each network contains basically three types of layers including neurons with similar characteristics. The first layer is called the input layer which connects the input neurons to other layer. The second layer, or hidden layer, is between the input and output layers and not connected externally. The third layer is the output layer which includes the output variables/neurons. Thus, generally, each neuron receives a number of inputs from either the original data or other neurons. The hidden layer can be more than one. Generally, connections are allowed from the input layer to the first hidden layer, from the first hidden layer to the second, and from the last hidden layer to the output layer. The strength of the connection between an input and a neuron is noted by the value of the weight. Negative weight values reflect inhibitory connections, while positive values designate excitatory connections. The neural network model can be considered as a simple subroutine of input–output computations in any descriptive or optimization procedure, because the input–output relation is easy to compute.

- **The multiplayer perception network**

Figure 7.1 shows a simple neural network with three types of layers where the input variables are in the neurons of the input layer and the output variables are in the neurons of the output layer. The number of the input and output variables are R and K, respectively. In order to provide a high level of flexibility in model formulation, the hidden layers are defined in the middle; its neurons correspond to hidden or artificial variables. Weights on all links specified by the rule set and the biases on units (b_i) corresponding to consequents are set so that the network responds in exactly the same manner as the rules upon which it is based.

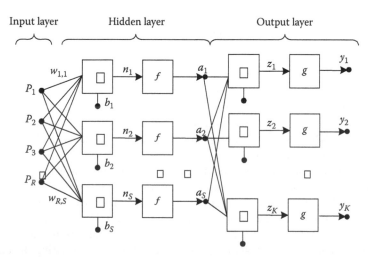

FIGURE 7.1   Illustration of a simple neural network.

Let a_j denote the artificial variables corresponding to the jth neuron of the hidden layer. This variable is calculated as follows:

$$a_j = f\left(\sum_{i=1}^{R} w_{ij} P_i + b_j\right) \tag{7.4}$$

where
R is the number of input variables
f is a transfer function

The coefficients w_{ij} are constants estimated by the training process and specified by the rule set.

The model output is calculated as follows:

$$y_j = g\left(\sum_{i=1}^{S} \overline{w}_{ij} a_i + b'_j\right) \tag{7.5}$$

where S is the number of hidden variables. The coefficients $\overline{w}_{jk}$ are also estimated by the training process. There are many transfer functions used in developing neural networks but here the three most popular ones are introduced. The hard-limit transfer function shown in Figure 7.2a limits the output of the neuron to either 0, if the net input argument n is less than 0; or 1, if n is greater than or equal to 0. This function is used to create neurons that make classification decisions (Demuth and Beale, 2002).

The linear transfer function is shown in Figure 7.2b. Neurons of this type are used as a linear approximation and also employed in the output layer. The sigmoid transfer function depicted in Figure 7.2c takes the input, which may have any value between plus and minus infinity, and squashes the output into the range 0–1. Also the tangent

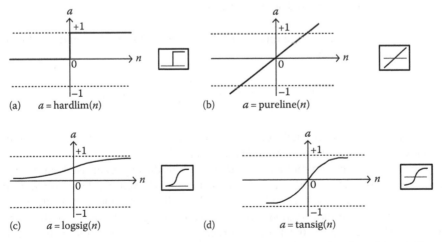

FIGURE 7.2 Transfer function types: (a) hard limit, (b) linear, (c) log sigmoid, and (d) tangent sigmoid.

sigmoid transfer function is shown in Figure 7.2d. The sigmoid transfer function can be formulated as

$$a = \frac{1}{1+e^{-cn}}, \quad c > 0 \tag{7.6}$$

This transfer function is commonly used in back propagation networks, in part because it is differentiable (Demuth and Beale, 2002).

A multilayer perceptron (MLP) network, which is usually called feed forward network, has one or more hidden layers. In this network, the information moves in only one direction, forward, from the input nodes through the hidden nodes to the output nodes. There are no cycles or loops in the network as illustrated in Figure 7.1. Once the network weights are fixed, the states of the units are totally determined by inputs, independent of initial and past states of the units. Thus, MLP is a static network in which information is transmitted through the connections between its neurons only in forward direction and the network has no feedback over its initial and past states. MLP can be trained with the standard back propagation algorithm (BPA).

The ANN models need the estimation of the model parameters including bias value in each neurons of hidden layer (b_j) and outputs layer (b_j') and also weights on all links w_{ij} and $\bar{w}_{ij}$. The most frequently applied methodology is training the network, which is an iteration procedure that successively improves the parameter values. The weights between the output layer and the hidden layer could be modified as

$$\bar{w}_{jk}(t+1) = \bar{w}_{jk}(t) + \beta E_k(t) y_k(t) \tag{7.7}$$

where
  β is the learning rate
  $y_k(t)$ is the actual output
  $E_k(t) = d_k(t) - y_k(t)$ is error of estimation and is calculated using the difference
    between the desired and actual outputs

Since the hidden variables are unknown, the weights between the input and hidden layers can be modified by

$$w_{ij}(t+1) = w_{ij}(t) + \beta a_j(t) \sum_k \bar{w}_{jk}(t) E_k(t) \tag{7.8}$$

This updating process is used iteratively for a large number of input–output pairs until the error terms $E_k(t)$ become sufficiently small.

The BPA is a common method of teaching ANNs during supervised learning for nonlinear multilayer networks. In this method of model training, the gradient of the error function is estimated using the chain rule of differentiation. In this algorithm, network weights are moved along the negative of the gradient of the performance function. The weights in the training process are updated through each iteration (which is usually called epoch) in the steepest descent direction based on the simulation error of the outputs.

Example 7.1

Assume that the groundwater table variation is a function of two variables including volumes of aquifer recharge and aquifer discharge. The monthly data of groundwater table variation and volumes of aquifer recharge and aquifer discharge for 12 years are given in Tables 7.1 through 7.3, respectively. The studies have also demonstrated that the relation of water table variation with these parameters is complicated and could not be formulated in regular form. Develop a MLP model to simulate the monthly groundwater table variation of this aquifer as a function of monthly volumes of aquifer recharge and aquifer discharge.

Solution

The water table and volumes of aquifer recharge and aquifer discharge data are standardized by subtracting the average of data and then dividing to its standard division. It should be noted that for using the log sigmoid function, the data should be between zero and one. In this example, a network with three layers and six neurons in a hidden layer is developed. The log sigmoid and linear functions in hidden and output layers are used as the transfer functions. The number of data for developing the network is 144. As shown in Figure 7.3, 100 series of data are used for model calibration (inputcal and outcal) and the remaining is considered for model validation (inputval and outval). The model is trained in 3500 steps (epochs) and the model goal is to achieve an error of less than 0.00001. The simulated and recorded monthly groundwater table variation data are compared in Figure 7.4. As shown in this figure, about 30% of data have been simulated with less than 5% error and 60% of data have been simulated with an error of less than 10%.

- **Temporal neural networks**

Temporal neural networks are an alternative neural network architecture whose primary purpose is to work on continuous data. The advantage of this architecture is to adapt to the network online and hence helpful in many real-time applications. In many cases such as hydrological modeling, the input pattern comprises one or more temporal signals such as time series prediction. Temporal perceptual learning relies on finding temporal relationships in sensory signal inputs. In an environment, statistically salient temporal correlations can be found by monitoring the arrival times of sensory signals.

Time delay operators, recurrent connections, and the hybrid method are different approaches, which are used to design temporal neural networks. Different types of temporal neural networks are introduced in the following subsection based on Karamouz et al. (2007).

- **Tapped delay line**

Tapped delay line (TDL) is based on a combination of time delay operators in a sequential order. The architecture of TDL corresponds to a buffer containing the N most recent inputs generated by a delay unit operator D. Given an input variable $p(t)$, D operating on $p(t)$ yields its past values $p(t-1)$, $p(t-2)$, ..., $p(t-N)$,

TABLE 7.1
Monthly Recharge to the Aquifer (MCM) of Example 7.1

Year	1995	1996	1997	1998	1999	2000	2001	2002	2003	2004	2005	2006
January	5.13	3.16	8.05	3.79	3.19	3.39	2.77	2.76	2.45	2.7	3.74	6.29
February	5.16	4.07	6.83	5.78	5.86	4.89	6.16	4.1	4.13	4.69	4.35	6.29
March	4.36	4.47	8.04	5.59	5.6	5.46	5.16	5.33	4.48	4.07	5.68	8.08
April	4.23	4.87	7.26	4.84	4.59	5.54	4.38	6.69	4.71	4.45	6.15	8.13
May	4.99	6.1	9.47	5.61	5.7	7.98	5.77	6.06	4.08	5.71	7.45	10.02
June	26.21	9.73	11.02	7.39	15.09	10.32	9.74	7.14	7.36	6.93	13.51	12.73
July	24.27	23.27	29.6	18.22	23.94	14.68	32.43	17.52	18.52	42.41	38.67	17.22
August	14.19	13.99	18.2	14.5	25.23	2.75	24.99	5.45	24.43	39.81	30.05	11.33
September	2.02	4.83	9.6	1.15	12.74	0.66	1.95	4.18	8.16	15.53	10.23	5.96
October	0.33	1.48	2.17	0.06	4.15	0.15	0.07	1.87	2.1	2.48	3.66	2.81
November	0.03	0.42	0.15	0.4	3.41	0.1	0	0.08	0.38	1.35	1.36	0.21
December	2.57	0.84	1.56	0.95	2.63	0.6	2.02	1.15	0.68	1.47	2.37	2.53

TABLE 7.2
Monthly Discharge of the Aquifer (MCM) of Example 7.1

Year	1995	1996	1997	1998	1999	2000	2001	2002	2003	2004	2005	2006
January	6.79	2.73	3.09	2.73	2.73	2.73	2.73	2.73	2.73	2.73	2.73	7.15
February	2.73	2.73	3.09	3.09	3.09	2.73	3.09	2.73	2.73	2.73	2.73	3.09
March	2.73	2.73	3.09	3.09	3.09	3.09	2.73	3.09	2.73	2.73	3.09	3.09
April	2.73	2.73	3.09	2.73	2.73	3.09	2.73	3.09	2.73	2.73	3.09	3.09
May	2.73	3.09	3.09	3.09	3.09	3.09	3.09	3.09	2.73	3.09	3.09	3.09
June	4.3	3.09	1.42	3.09	1.42	3.09	3.09	3.09	3.09	3.09	1.42	1.42
July	4.3	4.3	5.44	1.6	4.3	1.42	1.52	1.6	1.6	4.52	4.52	1.6
August	13.61	9.55	9.72	9.55	8.36	6.79	8.36	7.15	8.36	12.65	13.57	9.55
September	18.98	10.86	15.27	10.86	13.61	10.86	10.86	10.86	11.21	17.67	19.34	15.27
October	23.04	14.92	23.04	14.92	23.04	14.92	18.98	14.92	14.92	27.11	27.11	18.98
November	23.04	14.92	23.04	14.92	23.04	14.92	18.98	14.92	14.92	27.11	27.11	23.04
December	10.86	10.86	10.86	10.86	10.86	6.79	10.86	6.79	6.79	10.86	14.92	10.86

TABLE 7.3

Monthly Groundwater Table Variation of the Aquifer (cm) of Example 7.1

Year	1995	1996	1997	1998	1999	2000	2001	2002	2003	2004	2005	2006
January	71.5	49.13	43.85	51.56	39.33	52.09	35.09	43.51	31.79	38.72	53.3	57.81
February	69.84	49.55	48.82	52.62	39.78	52.75	35.13	43.54	31.51	38.69	54.31	56.95
March	72.26	50.89	52.56	55.31	42.56	54.9	38.2	44.91	32.91	40.64	55.93	60.15
April	73.89	52.63	57.51	57.81	45.07	57.28	40.63	47.15	34.66	41.98	58.52	65.15
May	75.39	54.77	61.68	59.92	46.93	59.73	42.27	50.75	36.63	43.7	61.58	70.19
June	77.65	57.78	68.07	62.44	49.54	64.62	44.96	53.73	37.98	46.32	65.95	77.12
July	99.56	64.42	77.67	66.75	63.21	71.86	51.61	57.78	42.25	50.16	78.03	88.43
August	119.53	83.39	101.82	83.37	82.85	85.11	82.52	73.7	59.18	88.05	112.18	104.05
September	120.11	87.84	110.3	88.32	99.72	81.07	99.15	72	75.24	115.22	128.66	105.84
October	103.14	81.81	104.62	78.61	98.85	70.87	90.24	65.33	72.19	113.07	119.56	96.52
November	80.43	68.37	83.75	63.75	79.95	56.1	71.33	52.28	59.37	88.45	96.11	80.35
December	57.41	53.87	60.85	49.23	60.32	41.28	52.35	37.44	44.83	62.69	70.36	57.52

```
File  Edit  Text  Cell  Tools  Debug  Desktop  Window  Help
  1
  2 -    load input;
  3 -    load out;
  4 -    input=input';
  5 -    out=out';
  6 -    inputcal=input(1:2,1:100);
  7 -    outcal=out(1,1:100);
  8 -    inputval=input(1:2,101:144);
  9 -    net=newff(minmax(input),[3,1],{'tansig' 'purelin'},'traingdx');
 10 -    net.performFcn='sse';
 11 -    net.trainParam.goal=1e-5;
 12 -    net.trainParam.show=20;
 13 -    net.trainParam.lr=0.01;
 14 -    net.trainParam.epochs=3000;
 15 -    [net,tr]=train(net,inputcal,outcal);
 16 -    simcal=sim(net,inputcal);
 17 -    simval=sim(net,inputval);
 18 -    w1=net.IW{1,1};
 19 -    w2=net.LW{2,1};
 20 -    b1=net.b{1,1};
 21 -    b2=net.b{2,1};
 22
 23
```

FIGURE 7.3 Structure of developed MLP model for simulation of monthly groundwater table.

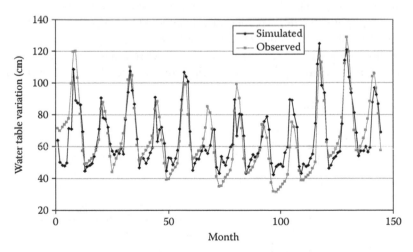

FIGURE 7.4 Comparison between the simulated groundwater table variation with MLP and the observed data.

where N is the tapped delay line memory length. Thus, the output of the TDL is an $(N + 1)$-dimensional vector, made up of the input signal at the current time and the previous input signals. The architecture of the TDL is illustrated in Figure 7.5. TDLs help networks to process dynamically through the flexibility in considering a sequential input.

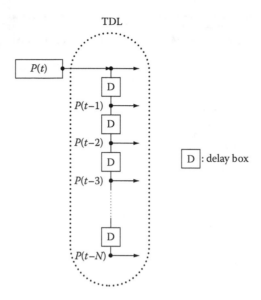

FIGURE 7.5   Tapped delay line in temporal neural networks model.

- **Adaptive linear filter**

The ADALINE (Adaptive Linear) network is similar to the perceptron, but its activation function is linear. The linear activation function allows ADALINE networks to take on any output value thus solving linearly separable problems. The architecture of ADALINE is shown in Figure 7.6. ADALINE has a counterpart in statistical modeling, in this case least square regression. The ADALINE network is a widely used ANN found in practical applications especially in adaptive filtering. In ADALINE,

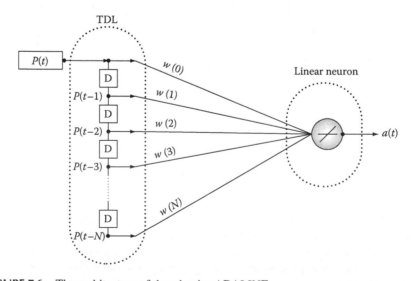

FIGURE 7.6   The architecture of the adaptive ADALINE.

the learning rule is based on the least mean squares method which minimizes the mean square error. The adaptive filter is produced by combining a tapped delay line and an ADALINE network. At each time step, this filter is adjusted and finds weights and biases that minimize the network's sum squared error for recent input and target vectors. In the ADALINE network, the output, $a(t)$ is obtained as follows:

$$a(t) = \sum_{k=0}^{N} w(k)P(t-k) + b \tag{7.9}$$

where
  $w(k)$ is the kth member of weight vector
  b is the bias of the linear neuron
  t denotes a discrete time

- **Recurrent neural network**

Recurrent neural networks (RNN) have feedback elements that enable signals from one layer to feedback to a previous layer. The recurrent scheme differs from MLP in the way that its outputs recur either from the output layer or the hidden layer back to its input layer. So a context unit consists of several time delay operators and represents time implicitly by its effects on processing, in contrast to a TDL component, which explicitly considers temporal processing. Attaching recurrent connections (context unit) to an MLP network results in an RNN. Thus, in a recurrent network, the weight matrix for each layer contains input weights from all other neurons in the network, not just neurons from the previous layer. The architecture of the RNN is illustrated in Figure 7.7. There are three general models of an RNN, depending

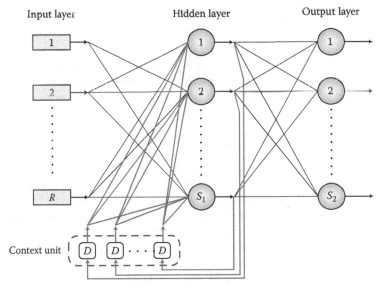

FIGURE 7.7 A typical architecture of Elman RNN. (From Karamouz, M. et al., *J. Am. Water Resourc. Assoc.*, 37(5), 1301, 2001.)

on the architecture of the recurrent connections: the Jordan RNN (Jordan, 1986), which has feedback connections from the output layer to input layer; the Elman RNN (Elman, 1990), which has feedback connections form the hidden layer to the input layer; and the local RNN (Frasconi et al., 1992), which uses only local feedback.

The network employs its output signals from the hidden layer to train the networks. The input layer is divided into two parts: the true input units and the context unit, which are a tapped delay line memory of S_1 neurons from the hidden layer. Elman RNN is formulated by the following equations:

$$a_j^1(t) = F\left(\sum_{i=1}^{R} w_{j,i}^1 P_i(t) + \sum_{c=1}^{S_1} w_{j,c}^C a_c^1(t-1) + b_j^1 \right), \quad 1 \le j \le S_1 \tag{7.10}$$

$$a_k^2(t) = G\left(\sum_{j=1}^{S_1} w_{k,j}^2 a_j^1(t) + b_k^2 \right), \quad 1 \le k \le S_2 \tag{7.11}$$

where
  t denotes a discrete time
  R is the number of input signals
  S_1 and S_2 are the numbers of hidden neurons and output neurons, respectively
  w^1 and w^C are the weight matrices of the hidden layer for real inputs and context unit
  w^2 is the output layer weight matrix
  b^1 and b^2 are the bias vectors of the hidden and output layers, respectively
  p is the input matrix
  a^1 and a^2 are the hidden and output layers' output vectors, respectively
  F and G are the hidden and output layers' activation functions, respectively

It is noteworthy that the modifiable connections are all feed forward, and the weights from the context units can be trained in exactly the same manner as the weights from the input units to the hidden units by the conventional back propagation method (Fauset, 1994).

• Input-delayed neural network

Input-delayed neural network (IDNN) can be used to design temporal neural networks as a capable alternative. The IDNN consists of two parts: the first part is a memory structure that saves latest information, and the second part is a nonlinear associator that processes information and predicts future. The memory structure is a time delay line that corresponds to a buffer containing the recent inputs generated by the delay unit operator D, while the associator is the conventional feed forward network with one hidden layer and one output layer (Coulibaly et al., 2001).

IDNN is a time delay neural network, where only the input layer owns the memory and this makes it different from the general time delay neural network. The structure of typical IDRNN (IDNN with recurrent connection) is shown in Figure 7.8 that

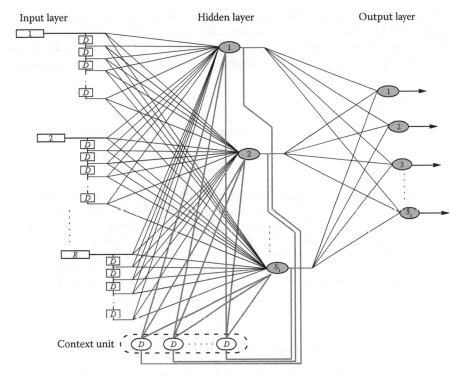

Input layer   Hidden layer   Output layer

Context unit

FIGURE 7.8 A typical architecture of IDRNN.

each input variables is delayed several time steps and fully connected to the hidden layer. The output of IDNN is given by

$$a_j^1(t) = F\left(\sum_{d=0}^{D}\sum_{i=1}^{R} w_{j,i,d}^1 P_{i,d+1}(t) + b_j^1\right), \quad 1 \leq j \leq S_1 \qquad (7.12)$$

$$a_k^2(t) = G\left(\sum_{j=1}^{S_1} w_{k,j}^2 a_j^1(t) + b_k^2\right), \quad 1 \leq k \leq S_2 \qquad (7.13)$$

where
  t denotes a discrete time
  D is the time delay memory order
  R is the number of input signals
  S_1 and S_2 are the numbers of hidden neurons and output neurons, respectively
  w^1 and w^2 are the weight matrices of hidden and output layers
  b^1 and b^2 are the bias vectors of the hidden and output layers
  p is the input matrix and a^1 and a^2 are the hidden and output layers' output vectors
  F and G are the hidden and output layers' activation functions, respectively

A TDRNN is presented using both mentioned components in the architecture of a neural network in order to perform dynamically. A major feature of this architecture is that the nonlinear hidden layer receives the contents of both the input time delays and the context unit, which makes it suitable for complex sequential input learning.

Example 7.2

For the aquifer of Example 7.1, simulate the groundwater table variation using the IDNN model.

Solution

The same procedure as the previous example is followed for preparing input data of the IDNN model. The developed structure of the IDNN model is shown in Figure 7.9. The tapped delay line memory of the developed model is 2 and there are 12 neurons in its hidden layer. The transfer function of hidden and output layers are tangent sigmoid and linear, respectively. The model is trained in 3000 steps (epochs) and the model goal is achieving an error less than 0.00001. The simulated and recorded monthly groundwater table variation data are compared in Figure 7.10. About 26% of data has been simulated with less than 5% error and 62% of data has been simulated with an error of less than 10%.

7.3.2 FUZZY SETS AND PARAMETER IMPRECISION

Two approaches of randomization or fuzzification are used for evaluating the parameter uncertainty in water resources and hydrologic modeling. Fuzzy rules could be developed to incorporate the uncertainty of the variables and it could be utilized as a simulation model or as a model to develop the operation policy in the real-time operation of the groundwater systems. In this section, the fundamentals of fuzzy sets and fuzzy decisions are introduced.

```
File  Edit  Text  Cell  Tools  Debug  Desktop  Window  Help

                                                              Stack: Base

1
2 -    load input;
3 -    load out;
4 -    input=input';
5 -    out=out';
6 -    inputcal=input(1:2,1:100);
7 -    outcal=out(1,1:100);
8 -    inputval=input(1:2,101:144);
9 -    net=newfftd(minmax(input),[0 1 2], [12,1],{'tansig' 'purelin'},'traingdx');
10 -   net.performFcn='sse';
11 -   net.trainParam.goal=1e-5;
12 -   net.trainParam.show=20;
13 -   net.trainParam.lr=0.01;
14 -   net.trainParam.epochs=3000;
15 -   [net,tr]=train(net,inputcal,outcal);
16 -   simcal=sim(net,inputcal);
17 -   simval=sim(net,inputval);
18 -   w1=net.IW{1,1};
19 -   w2=net.LW{2,1};
20 -   b1=net.b{1,1};
21 -   b2=net.b{2,1};
```

FIGURE 7.9 Developed IDNN model for simulation of groundwater table variation.

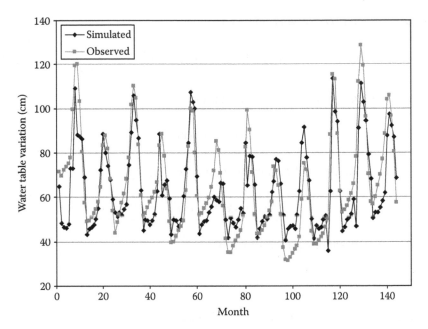

FIGURE 7.10 Comparison between the simulated groundwater table variation with IDNN and the observed data.

Let X be the set of certain objects. A *fuzzy set A* in X is a set of ordered pairs as follows:

$$A = \{(x, \mu_A(x))\}, \quad x \in X \tag{7.14}$$

where
$\mu_A : X \mapsto [0,1]$ is called the membership function
$\mu_A(x)$ is the grade of membership of x in A

In the classical set theory $\mu_A(x)$ equals 1 or 0, because x either belongs to A or does not.

The basic concepts of fuzzy sets are as follows:

A fuzzy set A is *empty*, if $\mu_A(x) = 0$ for all $x \in X$.
A fuzzy set A is called *normal*, if

$$\sup_x \mu_A(x) = 1 \tag{7.15}$$

For normalizing a nonempty fuzzy set, the normalized membership function is computed as follows:

$$\bar{\mu}_A(x) = \frac{\mu_A(x)}{\sup_x \mu_A(x)} \tag{7.16}$$

The *support* of a fuzzy set A is defined as

$$S(A) = \{x \,|\, x \in X, \mu_A(x) > 0\} \tag{7.17}$$

The fuzzy sets A and B are equal, if for all $x \in X$

$$\mu_A(x) = \mu_B(x) \tag{7.18}$$

A fuzzy set A is a *subset* of B ($A \subseteq B$), if for all $x \in X$,

$$\mu_A(x) \le \mu_B(x) \tag{7.19}$$

A' is the *complement* of A, if for all $x \in X$,

$$\mu_{A'}(x) = 1 - \mu_A(x) \tag{7.20}$$

The *intersection* of fuzzy sets A and B is defined as

$$\mu_{A \cap B}(x) = \min\{\mu_A(x);\ \mu_B(x)\} \quad \text{(for all } x \in X) \tag{7.21}$$

Notice that (v) and (vii) imply that $A \subseteq B$ if and only if $A \cap B = A$.
  The *union* of fuzzy sets A and B is given as

$$\mu_{A \cup B}(x) = \max\{\mu_A(x);\ \mu_B(x)\} \quad \text{(for all } x \in X) \tag{7.22}$$

The *algebraic product* of fuzzy sets A and B is denoted by AB and is defined by the following relation:

$$\mu_{AB}(x) = \mu_A(x)\mu_B(x) \quad \text{(for all } x \in X) \tag{7.23}$$

The *algebraic sum* of fuzzy sets A and B is denoted by $A + B$ and has the membership function

$$\mu_{A+B}(x) = \mu_A(x) + \mu_B(x) - \mu_A(x)\mu_B(x) \quad \text{(for all } x \in X) \tag{7.24}$$

A fuzzy set A is *convex*, if for all $x, y \in X$ and $\lambda \in [0,1]$,

$$\mu_A(\lambda x + (1 - \lambda)y) \ge \min\{\mu_A(x);\ \mu_A(y)\} \tag{7.25}$$

If A and B are convex, then it can be demonstrated that $A \cap B$ is also convex.
  A fuzzy set A is *concave*, if A' is convex. In the case that A and B are concave, the $A \cup B$ will also be concave.

Consider $f : X \mapsto Y$ a mapping from set X to set Y and A a fuzzy set in X. The fuzzy set B *induced* by mapping f is defined in Y with the following membership function:

$$\mu_B(y) = \sup_{x \in f^{-1}(y)} \mu_A(x) \tag{7.26}$$

where $f^{-1}(y) = \left\{ x \mid x \in X, f(x) = y \right\}$.

Example 7.3

Fuzzy sets A and B are defined in $X = [-\infty, \infty]$, by the membership functions as follows (Figure 7.11):

$$\mu_A(x) = \begin{cases} x - 1 & \text{if } 1 \le x \le 2 \\ 1 & \text{if } 2 < x \le 3 \\ -x + 4 & \text{if } 3 < x \le 4 \\ 0 & \text{otherwise} \end{cases}$$

$$\mu_B(x) = \begin{cases} x - 3 & \text{if } 3 \le x \le 4 \\ 1 & \text{if } 4 \le x \le 5 \\ -x + 6 & \text{if } 5 \le x \le 6 \\ 0 & \text{otherwise} \end{cases}$$

Determine the membership functions of $A \cap B$, $A \cup B$, AB, and $A + B$ fuzzy sets.

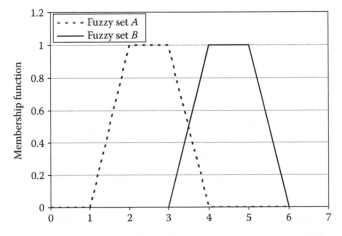

FIGURE 7.11   Membership function of A and B.

Solution

$\mu_{A\cap B}(x)$ is $\min\{\mu_A(x);\mu_B(x)\}$, therefore $A \cap B$ is

$$\mu_{A\cap B}(x) = \begin{cases} x-3 & \text{if } 3 \leq x \leq 3.5 \\ -x+4 & \text{if } 3.5 < x \leq 4 \\ 0 & \text{otherwise} \end{cases}$$

and $\mu_{A\cup B}(x) = \max\{\mu_A(x);\mu_B(x)\}$, thus $A \cup B$ is as follows :

$$\mu_{A\cup B}(x) = \begin{cases} x-1 & \text{if } 1 \leq x \leq 2 \\ 1 & \text{if } 2 \leq x \leq 3 \\ -x+4 & \text{if } 3 \leq x \leq 3.5 \\ x-3 & \text{if } 3.5 \leq x \leq 4 \\ 1 & \text{if } 4 \leq x \leq 5 \\ -x+6 & \text{if } 5 < x \leq 6 \\ 0 & \text{otherwise} \end{cases}$$

The product of A and B is determined as follows:

$$\mu_{AB}(x) = \begin{cases} (x-3)(-x+4) & \text{if } 3 \leq x \leq 4 \\ 0 & \text{otherwise} \end{cases}$$

Finally the summation of fuzzy sets is determined as follows:

$$\mu_{A+B}(x) = \begin{cases} x-1 & \text{if } 1 \leq x \leq 2 \\ 1 & \text{if } 2 \leq x \leq 3 \\ 1+(x-3)(x-4) & \text{if } 3 \leq x \leq 4 \\ 1 & \text{if } 4 \leq x \leq 5 \\ -x+6 & \text{if } 5 \leq x \leq 6 \\ 0 & \text{otherwise} \end{cases}$$

The membership functions of $A \cap B$, $A \cup B$, AB, and $A + B$ fuzzy sets are illustrated in Figure 7.12.

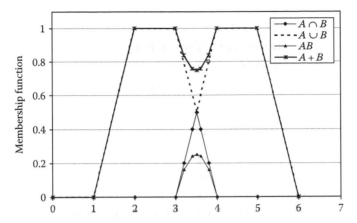

FIGURE 7.12 Operations of fuzzy sets $A \cap B$, $A \cup B$, AB, $A + B$.

Let X be the set of decision alternatives. A fuzzy constraint is defined by a fuzzy set C in X, and the membership function value $\mu_c(x)$ shows the degree an alternative satisfies the constraint C. Similarly, any objective function can be identified by a fuzzy set G, and $\mu_G(x)$ shows the degree an alternative represents an attainment of the goal represented by this objective function. Any multi-criteria decision-making problem can be defined in a fuzzy environment as the set of fuzzy goals G_1, ..., G_M and fuzzy constraints C_1, ..., C_N. The solution of the problem is the decision alternative that satisfies the constraints C_j in as high a level as possible and gives the best attainment of the goals possible. In other words, the intersection of the fuzzy sets G_1, ..., G_M, C_1, ..., C_N must have its highest membership value. Therefore we consider

$$\mu_D(x) = \min\{\mu_{G_1}(x),...,\mu_{G_M}(x); \quad \mu_{C_1}(x),...,\mu_{C_N}(x)\} \tag{7.27}$$

and find alternative x that maximizes this membership function. The interested reader may be referred to Bellman and Zadeh (1970), or to the section of fuzzy decision in Szidarovszky et al. (1986).

Example 7.4

Assume that there are five alternatives to select from for improving a groundwater system; there are three constraints including groundwater table fluctuation and supplying the minimum water demand with acceptable water quality. Two objective functions are considered including maximizing annual groundwater yield and minimizing the costs. The membership values for all alternatives with respect to the constraints and objectives are given in Table 7.4.

Solution

The last row gives the values of μ_D by selecting the smallest numbers of all columns. The largest value of μ_D appears in the third columns, so the second alternative is the best.

TABLE 7.4
Membership Values in Fuzzy Decision

	Alternatives				
	1	2	3	4	5
	0.85	0.95	0.80	0.85	0.73
μ_{G_2}	0.90	0.90	0.75	0.84	0.77
μ_{C_1}	0.89	0.87	0.83	0.9	0.95
μ_{C_2}	0.82	0.90	0.74	0.83	0.63
μ_{C_3}	0.73	0.87	0.78	0.81	0.66
μ_D	0.73	0.87	0.74	0.81	0.63

7.3.3 System Dynamics

System dynamics (SD) approach gives users a better understanding of the complex interrelationships among different elements within a system. SD provides insight into the feedback process for simulating the system behavior. Simulation of the model over time is considered essential to understanding the dynamics of the system. The development of an SD simulation model includes major steps such as understanding of the system and its boundaries, identifying the key variables, representation of the physical processes or variables through mathematical relationships, mapping the structure of the model, and simulating the model for understanding its behavior. The SD approach is based on the theory of feedback process using system thinking paradigm. Feedback systems can be classified into two different classes (Ahmad and Simonovic, 2000):

Negative feedback, which seeks a goal and responds as a consequence of failing to achieve the goal
Positive feedback, which generates growth processes where action builds a result that generates still greater action

Feedback loop indicates how a system might behave because of its internal feedback loops and the effects that positive and negative feedback loops have on a system.

It was originally rooted in the management and engineering sciences in the 1960s but has gradually developed into other fields. The advantages of this approach are the ease of constructing "what if" scenarios and tackling the physics of big, messy, real-world problems metallically. In addition, the general principles of the SD simulation tools could be applied to social, economic, and ecological and biological systems for system development (Karamouz et al., 2010).

A system includes a collection of elements which continually interact over time to form a unified body. These interactions among the elements of a system are called the structure of the system. Figure 7.13 shows the basic elements of the SD simulation model. It is similar to a reservoir system. As shown in this figure, the interactions among inflow, storage, outflow, and other site-specific issues such as

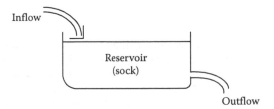

FIGURE 7.13 A schematic of an urban storage tank in system dynamic modeling.

release values forms the structure of the reservoir system. Stock represents accumulations and can be used as source, for example, water stored in the reservoir. The accumulation needs flow (inflow to the reservoir), which is modeled by flow. Flow and stock are inseparable and make the minimum set of elements needed to describe dynamics. Converters convert input to outputs. They can represent information or material quantities. Connectors link stocks to converters, stocks to flow regulators, and converters to other converters. It, representing the relationship between other objects, conveys information from one variable to the others. Stocks, flows, converters, and connectors tools are used for object-oriented programming.

The behavior and response of the system is simulated over time considering the dynamics of the system. In the reservoir example, the behavior is described by the dynamics of reservoir storage filling and draining. This behavior is due to the influences of inflow, outflow, losses, and environment, which are elements of the system. SD can also be used to analyze how structural changes in one part of a system might affect the behavior of the system as a whole. The visual effects of SD mapping help the decision makers to clearly evaluate the impacts of different scenarios. The model results are functionally transparent to all parties involved in the water management.

The modeling process is a cycling process arrived at by cycling back through the entire sequence several times whileworking on the model. There are some steps for defining the system including clearly stating the purpose of the model, developing a base behavior pattern, and developing a system diagram.

In the first step, the purposes of the modeling are clarified utilizing some tools such as reference behavior pattern (RBP). RBP is a graph over time for one or more variables which best capture the dynamic phenomenon seeking to understand and focus the efforts for the behavioral dimensions. In the second step of the modeling process, the hypotheses responsible for generating the behavior pattern are represented. A framework including dynamic organizing principles, based on stock/flow, feedback loop is developed as the core of the model. Once the flows associated with hypotheses have been characterized, the next step in mapping is to close loops. A system diagram is composed of the essential actors or sectors. The model will help to focus the effort on the structural dimension.

There are many tools used for implementing the object-oriented modeling approach. Computer software tools such as STELLA developed by High Performance Systems Inc., VENSIM developed by Ventana System Inc., and POWERSIM developed by POWERSIM Software help the execution of these processes.

Example 7.5

In a region, the water authority board considers a law to reduce the rate of groundwater and river withdrawals. It would be given the authority to set an initial charge price for consumed water. The water consumers are initially charged $0.1/m³ and then based on the quantity of produced wastewater. A group of researchers collected and analyzed data about the monthly effects that resulted from this regulation. For this period the following relationships are developed:

$$A_t = 10 - 4.2 * P_{st} \tag{7.28}$$

$$B_t = 0.9(0.3 + 2P_{rt})A_t \tag{7.29}$$

$$C_t = 0.38(A_t - B_t) \tag{7.30}$$

$$D_t = A_t - B_t - C_t \tag{7.31}$$

$$E_t = A_t - B_t \tag{7.32a}$$

$$P_{s,t+1} = \frac{0.88 - 0.006B_t}{A_t + P_{rt}} \tag{7.32b}$$

$$P_{n,t+1} = 0.4 - 0.03E_t \tag{7.33}$$

where
    A_t is the quantity of consumed water per month t (1000 m³/month)
    B_t is the quantity of treated wastewater per month t (1000 m³/month)
    C_t is the quantity of water loss without any usage per month t (1000 m³/month)
    D_t is the quantity of wastewater discharged to groundwater through absorption
        wells and septic tanks per month t (1000 m³/month)
    E_t is the quantity of supplied water per month t (1000 m³/month)
    $P_{s,t}$ is the sale price of 1 m³ of supplied water
    $P_{n,t}$ is the cost of supplying of 1 m³ of water
    $P_{r,t}$ is the initial charge for 1 m³ water usage

Based on the above information, explain the expected changes in water usage and wastewater production and the corresponding economic impacts through answering the following questions.

Explain above equations in the system structure. Draw a system graph which is primarily designed to support statements about the water flow in the city. Represent the water supply system of the city within a system boundary, SB, using three components called (1) water treatment plants, (2) water distribution system, and (3) wastewater treatment system. Relate these three components of the system to three systems in its environment including (1) water loss, (2) groundwater, and (3) surface water resources. Then show what components are associated with each of the above described prices. Finally, show in your graph two cash flows, namely, NB_t = monthly water supply budget (10³ $/month) and R_t = monthly water supplying revenue (10³ $/month).

Solve the problem with initial values of $A_0 = 9.1$, $B_0 = 3.7$, $C_0 = 0.9$, $D_0 = 5.08$, $E_0 = 8.36$, and $P_{rt} = 0.05$. Repeat the calculation with $P_{rt} = 0.1$, 0.2, 0.3. Draw the variation of treated water and discharged water to groundwater as a function of P_{rt}.

Plot the diagram of the effect of different P_{rt} on the monthly revenue of water authority board and water supplying budget.

Solution

1. (a) $A_t = 10 - 4.2*P_{st}$: As the sale price of water increases, the quantity of water consumption decreases proportionally from some consumption limitation and employing water conservation strategies. The maximum of the quantity of water consumption is equal to 10 (for $P_{s,t} = 0$).

   (b) $B_t = 0.9(0.3 + 2P_{rt})A_t$: The treated wastewater is a portion of consumed water and its quantity increases by increase in the initial charge of consumers.

   (c) $C_t = 0.38(A_t - B_t)$: 38% of nontreated water is lost because of problems in the water supply and distribution system.

   (d) $D_t = A_t - B_t - C_t$: The quantity of discharged wastewater to groundwater and river is obtained by mass continuity equation knowing the quantity of consumed, treated, and lost water quantities in the water supply system.

   (e) $E_t = A_t - B_t$: The required water supply of the month is obtained by using the continuity equation considering the water consumption and the treated wastewater.

   (f) $P_{s,t+1} = 0.88 - 0.006B_t/A_t + P_{rt}$: The water sale cost will decrease with the percentage of wastewater treated the month before, and will increase with the initial charge.

   (g) $P_{n,t+1} = 0.4 - 0.03E_t$: The price of water production in a month will decrease with an increase of quantity of water supply in the proceeding month.

2. System graph and associated costs are illustrated in Figure 7.14. Also, $NB_t = P_{n,t}E_t$ and $R_t = P_{s,t}A_t$

3. The results are tabulated in Tables 7.5 through 7.8.

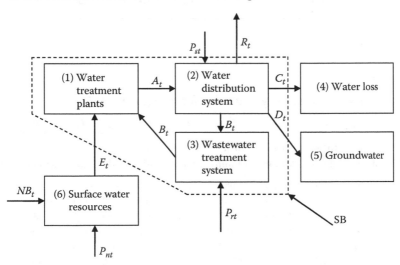

FIGURE 7.14　The diagram of water flow in water supply and distribution system.

TABLE 7.5

Variation of Water Quantity and Price for $P_{rt} = 0.05$

T	A_t	B_t	C_t	D_t	E_t	P_{rt}	P_{st}	P_{nt}	NB	R
Month	$\dfrac{10^3\,m^3}{Month}$	$\dfrac{10^3\,m^3}{Month}$	$\dfrac{10^3\,m^3}{Month}$	$\dfrac{10^3\,m^3}{Month}$	$\dfrac{10^3\,m^3}{Month}$	$\dfrac{\$}{m^3}$	$\dfrac{\$}{m^3}$	$\dfrac{\$}{m^3}$	$\dfrac{10^3\$}{Month}$	$\dfrac{10^3\$}{Month}$
0	9.100	3.700	0.900	5.080	8.360	0.050	—	—	—	—
1	6.104	2.198	1.485	2.422	3.907	0.050	0.928	0.149	0.583	5.662
2	6.103	2.197	1.484	2.422	3.906	0.050	0.928	0.283	1.105	5.663
3	6.103	2.197	1.484	2.422	3.906	0.050	0.928	0.283	1.105	5.663
4	6.103	2.197	1.484	2.422	3.906	0.050	0.928	0.283	1.105	5.663
5	6.103	2.197	1.484	2.422	3.906	0.050	0.928	0.283	1.105	5.663

TABLE 7.6

Variation of Water Quantity and Price for P_{rt} Equal to 0.1

T	A_t	B_t	C_t	D_t	E_t	P_{rt}	P_{st}	P_{nt}	NB	R
Month	$\dfrac{10^3\,m^3}{Month}$	$\dfrac{10^3\,m^3}{Month}$	$\dfrac{10^3\,m^3}{Month}$	$\dfrac{10^3\,m^3}{Month}$	$\dfrac{10^3\,m^3}{Month}$	$\dfrac{\$}{m^3}$	$\dfrac{\$}{m^3}$	$\dfrac{\$}{m^3}$	$\dfrac{10^3\$}{Month}$	$\dfrac{10^3\$}{Month}$
0	9.100	3.700	0.900	5.080	8.360	0.100	—	—	—	—
1	5.894	2.652	1.232	2.010	3.242	0.100	0.978	0.149	0.484	5.762
2	5.895	2.653	1.232	2.010	3.242	0.100	0.977	0.303	0.982	5.762
3	5.895	2.653	1.232	2.010	3.242	0.100	0.977	0.303	0.982	5.762
4	5.895	2.653	1.232	2.010	3.242	0.100	0.977	0.303	0.982	5.762
5	5.895	2.653	1.232	2.010	3.242	0.100	0.977	0.303	0.982	5.762

TABLE 7.7

Variation of Water Quantity and Price for P_{rt} Equal to 0.15

T	A_t	B_t	C_t	D_t	E_t	P_{rt}	P_{st}	P_{nt}	NB	R
Month	$\dfrac{10^3\,m^3}{Month}$	$\dfrac{10^3\,m^3}{Month}$	$\dfrac{10^3\,m^3}{Month}$	$\dfrac{10^3\,m^3}{Month}$	$\dfrac{10^3\,m^3}{Month}$	$\dfrac{\$}{m^3}$	$\dfrac{\$}{m^3}$	$\dfrac{\$}{m^3}$	$\dfrac{10^3\$}{Month}$	$\dfrac{10^3\$}{Month}$
0	9.100	3.700	0.900	5.080	8.360	0.200	—	—	—	—
1	5.474	3.449	0.770	1.256	2.025	0.200	1.078	0.149	0.302	5.899
2	5.480	3.452	0.770	1.257	2.028	0.200	1.076	0.339	0.688	5.898
3	5.480	3.452	0.770	1.257	2.028	0.200	1.076	0.339	0.688	5.898
4	5.480	3.452	0.770	1.257	2.028	0.200	1.076	0.339	0.688	5.898
5	5.480	3.452	0.770	1.257	2.028	0.200	1.076	0.339	0.688	5.898

TABLE 7.8

Variation of Water Quantity and Price for P_{rt} Equal to 0.25

T	A_t	B_t	C_t	D_t	E_t	P_{rt}	P_{st}	P_{nt}	NB	R
Month	$10^3 m^3$ Month	$10^3 m^3$ Month	$10^3 m^3$ Month	$10^3 m^3$ Month	$10^3 m^3$ Month	$\dfrac{\$}{m^3}$	$\dfrac{\$}{m^3}$	$\dfrac{\$}{m^3}$	$10^3 \$$ Month	$10^3 \$$ Month
0	9.100	3.700	0.900	5.080	8.360	0.300	—	—	—	—
1	5.054	4.094	0.365	0.595	0.960	0.300	1.178	0.149	0.143	5.952
2	5.064	4.102	0.366	0.597	0.962	0.300	1.175	0.371	0.357	5.951
3	5.064	4.102	0.366	0.597	0.962	0.300	1.175	0.371	0.357	5.951
4	5.064	4.102	0.366	0.597	0.962	0.300	1.175	0.371	0.357	5.951
5	5.064	4.102	0.366	0.597	0.962	0.300	1.175	0.371	0.357	5.951

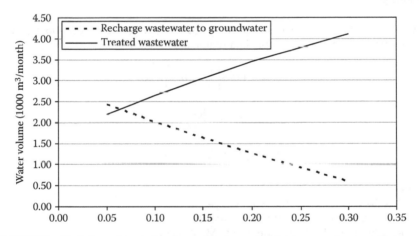

FIGURE 7.15 Variation of treated water and recharged water to groundwater with P_{rt}.

4. The variation of treated water and recharged water to groundwater with P_{rt} is plotted in Figure 7.15. By increasing the P_{rt}, the treated water quantity considerably increases and the recharged wastewater to groundwater decreases.

5. The diagram of effect of different P_{rt} on the monthly revenue and water supplying budget is illustrated in Figure 7.16. The figure shows the monthly revenue increases and the monthly water supplying budget decreases due to increasing P_{rt}. It is increasing the treated water that leads to decreasing water consumption.

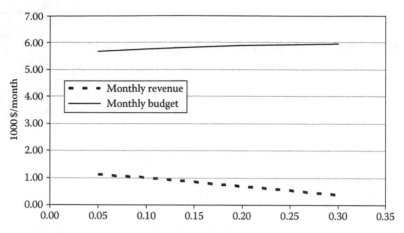

FIGURE 7.16   The effect of different P_{rt} on the monthly revenue and water supply budget.

Example 7.6

Assume that the relationships between population, water demands, and groundwater resources for water supply in a region are as follows:

$$P_{t+1} = P_t + 0.8\left(\frac{D_t}{P_t} - 0.5\right) \times 10^{10} \tag{7.34}$$

$$D_t = \frac{R_t P_t}{2 \times 10^9} \tag{7.35}$$

$$R_t = 10^9 + 0.2P_t - 0.02P_t^{1.168} \tag{7.36}$$

where
   P_t is the population in the region at year t
   D_t is the water demand (MCM) at year t
   R_t is the groundwater resources (MCM) at year t

Assuming that the city population in the first year is equal to 0.5×10^5, determine the maximum possible population of the region.

Solution

For determining the maximum possible population of the region, Equation 7.36 is replaced in Equation 7.35 and then it is replaced in Equation 7.34. It is solved based on the population factor.

$$D_t = \frac{(10^9 + 0.2P_t - 0.02P_t^{1.168})P_t}{2 \times 10^9} \tag{7.37}$$

Finally the obtained relation is substituted in the population relation.

$$P_{t+1} = P_t + 0.8\left(\frac{(10^9 + 0.2P_t - 0.02P_t^{1.168})P_t/2 \times 10^9}{P_t} - 0.5\right) \times 10^{10} \tag{7.38}$$

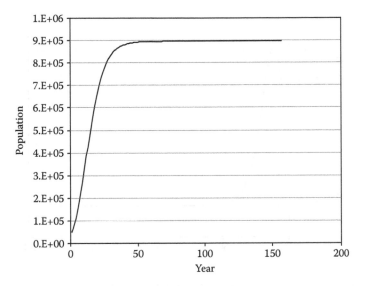

FIGURE 7.17 Variations of population over time.

The population of each year is more than the pervious year $P_{t+1} > P_t$, so that the expression in brackets should be more than zero.

$$0.8\left(\frac{(10^9 + 0.2P_t - 0.02P_t^{1.168})P_t/2 \times 10^9}{P_t} - 0.5\right) > 0$$

$$\rightarrow \frac{(10^9 + 0.2P_t - 0.02P_t^{1.168})}{2 \times 10^9} > 0.5 \qquad (7.39)$$

$$10^9 + 0.2P_t - 0.02P_t^{1.168} > 10^9 \Rightarrow 0.2P_t - 0.02P_t^{1.168} > 0$$

$$\Rightarrow P_t > 0.1P_t^{1.168} \qquad 10 > P_t^{0.168} \Rightarrow P_t < 9 \times 10^5 \qquad (7.40)$$

Therefore, the population cannot exceed 9×10^5 persons because of limitations of water resources. Once the population reaches this value, it will fluctuate around it. The variations of P_t, R_t, and D_t are shown in Figures 7.17 through 7.19, respectively. As it is shown in these figures, after about 70 years, the curves converge to straight lines, reaching their limits.

7.4 OPTIMIZATION MODELS FOR GROUNDWATER MANAGEMENT

Nowadays, due to increasing water demands and water pollution, a significant challenge is to determine the optimal amount of water withdrawal from the aquifer to supply the water demands and sustain native peoples and ecosystems. The optimal water supply to municipal, industrial, and agricultural water demands considering the physical, socioeconomic, and environmental constraints is the major objective

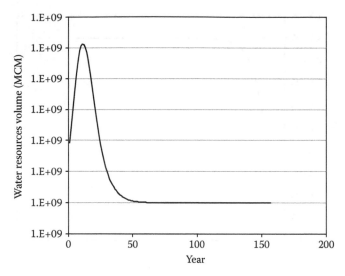

FIGURE 7.18   Variations of available water resources over time.

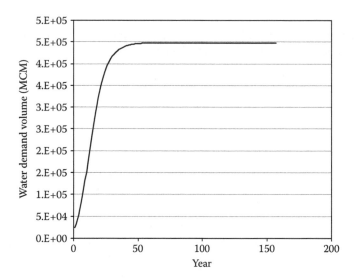

FIGURE 7.19   Variations of water demand over time.

of groundwater planning and management. The following planning problems are associated with groundwater supply management (Yeh, 1992):

1. The determination of an optimal pumping pattern such as location of pumping wells and their pumping rate to satisfy water demands
2. The timing and staging of well system development (capacity expansion) considering future water demands
3. The design of water transfer facilities for optimal allocation and distribution of water to demand points in the basin

In this section, some typical optimization models for groundwater operation management, capacity expansion, and conjunctive use of surface and groundwater resources are presented.

7.4.1 Optimization Model for Groundwater Operation

Optimal policies for groundwater operation such as optimal extraction, allocation, and optimal recharge of aquifer can be determined using the groundwater operation model considering physical, environmental, and socioeconomic constraints. The objective function of the optimization model could be to minimize the total discounted operational costs as follows:

$$Min \ Z = \sum_{t=1}^{T} \left(\frac{1}{1+\alpha} \right)^t \sum_{w \in W} C_w \left(Q_w^t, h_w^t \right) \tag{7.41}$$

where
  T is the number of time steps in the planning horizon
  α is the interest rate
  W is the set of well sites
  $C_w \left(Q_w^t, h_w^t \right)$ is the water extraction cost from the well site w, which is a nonlinear function of discharge Q_w^t and head h_w^t at each time step t

The water head at each time step can be determined using groundwater response equations and it can be considered as a system's constraint. The response equations for each planning period in a linear, distributed parameter groundwater system can be expressed as follows (Willis and Yeh, 1987):

$$h^t - A_1 h^{t-1} - A_2 g(Q^{t-1}) = 0 \quad \forall t \tag{7.42}$$

where
  h^t and h^{t-1} are vectors of groundwater head at time steps t and $t - 1$
  A_1 and A_2 are coefficient matrices vector

Q^{t-1} is defined as follows:

$$Q^{t-1} = (Q_1^{t-1}, \ Q_2^{t-1}, \ldots, \ Q_w^{t-1})^T \quad \forall t, \ w \in W \tag{7.43}$$

The following constraints can be considered in a groundwater operation optimization model:

- Supplying demands:

$$\sum_{w \in W} Q_w^t \geq D^t \quad \forall t \tag{7.44}$$

where D^t is the total water demand at time step t.

- The discharge capacity for each well:

$$0 \leq Q_w^t \leq Q_w^{\max} \quad \forall t, \, w \in W \tag{7.45}$$

where $Q_w^{\max}$ is the discharge capacity of well w.

The lower bound constraints on the head levels (h_w) are considered to ensure that excessive depletion of the aquifer does not occur:

$$h_w^t \geq h_w^{\min} \quad \forall t, \, w \in W \tag{7.46}$$

where $h_w^{\min}$ is the minimum groundwater head at well site w.

The nonnegativity of the decision variables is considered as

$$h^t, Q_w^t \geq 0 \quad \forall t, \, w \in W \tag{7.47}$$

The above planning model is the basic framework for groundwater systems planning and management.

Example 7.7

Groundwater pumping schedules includes determining the discharges from well sites and the lengths of the planning periods. The pumping schedule for three planning periods is shown in Figure 7.20. Develop a management model to minimize the drawdown for determining the optimal pumping pattern in the basin for the pumping schedule.

Solution

The objective function of the management model is expressed as

$$\min z = \sum_k \sum_{w \in \pi} s_w^k \tag{7.48}$$

where s_w^k is the drawdown at well w at the end of period k.

The hydraulic response equations could be developed from the Theis equation. The drawdown $s(r,t)$ occurring during the first planning period can be calculated as

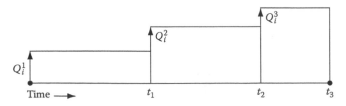

FIGURE 7.20 Pumping schedule during planning horizon.

$$s(r,t) = \sum_{w \in \pi} \frac{Q_w^1}{4\pi T} W\left(\frac{r_w^{-2}S}{4Tt}\right), \quad 0 \le t \le t_1 \tag{7.49}$$

where
Q_w^k as the discharge from well site i in period k
t, t_1 is the length of the planning period
π is number of wells
$\bar{r}_w$ is the distance from the wth well to any point r in the aquifer
S and T are the storage and transmissivity parameters, respectively

In the second time period, the drawdown is given as

$$s(r,t) = \sum_{w \in \pi} \frac{Q_w^1}{4\pi T} W\left(\frac{r_w^{-2}S}{4Tt}\right) + \sum_{w \in \pi} \left(\frac{Q_w^2 - Q_w^1}{4\pi T}\right) W\left(\frac{r_w^{-2}S}{4T(t - t_1)}\right), \quad t_1 \le t \le t_2 \tag{7.50}$$

Similarly, for the third period,

$$s(r,t) = \sum_{w \in \pi} \frac{Q_w^1}{4\pi T} W\left(\frac{r_w^{-2}S}{4Tt}\right) + \sum_{w \in \pi} \left(\frac{Q_w^2 - Q_w^1}{4\pi T}\right) W\left(\frac{r_w^{-2}S}{4T(t - t_1)}\right)$$

$$+ \sum_{w \in \pi} \left(\frac{Q_w^3 - Q_w^2}{4\pi T}\right) W\left(\frac{r_w^{-2}S}{4T(t - t_2)}\right), \quad t \ge t_3 \tag{7.51}$$

The water withdrawal target for each period can be considered as a model constrain

$$\sum_{w \in \pi} Q_w^k \ge D^k \quad \forall k \tag{7.52}$$

where D^k is the demand in period k.
The drawdown should be less than the maximum allowable drawdown at certain control locations in the aquifer

$$s_a^k \le s^*, \quad a \in \Delta \tag{7.53}$$

where $a \in \Delta$ are the control locations.

7.4.2 Optimization Model for Capacity Expansion

A full-scale network system is initially installed to use groundwater resources. However, the system capacity may exceed water demand in the early stages because water demand generally increases with time. Therefore, a capacity expansion planning capable of determining an optimal schedule is needed to expand system capacity according to increasing water demand. The target of capacity expansion is to increase the reliability of water supply by structural methods. The capacity expansion model provides information about how to expand the water supply system so that they meet increasing demand over time. This involves deciding the expansion size (sizing), expansion time (timing), and expansion capacity types. Construction of a new dam, increasing the height of an existing reservoir, well field development, and construction of tunnels or channels for interbasin water transfers are some common examples of expansion capacity types; however, they are somewhat limited by site-related, geophysical, geographic, economic, and institutional factors.

The capacity expansion or timing and staging of well field development are usual problems in groundwater systems planning. In these problems it is assumed that in conventional design, well sites have been developed and operation policies are developed for existing conditions. Therefore, the capital cost can be neglected in developing the operation policies and should be considered in capacity expansion models. The objective function of the capacity expansion model, which is the minimization of total discounted capital and operational costs, is as follows:

$$Min\ Z = \sum_{t=1}^{T} \left(\frac{1}{1+\alpha}\right)^t \sum_{w \in W}\left(KC(Q_w^{max})X_w^t + C_w(Q_w^t, h_w^t)Y_w^t\right) \qquad (7.54)$$

where
T is the number of time steps in the planning horizon
α is the interest rate
W is the set of well sites
$KC(Q_w^{max})$ is the capital cost of well site w, which is considered as a function of well capacity, (Q_w^{max})
$C_w(Q_w^t, h_w^t)$ are the water extraction costs from the well site w, which is a nonlinear function of discharge Q_w^t and head h_w^t at each time step t
X_w^t and Y_w^t are 0–1 integer variables
X_w^t indicates whether or not well site w is developed in period t
X_w^t is equal to 1, if well site w is developed in time period t and equal to zero otherwise

Similarly, Y_w^t is equal to 1 if well site w is in operation and equal to zero otherwise.

This objective function should be minimized considering groundwater response equations and the system's constraints.

- The response equations for each planning period is

$$h^t - A_1 h^{t-1} - A_2 g(Q^{t-1}) = 0 \quad \forall t \qquad (7.55)$$

- Supplying demands:

$$\sum_{w \in W} Y_w^t Q_w^t \geq D^t \quad \forall t \qquad (7.56)$$

where D^t is the total water demand in time step t.

- The constraint of well capacity for each well site w is

$$0 \leq Q_w^t \leq Q_w^{max} \quad \forall t, w \in W \tag{7.57}$$

where Q_w^{max} is the discharge capacity of well w.
- The lower bound constraints on head levels (h_w) are considered to ensure that the excessive depletion of the aquifer does not occur:

$$h_w^t \geq h_w^{min} \quad \forall t, w \in W \tag{7.58}$$

where h_w^{min} is the minimum groundwater head at well site w.
- The 0–1 integer variables are defined as follows:

$$X_w^t, Y_w^t = [0,1] \tag{7.59}$$

- The maximum number of developed well over the planning horizon is one, therefore,

$$\sum_{t=1}^{T} X_w^t \leq 1 \quad w \in W \tag{7.60}$$

- The limitation on total capital and operational cost at each planning period can be considered as follows:

$$\sum_{w \in W} \left(KC(Q_w^{max})X_w^t + C_w \left(Q_w^t, h_w^t \right) Y_w^t \right) \leq C_{total,t} \quad \forall t \tag{7.61}$$

where $C_{total,t}$ is the total capital and operational cost in planning period t.
- The nonnegativity of the decision variables:

$$h^t, Q_w^t \geq 0 \quad \forall t, w \in W \tag{7.62}$$

Because of the uncertainties in future operational and capital cost functions, and projected water demands, the developed optimal policies should be revised when more information becomes available.

Example 7.8

The water demand of a city is supplied from groundwater resources and it increases over time due to population growth as shown in Figure 7.21. Four well field development projects are studied for supplying future demands of this city in a 30 year planning time horizon. The current capacity of water supply system is 100 MCM. Assume that the construction cost stays constant. The capacity and capital costs of these projects are shown in Table 7.9. The interest rate is considered to be 5%. Find the optimal sequence of implementing projects.

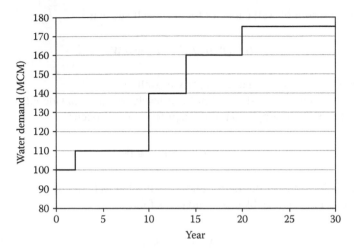

FIGURE 7.21 Water demand of a city over a 30 year period.

TABLE 7.9
Initial Investment and Capacity of Well Field Development

Well Field Development Project	Capital Costs ($10⁶)	Capacity (MCM)
A	2	15
B	2.5	20
C	1.8	10
D	4	30

Solution

Table 7.10 shows the present value of initial investment of the projects, which are estimated using $P = F/(1 + i)^t$ in the different years that the water demand has increased. In order to find the best timing for the construction of well field development projects, minimization of the present value of cost is considered as the objective function of a linear programming model as follows:

TABLE 7.10
Present Value of Capital Costs for Well Field Development Projects ($10⁶)

Project	$t = 2$ Years	$t = 10$ Years	$t = 14$ Years	$t = 20$ Years
A	1.81	1.23	1.01	0.75
B	2.27	1.53	1.26	0.94
C	1.63	1.11	0.91	0.68
D	3.63	2.46	2.02	1.51

Minimize $C = 1.81X_{11} + 1.23X_{12} + 1.01X_{13} + 0.75X_{14}$

$$+ 2.27X_{21} + 1.53X_{22} + 1.26X_{23} + 0.94X_{24}$$

$$+ 1.63X_{31} + 1.11X_{32} + 0.91X_{33} + 0.68X_{34}$$

$$+ 3.63X_{41} + 2.46X_{42} + 2.02X_{43} + 1.51X_{44}$$

Subject to

$15X_{11} + 20X_{21} + 10X_{31} + 30X_{41} \geq 10$

$15X_{11} + 20X_{21} + 10X_{31} + 30X_{41} + 15X_{12} + 20X_{22} + 10X_{32} + 30X_{42} \geq 40$

$15X_{11} + 20X_{21} + 10X_{31} + 30X_{41} + 15X_{12} + 20X_{22} + 10X_{32} + 30X_{42} +$

$15X_{13} + 20X_{23} + 10X_{33} + 30X_{43} \geq 60$

$15X_{11} + 20X_{21} + 10X_{31} + 30X_{41} + 15X_{12} + 20X_{22} + 10X_{32} + 30X_{42} +$

$15X_{13} + 20X_{23} + 10X_{33} + 30X_{43} + 15X_{14} + 20X_{24} + 10X_{34} + 30X_{44} \geq 75$

$X_{11} + X_{12} + X_{13} + X_{14} \leq 1$

$X_{21} + X_{22} + X_{23} + X_{24} \leq 1$

$X_{31} + X_{32} + X_{33} + X_{34} \leq 1$

$X_{41} + X_{42} + X_{43} + X_{44} \leq 1$

$$X_{ij} = \begin{cases} 1 & \text{if project } i \text{ constructed in the } j\text{th time period} \\ 0 & \text{otherwise} \end{cases}$$

In this problem, the constraints are defined to satisfy the additional demand in different years. They also assure that each project can only be constructed once. The developed program in Lingo 8 is shown in Figure 7.22. The optimal timing of the implementation of the projects is found to be as projects C at year 2, D at year 10, B at year 14, and A at year 20. The present value of costs of construction of the projects is also estimated to be 6.1×10^6.

7.4.3 OPTIMIZATION MODEL FOR WATER ALLOCATION

Policy instruments for groundwater allocation such as tradable extraction rights, and groundwater extraction charges determine both the economic efficiency of groundwater use and the distribution of groundwater among users. Groundwater allocation problem provides the optimal water distributed to different users such as agricultural, industrial, and municipal demands in a river basin, with a minimal effect on the environment and water users. The objective function of this model maximizes the net discounted benefit of water supply as follows:

```
 File  Edit  LINGO  Window  Help

MODEL:
SETS:
LANDFILLS/L1 L2 L3 L4/:CAPACITY, VALUE;
STEPS/S1 S2 S3 S4/: STEP;
TIMES/T1 T2 T3 T4/: DEMAND, YEAR;
LINKS( LANDFILLS, TIMES): COST, X;
ENDSETS

DATA:
CAPACITY=15 20 10 30;
VALUE=2.0 2.5 1.8 4.0;
STEP=1 2 3 4;
DEMAND=10 40 60 75;
YEAR=2 10 14 20;
ENDDATA

MIN=@SUM(LANDFILLS(I):
    @SUM(TIMES(J):
      COST(I,J)*X(I,J)));

@FOR(TIMES(T):
  @SUM(STEPS(J)|J#LE#T:
    @SUM(LANDFILLS(I):
     X(I,J)*CAPACITY(I))) >= DEMAND(T));

@FOR (TIMES(J):
 @FOR (LANDFILLS(I):
  COST(I,J)=VALUE(I)/1.05^(YEAR(J))));

@FOR (TIMES(T):
 @SUM(LANDFILLS(I):
  X(I,T))=1);

@FOR (LANDFILLS(I):
 @SUM(TIMES(T):
  X(I,T))=1);

@FOR (TIMES(T):
 @FOR (LANDFILLS(I):
   @BIN(X(I,T))));
END
```

FIGURE 7.22   Solution for Example 7.8 (Lingo 8 source code).

$$Max\ Z = \sum_{t=1}^{T}\left(\frac{1}{1+\alpha}\right)^t \sum_i \left\{ \int_0^{Q_i^t} D_i\left(Q_i^t\right)dQ - \sum_{w\in W} C_w\left(Q_w^t, h_w^t\right) \right\} \qquad (7.63)$$

where

T is the number of time steps in the planning horizon

α is the interest rate

W is the set of well sites

$D_i\left(Q_i^t\right)$ is the groundwater demand function of the subregion i, which depends on Q_i^t

Q_i^t is the water pumped from the subregion I

$C_w\left(Q_w^t, h_w^t\right)$ is the water extraction cost from the well site w, which is a nonlinear function of the discharge Q_w^t and head h_w^t at each time step t

This objective function should be maximized considering the groundwater response equations and the system's constraints:

- The response equation for each planning period is

$$h^t - A_1 h^{t-1} - A_2 g(Q^{t-1}) = 0 \quad \forall t \tag{7.64}$$

- The constraint of water balance in each subregion is

$$Q_i^t = \sum_{w \in W} Q_w^t \tag{7.65}$$

- The constraint of well capacity for each well site w is

$$0 \leq Q_w^t \leq Q_w^{max} \quad \forall t, w \in W \tag{7.66}$$

where Q_w^{max} is the discharge capacity of well w.
- The nonnegativity of the decision variables:

$$h^t, Q_w^t, Q_i^t \geq 0, \quad \forall t, w \in W \quad \forall i \tag{7.67}$$

The water allocation model can provide a schedule for optimal groundwater extraction.

Example 7.9

Develop a stochastic planning model to determine the optimal, temporal allocation of the groundwater supply of a confined aquifer. Assume (1) the dynamics of the groundwater system can be represented by the model

$$S_{t-1} = S_t - P_t + R_t \tag{7.68}$$

where
   S_t is the quantity of water in groundwater storage at the end of time period t, where t is numbered backward with respect to real time
   P_t and R_t are the groundwater pumping and recharge occurring during period t

and (2) the groundwater recharge is a random variable described by the probability density function, $h(R_t)dR_t$, which is independently distributed over time.

Solution

The stochastic planning problem can be formulated as a dynamic programming problem. More details about dynamic programming are presented in special methods for optimization techniques in this chapter. The stages of the model correspond to time; the state variable is the groundwater storage. Defining $B_t(S_t, P_t)$ as the net benefit resulting from pumping P_t then $f_t(S)$, the optimal return with t stages remaining, is given by the recursive equation (Burt, 1964)

$$f_t(S) = \max_{R \leq S} \left\{ B_t(S_t, P_t) + \left(\frac{1}{1+\rho} \right) E[f_{t-1}(S_t + R_t - P_t)] \right\} \tag{7.69}$$

$$f_t(S) = \max_{R \leq S} \left\{ B_t(S_t, P_t) + \left(\frac{1}{1+\rho} \right) \int_0^\infty f_{t-1}(S_t + R_t - P_t) h(R_t) dR_t \right\} \tag{7.70}$$

where
  E is the mathematical expectation operator
  ρ is the discount rate

The functional equation can be interpreted as the maximization with respect to the water consumption at stage t of the immediate net benefit plus the expected return in the $t - 1$ remaining stages, given that an optimal policy will be used during the remaining $t - 1$ stages.

Solution of the dynamic programming model is usually accomplished by discretizing the groundwater storage, recharge, and pumping. The dynamic programming model then becomes a finite Markovian decision process. Defining N discrete levels for groundwater storage ($j = 1, ..., N$), m values for R_t, and n values for P, ($k = 1, ..., n$), the functional equation can be expressed for a particular groundwater storage level, S_i, as

$$f_t(S_i) = \max_k \left\{ B_t(S_{ti}, P_{tk}) + \left(\frac{1}{1+\rho} \right) \sum_{j=1}^N f_{t-1}(S_{ti} + R_{tj} - P_{tk}) \cdot P\left(R_t = R_t^j\right) \right\}, \quad p_{tk} \leq S_i \tag{7.71}$$

For each groundwater storage level, S_{ti}, we can also define a probability, P_{ij}^k for each P_{tk} as the probability of going to S_{ti}, given that the groundwater storage was S_{ti}, at the beginning of the period and P_{tk} is the withdrawal. The recursive equation can then be expressed in terms of these conditional probabilities as

$$f_t(S_i) = \max_k \left\{ B_t^k(S_i) + \left(\frac{1}{1+\rho} \right) \sum_{j=1}^n p_{ij}^k f_{t-1}(S_i) \right\}, \quad t = 0,1,2, \quad B_t^k(S_i) = B_t(S_{ti}, P_{tk}) \tag{7.72}$$

An interesting property of the dynamic programming model occurs when the system benefit function is constant over the planning horizon. The solution of the programming model will then converge to an optimal policy that is independent of the stage when the number of stages becomes very large. The optimal policies are found from the solution of the equation

$$(I - \beta P)x = b \tag{7.73}$$

where
  I is the identity matrix, $\beta = (1 + \rho)^{-1}$
  P is an $n \times n$ matrix containing p_{ij}^k, $b = \left[B_t^k(S_i), ..., B_t^k(S_i) \right]^T$
  $x = [f_1, f_2, ..., f_n]^T$ the optimal return vector

7.4.4 OPTIMIZATION MODEL FOR CONJUNCTIVE WATER USE

The conjunctive water use model provides an optimal allocation schedule from groundwater and surface water resources in a region to meet different water demands. The objective function maximizes the net discounted economic benefit as follows:

$$Max\ Z = \sum_{t=1}^{T} \left(\frac{1}{1+\alpha}\right)^t \left\{ \sum_l \left[\int_0^{Q_l^t} D_l\left(Q_l^t\right) dQ - \sum_m g_{ml}^t\left(G_{ml}^t\right) - \sum_n s_{nl}^t\left(S_{nl}^t\right) \right] \right\} \quad (7.74)$$

where
  $l/m/n$ are indexes that define different demands, groundwater resources, surface water resources in the region
  $D_l\left(Q_l^t\right)$ is the demand function, which depends on the water allocated to demand l, Q_l^t
  $g_{ml}^t\left(G_{ml}^t\right)$ is the cost function of groundwater allocation from resource m, which depends on the amount of groundwater allocated to that demand, G_{ml}^t
  $s_{nl}^t\left(S_{nl}^t\right)$ is the cost function of surface water allocation from resource n, which depends on the amount of surface water allocated to that demand, S_{nl}^t

The model is constrained to response equations, and physical limitations.

- The response equations for each groundwater resource is

$$h_g^t - A_1^m h_g^{t-1} - A_2^m g\left(\sum_m G_{ml}^{t-1}\right) = 0 \quad (7.75)$$

- Surface water balance equation is

$$\bar{S}_n^t = \bar{S}_n^{t-1} + R_j^t - \sum_l S_{nl}^t \quad (7.76)$$

- Related groundwater allocation and pumping schedule is

$$\sum_l G_{ml} = \sum_{w \in W} Q_{w,m}^t \quad \forall m \quad (7.77)$$

where
  $Q_{w,m}^t$ is the pumping rate from well w in resource m
  W defines the well site of the region

- Well capacity limitation:

$$Q_{w,m}^t \leq Q_{w,m}^{max} \quad \forall m \quad (7.78)$$

- The lower bound constraints on the head levels (h_w) are considered to ensure the excessive depletion of the aquifer does not occur:

$$h_w^t \geq h_w^{min} \quad \forall t, \; w \in W \tag{7.79}$$

- The nonnegativity of the decision variables:

$$h_l^t, Q_w^t, Q_l^t, G_{ml}^t, S_{nl}^t \geq 0 \quad \forall t, \; w \in W, \quad \forall l, m, n \tag{7.80}$$

This optimization problem can be solved using different methods such as nonlinear, dynamic programming, and genetic algorithm. For more details and examples of this management model refer to Chapter 8.

7.4.5 OPTIMAL GROUNDWATER QUALITY MANAGEMENT MODEL

The management of groundwater with water quality issue in aquifer systems is a multi-objective planning problem, which is solved with trade-offs between different objectives. Wastewater can be injected to control seawater intrusion, decrease the land subsidence, and increase the groundwater level. In the groundwater quality management model, the objective function is to minimize the possible contamination of the aquifer or the cost of surface waste treatment and/or waste storage. The decision variables of the optimization model are the well location and pumping/injection rates to satisfy the water demand and the maximum waste input concentration to satisfy the groundwater quality requirements. The objective function of the optimization model defined as the weighted sum of the water quantity and quality objectives such as minimization of the possible contamination of the aquifer or the cost of surface waste treatment can be expressed as follows:

If we consider a dynamic model and assume (1) a single waste constituent and (2) that the planning horizon consists of T^* discrete time periods, or

$$Max \; z = \sum_{k=1}^{T^*} \sum_1 \lambda_1 f_1 \left(h^k, c_s^k, Q_w^k, Q_r^k \right) \tag{7.81}$$

f_l is the lth objective of the planning problem and we have assumed that a particular objective is a function of the state variables (the head and mass concentrations in planning period k) and the pumping (Q_w^k) and injection (Q_r^k) policies. The λ_1 represents the preferences or weights associated with planning objective.

The constraints of the model are meeting water demand and waste load disposal target, expressed as follows:

$$\sum_{W \in \pi} Q_w^k \geq D^k \quad \forall k \tag{7.82}$$

$$\sum_{r \in \psi} Q_r^k \geq WL^k \quad \forall k \tag{7.83}$$

where

Q_w^k is the pumping rate

Q_r^k is the injection rate

D_k is the water demand

WL_k is the magnitude of the waste load in planning period k

The index sets π and ψ define the feasible pumping and injection sites in the basin

Others constrains are on the maximum pumping and injection rates as well, as follows:

$$Q_w^k \leq Q_w^*, \quad w \in \pi \quad \forall k \tag{7.84}$$

$$Q_r^k \leq Q_r^*, \quad r \subset \psi \quad \forall k \tag{7.85}$$

Also there are constrains on the concentrations of water withdrawal for demand supplying and injection as follows:

$$c_{s,w}^k \leq \bar{c}_{s,w}^*, \quad w \in \pi \quad \forall k \tag{7.86}$$

$$c_{s,r}^k \leq \bar{c}_{s,r}^*, \quad r \in \psi \quad \forall k \tag{7.87}$$

The simulation model can be utilized for the estimation of concentration and water head as the state variables. It can be linked with the optimization model for developing control policies. The decision variables are the waste injection concentrations and the pumping and injection rates.

7.4.6 OPTIMIZATION MODEL FOR PARAMETERS OF GROUNDWATER MODEL

The system parameters in groundwater flow and transport simulation are estimated for precise modeling. These parameters are transmissivity, storage coefficient, conductivity, dispersivity, retardation factor, and reaction rate. The techniques based on trial and error and graphical matching are traditional methods. The parameters values of groundwater modeling are estimated from observations of the system response through the inverse problem. In this method, the parameters are determined to match the observations through a systematic adjustment in the spatial and time domains. In the inverse problem, an optimization model is developed to determine the system's hydraulic and mass transport parameters.

Yeh (1986) classified the solution algorithms of parameter identification in groundwater systems as being part of either the *output error criterion method* or *the equation error criterion method*. In the equation error criterion method, unknown parameters are determined based on analytical solutions of the governing equations of the aquifer system such as finite-element or finite-difference approximation. In this method, an optimization model is developed to minimize the sum of squares of errors for the equation, to obtain the unknown parameters. The error is estimated using the observed or interpolated data at each node of the domain. In output error criterion method, an optimization model is developed to minimize a given output error criterion in the identification of the flow model parameters. The parameter estimates are required to satisfy the aquifer system's hydraulic or water duality equations and possible upper or lower bounds (Willis and Yeh, 1987).

7.5   OPTIMIZATION TECHNIQUES

Decision making is the science of choice. In a particular problem we might have a single person who is in charge of deciding what to do, or there are several people or organizations which are involved in the decision-making process who are termed as *decision makers*. The options from which the selection is made are called *alternatives* and the set of all possible alternatives is called the *decision space*. If the decision space is finite, then the decision-making problem is called *discrete*, and if the decision alternatives are characterized by continuous variables, then the problem is called *continuous*.

All decision-making problems are faced with limitations which result from financial constraints, technological, social, and environmental restrictions among others. These limitations are called the *constraints* of the problem. The alternatives satisfying all constraints are called *feasible*, and the set of all feasible alternatives is called the *feasible decision space*.

The goodness of any alternative can be characterized by several ways, which are called the *attributes*. Most of them can be described verbally such as "how clean the water is," or "how does it satisfy the customers." In mathematical modeling we need measurable quantities describing the goodness of each alternative with respects to the attributes. Such measures are called the *criteria*. For example, water quality can be measured by the pollutants' concentration, consumer satisfaction by supplies, and so on.

If there is only one criterion, the problem is called the *single-criterion* optimization problem. In the presence of more than one criterion, we have a *multiple-criteria* decision problem. In the case of multiple decision makers, we might consider the problem as multiple-criteria decision making when the criteria of all decision makers are considered as the criteria of the problem (Karamouz et al., 2003).

A set of constraints considering the technical, economic, legal, or political limitations of the project, should be considered in evaluating the optimal solution. There may also be some constraints in the forms of either equalities or inequalities on decision and state variables.

The optimization problem can be solved through a manual trial-and-error adjustment or using a formal optimization technique. Although the trial-and-error method is simple it is tedious and also does not guarantee reaching the optimal solution. On the contrary, an optimization technique can be used to identify the optimal solution which is feasible in terms of satisfying all the constraints.

The commonly used optimization techniques are

1. Linear programming (LP) (e.g., Lefkoff and Gorelick, 1987)
2. Nonlinear programming (NLP) (e.g., Ahlfeld et al., 1988; Wagner and Gorelick, 1987)

3. Mixed integer linear programming (MILP) (e.g., Willis, 1976, 1979)
4. Mixed integer nonlinear programming (MINLP) (e.g., McKinney and Lin, 1995)
5. Differential dynamic programming (DDP) (e.g., Chang et al., 1992; Culver and Shoemaker, 1992; Sun and Zheng, 1999)

These methods are considered as "gradient" methods because the gradients of the objective function regarding the variables to be optimized are repetitively calculated in the process of finding the optimal solution. Beside the computational efficiency of these techniques, there are some significant limitations. First, when the objective function is highly complex and nonlinear, with multiple optimal points, a gradient method may be trapped in one of the local optimal and cannot reach the globally optimal solution. Second, sometimes numerical difficulties in gradient methods result in instability and convergence problems.

More recently, evolutionary optimization methods such as simulated annealing (SA), genetic algorithms (GA), and tabu search (TS) which are based on heuristic search techniques have been introduced. These methods imitate natural systems, such as biological evolution in the case of genetic algorithms, to identify the optimal solution, instead of using the gradients of the objective function. The evolutionary methods generally require intensive computational efforts; however, they are increasingly being used because of their ability to identify global optimum, their efficiency in handling discrete decision variables, and their easy application in different simulation models.

7.5.1 SINGLE-CRITERION OPTIMIZATION

Let X denote the feasible decision space, and let be any feasible alternative. If X is a finite set, then the value of X can be any one of the number 1, 2, ..., N when N is the number of the elements of X. If X is a continuous set characterized by continuous decision variables, then x is a vector, $x = (x_1, ..., x_n)$ where n is the number of the decision variables and $x_1, x_2, ..., x_n$ are the decision variables. If f denotes the single criterion, then we assume that f is defined on X and for all $x \subset X, f(x)$ is a real number. This condition is usually defined as $f: X \to R$. Since f is real valued, all possible values of $f(x)$ can be represented on the real line, and the set

$$H = \left\{ f(x) \big| x \in X \right\} \tag{7.88}$$

is called the *criterion space*. Notice that the elements of X show the decision choices and the elements of H give the consequence of the choices in terms of the values of the evaluation criterion. We can simply say that X shows what we can do and H shows what we can get.

If X is finite and the number of its elements is small then by comparing the $f(x)$ values, the best alternative can be easily identified. If the number of elements of X is large or even infinite but X is a discrete set (for example, when it is defined by decision variables with integer values), then the methods of discrete and combinatorial optimization are used. This methodology can be found in any textbook of advanced optimization techniques.

The most simple single-criterion optimization problem with continuous variables is *linear programming*. In order to give a general formulation, let n be the number of the decision variables, and $x_1, x_2, ..., x_n$ the decision variables. It is assumed that the criterion and all constraints are linear. The criterion is sometimes also called *the objective function* and it is assumed to have the form

$$f(x) = c_1 x_1 + c_2 x_2 + \cdots + c_n x_n = \underline{c}^T \underline{x} \tag{7.89}$$

where $c_1, c_2, ..., c_n$ are real numbers, $\underline{c} = (c_1, c_2, ..., c_n)$ and

$$\underline{x} = \begin{bmatrix} x_1 \\ x_2 \\ \vdots \\ x_n \end{bmatrix} \tag{7.90}$$

Linear constraints are equalities or inequalities. Equality constraints have the general form

$$a_{i1} x_1 + a_{i2} x_2 + \cdots + a_{in} x_n = b_i \tag{7.91}$$

and inequality constraints can be written as

$$a_{i1} x_1 + a_{i2} x_2 + \cdots + a_{in} x_n \underset{\leq}{\overset{\geq}{}} b_i \tag{7.92}$$

By repeating the above procedure to all variables and constraints, if necessary, the resulting problem will have only nonnegative variables and $\leq$ type inequalities in the constraints. Hence any linear programming problem can be rewritten in this way into its *primal form*:

$$\text{Maximize} \quad c_1 x_1 + c_2 x_2 + \cdots + c_n x_n$$

$$\text{Subject to}$$

$$x_1, x_2, ..., x_n \geq 0$$

$$a_{11} x_1 + a_{12} x_2 + \cdots + a_{1n} x_n \leq b_1$$

$$a_{21} x_1 + a_{22} x_2 + \cdots + a_{2n} x_n \leq b_2 \tag{7.93}$$

$$\vdots$$

$$a_{m1} x_1 + a_{m2} x_2 + \cdots + a_{mn} x_n \leq b_m$$

Note that the objective function may represent cost, contained pollutant amount, and so on, when smaller objective values represent better choices. In such cases the objective function is minimized. In the above formulation we assume maximization, since minimization problems can be rewritten into maximization problems by multiplying the objective function by (−1).

In the very simple case of two decision variables, Equation 7.93 can be solved by the graphical approach. As an illustration consider the following example. The most common procedure to solve this kind of problems is the *Simplex Method*. For more details please see Karamouz et al., 2003.

7.5.2 MULTI-CRITERIA OPTIMIZATION

As in the previous session, let X denote the feasible decision space, and let $f_1, ..., f_I$ be the criteria. Set X shows which decisions can be made; however, in the presence of multiple criteria we need to represent the set of all possible outcomes. The set

$$H = \left\{ (f_1(x),..., f_I(x)) \middle| x \in X \right\} \tag{7.94}$$

is called the *criteria space* that shows what we can get by selecting different alternatives.

Example 7.10

Assume that selection has to be made from four available technologies to perform a certain groundwater cleaning restoration task. These technologies are evaluated by their costs and convenience of usage. The cost data are given in $1000 units and the convenience of usage is measured in a subjective scale between 0 and 100, where 100 is the best possible measure. The data are given in Table 7.11. Which technique is the acceptable solution?

Solution

The cost data is multiplied by (−1) in order to maximize both criteria. Similarly to the single-criterion case, the criteria are sometimes called *the objective function*. The criteria space has four isolated points in this case as shown in Figure 7.23. It is clear that

TABLE 7.11

Costs and Convenience Numbers of Four Projects

Technology	Cost	Convenience
1	−25	85
2	−18	65
3	−27	93
4	−15	50

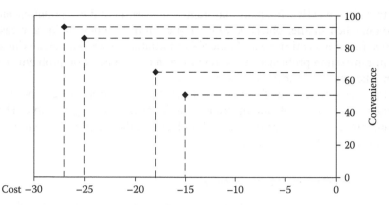

FIGURE 7.23   Criteria space in the discrete case.

none of the points dominates any other point, since lower cost is always accompanied by lower convenience, that is, increasing any objective should result in decreasing the other one. Therefore, any one of the four technologies could be accepted as the solution.

- **Sequential optimization**

This method is based on the ordinal preference order of the objectives. Assume that the objectives are numbered so that f_1 is the most important, f_2 is the second most important, and so on, and f_I is the least important. In applying the method, first f_1 is optimized. If the optimal solution is unique, then the procedure terminates regardless of the values of the less important other objectives. Otherwise, f_1 is kept on its optimal level and f_2 is optimized. If there is a unique optimal solution, then it is selected as the solution of the problem, otherwise both f_1 and f_2 are kept optimal and f_3 is optimized, and so on, until there is a unique optimal solution or f_I is already optimized.

- **The ε-constraint method**

In the case of sequential optimization, the procedure often terminates before all objectives are optimized. Therefore, less important objectives may have very unfavorable values. In order to avoid such possibilities the following method can be suggested. Assume that the most important objective is specified, and minimal acceptable levels are given for all other objectives. If f_1 is the most important objective, and $\varepsilon_2, \varepsilon_3, \ldots, \varepsilon_I$ are the minimum acceptable levels, then the solution is obtained by solving the following single-objective optimization problem:

$$\text{Maximize} \quad f_1(x)$$

$$\text{Subject to} \quad x \in X$$

$$f_2(x) \geq \varepsilon_2$$

$$\vdots \qquad\qquad\qquad (7.95)$$

$$f_I(x) \geq \varepsilon_I$$

Example 7.11

Assume that in Example 7.10, the cost is more important than convenience, but we want to have at least 82 in convenience. Which technique is the acceptable solution?

Solution

This condition makes technology variants 2 and 4 infeasible. Both alternatives 1 and 3 satisfy these minimal requirements. The less expensive among these two remaining alternatives is variant 1, so it is the choice.

- **The weighting method**

In the application of this method all objectives are taken into account by importance weights supplied by the decision makers. Assume that $c_1, c_2, ..., c_I$ are the relative importance factors of the objectives, which are specified by the decision makers on some subjective basis, or are obtained by some rigorous methods such as pair-wise comparisons. It is always assumed that $c_i > 0$ for all i, and $\sum_{i=1}^{I} c_i = 1$. In applying the weighting method, the composite objective is a weighted average of all objectives, where the weights are the relative importance factors $c_1, ..., c_I$. Hence, the single-objective optimization problem

$$\text{Maximize} \quad c_1 f_1(x) + c_2 f_2(x) + \cdots + c_I f_I(x)$$

$$\text{Subject to} \quad x \in X \tag{7.96}$$

is solved. A transformation called the *normalizing* procedure is needed that is a simple linear transformation. The procedure is outlined as follows:

Let f_i denote an objective function. Let m_i and M_i denote its possible smallest and largest values, respectively. These values may be computed by simply minimizing and maximizing f_i over the feasible decision space X or their values also can be supplied by the decision makers based on subjective judgments. Then the normalized objective function is obtained as

$$\overline{f_i}(x) = \frac{f_i(x) - m_i}{M_i - m_i} \tag{7.97}$$

In many practical cases, the decision makers are able to provide a utility function to each objective that shows his/her satisfaction level for the different values of that objective. Such a function is illustrated in Figure 7.24 where m_i gives the least 0 satisfaction level and M_i gives the largest 1 (=100%) satisfaction level. Then objective f_i is replaced by the corresponding utility value

$$\overline{f_i}(x) = u_i(f_i(x)) \tag{7.98}$$

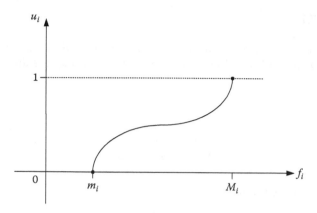

FIGURE 7.24 Utility function of objective f_i.

Notice that Equation 7.97 corresponds to the linear utility function

$$u_i(f_i) = \frac{(f_i - m_i)}{(M_i - m_i)} \tag{7.99}$$

We always require that the lowest possible utility value is 0 and the largest is 1, so $\overline{f}_i(x) \in [0,1]$ for all $x \in X$. Then Equation 7.96 is modified as

$$\text{Maximize} \quad c_1\overline{f}_1(x) + c_2\overline{f}_2(x) + \cdots + c_I\overline{f}_I(x)$$
$$\text{Subject to} \tag{7.100}$$
$$x \in X$$

In this new composite objective only dimensionless values are added, and the linear combination of the normalized objectives can be interpreted as an average satisfaction level based on all objectives.

Example 7.12

Assume that in Example 7.10, the decision makers consider \$30,000 as the worst possible cost, and they believe that a cost of \$15,000 would give them 100% satisfaction.

Solution

So, $m_1 = -30$ and $M_1 = -15$ and if linear utility function is assumed between these values, then the linearization is based on the linear transformation as: $\overline{f}_1 = (f_1 + 30)/(-15 + 30)$, so the cost values in Table 7.11 have to be modified accordingly. The convenience data are already given in satisfaction levels (in %), so these numbers

---

TABLE 7.12

Normalized Objectives of Four Projects

Technology	Cost	Convenience
1	0.33	0.85
2	0.8	0.65
3	0.2	0.93
4	1	0.50

---

are divided by 100 in order to transform them into the unit interval [0,1]. The normalized values are shown in Table 7.12.

Assume that equal importance is given to the objectives, that is, $c_1 = 0.3$ and $c_2 = 0.7$. The composite objective values for the four technology variants are as follows:

$$0.3(0.33) + 0.7(0.85) = 0.694$$

$$0.3(0.8) + 0.7(0.65) = 0.695$$

$$0.3(0.2) + 0.7(0.93) = 0.711$$

$$0.3(1) + 0.7(0.5) = 0.65$$

Therefore, the third alternative is the choice.

- **Interactive fuzzy approach**

The fuzzy application became possible when Bellman and Zadeh (1970) and, a few years later, Zimmermann (1978) introduced fuzzy sets into the field of multiple criteria analysis. In recent years, an interactive approach in fuzzy environment is used as one of the most effective approaches for solving multi-objective optimization problems. For fuzzy multi-objective optimization, the concept of membership-Pareto (M-Pareto) optimal solutions, which is defined in terms of membership functions instead of objective functions, is introduced. The decision maker must specify the reference membership levels to express the preferences for the objective functions and to generate a candidate for the satisfying solution (also called the M-Pareto optimal solution). For the decision maker's reference membership levels, $\bar{\mu}_i^r$, the corresponding optimal solution which is the nearest to the requirements in the mini-max sense or better than that if the reference membership levels are attainable, is obtained by solving the following problem:

$$\underset{x \in X}{Min} \ \underset{i=1,\ldots,p}{Max} \left[\bar{\mu}_i^r - \mu_i(Z_i(x)) \right] \tag{7.101}$$

or equivalently

$$Min \quad \lambda$$

$$s.t. \quad \overline{\mu}_i^r - \mu_i(Z_i(x)) \leq \lambda, \quad i = 1, \ldots, p \quad (7.102)$$

$$x \in X$$

where $\mu_i(Z_i(x))$ is the membership function for each of the objective functions $Z_i(x)$, $i = 1, \ldots, p$. This method for obtaining the trade-off information between p objectives can be summarized as follows:

1. Solve the p individual optimization problems to find the optimal solution for each of the individual p objectives.
2. Compute the value of each of the objectives and determine the individual minimum Z_i^{min} and individual maximum Z_i^{max} of each objective function under the given constraints.
3. Derive a membership function for each of the objectives and the associated extreme values. The membership function $\mu_i(Z_i(x))$ for the fuzzy-max objective function $Z_i(x)$ can be defined as

$$\mu_i(Z_i(x)) = \begin{cases} 0 & Z_i(x) \leq Z_i^{min} \\ \dfrac{(Z_i(x) - Z_i^{min})}{(Z_i^{max} - Z_i^{min})} & Z_i^{min} < Z_i(x) \leq Z_i^{max} \\ 1 & Z_i(x) > Z_i^{max} \end{cases} \quad (7.103)$$

4. Set the initial reference membership levels to 1 for incorporating the initial opinion of the decision maker.
5. Solve min–max problem, then the optimal solution and membership functions are determined based on the results.
6. The reference membership levels and membership functions of the objectives can be revised considering the results of trade-off between objectives and repeat from step 5.

7.5.3 SPECIAL METHODS FOR GROUNDWATER OPTIMIZATION

Some special methodologies, which are useful in the application in groundwater optimization, are presented in this section.

- **Dynamic programming**

In the water resources planning, dynamism of the system is a major concern, which has to be taken into account in modeling and optimization. Dynamic programming (DP) is the most popular tool in optimizing dynamic systems. DP is used extensively in the optimization of groundwater resource systems (Buras, 1966). The popularity and success of this technique can be attributed in part to its efficiency in incorporating nonlinear constraints and objectives and stochastic or random variables in the DP formulation of the management or planning problem. In this method, highly complex problems with a large number of decision variables are decomposed into a series of subproblems that can be solved recursively (Willis and Yeh, 1987). In the DP formulation of the planning or operational model, the optimization model is described or characterized by the state variables of the system, the system stages, and the control or decision variables. In groundwater systems, the state variables may represent the hydraulic head or mass concentrations, or the amount of groundwater in storage at any time. The decision variables control, in any time period, the pumping, recharge, or waste injection schedules. The time element of the planning problem defines the stages of the DP model.

An important element of any DP model is the state transition equations of the system. The response or transfer equations, which can be developed using finite-difference or finite-element methods, define how the system state variables change over successive stages or time periods. Consider the linear groundwater system which is homogeneous and isotropic. The stage-to-stage transformation is then described by the continuity equation, which for a lumped parameter the linear system can be expressed as

$$S_{t+1} - S_t + R_t = I_t - E_t \tag{7.104}$$

where
S_t is the storage at the beginning of time period or stage t
R_t is the groundwater extraction during stage t
I_t is the groundwater recharge
E_t is the subsurface outflow, leakage, or evaporation that occurs during stage t

Assuming that the system objective, $f(R_t, S_t)$ is a function of the state variable, the groundwater storage, and the pumping occurring in any stage or time period,

$$Min \; Z = \sum_{t=1}^{T} f(R_t, S_t) \tag{7.105}$$

where T is the time horizon. Additional constraints of the model include maximum and minimum storage bounds to limit the groundwater extractions or to prevent possible flooding of the land surface, or

$$S_t^{max} \geq S_t \geq S_t^{min}, \quad t = 1,\ldots,T \tag{7.106}$$

$$R_t^{max} \geq R_t \geq R_t^{min}, \quad t = 1,\ldots,T \tag{7.107}$$

$$\left| S_{t+1} - S_t \right| \leq SC_s, \quad t = 1,\ldots,T \tag{7.108}$$

where SC_s is the maximum allowable change in storage of reservoir s within each month considering dam stability and safety conditions. Because of many decision variables and constraints, finding the direct solution is computationally difficult. Through the DP procedure, this unique optimization problem is changed to solve many smaller sized problems in the form of a recursive function. The DP recursive equation can be expressed as

$$V_t(S_t) = min\{ f_t(R_t, S_t) + V_{t+1}(f_{t+1}(R_{t+1}, S_{t+1}))\} \tag{7.109}$$

In most applications, the optimization Equation 7.109 cannot be solved analytically and recursive functions are determined only numerically. In such cases, discrete dynamic programming (DDP) is used, where the state and decision spaces are discretized. The main problem of using this method is known as the "curse of dimensionality." This is because the number of points that should be considered in the evaluation of different possible situations drastically increases by the number of discretizations. Therefore, DDP is used only up to four or five decision and state variables.

Different successive approximation algorithms such as differential dynamic programming (DIFF DP), discrete differential dynamic programming (DDDP), and state incremental dynamic programming (IDP) are developed to overcome this problem. In applying any of these methods, the user should provide an initial estimation of the optimal policy. This estimation is improved in each step until convergence occurs. However, the limit might be only a local optimum. The survey paper by Yakowitz (1982) gives a summary of the different versions of DP with applications to water resources planning and management.

Example 7.13

Assume an aquifer which supplies part of the water demand for a city. The monthly water demand (D_t) is about 25 million cubic meters. The maximum storage bond of the aquifer is 30 million cubic meters. Let S_t (aquifer storage at the beginning of month t) take the discrete values, 0, 10, 20 and 30. The cost of operation is estimated as the total seem of squared deviations from storage (ST_t) and release target (D_t) as follows:

$$C_t(S_t, R_t) = (S_t - ST_t)^2 - (R_t - D_t)^2 \tag{7.110}$$

Formulate a forward-moving deterministic DP model for finding the optimal release in the next 3 months.

Formulate a backward-moving deterministic DP model for finding the optimal release in the next 3 months.

Solve the DP model developed in part (a), assuming that the inflows to the aquifer in the next 3 months ($t = 1, 2, 3$) are forecasted to be 10, 30, and 20, respectively. The aquifer storage at the current month is 20 million cubic meters.

Solution

(a) Forward deterministic DP model is formulated as follows:

$$f_t(S_{t+1}) = Min(C_t(S_t, R_t) + f_{t-1}(S_t)) \qquad t = 1, 2, 3, \quad S_t \in \Omega_t$$

$$S_t \leq S_{max}, \quad S_{max} = 30\,MCM$$

$$C_t(S_t, R_t) = (S_t - ST_t)^2 - (R_t - D_t)^2 \tag{7.111}$$

$$S_{t+1} = S_t + I_t - R_t$$

where

Ω_t is the set of discrete storage volumes (states) that will be considered for the beginning of month t at site s

S_{t+1} is the storage at the beginning of month $t + 1$

(b) The backward deterministic DP model is formulated as follows:

$$f_t(S_t) = Min(C_t(S_t, R_t) + f_{t+1}(S_{t+1})) \qquad t = 1, 2, 3 \tag{7.112}$$

The constraints are as mentioned above.

(c) Loss is calculated using the forward formulation for 3 months. Results are shown in Table 7.13.

Therefore, the optimal water withdrawals from the aquifer are $R_t^* = 20$, $R_2^* = 20$ and $R_3^* = 20, 30$.

TABLE 7.13
DP Model Solution for Example 7.13

t	I_t	S_t	$R_t : 10$	20	30	40	$f_t(S_t, R_t)$	R_t^*
$t = 3$	20	0	625	425	425	625	425	20
		10	325	125	125	325	125	20,30
		20	225	25	25	225	25	20,30
		30	325	125	125	325	125	20,30
$t = 2$	30	0	625+25	425+125	425+425	—	550	20
		10	325+125	125+25	125+125	325+425	150	20
		20	—	25+125	25+25	225+125	50	30
		30	—	—	125+125	325+25	25	30
$t = 1$	10	20	225+50	25+150	25+550	—	175	20

R_t^* is the optimal policy at each month.

- **Genetic algorithm**

Genetic algorithms are stochastic search methods that imitate the process of natural selection and mechanism of population genetics. They were first introduced by Holland (1975) and later further developed and popularized by Goldberg (1989). Genetic algorithms are used in a number of different application areas ranging from function optimization to solving large combinatorial optimization problems. The GA begins with the creation of a set of solutions referred to as a population of individuals. Each individual in a population consists of a set of parameter values that completely describes a solution. A solution is encoded in a string called a chromosome, which consists of genes that can take a number of values. A general form of three genes (which are the string elements, for example, bits) in one chromosome with 4-bit representation is shown in Figure 7.25. Initially, the collection of solutions (population) is generated randomly and at each iteration (also called generation) a new generation of solutions is formed by applying genetic operators including selection, crossover, and mutation, analogous to ones from natural evolution. Each solution is evaluated using an objective function (called a fitness function), and this process is repeated until some form of convergence in fitness is achieved. The objective of the optimization process is to minimize or maximize the fitness.

In *creating* chromosomes, we have to map the decision space into a set of encoded strings. The encoding mechanism depends on the problem being solved. One possible way is to use the binary representation of the decision variables. We have to decide also on the size of the breeding pool, that is, on the number of chromosomes being involved in the process. Larger number of chromosomes increases diversity, but the algorithm becomes slower. First, an initial population is selected randomly.

The *fitness* of each string is often evaluated based on the objective functions, the constraints, and the encoding mechanism. A usual way is to add a penalty term to the objective function if any chromosome becomes infeasible by violating one or more constraints.

Pairs of chromosomes are formed randomly and they are then subjected to certain genetic manipulations. They modify the genes in the parent chromosomes. A typical procedure is called *crossover*, which swaps a part of the genetic information contained in the two chromosomes. Usually a sub-string portion is selected randomly in the chromosomes, and the genes within that sub-string are exchanged. By this way new offspring are created to replace the parent chromosomes. An example of using crossover in generation-two chromosomes is shown in Figure 7.26. The particular

$$\underline{1\,0\,1\,1\,0\,0} \quad \underline{1\,0\,0\,1\,1\,1} \quad \underline{0\,0\,1\,0\,0\,0} \cdots \cdots$$
$$H1 \qquad\qquad H2 \qquad\qquad H3 \cdots \cdots$$

FIGURE 7.25   A general framework of a chromosome including genes.

A: 0 1 0 1 0 1 0 1 0 1 0 1 1 | 1 1 0 0 0   C: 0 1 0 1 0 1 0 1 0 1 0 1 1 | 0 1 1 0 1

B: 1 0 1 0 1 0 1 0 1 0 1 0 0 | 0 1 1 0 1   D: 1 0 1 0 1 0 1 0 1 0 1 0 0 | 1 1 0 0 0

FIGURE 7.26   An example of using crossover in generation of chromosomes C and D from chromosomes A and B.

A: 0 1 0 1 0 1 0 1 0 1 0 1 1 1 1 0 0 0   C: 0 0 0 1 0 0 1 1 0 1 1 1 0 1 1 0 0 1

B: 1 0 1 0 1 0 1 0 1 0 1 0 0 0 1 1 0 1   D: 1 1 1 0 1 1 0 0 1 0 0 0 1 0 1 1 0 0

FIGURE 7.27   An example of using mutation in generation of chromosomes C and D from chromosomes A and B.

structure of the crossover mechanism is usually problem specific and has to result in meaningful and feasible solutions.

There is no guarantee that the new pairs of chromosomes will be better than the parent chromosomes. To overcome this difficulty *mutations* are allowed to occur, which reverse one or more genes in a chromosome. This simple step may reintroduce certain genes into the solution that may be essential for the optimal solution and being lost from the breeding population in previous stages. An example of using mutation in generation-two chromosomes is shown in Figure 7.27. Using mutation more frequently makes the search more broadly random, resulting in slower convergence of the algorithm.

By repeating the above steps, we sequentially replace either the entire previous population or only the less fit members of it. The cycle of creation, evaluation, selection, and manipulation is iterated until an optimal or an acceptable solution is reached.

GA again raises a couple of important features. First, it is a stochastic algorithm; randomness has an essential role in genetic algorithms. Both selection and reproduction need random procedures. A second very important point is that genetic algorithms always consider a population of solutions. Keeping in memory more than a single solution at each iteration offers a lot of advantages. The algorithm can recombine different solutions to get better answers and so it can use the benefits of assortment. The robustness of the algorithm should also be mentioned as something essential for the algorithm success. Robustness refers to the ability to perform consistently well on a broad range of problem types.

Example 7.14

In an area, the cost function of aquifer restoration is the function of treatment and pumping operation. Find the maximum point of function $f(x) = -x^2 + x + 1$ in the [0,1] interval using the genetic algorithm where $f(x)$ is cost function, and x is injection concentrations (oxygen and nutrient).

Solution

First the decision space is discretized and each is encoded. For illustration purposes, we have 4-bit representations (musk). The population in musk and corresponding value is provided in Table 7.14. Next the initial population is selected. We have chosen six chromosomes; they are in Table 7.15, where their actual values and the objective function values are also presented.

The best objective function values are obtained at 0101 and 0100. Assume mutation probability P_m is 0.01 and crossover probability is considered as 1. Also assume that four random numbers are generated to compare with mutation probability for

TABLE 7.14

Population of Chromosomes

Musk	X	Musk	X
1001	0.6	0000	0
1010	0.67	0001	0.07
0101	0.34	0010	0.13
0111	0.47	0100	0.27
1110	0.93	1000	0.53
1101	0.87	0011	0.2
1011	0.73	0110	0.4
1111	1	1100	0.8

TABLE 7.15

Initial Population in GA Solution

Chromosomes	Value	Objective Function
0000	0	1
0101	0.34	1.224
0100	0.27	1.197
1101	0.87	1.113
1111	1	1
0010	0.13	1.113

TABLE 7.16

Second Population in GA Solution

Chromosomes	Value	Objective Function
0000	0	1
0101	0.34	1.224
0100	0.27	1.197
1101	0.87	1.113
1111	1	1
0010	0.13	1.113

each bit as [0.213, 0.008, 0.853, 0.462] of chromosome 0001. A simple crossover procedure is performed by interchanging the two middle bits to obtain 0000 and 0100. The new population is presented in Table 7.16. The next populations are generated and presented in Tables 7.17 and 7.18.

The results show how the best objective function evolves from population to population. Starting from the initial table, the value of 1.224 increased to 1.249, then remained the same. Since this value is the global optimum, it will not increase in further steps. In practical application we stop the procedure if no, or very small, improvement occurs after a certain (user-specified) number of iterations.

TABLE 7.17
Third Population in GA Solution

Chromosomes	Value	Objective Function
1010	0.67	1.221
0111	0.47	1.249
0110	0.4	1.240
1111	1	1
0101	0.34	1.224
1000	0.53	1.249

TABLE 7.18
Fourth Population in GA Solution

Chromosomes	Value	Objective Function
1001	0.6	1.24
0111	0.47	1.249
0000	0	1
0011	0.2	1.16
1100	0.8	1.16
1000	0.53	1.249

- **Simulation annealing**

Simulation annealing (SA) is developed based on the analogy between the physical annealing process of solids and optimization problems. In SA, one can simulate the behavior of a system of particles in thermal equilibrium using a stochastic relaxation technique developed by Metropolis et al. (1953). This stochastic relaxation technique helps the SA procedure escape from the local minimum. SA starts at a feasible solution, q_0 (a real-valued vector representing all decision variables), and objective function $J_0 = J(q_0)$. A new solution q_1 is randomly selected from the neighbors of the initial solution and the objective function $J_1 = J(q_1)$ is evaluated. If the new solution has a smaller objective function value $J_1 < J_0$ (in minimization problems), the new solution is definitely better than the old one and therefore it is accepted and the search moves to q_1 and continues from there. On the other hand, if the new solution is not better than the current one, $J_1 < J_0$, the new solution may be accepted, depending on the acceptance probability defined as

$$P_r(accept) = \exp\left\{\frac{-(J_1 - J_0)}{T}\right\}$$

(7.113)

where T is a positive number that will be discussed later.

For a given value of T, the acceptance probability is high when the difference between the objective function values is small. To decide whether or not to move to the new solution q_1 or stay with the old solution q_0, SA generates a random number U between 0 and 1. If $U < P_r(accept)$, SA moves to q_1 and resumes the search. Otherwise, SA stays with q_0 and selects another neighboring solution. At this step one iteration has been completed.

The control parameter T plays an important role in reaching the optimal solution through SA because the acceptance probability is strongly influenced by the choice of T. When the T is high the acceptance probability calculated is close to 1, and the new solution will very likely be accepted, even if its objective function value is considerably worse than that of the old solution. On the other hand, if T is low, approaching zero, then the acceptance probability is also low, approaching zero. This effectively precludes selection of the new solution even if its objective function value is only slightly worse than that of the old solution. Therefore, there is no new solution for which $J_1 > J_0$, is likely to be selected. Rather, only "downhill" points ($J_1 < J_0$) are accepted; therefore, a low value of T leads to a descending search.

The successful performance of the SA method requires that T be assigned a high initial value, and that it be reduced gradually throughout the sequence of calculation by the reduction factor λ. The number of iterations under each constant T should be sufficiently large. But higher initial T and larger number of iterations mean longer simulation time and may severely limit computational efficiency. SA stops either when it has completed all the iterations or when the termination criterion (a specific value of objective function) is reached.

As a rule of thumb to balance the computational efficiency and quality of the SA method, the initial T should be chosen in such a way that the acceptance ratio is somewhere between 70% and 80%. It is recommended that a number of simulation runs should be executed first to estimate the acceptance ratio. Second, the number of iterations under a constant T is about ten times the size of the decision variables. In literature the amount of 0.9 is suggested as a reasonable number for λ, the reduction factor (Egles, 1990).

SA has many advantages including easy implementation, not requiring much computer memory and coding and providing guarantee for identification of an optimal solution if an appropriate schedule is selected. The last point becomes much more important in the cases when the solution space is large and the objective function has several local minima or changes dramatically with small changes in the parameter values. Examples of the SA application in the hydrologic literature include Dougherty and Marryott (1991) and Wang and Zheng (1998).

7.6 CONFLICT RESOLUTION

The operation of groundwater systems is mostly a multi-objective problem and there are some economic, hydraulic, water quality, or environmental objectives that are usually in conflict, for example

- There are several demand points and the water which is supplied to one of these demand points, cannot be used by others. Therefore, the major conflict issue in groundwater operation happens when the groundwater storage is not capable of supplying all of the demands.

- Extra discharge of aquifer or high variation of groundwater table can cause some problems such as settlement of buildings and ground surface.
- Recharge of aquifer with polluted water such as infiltration of agricultural return flows, disposal of sewage or water treatment plant sludge, and sanitary landfills can produce some environmental problems.

This topic is discussed by Karamouz et al. (2003). There are many interbasin issues that could be treated in the same fashion.

The optimal operation policies of groundwater systems can be developed using conflict resolution models considering groundwater systems hydraulic, water quality response equation, possible well capacity, hydraulic gradient, and water demand requirements. The common conflict issue in groundwater systems planning and operation happens when the aquifer should supply water to different demand points for different purposes and there are some constraints on groundwater table variations and groundwater quality. Consider an aquifer system which supplies the following demands:

- Domestic water demand
- Agricultural water demand
- Industrial water demand

The utility function for each of these agencies should be considered when formulating conflict resolution problems.

There are several alternative ways to solve conflict situations. One might consider the problem as a multi-objective optimization problem with the objectives of the different decision makers. Conflict situations can also be modeled as social choice problems in which the rankings of the decision makers are taken into account in the final decision. A third way of resolving conflicts was offered by Nash, who considered a certain set of conditions that the solution has to satisfy, and proved that exactly one solution satisfies his "fairness" requirements. In this section, the Nash bargaining solution will be outlined:

Assume that these are I decision makers. Let X be the decision space and $f_i : X \mapsto R$ the objective function of decision maker i. The criteria space is defined as

$$H = \left\{ \underline{u} \,\middle|\, \underline{u} \in R^I, \quad \underline{u} = (u_i), \quad u_i = f_i(x) \text{ with some } x \in X \right\} \qquad (7.114)$$

It is also assumed that in the case when decision makers are unable to reach an agreement, all decision makers will get low objective function values. Let d_i denote this value for decision maker i, and let $\underline{d} = (d_1, d_2, \ldots, d_I)$. Therefore, the conflict is completely defined by the pair $(H, \underline{d})$, here H shows the set of all possible outcomes and $\underline{d}$ shows the outcomes if no agreement is reached. Therefore, any solution of the conflict depends on both H and $\underline{d}$. Let the solution be therefore denoted as a function of H and $\underline{d}$: $\phi(H, \underline{d})$. It is assumed that the solution function satisfies the following conditions:

1. The solution has to be feasible: $\phi(H, \underline{d}) \in H$.
2. The solution has to at least provide the disagreement outcome to all decision makers, $\phi(H, \underline{d}) \geq \underline{d}$.

3. The solution has to be non-dominated. That is, there is no $f \in H$ such that $\underline{f} \neq \varphi(H, \underline{d})$ and $\underline{f} \geq \varphi(H, \underline{d})$.
4. The solution must not depend on unfavorable alternatives. That is, if $H_1 \subset H$ is a subset of H such that $\varphi(H, \underline{d}) \in H_1$ then $\varphi(H, \underline{d}) = \varphi(H_1, \underline{d})$.
5. Increasing linear transformation should not alter the solution. Let T be a linear transformation on H such that $T(\underline{f}) = (\alpha_1 f_1 + \beta_1, \ldots, \alpha_I f_I + \beta_I)$ with $\alpha_i > 0$ for all i, then $\varphi(T(H), T(\underline{d})) = T(\varphi(\overline{H}, \underline{d}))$.
6. If two decision makers have equal positions in the definition of the conflict then they must get equal objective values at the solution. Decision makers i and j have an equal position if $d_i = d_j$ and any vector $\underline{f} = (f_1, \ldots, f_I) \in H$ if and only if $(\overline{f}_1, \ldots, \overline{f}_I) \in H$ with $\overline{f}_i = f_j, \overline{f}_j = f_i, \overline{f}_l = f_l$ for $l \neq i, j$. Then we require that $\varphi_i(H, \underline{d}) = \varphi_j(H, \underline{d})$.

Therefore, all of the conflict issues and the responsible agencies should be recognized in the first step in conflict resolution studies. The utility function typically consists of a range of full satisfaction (utility = 1.0) and point(s) of no satisfaction (utility = 0).

Before showing the method to find the solution satisfying these properties some conditions must be followed. The feasibility condition requires that the decision makers cannot get more than the amount available. No decision maker would agree in an outcome that is worse than the amount he/she would get anyway without the agreement. This property is given in condition 2. The requirement that the solution is non-dominated shows that there is no better possibility available for all. The fourth requirement states that if certain possibilities become infeasible but the solution remains feasible, then the solution must not change. If any of the decision makers changes the unit of his/her objective, then a linear transformation is performed on H. The fifth property requires that the solution must remain the same. The last requirement shows a certain kind of fairness stating that if two decision makers have the same outcome possibilities and the same disagreement outcome, then there is no reason to distinguish among them in the final solution.

If H is convex, closed, and bounded, and there is at least one $\underline{f} \in H$ such that $\underline{f} > \underline{d}$, then there is a unique solution $\underline{f}^* = \varphi(H, \underline{d})$, which can be obtained as the unique solution of the following optimization problem:

$$\text{Maximize} \quad (f_1 - d_1)(f_2 - d_2) \ldots (f_I - d_I)$$

$$\text{Subject to}$$

$$f_i \geq d_i \quad (i = 1, 2, \ldots, I) \tag{7.115}$$

$$\underline{f} = (f_1, \ldots, f_I) \in H$$

The objective function is called the *Nash product*. Notice that this method can be considered as a special distance-based method, when the geometric distance is maximized from the disagreement point $\underline{d}$.

Example 7.15

Consider Example 7.10 again. Assume that the cost and convenience are the objectives of two different decision makers. Assume that the disagreement point for cost and continence is $d = (-30, 50)$. Choose the best technology using Nash product.

Solution

Assume that $d = (-30, 40)$ when -30 is considered as a worst possibility of cost and 50 as the same for convenience. The value of the Nash product for the four alternatives are as follows:

$$(-25 + 30)(85 - 40) = 225$$

$$(-18 + 30)(65 - 40) = 300$$

$$(-27 + 30)(93 - 40) = 159$$

$$(-15 + 30)(50 - 40) = 150$$

Since the second alternative gives the largest value, the second technology variant is considered as the solution.

Example 7.16

In an unconfined aquifer system, the following agencies are affected by the decisions made to discharge from aquifer to supply water demands:

Agency 1: Department of Water Supply
Agency 2: Department of Agriculture
Agency 3: Industries
Agency 4: Department of Environmental Protection
Agency 5: Department of Domestic Water Use

Department of water supply has a twofold role, namely, to allocate water to different purposes and control the groundwater table variations. The decision makers in different agencies are asked to set their utility functions, which are shown in Figures 7.28 through 7.30.

The analyst should set the weights shown in Table 7.19 on the role and authority of different agencies in the political climate of that region.

The initial surface, volume, and total dissolved solids (TDS) concentration of water content of the aquifer are $600 \, km^2$, 1140 million cubic meters, and $1250 \, mg/L$, respectively. The net underground inflow is 1100 million cubic meters per year with a TDS concentration equal to $1250 \, mg/L$. Assume that 60% of allocated water returns to the aquifer as return flow and the average TDS concentration of return flow is $2000 \, mg/L$, and the average storage coefficient of the aquifer is 0.06. Find the most appropriate water allocation scheme for this year using the Nash's bargaining theory.

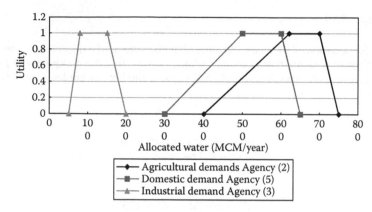

FIGURE 7.28   Utility function of different agencies for the water allocated to different demands.

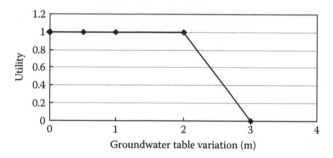

FIGURE 7.29   Utility function of Agency 1 for the average groundwater variation.

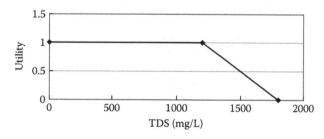

FIGURE 7.30   Utility function of Agency 4 for the TDS concentration.

Solution

The nonsymmetric Nash solution of the problem is the unique optimal solution of the following problem:

$$\text{Maximize} \prod_{i=1}^{5} (f - d_i)_i^{w_i} \tag{7.116}$$

Subject to

$$d_i \le f_i \le f_i^* \quad \forall i \tag{7.117}$$

TABLE 7.19

Relative Weights or Relative Authority of Agencies

Agencies	Relative Weight (Case 1)	Relative Weight (Case 2)
Department of agriculture	0.133	0.17
Department of domestic water	0.33	0.2
Department of water supply	0.2	0.23
Industries	0.133	0.1
Department of environmental protection	0.2	0.3

where

w_i is the relative weight

t_i is the utility function

d_i is the disagreement point

f_i^* is the ideal point of player (agency) i

This objective function should be maximized considering the constraints of groundwater surface variation and groundwater quality that are as follows:

- The average groundwater table variation:

$$\Delta I \approx (q_1 + q_2 + q_3 - IN) \times \left(\frac{1}{S}\right) \times \left(\frac{1}{A}\right) \qquad (7.118)$$

where

q_1, q_2, q_3 are the annual agricultural, domestic, and industrial groundwater withdrawal (MCM)

IN is the net annual underground inflow (MCM)

S is the storage coefficient

A is the area of aquifer (km²)

Therefore,

$$\Delta I \approx \frac{(q_1 + q_2 + q_3 - 1100)}{0.06 \times 600} \qquad (7.119)$$

The average groundwater quality:

Considering the TDS as the indicator water quality variable, the average TDS concentration of groundwater at the end of the year can be estimated as follows:

$$C_{new} = \frac{(V_{in.} \times C_{in.} + IN \times C_{IN} + \alpha_1 \times C_1 \times q_1 + \alpha_2 \times C_2 \times q_2 + \alpha_{13} \times C_3 \times q_3 - C_{mean}(q_1 + q_2 + q_3))}{(V_{in.} + IN + (\alpha_{11} - 1) \times q_1 + (\alpha_2 - 1) \times q_2 + (\alpha_3 - 1) \times q_3)}$$

$$(7.120)$$

and

$$C_{mean} = \frac{(C_{in.} + C_{new})}{2} \qquad (7.121)$$

TABLE 7.20
Results of Conflict Resolution Method

Variable	Result (Case 1)	Result (Case 1)
Allocated water to agricultural demands (MCM)	492	533
Allocated water to domestic demands (MCM)	500	460
Allocated water to industries (MCM)	80	80
Groundwater table drawdown (m)	2.64	2
The final average TDS concentration of groundwater	1440	1439

where

$V_{in.}/C_{in.}$ is the initial volume and TDS concentration of water content of aquifer

C_{IN} is the TDS concentration of net groundwater inflow

$\alpha_1/\alpha_2/\alpha_3$ is the return flow percent of annual agricultural, domestic, and industrial groundwater withdrawal

$C_1/C_2/C_3$ is the average TDS concentration of agricultural, domestic, and industrial groundwater withdrawal

C_{mean} is the average TDS concentration of groundwater

C_{new} is the TDS concentration of aquifer at the end of the year

Therefore,

$$C_{new} = \frac{\begin{array}{c}(1440 \times 1250 + 1100 \times 1250 + 0.6 \times 2000 \times q_1 + 0.6 \\ \times 2000 \times q_2 + 0.6 \times 2000 \times q_3 - C_{mean}(q_1 + q_2 + q_3))\end{array}}{1440 + 1100 + (0.6 - 1)(q_1 + q_2 + q_3)}$$

and

$$C_{mean} = \frac{(1250 + C_{new})}{2}$$

This nonlinear optimization can be solved using different NLP solvers. The allocated water considering different relative weights for agencies is presented in Table 7.20.

As it can be seen in Table 7.20 in case 1, domestic demands which have the highest priority and industrial demands that have less volume, have been completely supplied but utilities of other agencies are less than 1. The comparison between cases 1 and 2 shows the effect of relative weights on allocated water. Increasing the relative weights of the Department of Water Supply, the Department of Environmental Protection, and the Department of Agriculture, has improved the allocated water to agricultural demands and groundwater table variation but it does not have any significant effect on the TDS concentration of groundwater.

7.7 GROUNDWATER SYSTEMS ECONOMICS

Economics includes analytical tools that are used to determine the allocations of scarce resources by balancing the competing objectives. Different important and complex decisions should be made in the process of planning, design, construction, operation, and maintenance of groundwater resources systems. Economic considerations beside the technological and environmental concerns play an important role in decision making about groundwater resources planning and management.

The economical analysis of groundwater development projects reveals two sides. They create value and they also encounter costs. The preferences of individuals for goods and services are considered in the value side of the analysis. The value of goods or services is related to the willingness to pay. The costs of economic activities are classified as fixed and variable costs. The range of operation or activity level does not affect the fixed costs such as general management and administrative salaries and taxes on facilities, but variable costs are determined based on the quantity of output or other measures of activity level.

Costs are also classified as private costs and social costs. The private costs are directly experienced by the decision makers but all costs of an action are considered in the social costs without considering who is affected by them (Field, 1997). The difference between social and private costs is called external costs or usually environmental costs. Since the firms or agencies do not normally consider these costs in decision making, they are called "external." However, it should be noted that these costs may be real costs for some members of society. Open access resources such as reservoirs in urban areas are the main source of external or environmental costs.

Example 7.17

There are two choices, A and B alternative, for well development in a region; $25,000 and $20,000 per week should be paid to provide the needed materials for project application, respectively. The other costs of two trenchless replacement methods are summarized in Table 7.21. Compare the two methods in terms of their fixed, variable, and total costs when 20 km of mains should be replaced. For the selected method, how many meters of the mains should be replaced before starting to make a profit, if there is a payment of $30/m of main replacement?

Solution

Choice A
As it is depicted in the Table 7.22, the plant setup is a fixed cost, whereas the payment for traffic jams and rent are variable costs.

Choice B
As can be seen in Table 7.22, although choice B has higher fixed costs, it has less total cost than choice A. The project will make profit at the point where total

TABLE 7.21
Costs of Choices A and B for Trenchless Replacement

Cost	Method A	Method B
Distance from construction site	12 week	8 week
Weekly rental of site	6,500	$1,500
Cost to setup and remove equipment	$175,000	$225,000
Transferring expenses	$25,000/km	$20,000/km

TABLE 7.22

Analysis of Costs of Two Choices for Main Replacement

Cost	Fixed	Variable	Choice A	Choice B
Rent		✓	$78,000	$12,000
Plant setup	✓		$1750,000	$225,000
Transferring		✓	12($25,000) = 300,000	8($20,000) = 160,000
Sum			$553,000	$505,000

revenue equals total cost of the kilometers of mains replaced. For choice B, the variable cost per meter of replacement is calculated as follows:

$$\frac{(\$12,000 + \$160,000)}{20,000} = \$8.6$$

Total cost = Total revenue when the project starts to make benefit. Considering the meters of the mains that should be replaced before starting to make a profit as X, we will, therefore, have

$$\$225,000 + \$8.6X = \$30X \text{ and } X \text{ is determined as } 10,514\,m$$

Therefore, if choice B is selected, the project will begin to make a profit after replacing 10,514 m of the mains.

Similarly external benefits are defined. An external benefit is a benefit that is experienced by somebody outside the decision about consuming resources or using services. For example, consider that a private power company constructed a dam for producing hydropower. The benefits experienced by the people downstream such as mitigating both floods and low flows are classified as external benefits.

Many values of an action cannot be measured in commensurable monetary units as is desired for economists. For example, the effects on human beings physically (through loss of health or life), emotionally (through loss of national prestige or personal integrity), and psychologically (through environmental changes) cannot be measured by monetary units. These sorts of values are called intangible or irreducible (James and Lee, 1971).

In most engineering activities, the capital investment is committed for a long period; therefore, the effects of time should be considered in economical analysis. This is because of the different values of the same amount of money spent or received at different times. The future value of a present amount of money will be larger than the existing amount due to opportunities that are available to invest the money in various enterprises to produce a return over a period. The interest rate is the rate at which the value of money increases from present to future. In contrast, the discount rate is the rate by which the value of money is discounted from future to present.

A summary of relations between general cash flow elements using discrete compounding interest factors is presented in Table 7.23 using the following notation:

i: interest rate

N: number of periods (years)

TABLE 7.23

Relations between General Cash Flow Elements Using Discrete Compounding Interest Factors

To Find:	Given:	Factor by which to Multiply "Given"	Factor Name	Factor Function Symbol
F	P	$(1+i)^N$	Single-payment compound amount	$\left(\dfrac{F}{P}, i\%, N\right)$
P	F	$\dfrac{1}{(1+i)^N}$	Single-payment present worth	$\left(\dfrac{P}{F}, i\%, N\right)$
F	A	$\dfrac{(1+i)^N - 1}{i}$	Uniform series compound amount	$\left(\dfrac{F}{A}, i\%, N\right)$
P	A	$\dfrac{(1+i)^N - 1}{i(1+i)^N}$	Uniform series present worth	$\left(\dfrac{P}{A}, i\%, N\right)$
A	F	$\dfrac{i}{(1+i)^N - 1}$	Sinking fund	$\left(\dfrac{A}{F}, i\%, N\right)$
A	P	$\dfrac{i(1+i)^N}{(1+i)^N - 1}$	Capital recovery	$\left(\dfrac{A}{P}, i\%, N\right)$
P	G	$\dfrac{(1+i)^N - 1 - Ni}{i^2(1+i)^N}$	Discount gradient	$\left(\dfrac{P}{G}, i\%, N\right)$
A	G	$\dfrac{(1+i)^N - 1 - Ni}{i(1+i)^N - i}$	Uniform series gradient	$\left(\dfrac{A}{G}, i\%, N\right)$

Source: DeGarmo, et al., *Engineering Economy*, Prentice-Hall, Upper Saddle River, NJ, 647 p., 1997.

P: present value of money
F: future value of money
A: end of cash flow period in a uniform series that continues for a specific number of periods
G: End-of-period uniform gradient cash flows

More details about time-dependent interest rates and interest formulas for continuous compounding are given in DeGarmo et al. (1997) and Au and Au (1983).

Example 7.18

A well development for pumping water to supply agricultural demand in a region produces benefits, as expressed in Figure 7.31: $125,000 benefit in year 1 is increased in 10 years on a uniform gradient to $1,250,000. Then it reaches $1,640,000 in year 25 with a uniform gradient of $26,000/year and then it remains

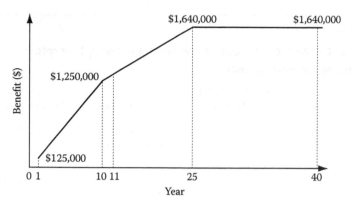

FIGURE 7.31　Cash flow diagram.

constant at $1,640,000 each year until year 40, the end of the project life span. Assume an interest rate of 5%. What is the present worth of this project?

Solution

First, the present value of the project benefits in years 1 through 10 are calculated as follows:

$$125,000\left(\frac{P}{G},\ 5\%,\ 10\right)+125,000\left(\frac{P}{A},\ 5\%,\ 10\right)=\$4,921,723$$

The same procedure is repeated in years 11 through 25:

$$1,276,000\left(\frac{P}{A},\ 5\%,15\right)\left(\frac{P}{F},\ 5\%,10\right)+26,000\left(\frac{P}{G},\ 5\%,15\right)\left(\frac{P}{F},\ 5\%,10\right)$$

$$=1,276,000\times10.379\times0.613+26,000\times63.28\times0.613=9,126,886$$

Present value of benefits in years 26 through 40:

$$1,640,000\left(\frac{P}{A},\ 5\%,15\right)\left(\frac{P}{F},\ 5\%,25\right)=1,640,000\times10.379\times0.295=5,021,360$$

The total present worth is equal to $19,069,969, the summation of the above three values.

7.7.1　ECONOMIC ANALYSIS OF MULTIPLE ALTERNATIVES

Through economic analysis it is determined whether a capital investment and the cost associated with the project within the project lifetime can be recovered by revenues or not. It should also be determined how attractive the savings are regarding the involved risks and the potential alternative uses. The five most common methods including the present worth, future worth, annual worth, internal rate of return, and external rate of return methods are briefly explained in this section. The life span of different projects should be incorporated in economic analysis through repeatability assumption.

In the present worth method, the alternative with the maximum present worth (PW) of the discounted sum of benefits minus costs over the project life span, Equation 7.122, is selected.

$$PW = \sum_{t=1}^{N} (B_t - C_t)\left(\frac{P}{F}, i\%, t\right) \tag{7.122}$$

where
C_t is the cost of the alternative at year t
B_t is the benefit of the alternative at year t
N is the study period
i is the interest rate

In the case of cost alternatives, the one with the least negative equivalent worth is selected. Cost alternatives are all negative cash flows except for a possible positive cash flow element from disposal of assets at the end of the projects useful life (DeGarmo et al., 1997).

In the Future Worth method, all benefits and costs of alternatives are converted into their future worth and then the alternative with the greatest future worth or the least negative future worth is selected. Two choices are available for selecting the study period including the repeatability assumption, and the co-terminated assumption. Based on the last assumption, for the alternatives with a life span shorter than the study period, the estimated annual cost of the activities is used during the remaining years. The same for the alternatives with a life span longer than the study period, a reestimated market value is normally used as a terminal cash flow at the end of the project's co-terminated life as stated by DeGarmo et al. (1997).

In the annual worth (AW) method, the costs and benefits of all alternatives are considered in the annual uniform and similar to the other methods, the alternative with the greatest annual worth or the least negative worth is selected. As in this method the worth of alternatives is evaluated annually, the projects with different economic lives are compared without any change in the study period.

The internal rate of return (IRR) method is commonly used in economic analysis of different alternatives. IRR is defined as the discount rate that when considered, the net present value or the net future value of the cash flow profile will be equal to zero. The PW and AW methods are usually used for finding the IRR. In this method, the minimum attractive rate of return (MARR) is defined and projects with an IRR less than the considered threshold are rejected.

The main assumption of this method is that the recovered funds, which are not consumed in each time, are reinvested in IRR. In such cases that it is not possible to reinvest the money in the IRR rather than the (MARR), the external rate of return (ERR) method should be used. The interest rate, external to a project at which net cash flows produced by the project over its life span can be reinvested, is considered in this method.

TABLE 7.24

Different Characteristics of Pumps

Economic Characteristics	Pump A	Pump B
Initial investment	$200	$700
Annual benefits	$95	$120
Salvage value	$50	$150
Useful life	6	12

Example 7.19

Two different types of pumps (A, B) can be used in a pumping station for removing water from an aquifer. The costs and benefits of each of the pumps are displayed in Table 7.24. Compare the two pumps economically using the MARR = 10% and considering the time of study equal to 12 year.

Solution

The study period is considered 12 years. By using the repeatability assumption, it can be written that

PW of costs of A = $200 + ($200 − $50)(P/F, 10%, 6) − $50 (P/F, %10, 12) = $269

PW of benefits of A = $95 (P/A, 10%, 12) = $647

PW of costs of B = $700 − $150 (P/F, 10%, 12) = $652

PW of benefits of B = $120 (P/A, 10%, 12) = $818

For selecting the economically best alternative, the graphical method is employed in this example. The location of projects on a graph showing present value of benefits versus present value of costs is determined as points A and B and then the slope of the line connecting the points is estimated and compared with a 45° slope line where the points on it have equal present value of costs and benefits (Figure 7.32). When the slope of the line, connecting two projects, is less (more) than 45°, the project with less (more) initial cost would be selected. As shown in Figure 7.32, pump A, which has less initial cost, should be selected.

7.7.2 Economic Evaluation of Projects Using Benefit–Cost Ratio Method

The benefit–cost ratio method which considers the ratio of benefits to costs has been widely used in the economic analysis of water resources projects. The ratio is estimated based on the equivalent worth of discounted benefits and costs in the form of annual worth, present worth, and/or future worth, to consider the time value of money. Equations 7.123 and 7.124 are basic formulations of the benefit–cost ratio method.

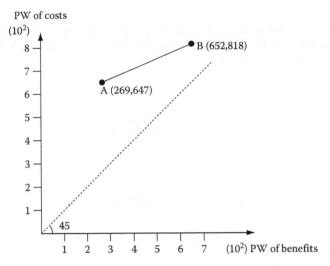

FIGURE 7.32 Comparing two pumps in Example 7.19 by the rate-of-return method.

$$\frac{B}{C} = \frac{B}{I+C} \tag{7.123}$$

$$\frac{B}{C} = \frac{B-C}{I} \tag{7.124}$$

where
  B is net equivalent benefits
  C is net equivalent annual cost including operation and maintenance costs
  I is initial investment

If the benefit–cost ratio of a project is less or equal to one, it is not considered. Through the following equation the associated salvage value of the investment is considered in the benefit–cost method:

$$\frac{B}{C} = \frac{B-C}{I-S} \tag{7.125}$$

where S is the salvage value of investment.

Example 7.20

Assume that the annual costs of operation and maintenance of pumps A and B are equal to $5500 and $7000, respectively. Compare two pumps from an economic aspect, using the benefit–cost ratio method.

Solution

For considering the salvage value of the pumps, Equation 7.125 is employed. Because of different life spans of the pumps, the annual worth method is employed for more simplicity as follows:

Pump A:

$$\frac{B}{C} = \frac{12,000 - 5,500}{30,000(A/P, 8\%, 7) - 600(A/F, 8\%, 7)} = 1.1358$$

Pump B:

$$\frac{B}{C} = \frac{14,000 - 7,000}{75,000(A/P, 8\%, 12) - 1,200(A/F, 8\%, 12)} = 0.711$$

As can be seen, the benefit–cost ratio for pump B is less than one. Therefore, it is rejected and pump A is selected.

7.8 CASE STUDIES

In this section, the application of mythologies is discussed as two practical case studies conducted by Karamouz et al. (2001), Karamouz et al. (2004), and Karamouz et al. (2007). The first case includes the application of object-oriented programming based on conflict resolution methods for water resource planning in the study area. The second case includes water pollution control model based on Nash bargaining theory (NBT) to resolve the existing conflicts for aquifer restoration.

7.8.1 CASE 1: DEVELOPMENT OF A SYSTEM DYNAMICS MODEL FOR WATER TRANSFER

In this case, a system approach to water transfer from southeast to southwest of Tehran, the capital of Iran, is discussed. By the construction of this channel, up to $7\,\text{m}^3/\text{s}$ of mostly urban drainage water can be transferred from east to west in the region to overcome the negative impacts of water rise in the east and to utilize the urban water drainage in the cultivated lands and for other municipal applications in the west. The channel intercepts several local rivers and drainage channels and could partially collect water from these outlets. In order to simulate the state of the system under different water transfer scenarios and water allocation schemes, an object-oriented simulation model has been developed. Using different objects, the user can change the allocation schemes for any give demand points. An economic model based on conflict resolution methods is developed for resource planning in the study area.

- **Area characteristics**

Distribution of land with different capabilities for agricultural development may significantly affect the water supply planning in the southern part of Tehran. Figure 7.33 shows a system of local rivers which includes Sorkheh-Hesar River, Absiah, and

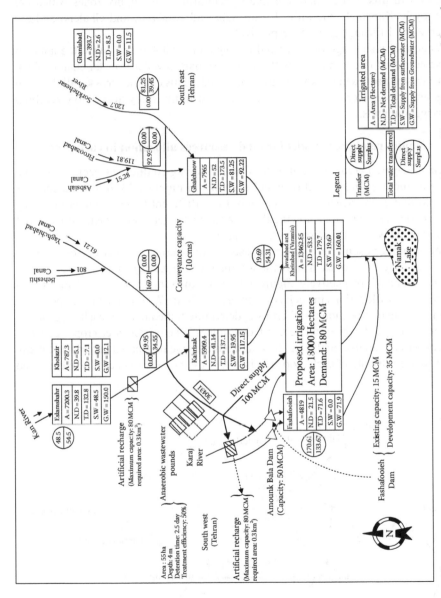

FIGURE 7.33 Schematic map of Tehran metropolitan area water supply and demand in the study area. (From Karamouz, M. et al., *J. Am. Water Resourc. Assoc.*, 37(5), 1301, 2001.)

Firouz-abad drainage channels in the east and Beheshti and Yakhchi-abad drainage channels and Kan River in the west. This collection of rivers and channels drains sanitary and industrial wastewater as well as urban surface water of different parts of the region and supplies water for irrigation zones in the southern part of Tehran. As shown in this figure, there are five agricultural water supply zones which are Fashfooyeh, Kahrizak, Varamin, Eslamshahr and Khalazir, and Ghaleno and are named Zone 1 through Zone 5, respectively.

Because of a shortage of water and having more suitable and cultivated lands in the western part in comparison with the eastern part, a water transfer channel from east to west is proposed by Karamouz et al. (2005). This channel crosses all the local rivers and wastewater channels from east to west in the study area before reaching the irrigation lands.

- **Conflict resolution model for land resources allocation in each zone**

For each irrigated area, an SD model of water sharing conflict has been developed. The dynamic hypothesis of this model is shown in Figure 7.34. In this model, it is assumed that each irrigated area has only three different types of crops including wheat, barley, and tomato. Knowing the area and water demand of each crop, the first estimate of its water allocation has been made.

The weight of each crop is defined based on their usage and price. According to the demand and price of each crop, the weight of wheat, barley, and tomato is considered as 3, 1, and 2, respectively. Figure 7.35 shows the schematic diagram for finding land resources allocations in each zone, based on Figure 7.34 of dynamic hypothesis.

Finally, when allocation of all crops is determined, the area that each crop can be planted with regard to its water demand, determines in converters area crop 1, 2, and 3. The converter available land refers to total lands that can be planted in a specific plain. The ultimate estimate of each crop area is made based on the total available lands and

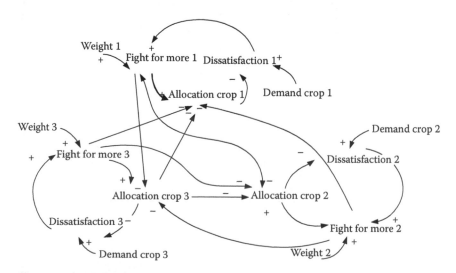

FIGURE 7.34 Dynamic hypothesis of land resources allocations in each zone.

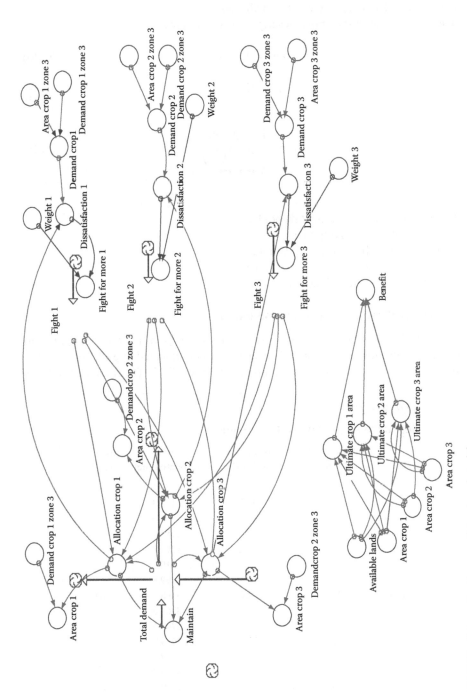

FIGURE 7.35 Schematic diagram of developed model.

crop area that has been determined before. The converter benefit represents the total benefit of plain due to its land resources allocations.

The formulation of the above descriptions is as follows:

$$Ds_i = As_i - Al_i \tag{7.126}$$

$$F_i = \sqrt{Ds_i} \times w_i \tag{7.127}$$

$$Al_i = f\left(\frac{\sum_t F}{\sum_i \sum_t F}\right) \tag{7.128}$$

where
   Ds_i is the dissatisfaction of stakeholder i
   As_i is aspiration of stakeholder i
   Al_i is the water allocation to stakeholder i
   F_i is fight for more water of stakeholder i
   w_i is the weight of stakeholder i

- **Results of the conflict resolution model**

As groundwater withdrawal is expensive in the study area, all residents prefer to use surface water for water supply. The surface water allocated to each zone and the land associated with each zone are presented in Table 7.25. The allocated water is

TABLE 7.25
Resource Allocation to Each Zone

Zone	Water Demand (MCM)	Allocated Water (MCM)	Area Allocated (ha)	Income ($10^6$$)
			1754.1	
Zone 1	53.8	40.3	2113.3	1.6
			951.6	
			2200.9	
Zone 2	102.9	75.8	1297.8	2.6
			2410.7	
			6299.8	
Zone 3	135.0	102.6	3198.8	5.4
			3964.3	
			2745.5	
Zone 4	112.3	79.3	3624.9	2.6
			1619.6	
			6296.4	
Zone 5	130.0	101.7	1280.6	2.7
			379.0	

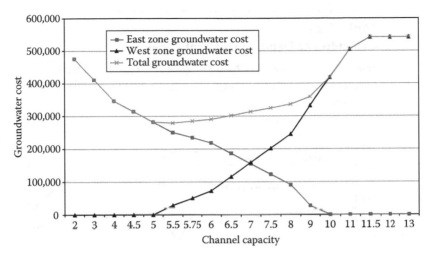

FIGURE 7.36 Variation of groundwater withdrawal cost at the western and eastern plains due to different water transfer channel capacity.

calculated, based on each zone's demand. The relative weight of each zone is also related to the condition of groundwater table in that zone. The land resource allocation in each zone is calculated from its weight and water demand. The relative weight of each crop is considered, based on its price.

- **Optimal groundwater withdrawal in each zone**

Due to the shortage of surface water in the study area, the remaining water demand of each zone is allocated from groundwater storage. Figure 7.36 shows the variation of groundwater withdrawal cost in the eastern and western plains. The cost of groundwater withdrawal is calculated by

$$\text{Cost} = \frac{h \times R}{\eta \times E} \tag{7.129}$$

where
R is the cost of the each KW/h power used for water purpose that is equal 0.004$
h is the depth of groundwater
E is the pumping efficiency that is assumed as 90%
η is equal to 0.102

As the groundwater table is down in the western plains, it is economical that more surface water is allocated to the western plains because of their high cost of groundwater withdrawal. The groundwater depth in different zones is shown in Table 7.26.

- **Sizing channel capacity**

As discussed before, groundwater withdrawal is economical in the eastern plains, because of higher groundwater table in these zones than western plains. So,

TABLE 7.26

Groundwater Depth in Different Zones

Zone	Zone 1	Zone 2	Zone 3	Zone 4	Zone 5
Groundwater depth (m)	50	14.5	21	20	8.6

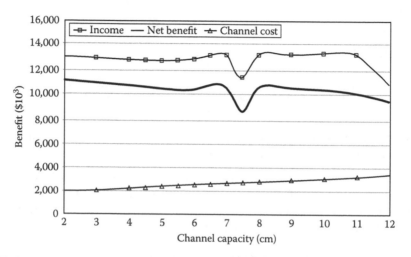

FIGURE 7.37   Variation of agricultural income and conveying capacity of the channel.

construction of a transfer channel, 15 km long from east to west of study area, will be economical. For finding the optimal channel capacity, different scenarios are considered and each scenarios net benefit of agricultural land is calculated.

The net benefit of each zone is equal benefit resulting from optimal land resource allocation benefit less groundwater withdrawal cost. Figure 7.37 shows the variation of agricultural income and channel cost in the study area. As presented in Figure 7.37 the channel should be designed to carry $7\,m^3$ of water per second.

- **Conclusion—making technologies work**

Many communities around the world are searching for solutions for water problems. Their ability to solve them is limited by scarce resources and data and, in some instances, undefined/unsettled water governance and by their planning schemes affected by politics and subjective decisions.

As a result, regions and municipalities are looking for quick results and planning schemes for the period of a decision-maker term in his or her office. In developing countries, there is a misrepresentation that developed countries have access to better techniques and hardware/software and therefore utilizing technologies from those countries will bring them the tools that could make magic. They ended up with black boxes with little or no adaptation capability and without the environment and experts which could utilize them. So reliance will be shifted to in-house models.

Again some of the in-house models soon become obsolete because of many assumptions and simplifications that were necessary to make in order to build the models with scarce data and also with substandard software support systems. So the new challenge in planning, more than the development of models and tool boxes, is geared toward making transparent data and algorithms that are adaptable to regional and native characteristics and could be used to bring different decision makers and stakeholders together and create a consensus. A shared vision planning is needed to make the selected techniques and allocation schemes useable, adaptable, and expandable. Participatory planning is the key to sound water management. SD and conflict resolution techniques can be used to formulate the problems with the intention of bringing stakeholders into the decision-making process.

7.8.2   CASE 2: CONFLICT RESOLUTION IN WATER POLLUTION CONTROL FOR AN AQUIFER

In this case, a conflict resolution approach to water pollution control in the Tehran metropolitan area, with its complex system of water supply, demand, and water pollution, is discussed. Tehran's annual domestic consumption is close to one billion cubic meters. Water resources in this region include water storage in three reservoirs, the Tehran Aquifer, as well as local rivers and channels which are mainly supplied by urban runoff and wastewater. The sewer system mainly consists of traditional absorption wells. Therefore, the return flow from the domestic consumption has been one of the main sources of groundwater recharge and pollution. Some parts of this sewage are drained into local rivers and drainage channels and partially contaminate the surface runoff and local flows. These polluted surface waters are used in conjunction with groundwater for irrigation purposes in the southern part of Tehran. Different decision makers and stakeholders are involved in water pollution control in the study area, which usually has conflicting interests. In this study, the Nash bargaining theory (NBT) is used to resolve the existing conflicts and provide the surface and groundwater pollution control policies considering different scenarios for the development projects. Results show how significant an integrated approach is for water pollution control of surface and groundwater resources in the Tehran region.

- **Water resources characteristics in the study area**

The Tehran plain lies between 35° and 36° 35′ Northern latitude and 50° 20′ and 51° 51′ Eastern longitude, in the south of the Alborz mountain ranges. About 800 million cubic meters of water per year is provided for domestic consumption by over eight million inhabitants. More than 60% of water consumption in Tehran returns to the Tehran Aquifer via traditional absorption wells. Some of the sewage is also drained into local rivers and is used for irrigation in the southern part of Tehran (Karamouz et al., 2001).

The groundwater in the Tehran Aquifer, which is used as drinking water, is polluted due to return flow from domestic absorption wells. In order to overcome the current problems, several development plans are being investigated and implemented. These projects will change the current balance of recharge and water use in the region (Karamouz et al., 2004).

TABLE 7.27

Forecasted Water Demand, Wastewater, and Aquifer Pollution Load—Case Study 1

Year	Population	Water Demand (MCM)	Wastewater (MCM)	Recharge to the Groundwater (MCM)	Aquifer Pollution Load (Nitrate) (kg/Year)
2006	8,120,402	889.184	693.564	522.744	40,251,288
2011	9,042,420	990.145	772.313	601.493	46,314,961
2016	9,956,493	1090.236	850.384	679.528	52,323,656
2021	10,786,922	1181.168	921.311	750.491	57,787,807

TABLE 7.28

Main Characteristics of Tehran Wastewater Collection Project

Project Phase	Population Covered by TWCP	Construction Cost (Cumulative) (US Dollars)	Nitrate Load Reduction (kg/year)	Reduction in Discharged Wastewater (m³/year)
1	2,100,000	13,650,000	13,810,797	179,361,000
2	4,200,000	27,300,000	27,621,594	358,722,000
3	6,300,000	40,950,000	41,432,391	538,083,000
4	8,400,000	54,600,000	55,243,188	717,444,000
Total	10,500,000	68,250,000	69,053,985	896,805,000

Table 7.27 presents the forecasted Tehran Aquifer recharge due to return flow from domestic water use (without considering the effects of TWCP) and the corresponding nitrate load as a water quality indicator. The Tehran wastewater collection project (TWCP) is the most important ongoing project for solving the current quantity and quality problems of sewage disposal in the study area. The initial study of the TWCP was performed with the aid of the World Health Organization (WHO) and the United Nations (UN). Table 7.28 and Figure 7.38 present the main characteristics of Tehran's wastewater collection system.

Another completed project is a network of drainage wells to lower the groundwater table in the southern part of the city. In this project, more than 100 drainage wells have been constructed. The pumped water is discharged to the local streams and channels and contributes to the surface water in the southern part of the city. For more detailed information about surface water resources in the study area see Karamouz et al. (2004).

In this study, optimal operating policies for drainage wells as well as the optimal coverage of the TWCP are developed for different scenarios in the development stages considering the objectives and utility functions of the decision makers and the stakeholders of the system.

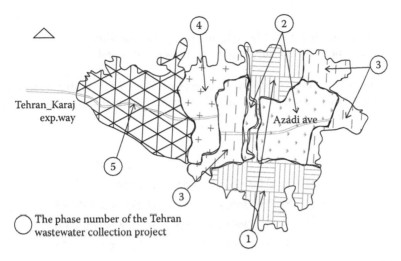

FIGURE 7.38 Area covered by different phases of the Tehran Wastewater Collection Project.

- **Conflict resolution model**

Based on the available data and the results of some brainstorming sessions, the main decision makers and stakeholders of the system and their utilities are as follows:

Department of Environment: Water quality in the Tehran Aquifer is the main concern of this department. The available data show that the concentration of some water quality variables such as nitrate, TDS, and coliform bacteria deviate from groundwater and drinking water quality standards. Nitrate is considered as the representative water quality variable in this study. As the suggested nitrate concentration for groundwater is usually between 25 and 50 mg/L, the utility function of this department ($f_{env.}$) is formulated as follows:

$$f_{env.} = \begin{cases} 1 & \text{if } c < 25 \text{ mg/L} \\ 1 - 0.04(c - 25) & \text{if } 25 \leq c < 50 \text{ mg/L} \\ 0 & \text{if } c \geq 50 \text{ mg/L} \end{cases} \tag{7.130}$$

where c is the average nitrate concentration in the Tehran Aquifer.

Tehran Water and Wastewater Company: The main objective of this company is water supply to domestic demands and wastewater collection and disposal. Considering the importance of the water supply reliability in urban areas, its most favorite range in the study area is more than 94%. Therefore, the utility function of the decision makers in this company for reliability of water supply is assumed as

$$f_{wws.} = \begin{cases} 1 & \text{if } s > 94 \\ 1 - 0.0156(s - 30) & \text{if } 30 < s < 94 \\ 0 & \text{if } 0 < s \leq 30 \end{cases} \tag{7.131}$$

where

$f_{wws.}$ is the utility function related to water supply reliability

s is the percentage of the supplied domestic water demand

As the construction cost of different phases of the TWCP is different, the utility function of the Tehran Water and Wastewater Company is assumed to be as follows:

$$f_{wwx.} = \begin{cases} 1 & \text{if } x \leq 13.65 \\ 0.95 - 0.00366(x - 13.65) & \text{if } 13.65 < x \leq 27.3 \\ 0.85 - 0.00366(x - 27.3) & \text{if } 27.3 < x \leq 40.95 \\ 0.75 - 0.00366(x - 40.95) & \text{if } 40.95 < x \leq 54.6 \\ 0.65 - 0.00366(x - 54.6) & \text{if } 54.6 < x \leq 68.25 \end{cases} \tag{7.132}$$

where

$f_{wwx.}$ is the construction cost of the TWCP

x is the total allocated budget to TWCP (million \$)

The utility function of the construction cost of the TWCP is evaluated based on the estimated cost of the different phases of the TWCP.

Tehran Health Department: The main objective of this department is to make sure that the water supplied to the domestic sector meets the drinking water quality standards. As the recommended nitrate concentration in the allocated water is less than 10 mg/L, the utility function of the decision makers of this department ($f_{hd.}$) is assumed to be as follows:

$$f_{hd.} = \begin{cases} 1 & \text{if } c \leq 10 \text{ mg/L} \\ 1 - 0.02(c - 10) & \text{if } 10 < c \leq 25 \text{ mg/L} \\ 0.7 - 0.028(c - 25) & \text{if } 25 < c \leq 50 \text{ mg/L} \\ 0 & \text{if } c > 50 \text{ mg/L} \end{cases} \tag{7.133}$$

where c is the average nitrate concentration in the Tehran Aquifer.

Water Supply Authority (Water and Wastewater Company): The main objective of this organization is to supply water in order to meet different water demands in the study area. Controlling the variations of the groundwater table elevation in the Tehran Aquifer and reducing the pumping costs of drainage wells, which are used to drawdown the rising groundwater table, are the main utilities of this company. The most favorite range of average groundwater table level variation in the Tehran Aquifer is about 2 cm due to its impacts on the subsidence of building foundations, especially in the southern part of the city which has experienced considerable groundwater table variations. The utility function of the Tehran Water Supply Authority for groundwater table variations is assumed to be as follows:

$$
f_{wsl.} = \begin{cases} 1 & \text{if } 0 \le l < 2 \text{ cm} \\ 1 - 0.0556(l-2) & \text{if } 2 \le l < 20 \text{ cm} \\ 0 & \text{if } l \ge 20 \text{ cm} \end{cases} \quad (7.134)
$$

where
$f_{wsl.}$ is the utility function related to the variation of the groundwater table level
l is the average annual variation of the groundwater table in the Tehran Aquifer

The existing pumping capacity of the drainage wells located in the southern part of Tehran is about 100 million cubic meters. The utility function of the Tehran Water Supply Authority for the total pumping capacity of drainage wells is as follows:

$$
f_{wsd.} = \begin{cases} 1 & \text{if } 0 \le d < 100 \text{ MCM} \\ 1 - 0.033(d-100) & \text{if } 100 \le d < 400 \text{ MCM} \\ 0 & \text{if } d \ge 400 \text{ MCM} \end{cases} \quad (7.135)
$$

$f_{wsd.}$ is the utility function related to the discharge volume from the drainage wells, respectively
d is the total annual discharge from drainage wells located in the southern part of Tehran (million cubic meters).

Considering the Nash product as the objective function, the following model is formulated for water pollution control in the study area:

$$
Max\ Z = (f_{env.} - d_{env.})^{w_1} \cdot (f_{wws.} - d_{wws.})^{w_2} \cdot (f_{wwx.} - d_{wwx.})^{w_3} \cdot
$$
$$
(f_{hd.} - d_{hd.})^{w_4} \cdot (f_{wsl.} - d_{wsl.})^{w_5} \cdot (f_{wsd.} - d_{wsd.})^{w_6} \quad (7.136)
$$

Subject to

$$
\Delta V_A = (R+I) - (s+d) \quad (7.137)
$$

$$
L = \frac{S_A}{A_A} \Delta V_A \quad (7.138)
$$

$$
c = \frac{c_R R + c_0 V_A}{V_A + R + I} \quad (7.139)
$$

$$
x = f(A) \quad (7.140)
$$

$$
R = f(A, P) \quad (7.141)
$$

where

$d_{env.}/d_{wws.}/d_{wwx.}/d_{hd.}/d_{wsl.}/d_{wsd.}$ are the disagreement points of each decision maker/agency in the study area

S_A is the storage coefficient

A_A is the aquifer area

ΔV_A is the variation in aquifer water storage

I is the aquifer recharge due to infiltration from precipitation, infiltration from streams, canals, as well as underground inflow at the boundaries

R is the infiltration from absorption wells

c_R is the concentration of water quality variables in the discharging flow

c_0 is the initial average concentration of water quality variable in the Tehran Aquifer

A is the area covered by the TWCP

L is the variation in the water table elevation

d is the discharge from drainage wells

s is the allocated groundwater to domestic demand

P is the population

x is the TWCP construction cost

w_i is the relative weight/authority of decision maker/stakeholder i

d_j is the disagreement point of decision makers/stakeholders corresponding to the utility function j

The results of applying this model for water pollution control in the Tehran Aquifer are presented in the next section.

- **Results and discussion**

The conflict resolution model presented in this study is applied for determining the optimal area that should be covered by the TWCP as well as the pumping discharge of drainage wells in the development stages of the project. In the water balance of the Tehran Aquifer, it is assumed that only the domestic water supply, the return flows from absorption wells, and the water discharge from drainage wells are variables, and the other terms in the water balance equation are constant. The discharges from the drainage wells usually control the annual groundwater table variation.

As mentioned before, the aquifer is polluted mainly due to wastewater discharge through absorption wells. Based on the available data, nitrate can be considered as the water quality indicator. The average concentration of nitrate and the total nitrogen in raw domestic sewage are 90 and 5 mg/L, respectively. Close to 86% of the total nitrogen can be converted to nitrate through nitrification process. Therefore, it is assumed that the ultimate nitrate concentration in wastewater discharged to the Tehran Aquifer is equal to 77 mg/L. As presented in Table 7.29, the per capita water use and the corresponding wastewater have been estimated. Therefore, the aquifer pollution load can be estimated considering different scenarios for Tehran's population and the subregions covered by the TWCP.

The relative authority of utility functions w_1, w_2, ..., w_6 are assumed as 0.18, 0.27, 0.13, 0.2, 0.13, and 0.09, respectively. Based on the form of utility function,

TABLE 7.29

Results of the Conflict Resolution Model for Water and Wastewater Management—Case Study 1

	Year			
Decision Variables	2006	2011	2016	2021
Average nitrate concentration (mg/L)	9.65	9.68	12.21	9.5
Reliability of domestic water supply	94.7	95.9	94.6	95.3
Average rising in groundwater table level (m)	0.077	0.072	0.0696	0.0652
Average discharge from drainage wells (MCM)	50	50	50	50
Total cost of TWCP (million $)	48.1	53.3	54.6	62.4
Population covered by TWCP (million persons)	7.4	8.2	8.4	9.6
Population covered by TWCP based on the timetable of the project (million persons)	2.5	4.6	6.7	8.8

the disagreement points of all decision makers are set to zero. The results of the proposed model for different planning horizons are presented in Table 7.29. As seen in this table, the population covered by the TWCP will reach 7.4 million people in the year 2006. In this year, the nitrate concentration will be less than the drinking water standard (10 mg/L). The results show that the discharge capacity of the drainage wells should be maintained as 50 million cubic meters per year. Based on the derived operating policies, the quality of groundwater and also the groundwater table variation can be effectively controlled during the planning horizon.

PROBLEMS

7.1 The studies in an aquifer have shown that groundwater TDS concentration of the aquifer is a function of TDS concentration and the volume of aquifer recharge. The monthly data of groundwater TDS concentration and the volume of aquifer recharge are given in Tables 7.30 to 7.32, respectively. Develop MLP and IDNN models to simulate the monthly TDS concentration of this aquifer as a function of monthly TDS concentration and the volume of aquifer recharge.

7.2 The variation of annual water demand from groundwater over a 30 years planning time horizon is exhibited in Table 7.33. The current capacity of water supply system is 100 MCM. It is assumed that the aquifer will supply 80% of the water demand in each year. Three projects are studied in order to supply demands of this area. Considering the cost of each project given in Table 7.34 and 6% rate of return, find the optimal sequence of implementing groundwater projects using Linear Programming.

7.3 Assume that there are six alternatives to select from for improving a groundwater system, there are two constraints including groundwater table fluctuation and supplying the minimum water demand. Two objective functions are

TABLE 7.30

TDS Concentration of the Aquifer (mg/l)

Year	1998	1999	2000	2001	2002	2003	2004	2005
Jan	355	275	391	387	470	347	428	622
Feb	255	402	297	303	353	472	414	382
Mar	357	490	498	451	445	613	424	508
Apr	301	346	325	118	390	405	467	252
May	405	354	293	317	439	425	426	497
June	285	522	328	103	371	411	488	331
July	439	361	338	461	517	449	351	312
Aug	380	650	429	779	414	734	343	787
Sep	269	326	284	364	383	447	427	670
Oct	341	308	457	239	490	396	443	352
Nov	301	381	283	257	382	499	402	233
Dec	303	339	405	785	392	430	759	555

TABLE 7.31

Recharged Water Volume to the Aquifer (MCM)

Year	1998	1999	2000	2001	2002	2003	2004	2005
Jan	36	27	40	27	10	8	10	12
Feb	25	41	30	4	8	10	9	5
Mar	36	48	48	46	10	13	10	10
Apr	30	35	33	0	9	9	10	7
May	41	36	29	16	10	10	10	16
June	28	50	33	5	9	9	11	7
July	44	37	34	21	11	10	8	9
Aug	39	56	43	95	9	15	8	60
Sep	26	33	28	83	9	10	10	103
Oct	35	31	45	29	11	9	10	51
Nov	30	39	19	9	9	11	9	7
Dec	30	34	19	15	9	10	9	17

considered including maximizing annual groundwater yield and minimizing the costs. The membership values for all alternatives with respect to the constraints and objectives are given in Table 7.35.

7.4 Assume that a central authority in an irrigation district has the authority to allocate the groundwater so as to maximize the net return from agricultural production. The district is also responsible for determining the optimal cropping pattern for the region, the amount of acreage devoted to the various crops in the area. Develop an optimization model to determine the optimal groundwater extraction pattern and the economic trade-offs associated with increasing pumping capacity or reducing head lower restriction.

TABLE 7.32

TDS Concentration of Charging Water to the Aquifer (mg/lit)

Year	1998	1999	2000	2001	2002	2003	2004	2005
Jan	853	685	921	680	1006	782	929	1105
Feb	637	940	735	518	793	1009	903	680
Mar	858	1075	1085	973	959	1266	922	1070
Apr	743	836	793	102	859	887	1001	548
May	946	852	724	933	950	924	925	1311
June	707	1116	800	303	826	899	1039	654
July	1002	866	820	912	1091	967	788	675
Aug	901	1235	985	1490	904	1487	773	1583
Sep	670	795	704	1122	847	964	927	1444
Oct	825	758	1029	423	1042	870	957	643
Nov	743	903	530	1274	846	1059	906	1301
Dec	747	822	522	1093	864	932	869	1025

TABLE 7.33

Annual Water Demand for Planning Horizon

	Year			
	2020	**2025**	**2035**	**2040**
Water Demand (MCM)	137.5	156.3	181.3	212.5

TABLE 7.34

Initial Investment and Capacity of Well Field Development

Well Field Development Project	Capital Costs (10^6)	Capacity (MCM)
A	2	15
B	3	20
C	1.5	10
D	5	25

7.5 Consider an aquifer with subsidence due to over extraction of groundwater. Develop an optimization model for subsidence control at critical points.

7.6 Consider an aquifer with excessive groundwater pumping as shown in Figure 7.39. Develop an optimization model to generate the pumping strategies that reverse the intrusion of saltwater while continuing to meet demands for fresh water.

7.7 Consider an aquifer that discharges to a river under steady flow conditions. Two water supply wells are to be operated in a way to maximize the total rate of water withdrawal under steady condition and minimize the amount

TABLE 7.35

Membership Values in Fuzzy Decision

	Alternatives					
	1	**2**	**3**	**4**	**5**	**6**
μ_{G_1}	0.8	0.65	0.7	0.9	0.9	0.85
μ_{G_2}	0.75	0.72	0.69	0.8	0.8	0.9
μ_{C_1}	0.7	0.68	0.65	0.75	0.8	0.8
μ_{C_2}	0.81	0.8	0.6	0.85	0.6	0.95

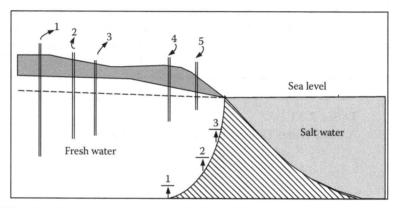

FIGURE 7.39 The salt water intrusion in Problem 7.6.

of water that is drawn from the river. Develop an optimization model to find pumping strategies.

7.8 Assume the net benefits of withdrawal from three wells are $f_1 = 6x_1 - x_1^2$, $f_2 = 7x_2 - 1.5x_2^2$, $f_3 = 8x_3 - 0.5x_3^2$, respectively. If the allowed total withdrawal from the aquifer is subject to 8 MCM and at least 2 MCM needs to be removed for environmental water requirement, find the optimal withdrawal from each well to maximize the net benefits using dynamic programming.

7.9 In order to supply the annual water demand of 50 MCM In a region, water is withdrawn from a river and a well. The cost of removing water from well is $0.5 per MCM per meter of lift and the cost for water allocation from the river is $35 per MCM. The minimum depth to water in the well is 50 m and the well depth is such that it is not possible to extract water below a depth of 90 m. Recharge to the aquifer is 10 MCM and due to 1 MCM of storage removal, the water level decreases as one meter. If the interest rate is 6 percent, develop and optimization model to minimize the costs during 3 years of planning horizon. Solve it using dynamic programming.

7.10 Find the maximum net benefit of water withdraw from an aquifer as a function of $f(x) = 2x - x^2$ in the [0,2] interval using the genetic algorithm.

TABLE 7.36

The Costs of Considered Three Projects

Costs	Timing	Project A	Project B	Project C
Initial investment	Year 0	$2,000,000	$1,000,000	$1,500,000
	Year 10	0	$1,000,000	$1,200,000
	Year 20	0	$1,000,000	0
Operation and Maintenance costs	Years 1–10	$7,000	$4,000	$6,000
	Years 10–20	$8,000	$7,000	$8,000
	Years 20–25	$9,000	$9,000	$9,000

TABLE 7.37

The Characteristics of Considered Projects

	Project A	Project B
Capital investment	$160,000	$240,000
Annual revenues	$80,000	$112,000
Annual expenses	$35,200	$68,800
Useful life (years)	8	12

7.11 The three alternatives described above are available for supplying a community water supply for the next 25 years when the economic lives and the period of analysis terminates. Using the costs presented in Table 7.36 and 5 percent discount rate, compare the three projects with present worth method.

7.12 The two alternatives including improving irrigation efficiency and well development in a region are available for supplying a community water supply for the next 15 years. Using the costs presented in Table 7.37, compare the three projects with present worth method.

REFERENCES

Ahlfeld, D.P., Mulvey, J.M., Pinder, G.F., and Wood, E.F. 1988. Contaminated groundwater remediation design using simulation, optimization and sensitivity theory, 1. Model development, *Water Resources Research*, 24(5), 431–441.

Ahmad, S. and Simonovic, S.P. 2000. Modeling reservoir operations using system dynamics. *Journal of Computing in Civil Engineering*, 14(3), 190–198.

Au, T. and Au, T.P. 1983. *Engineering Economics for Capital Investment Analysis*, Allyn and Bacon, Boston, MA, 540 p.

Bellman, R.E. and Zadeh, L.A. 1970. Decision making in a fuzzy environment. *Journal of Management Science*, 17, 141–164.

Buras, N. 1966. Dynamic programming in water resources development. *Journal of Advances in Hydroscience*, 3, 367–412.

Burt, O.R. 1964. *The Economics of Conjunctive Use of Groundwater and Surface Water*, Hilgardia, University of California, Division of Agricultural, Davis.

Chang, L.C., Shoemaker, C.A., and Liu, P.L.-F. 1992. Optimal time-varying pumping rates for groundwater remediation: Application of a constrained optimal control theory. *Water Resources Research*, 28(12), 3157–3174.

Coulibaly, P., Anctil, F., and Bobe, E.B. 2001. Multivariate reservoir inflow forecasting using temporal neural networks. *Journal of Hydrologic Engineering ASCE*, 6(5), 367–376.

Culver, T.B. and Shoemaker, C.A. 1992. Dynamic optimal control for groundwater remediation with flexible management periods. *Water Resources Research*, 28(3), 629–641.

DeGarmo, E.P., Sullivan, W.G., Bontadelli, J.A., and Wicks, E.M. 1997. *Engineering Economy*, Prentice-Hall, Upper Saddle River, NJ, 647 p.

Delleur, J.W. 1999. *The Handbook of Groundwater Engineering*. CRC Press/Springer-Verlag, Boca Raton, FL, 992 p.

Demuth, H. and Beale, M. 2002. *Neural Network Toolbox for MATLAB, User's Guide*. http://www.mathworks.com//.

Dougherty, D.E. and Marryott, R.A. 1991. Optimal groundwater management 1. Simulated annealing. *Water Resources Research*, 27(10), 2493–2508.

Egles, R.W. 1990. Simulated annealing: A tool for operational research. *European Journal of Operational Research*, 46, 271–281.

Elman, J.L. 1990. Finding structure in time. *Cognitive Science*, 14, 179–211.

Fauset, L.V. 1994. *Fundamentals of Neural Networks, Architecture, Algorithms, and Application*. Prentice Hall, NJ, 461 p.

Field, B.C. 1997. *Environmental Economics: An Introduction*. McGraw-Hill, New York.

Fransconi, P., Gori, M., and Soda, G. 1992. Local feedback multilayered networks. *Neural Computation*, 4, 120–130.

Goldberg, D.E. 1989. *Genetic Algorithms in Search, Optimization, and Machine Learning*. Addison-Wesley, Reading, MA, 412 p.

Holland, J. 1975. *Adaptation in Natural and Artificial Systems*. The University of Michigan Press, Ann Arbour, MI, 228 p.

James, L.D. and Lee, R.R. 1971. *Economics of Water Resources Planning*. McGraw-Hill, New York, 640 p.

Jordan, M.I. 1986. Attractor dynamics and parallelism in a connectionist sequential machine. *8th Annual Conf*erence, 15–17 August Cognitive Science Society, MIT Press, Amherst, MA, pp. 531–546.

Karamouz, M., Szidarovszky, F., and Zahraie, B. 2003. *Water Resources Systems Analysis*. Lewis Publishers, Boca Raton, FL, 590 p.

Karamouz, M., Kerachian, R., and Moridi, M. 2004. Conflict resolution in water pollution control in urban areas: A case study. In *Proceeding of the 4th International Conference on Decision Making in Urban and Civil Engineering*, Porto, Portugal, October 28–30, 2004.

Karamouz, M., Moridi, A., and Nazif, S. 2005. Development of an object oriented programming model for water transfer in Tehran metropolitan area. In *Proceedings of the 10th International Conference on Urban Drainage*, Copenhagen, Denmark, June 10–13.

Karamouz, M., Razavi, S., and Araghinejad, Sh. 2007. Long-lead seasonal rainfall forecasting using time-delay recurrent neural networks: a case study. *Journal of Hydrological Processes*, 22(2), 229–238.

Karamouz, M., Moridi, M., and Nazif, N. 2010. *Urban Water Engineering and Management*, CRC Press, Boca Raton, FL, 602 p.

Kresic, N. 2008. *Groundwater Resources: Sustainability, Management, and Restoration*. McGraw-Hill Professional, 852 p.

Lefkoff, L.J. and Gorelick, S.M. 1987. AQMAN: Linear and quadratic programming matrix generator using two-dimensional groundwater flow simulation for aquifer Management Modeling, U.S. Geological Survey Water Resources Investigations Report 87-4061.

McKinney, D.C. and Lin, M.D. 1995. Approximate mixed-integer nonlinear programming methods for optimal aquifer remediation design. *Water Resources Research*, 31(3), 731–740.

Metropolis, N., Rosenbluth, A., Rosenbluth, M., Teller, A., and Teller, E. 1953. Equation of state calculations by fast computing machines. *Journal of Chemical Physics*, 21, 1087–1092.

ReVelle, C. and McGarity, A.E. eds., 1997. *Design and Operation of Civil and Environmental Engineering Systems*. John Wiley & Sons, New York, 752 p.

StatSoft. *Electronic Statistics Textbook*. 2002. URL. http://www.statsoftinc.com//textbook// stathome.html.

Sun, M. and Zheng, C. 1999. Long-term groundwater management by a MODFLOW based dynamic groundwater optimization tool. *Journal of AWRA*, 35(1), 99–111.

Szidarovszky, F., Gershon, M., and Duckstein, L. 1986. *Techniques of Multi-Objective Decision Making in Systems Management*. Elsevier, Amsterdam, the Netherlands.

Todd, D.K. 1980. *Groundwater Hydrology*. John Wiley & Sons, Inc., New York, 535 p.

Wagner, B.J. and Gorelick, S.M. 1987. Optimal groundwater quality management under parameter uncertainty. *Water Resources Research*, 23(7), 1162–1174.

Wang, M. and Zheng, C. 1998. Application of genetic algorithms and simulated annealing in groundwater management: Formulation and comparison. *Journal of American Water Resources Association*, 34(3), 519–530.

Willis, R. and Yeh, W.W.-G. 1987. *Groundwater Systems Planning and Management*. Prentice Hall, Englewood Cliffs, NJ, 416 p.

Willis, R.L. 1976. Optimal groundwater quality management, well injection of waste waters. *Water Resources Research*, 12, 47–53.

Willis, R.L. 1979. A planning model for the management of groundwater quality. *Water Resources Research*, 15, 1305–1313.

Yakowitz, S. 1982. Dynamic programming applications in water resources. *Water Resources Research*, 18, 673–696.

Yeh, W.W.-G. 1986. Review of parameter identification procedures in groundwater hydrology: The inverse problem, Water Resources Research, 22(1), 95–108.

Yeh, W.W.-G. 1992. System analysis in groundwater planning and management. *Journal of Water Resources Planning and Management*, 118(3), 224–237.

Zimmermann, H.J. 1978. Fuzzy programming and linear programming withseveral objective functions. *Journal of Fuzzy Sets and System*, 1, 45–55.

8 Surface Water and Groundwater Interaction

8.1 INTRODUCTION

Surface and groundwater systems are in continuous dynamic interaction. They are the two components of hydrological cycle that are not isolated from the interaction. Surface water and groundwater resources are traditionally managed as independent systems. This can lead to over-allocation of resources and adverse environmental impacts. Development of land and water resources affects the quantity and quality; therefore, understanding the connectivity between surface water and groundwater systems is required to plan and manage surface water and groundwater resources accordingly.

Effective policies and management practices should be proposed based on a foundation that recognizes that surface water and groundwater are simply two components of a single integrated resource. The river basin can be considered as a unit for analyzing the interactions and interrelations between the various components of the hydrologic system and for practicing the water management policy. Due to increasing concerns over water resources and the environment, considering groundwater and surface as a single resource has become important.

Watershed planners formulate and practice water management strategies to integrate important issues such as water resource conservation, economic, social, environmental, water resource allocation, and water demand management. Issues related to water supply, water quality, and degradation of aquatic environments can be practiced.

One of the strategies of water supply management is the conjunctive use of surface and groundwater. It is applied to optimize the water resources development, management, and conservation within a watershed. Conjunctive use is the application of programming techniques for optimum utilization of water resources in a regional scale. The artificial recharge of aquifers is also used effectively for water resources development.

In this chapter, for better understanding the river-aquifer systems, the interaction of surface and groundwater followed by the conjunctive water use, and, finally, conjunctive use modeling with some example and case studies are explained.

8.2 INTERACTION BETWEEN SURFACE AND GROUNDWATER

The water that flows directly on top of the ground is called surface water such as streams, lakes, springs, and reservoirs. The overland flow can be generated when the rate of rainfall exceeds the infiltration capacity of the soil.

Surface water includes streams, lakes, springs and reservoirs. Baseflow comprises of streams under the ground surface that have been infiltrated from the surface water.

Most of this water returns to the stream at the sharp steps to sustain flows during extended dry periods. In the following, the infiltration is discussed as the process for groundwater and surface water.

8.2.1 INFILTRATION

The source of surface water for groundwater recharge is infiltration of precipitation through the unsaturated zone. Rainwater chemistry, soil chemistry, vegetated land, soil moisture, and hydraulic conductivity are effective in soil infiltration. The unsaturated zone consists of soil, air, and water that control's the infiltration rates. Groundwater flow quantifying in this zone is more complicated than the saturated zone. Water seeps from the surface into the subsurface depending upon the gravity and moisture potential, which are called hydraulic and pressure heads, respectively. Moisture potential varies with the moisture content and pore size of the medium. The total potential h in unsaturated flow is the summation of moisture potential $\psi(\theta)$ and gravity potential Z.

Any change in the infiltration rate requires a change in moisture content θ. Regarding the infiltration in the unsaturated zone, which is controlled by moisture potential $\psi(\theta)$ as well as hydraulic conductivity $K(\theta_v)$, Richard's equation is used to describe the 1D flow in the unsaturated zone:

$$\frac{\partial}{\partial z}\left[K(\theta_v)\left(\frac{\partial h}{\partial z}\right)\right] = \frac{\partial \theta}{\partial t} \tag{8.1}$$

$$\frac{\partial}{\partial z}\left[K(\theta_v)\left(\frac{\partial \psi}{\partial z} + \frac{\partial z}{\partial z}\right)\right] = \frac{\partial \theta}{\partial t} \tag{8.2}$$

Based on Richard's equation, a decrease in the moisture potential decreases cumulative infiltration. When the moisture potential approaches zero, the infiltration rate decreases to a rate equivalent to saturated vertical hydraulic conductivity, K_v. The infiltration rate is shown in Figure 8.1. The moisture content of the soil controls the

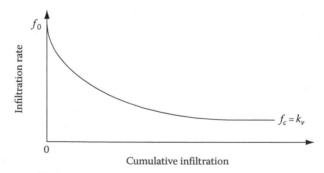

FIGURE 8.1   Infiltration capacity curve. (From U.S. Army Corps of Engineers, *Engineering and Design, Groundwater Hydrology*, 1999.)

initial infiltration capacity f_0. The saturated vertical hydraulic conductivity of the soil K_v controls the final infiltration capacity.

8.2.2 Concepts of Interaction between Surface Water and Groundwater

The interaction between streams and groundwater and the direction of flow are controlled based on the seasonal conditions. Groundwater discharges to the stream when the hydraulic gradient of the aquifer is toward the stream and the stream is a gaining or effluent stream. Figure 8.2 shows the gaining streams receives water from the groundwater system.

Stream loses water to groundwater by outflow through the streambed when the hydraulic gradient of the aquifer is away from the stream and the stream is losing or influent. The depth of water and the hydraulic gradient are the effective factors on the interaction flow. Figure 8.3 shows the losing streams lose water to the groundwater system.

In some conditions, the flow direction can vary in two sides of a stream—one side receives groundwater, while the other side loses water to groundwater. In some environments, the flow direction can vary along a stream; some reaches receive groundwater or lose water to groundwater, or the reach loses or gains water at different times of the year.

Losing streams can be connected to groundwater by a continuous saturated zone directly, as shown in Figure 8.3, or it may be disconnected from the groundwater

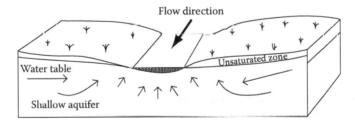

FIGURE 8.2 Groundwater discharges to the stream. (From Winter et al., 1998.)

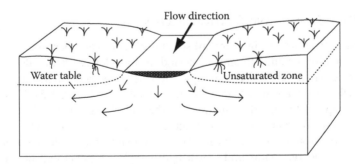

FIGURE 8.3 Stream loses water to groundwater. (From Winter et al., 1998.)

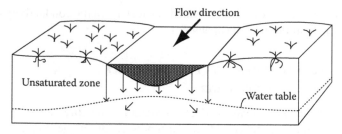

FIGURE 8.4   Disconnected streams. (From Winter et al., 1998.)

system by an unsaturated zone. In this case, the water seeps through a streambed of streams to a deep water table, as illustrated in Figure 8.4.

In order to assess the interaction between a stream and the groundwater, it is assumed that water moves vertically through the streambed. The rate of flow between the stream and the aquifer can be estimated based on Darcy's equation:

$$q = -KLWi \qquad (8.3)$$

where
  K is the vertical hydraulic conductivity of the steam bed
  L is the length of the stream reach over which seepage is assessed
  W is the average width of the stream over the reach
  i is the hydraulic gradient between the stream and the groundwater, which can be
    defined as

$$i = \frac{h_{aquifer} - h_{stream}}{M} \qquad (8.4)$$

where
  $h_{aquifer}$ is the elevation of the groundwater
  h_{stream} is the elevation of the stream water surface
  M is the thickness of the streambed

8.2.3   BANK STORAGE AND BASEFLOW RECESSION

A rapid rise in stream stage in nearly all streams generates another type of interaction between groundwater and streams. A period of high flows, such as floods, release from a dam, storm precipitation, and rapid snowmelt may cause tremendous amounts of water to flow from the stream into the streambanks (see Figure 8.5). This flow will contribute seepage through the streambed to the underlying water table, thus saturating the depleted stream-side deposits. It may require a few days or weeks of high flow to replenish the groundwater reservoirs and streambanks returns to the stream.

The rising and recession limbs of a hydrograph are generated in a rapid rise in stream stage. Discharge during this decay period is due to the groundwater

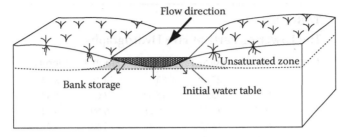

FIGURE 8.5 Movement of water into stream banks as bank storage. (From Winter et al., 1998.)

contributions as the stream drains water from the declining groundwater reservoir. The baseflow recession is the lower part of the falling limb. Since the baseflow is principally groundwater effluent, the baseflow recession curves represent withdrawal from groundwater storage. This baseflow recession for a drainage basin is the function of the overall topography, drainage patterns, soils, and geology of the watershed.

The term hydrograph separation refers to the process of separating the time distribution of baseflow from the total runoff hydrograph. The baseflow recession equation is

$$Q = Q_0 e^{-at} \tag{8.5}$$

where
  Q is the flow at time t in recession limbs of a hydrograph
  Q_0 is the flow at the start of the recession (m³/s)
  a is recession constant for the watershed
  t is the time from the start of the recession

Initially, the rate of bank storage discharge is high because of the steep water-level gradient, but as the gradient decreases, so does the groundwater runoff. The recession segment of the stream hydrograph gradually tapers off into what is called a depletion curve. There are some methods for hydrograph separation such as master-depletion curve method, constant-discharge baseflow separation, constant-slope baseflow separation, and concave-baseflow separation. This method needs some hydrographs for different events in the watershed. If adequate data are not available, then other separating hydrograph methods are used.

8.2.3.1 Master-Depletion Curve Method

Depletion curves are plotted as a combination of several arcs of recession limbs of hydrographs with overlapping in their lower parts. In order to plot master-depletion curve, a hydrograph is plotted on a tracing paper, which is a semilog paper. Then, another hydrograph is plotted on the tracing paper, which is moved horizontally until the arc coincides in its lower part with the arc already traced. The process is continued until all the available arcs are plotted.

TABLE 8.1

The Lowering Parts of Five Hydrographs

Time	Storm 1	Storm 2	Storm 3	Storm 4	Storm 5
1	36.7	27.92	83	14.25	8.93
2	29	23.6	15.34	8.6	5.43
3	23.7	19.32	10.28	6.69	4.24
4	20.7	16.67	8.49	5.68	3.91
5	17.7	14.29	7	4.94	3.46
6	14.38	12.16	5.9	4.32	2.86
7	12.88	10.71	5.43	3.82	2.52
8	11.64	9.33	5	3.38	2.3
9	10.52	8.22	4.8	3	2.1
10	10	6.95	4.6	2.9	2.05
11	9	6.3	4.5	2.8	2
12	8.7	6.1	4.46	2.7	
13	8.5	6	4.35		

Example 8.1

The lowering parts of five hydrographs are presented in Table 8.1. Find the master depletion curve for the watershed.

Solution

Using semi-log paper, log Q versus time, the recession curves are plot as shown in Figure 8.6. In plotting the figure, the recession curve for the smaller storm is plotted and then recession curve for the next event with larger magnitude of Q is plotted; hence, the master-depletion curve appears to extend along a line coincident with

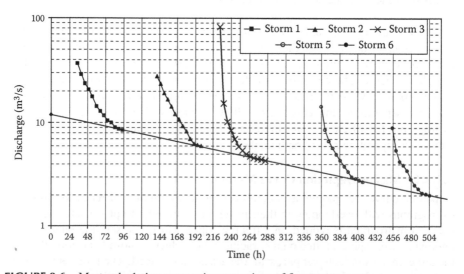

FIGURE 8.6 Master depletion curve using recessions of five storms events.

the recession curve of the first event. The master-depletion curve is constructed in such a way that it extends through the recessions of the observed events. This curve is presented by the form of Equation 8.5 using any two points. With points $t = 0$, $Q_0 = 12$, and $t = 528$, $Q_t = 1.85$, recession constant for the watershed is calculated using Equation 8.5 as follows:

$$1.85 = 12e^{-a(528-0)} \quad a = 0.00354$$

So, the form of master depletion curve for the watershed is

$$Q = 12e^{-0.00354t}$$

8.2.3.2 Constant-Discharge Baseflow Separation

In this method, the line of separating baseflow and direct runoff begins from minimum value immediately prior to beginning of storm hydrograph (A) to a constant discharge rate until it intersects the recession limb of the hydrograph (B). This method is illustrated in Figure 8.7, which is easy to apply.

8.2.3.3 Constant-Slope Baseflow Separation

This method allows the estimation of baseflow recession in streams with limited continuous data. The constant slope method connects the start of the rising limb (A) to a point directly under the hydrograph peak with the same slope of the hydrograph starting line (C) and then extends to the point of end-of-surface runoff (D). Point D can be estimated by the following empirical equation:

$$N = 0.83A^{0.2} \tag{8.6}$$

where
   N is the number of days after a peak when surface runoff ends
   A is the watershed area (km^2)

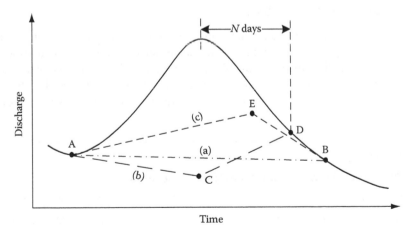

FIGURE 8.7 Graphics methods for separating the baseflow from direct runoff: (a) constant-discharge baseflow method, (b) constant-slope baseflow method, (c) concave-baseflow method.

In this method, it is assumed that the baseflow decreases in the rising limb due to movement of water into streambanks as bank storage and then increases in the recession limb.

8.2.3.4   Concave-Baseflow Separation

The concave method connects the start of the rising limb (A) to a point directly under the turning point of hydrograph recession limb (E) and then extends to the point of end-of-surface runoff (B). Point C is derived by intersecting the vertical line from the turning point of hydrograph recession limb and the extended back line of recession curve.

Example 8.2

The observed discharges in a watershed outlet during a rainfall event are presented in Table 8.2. The area of the watershed is about 20,000 km². Separate the baseflow from the direct runoff.

Solution

Points A, B, C, D, and E are determined using graphical methods, which include the constant-discharge, constant-slope, and concave-baseflow separation methods as shown in Figure 8.8. The number of days after a peak when surface runoff ends in constant-slope method is calculated using Equation 8.6 as follows:

$$N = 0.83 \times 2000^{0.2} \approx 6\,\text{days}$$

The baseflow and the direct runoff in each method are presented in Table 8.3.

8.2.4   GROUNDWATER AND LAKES

Groundwater interacts with lakes the same way as it interacts with streams. Lakes interact with the groundwater with a much larger surface water and bed area than streams. Due to sediments in the lake floor, the permeability of the lake bed is lower than that of streams. Lakes can receive groundwater inflow throughout their bed and have seepage loss throughout the lake bed, or they receive groundwater inflow in

---

TABLE 8.2

Watershed Flows During a Rainfall Event (m³/s)

Time (Day)	1	2	3	4	5	6	7	8	9
Flow (m³/s)	1200	1150	1350	1720	2200	2620	3200	2920	2510
Time (Day)	10	11	12	13	14	15	16	17	
Flow (m³/s)	2200	1850	1500	1320	1160	1060	1040	1030	

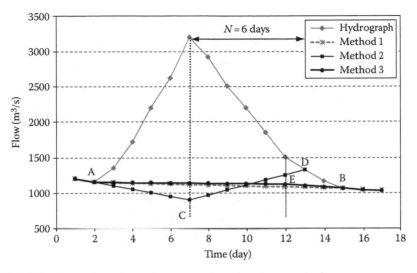

FIGURE 8.8　Illustration of base flow separation using three methods.

TABLE 8.3
The Direct Runoff Obtained from Three Methods (m³/s)

Time (Days)	Method 1	Method 2	Method 3
1	0	0	0
2	0	0	0
3	207	250	203
4	584	670	576
5	1071	1200	1059
6	1498	1670	1482
7	2085	2300	2065
8	1812	1950	1788
9	1408	1470	1381
10	1105	1090	1074
11	762	670	727
12	419	250	380
13	246	0	220
14	93	0	80
15	0	0	0
16	0	0	0
17	0	0	0

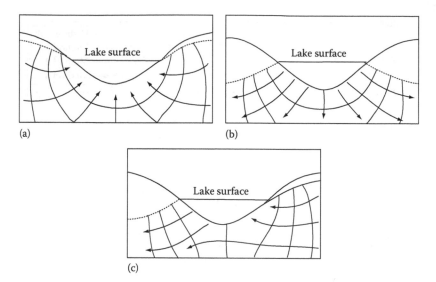

FIGURE 8.9 (a) Lakes receiving groundwater inflow; (b) lakes losing water as seepage to groundwater or (c) both at the same time. (From U.S. Geological Survey, 1998.)

some parts of the bed and lose groundwater inflow in other parts. Figure 8.9 shows the basin ways of groundwater and lakes interaction.

In order to estimate effects on lakes, wetlands, or any surface water feature on groundwater, the same approach of the stream interaction can be applied.

8.3   CONJUNCTIVE USE OF SURFACE AND GROUNDWATER

Water resources of surface water and groundwater are managed separately. However, surface water including streams, lakes, reservoirs, and wetlands interact with groundwater. Withdrawal of water from streams can deplete groundwater or, conversely, water in streams, lakes, or wetlands can be depleted by pumping of groundwater. Also pollution of surface water can cause degradation of groundwater quality and vice versa. Considering quantity and quality effects of development of one on the other and using integrated water resources management approach for sustainable development, the surface water and groundwater should be managed simultaneously.

Conjunctive use of surface and groundwater consists of harmoniously combining the use of both sources of water in order to minimize the undesirable physical, environmental, and economical effects of each solution and to optimize the water demand/supply balance. Surface water and groundwater are used simultaneously for planning the usage of water. It is applied to supply water demands and quality management of water resources for obtaining the sustainable development.

Surface water operation requires reservoirs for water storage in suitable locations. But groundwater can be stored naturally in aquifer without the need of storage construction. With conjunctive use planning, water can be temporarily stored in the aquifers and water demand can be supplied from surface water. During water scarcity, water can be withdrawn to meet water demands. The characteristics and extent of the interactions of ground water and surface water affect the success of such planning of conjunctive use of surface and groundwater resources.

8.3.1 ADVANTAGES OF CONJUNCTIVE USE

Conjunctive use of surface and groundwater resources increases the management efficiency for supplying water by storing the existing water resources to supply future water demands. Groundwater resources can be considered as a reliable resource for supplying water demands. Usually, periods of lowest streamflow and recharge coincide with the largest demand, or vice versa. Conjunctive water management could reduce the agricultural damages in dry periods. During wet periods, surface water can be utilized to supply the water demands and to recharge the aquifers. Generally, it could be a reliable alternative for water supply in the basin during all seasons. Also using the groundwater resources for supplying the water demand can decrease the allocation streamflow volume that leads to the reduction in the construction costs of large scale canals.

On the other hand, groundwater exploitation as a single water resource leads to groundwater depletion and inordinate fluctuations of groundwater level. So, the conjunctive use can replenish aquifers in terms of quality and quantity. Conjunctive water management is utilized in many areas to assure the sustainable availability of water, particularly for agricultural sector.

Justification of conjunctive use plans can be based on benefit/cost analysis. For this purpose, water balance and the hydrologic conditions in the plain are evaluated. The data for evaluation consist of land use, crop pattern, water demand, meteorological, and hydrologic conditions. Based on the analysis, surface and groundwater availability and water demand are determined. In order to evaluate the management plans in the case study, some scenarios are generated based on conjunctive use planning and their impacts on water demand and supply are assessed. For this purpose, a watershed model is needed to simulate the scenarios in different hydrologic conditions and then optimal policies for utilization of groundwater to conjunct with surface waters are developed. The optimal conjunctive water management strategies can be developed using simulation/optimization models. These models can help determine the best spatially distributed balance between using surface and underground water. They can be used for planning, management, and operations perspectives. Finally, benefits and costs of applying the generated scenarios are determined and the best scenario for executive conjunctive use planning is selected based on benefits/costs analysis.

However, the best conjunctive water management involves satisfying hydrogeologic, technical, sociologic, ecologic, economic, and institutional constraints. Water laws and regulations can help conjunctive water management.

8.3.2 SURFACE AND GROUNDWATER CONJUNCTIVE USE MODELING

The conjunctive water use model provides an optimal allocation schedule for surface and groundwater resources in a region to meet different water demands. Conjunctive water use planning should increase the efficiency, reliability, and cost-effectiveness of water use. Conjunctive water use can be simulated by coupled stream-aquifer models. In order to develop optimal policies, the conjunctive use planning can be formulated as an optimization model of the water resources system. The decision variables of the model are water allocations of the groundwater and surface in each planning period. The objective functions of optimization model is to maximize the objectives of the water resource system while satisfying the hydraulic response equations of the surface and groundwater systems, and any constraints limiting the head variations and the surface water availability (Willis and Yeh, 1987).

The economic based objective function of optimization model can be considered as maximizing the net benefit of water use. The economic benefit can be achieved by the sale of the water to different sectors and sale of the agricultural and industrial productions due to water allocation. The cost includes capital, operation, and maintenance to extract and convey the allocated water.

The objective function of conjunctive use planning can be as follows:

$$Max\ Z = \sum_{t=1}^{T} B_t - C_t \tag{8.7}$$

where

$$B_t = fb_t(GW_t, SW_t, D_t) \tag{8.8}$$

$$C_t = fc_t(GW_t, SW_t) \tag{8.9}$$

where B_t is the total benefits of surface water (SW_t) and groundwater (GW_t) allocations generated through the revenue of water sale and benefit fraction from water productions, which is function (fb_t) of surface water, groundwater, and water demands D_t in time t. C_t is the total costs of surface water and groundwater allocations based on the costs of capital, operation, and maintenance, which is function (fc_t) of surface water and groundwater volumes in time t.

The optimization model can be constrained by the continuity equation, the hydraulic response equations, possible limitations on river runoff, pumping discharge volume from groundwater, allocated water discharge volume from surface water, groundwater table fluctuations, and appropriate well pumping rates at each planning period. In order to prevent excessive groundwater overdraft or water logging of the agricultural areas of the basin, constraints can be introduced to limit the head variation in each aquifer system. The hydraulic head in the aquifers can be obtained utilizing the hydraulic response equations described in Chapters 4 and 6. For a groundwater system, the response equations can be expressed functionally as

$$h_t = f(h_{t-1}, GW_t, SW_t) \tag{8.10}$$

where h is hydraulic heads in groundwater system at the period t. In recent years, more complex and user-friendly simulation packages have been used to develop and calibrate the groundwater response equations considering the principle of superposition.

The optimization model can be linked with the water head simulation model to obtain the optimal water allocation policies. Also, the water quality simulation model for groundwater system can be utilized to evaluate the water allocation policies on the objective function. The structure of the optimization model, which is a nonlinear programming problem, can be developed based on target objective functions and the possible constraints in the study region.

Example 8.3

Develop a conjunctive use model to determine the optimal allocation of surface and groundwater in the river basin as shown in Figure 8.10. There are two agricultural plains above unconfined aquifers A and B.

Solution

Assuming that the system objective is to minimize the costs of surface water and groundwater allocation. The objective function optimization model is expressed by

$$Min\ Z = \sum_{t=1}^{T} fc_1(GW_{At}) + fc_2(GW_{Bt}) + fc_3(SW_{At}) + fc_4(GW_{At}, WD_{At})$$

$$+ fc_5(GW_{At}, WD_{At}) + fc_6(RE_t, ED_t)$$

where
GW_{At}, GW_{Bt}, are groundwater extraction from aquifers A and B
SW_{At} is diversion water from river to the aquifer A at each time period
fc_1 and fc_2 are the cost function related to well pumping of aquifers A and B
fc_3 is the cost function associated with canal diversion construction
WD_{At}, WD_{Bt} are water demands of agricultural zones A and B
fc_4 and fc_5 are the cost functions related to agricultural damages due to water scarcity
ED_t is the environmental water demand and RE_t is the instream flow in the downstream of the river
fc_6 is the cost function related to un-supplying the environmental water demand.

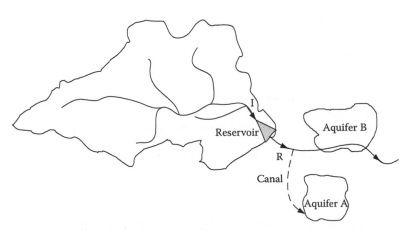

FIGURE 8.10   Schematic map of the problem river basin.

This problem is constrained by the continuity equation for the reservoir as follows:

$$S_t = S_{t-1} + I_t - E_t - R_t$$

where
S_t is the reservoir storage
I_t is inflow to the reservoir
E_t is the evaporation loss
R_t is the reservoir release at time t

The reservoir storage should be in a feasible range, between minimum and maximum storage volumes S_{min} and S_{max} as follows:

$$S_{min} \leq S_t \leq S_{max}$$

Writing water balance in the river leads to the following equation:

$$R_t - SW_{At} - SW_{Bt} = RE_t$$

There are limitations on capacity of diversion canal

$$SW_{At} \leq Cap_A$$

where Cap_A is the capacity of the canal for water convey.
For unconfined aquifer A, the hydraulic head can be calculated using groundwater flow equations as follows:

$$\frac{\partial}{\partial x}\left(T\frac{\partial h}{\partial x}\right) + \frac{\partial}{\partial y}\left(T\frac{\partial h}{\partial y}\right) + RW_{At} - GW_{At} = S_y\frac{\partial h}{\partial t}$$

where
h is the hydraulic head
RW_{At} is the groundwater recharge
S_S is the specific storage
T is transmissivity for the aquifer

The hydraulic heads are obtained by solving the differential equations utilizing numerical methods introduced in Chapter 6.
A constraint can be applied for limiting the maximum and minimum groundwater levels.
(h_{min} and h_{max}) in each planning period as follows:

$$h_{min} \leq h_t \leq h_{max}$$

The optimization model can be solved using the optimization techniques introduced in Chapter 7.

Example 8.4

Determine the average monthly allocated groundwater and surface water to four agricultural zones shown in Figure 8.11. As it can be seen in this figure, a channel can transfer surface water to Zone 4. The monthly discharge of rivers and the volume of discharge by precipitation during the last 2 years and gross water demands of the agricultural zones are presented in Tables 8.4 and 8.5, respectively. The equations of the average groundwater table fluctuations, based on the discharge from agricultural wells and recharge from precipitation, are presented in Table 8.6. In this table, $G(i)$ is the groundwater discharge from agricultural zone i (MCM), $Q(3)$ is the transferred surface water to zone (4), and P is the groundwater recharge from precipitation (MCM). The effects of other parameters such as recharge of precipitation and underground inflow and outflow are considered to be a monthly constant. The acceptable (without any cost) range for cumulative groundwater table variation in each zone is $\perp 4\,m$. The maximum capacity of water transfer channel is $4\,m^3/s$, and it is assumed that 90% of transferred water is supplied by River 1. The initial groundwater table depth in Zones 1, 2, 3, and 4 are considered to be 17, 8.1, 12.6, and 67 m, respectively.

The objectives of the developed model are water supply to agricultural lands, minimizing the pumping cost, and controlling the average groundwater table fluctuations in agricultural zones.

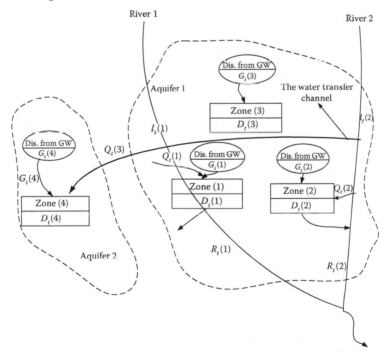

FIGURE 8.11 I_t (1): The inflow surface water to Zone (1) in month t, (MCM), I_t (2): The inflow surface water to Zone (2) in month t, (MCM), R_t (2): The outflow surface water from Zone (1) in month t, (MCM), R_t (1): The outflow surface water from Zone (2) in month t, (MCM), $G_t(i)$: The extracted groundwater from Zone (i) in month t, (MCM), $Q_t(i)$: Allocated surface water to Zone (i) in month t, (MCM), $D_t(i)$ Agricultural water demand in Zone (i) in month t, Components of surface and groundwater resources for Example 8.3.

TABLE 8.4
The Values of Monthly River Discharge and Precipitation

Water Year	Month	Discharge from River (1) (MCM)	Discharge from River (2) (MCM)	Recharge of Aquifer (1) by Precipitation (MCM)	Recharge of Aquifer (2) by Precipitation (MCM)
1	October	13.48	15.49	0.7	0.05
1	November	6.16	16.51	1.2	0.2
1	December	9.38	22.78	2	0.18
1	January	14.34	20.19	2.5	0.19
1	February	15.9	27.18	3	0.3
1	March	18.17	26.25	3	0.31
1	April	10.08	26.99	2.5	0.3
1	May	7.11	26.16	2	0.15
1	June	5.3	23.14	0.5	0.04
1	July	3.76	20.29	0	0.01
1	August	3.23	19.65	0	0.2
1	September	2.85	17.43	0.5	0.05
2	October	18.83	30.18	0.6	0.06
2	November	7.44	26.23	1.2	0.2
2	December	8.58	26.64	2.2	2.2
2	January	10.89	26.45	2.3	2.3
2	February	14.72	25.61	3.2	3.2
2	March	25.79	20.22	2.9	2.9
2	April	35.17	48.29	2.6	2.6
2	May	15.07	35.09	2	2
2	June	9.47	28.07	0.56	0.56
2	July	19.08	23.16	0.2	0.2
2	August	37.51	19.16	0	0
2	September	30.25	16.47	0.51	0.51

Solution

The pumping cost can be considered as a linear function of pumping power (P_{pump}) as follows:

$$P_{pump} = \frac{\gamma G H}{75 * \eta}$$

where
   γ is the specific gravity of water (kg/m³)
   G is the pumping discharge (m³/s)
   H is the groundwater table depth (m)
   η is the pumping efficiency

Since the objective function and the constraints are nonlinear, the dynamic programming (DP) method can be effectively used in this study. The recursive function of the DP model is developed as follows:

TABLE 8.5
Monthly Agricultural Water Demands and Downstream Water Right (MCM)

Month	Zone (1)	Zone (2)	Zone (3)	Zone (4)	Downstream Water Right
October	8.536	10.07	0.64	3.01	1.86
November	8.494	8.59	0.48	3.82	3.43
December	3.45	2.57	0.2	1.41	1.045
January	2.448	2.03	0.153	1.24	0.965
February	2.371	1.79	0.133	0.81	0.37
March	4.5	3.33	0.353	1.32	0.54
April	11.44	9.75	1.176	6.15	6.92
May	34.64	31.76	3.44	20.58	28.57
June	36.6	35.22	4.02	19.12	27.6
July	19.45	18.83	2.53	4.93	8.92
August	23.9	25.86	2.66	4.97	5.78
September	21.31	23.49	1.68	4.26	3.76

TABLE 8.6
The Equations of Average Groundwater Table Fluctuation (m)

Zone (1) $\Delta L_1 = 1.9 \times 10^{-4}\, G(1) + 1.2e^{-3}\, G(2) - 2e^{-5}\, G(3)^2 + 7.8e^{-4}\, G(3) - 0.02P - 0.03$

Zone (2) $\Delta L_2 = 1 \times 10^{-6}\, G(1) + 3.7e^{-3}\, G(2) - 4e^{-5}\, G(3)^2 + 1.9e^{-4}\, G(3) - 0.018P - 0.07$

Zone (3) $\Delta L_3 = 1 \times 10^{-7}\, G(1) + 1.1e^{-3}\, G(2) - 4.1e^{-4}\, G(3) - 0.01P - 0.02$

Zone (4) $\Delta L_4 = -0.0149G(4)^2 + 0.41G(4) - 7e^{-5}\, Q(3)^2 - 0.028Q(3) - 0.12$

Note: The positive values show the groundwater table drawdown.

$$f_t(G_t(i), Q_t(i)) = Min\left\{ C_t(G_t(i), Q_t(i)) + f_t^*(G_{t-1}(i), Q_{t-1}(i)) \right\}$$

where

   i is index of Zone $i = 1, \ldots, 4$

   $G_t(i)$ is the amount of groundwater extracted from zone i, in month t(MCM)

   $Q_t(i)$ is the allocated surface water to zone i, in month t(MCM)

   $C_t(G_t(i), Q_t(i))$ is the cost of operation during time period t

   $f_t^*(G_{t-1}(i), Q_{t-1}(i))$ is the minimum of the total operational cost till the end of time period $t-1$

The cost of operation in each period is estimated as follows:

$$C_t(G_t(i), Q_t(i)) = \sum_{i=1}^{4} Loss_t(i)$$

$$Loss_t(i) = \alpha(D_t(i) - Q_t(i) - G_t(i))^2 + \beta(G_t(i)H_t(i)) + \gamma\left(\left|L_t(i)\right| - L_{max}(i)\right)^2$$

$$\alpha = 0 \quad \text{if } D_t(i) \leq (Q_t(i) + G_t(i))$$

$$\gamma = 0 \quad \text{if } \left|L_t(i)\right| \leq L_{max}(i)$$

where
  $D_t(i)$ is the agricultural water demand in zone i and time period t
  $L_t(i)$ is the total variation of water table level until the end of time period t
  $L_{max}(i)$ is the maximum allowable cumulative groundwater table fluctuation in
    agricultural zone i
  $H_t(i)$ is the depth of groundwater table in agricultural zone i at the end of month t
  α, β, γ are weights of objectives (constant)

The constraints of the model are as follows:

Control of groundwater level fluctuations

$$H_t(i) = H_0(i) + \sum_{t=1}^{T} \Delta L_t(i), \quad i = 1, ..., 4$$

$$L_t(i) = \sum_{t=1}^{T} \Delta L_t(i), \quad i = 1, ..., 4$$

$$\Delta L_t(i) = f_i(G_t(i), Q_t(i), I, O)$$

where
  $H_0(i)$ is the initial depth of groundwater table in the agricultural zone i
  $\Delta L_t(i)$ is the variation of water table level in time period t in agricultural zone i
    (draw down is considered to be positive)
  $f_i(G_t(i), Q_t(i), I, O)$ is the groundwater table fluctuation in agricultural zone i and
    time period t, which is the function of total discharge(O), and the total
    recharge (I) of the aquifers

Estimation of surface water outflow from the agricultural Zones 1 and 2 ($R_t(i)$):

$$R_t(i) = I_t(i) - Q_t(i) - \left(1 - \frac{\psi}{100}\right)Q_t(3), \quad i = 1, 2$$

where
  ψ is the percentage of transferred water to Zone 4 from surface water resources
    of Zone 2
  $I_t(i)$ is the surface water inflow to zone i in the time period t(MCM)

Minimum instream flow requirements downstream of agricultural zones:

$$\sum_{i=1}^{2} R_t(i) \geq R_{t,min}$$

TABLE 8.7
Average Allocated Surface and Groundwater to Different Zones in July (MCM)

Source	Zone (1)	Zone (2)	Zone (3)	Zone (4)
Surface water	5.8	15.6	—	—
Groundwater	21.2	3.0	3.0	1.0
Transferred water	—	—	—	3

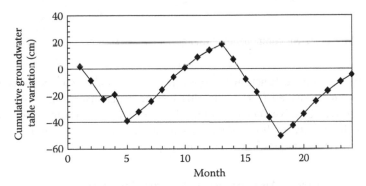

FIGURE 8.12 Cumulative groundwater table variation in zone 1 for Example 8.3.

where
$R_{t,min}$ is the minimum instream flow required in downstream in time period t
$R_t(i)$ is the surface water outflow from zone i in time period t in MCM

Maximum capacity of canal:

$$Q_t(4) \le Q_{t,max}$$

As the problem is multiobjective, the coefficients α, β, and γ are the relative weights of objectives, which should be determined in order to make these objectives comparable. The allocated water to different agricultural zones in July (a dry month) for $\alpha = 1$, $\beta = 0.005$, and $\gamma = 120$ and the corresponding cumulative groundwater table are presented in Table 8.7 and Figure 8.12 (Karamouz et al., 2002).

8.3.3 MULTIOBJECTIVE CONJUNCTIVE USE OPTIMIZATION

In a basin system, the components of water resources and water users have interaction with others that lead to complex system. The basin could include the reservoir, stream, river, aquifer, irrigated area, water demands, pumping wells, and diversion canal. Figure 8.13 illustrates different components of a basin. There are interactions between groundwater and water storage reservoir, irrigated field, and lakes. Also return flows from the allocated water to different sectors for supplying the agricultural, municipal, and industrial water demands recharge the aquifers. Water can be delivered to users through pipes and canals.

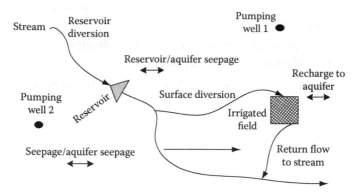

FIGURE 8.13   Different components of an aquifer-stream system.

The major objective function for optimization of models can be considered to maximize water provided from wells plus surface water diversions to supply water demands. Other objective functions could be considered to minimize the total cost or improve the quality of allocated water. The multiobjective optimization model can be solved by the techniques introduced in Chapter 7.

8.3.4   CASE STUDY 1: APPLICATION OF GENETIC ALGORITHMS AND ARTIFICIAL NEURAL NETWORKS IN CONJUNCTIVE USE OF SURFACE AND GROUNDWATER RESOURCES

The proposed model is applied to the conjunctive use of the surface and groundwater resources in the southern part of Tehran. Tehran plain is bounded by the Kan River in the West and the Sorkhehesar River in the East. About one billion cubic meters of water per year is provided for domestic consumption of over 8 million people. Figure 8.14 shows a map of the surface water resources of Tehran Plain.

More than 60% of the water consumption in Tehran returns to the Tehran aquifer via traditional absorption wells. Three major agricultural zones namely Eslamshahr–Kahrizak (Zone 1), Ghaleno (Zone 2), and Khalazir (Zone 3) are the main users of surface and groundwater resources in the study area (see Figure 8.14). Tehran aquifer is mainly recharged by inflow at the boundaries, precipitation, local rivers, and return flows from domestic, industrial, and agricultural sectors. The discharge from the aquifer is through water extraction from wells, springs, and as well as groundwater outflow and evapotranspiration.

8.3.4.1   Simulation Model

To simulate the groundwater table elevation in this study area, the PMWIN model is used. Tehran aquifer is considered as a single-layered aquifer, and therefore, only the horizontal hydraulic conductivity is estimated. Figure 8.15 shows the generated grid as well as the boundary conditions in the PMWIN model. Figure 8.16 presents the comparison between computed and historical groundwater table elevation in the

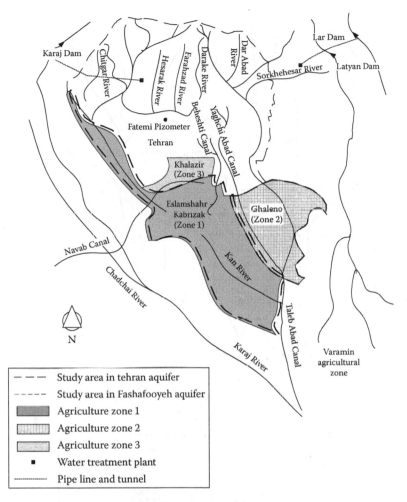

FIGURE 8.14 Surface water resources and the agricultural zones in the study.

study area. This figure shows how closely the model can reproduce the monthly water table variations.

The response functions in the optimization model shows the monthly groundwater table variations in Tehran aquifer. They are functions of discharge, recharge, inflow, and outflow at the boundaries as well as physical characteristics of the aquifer. Based on the available water quality data, NO_3 that usually deviates from the standards is selected as indicator water quality variables in the simulation model. The linkage of the optimization and simulation models will significantly increase the computational problems and also the time needed to achieve the optimal solution of the model. In the proposed methodology, the results of the (Processing Modflow) groundwater simulation model are used to train the artificial neural networks (ANN) based simulation model.

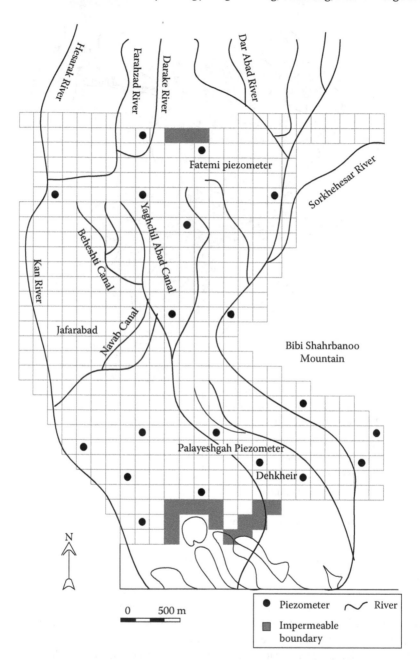

FIGURE 8.15   The generated grid and boundary condition in the PMWIN model.

The results of the frequent execution of the groundwater simulation model (PMWIN) for different sets of recharge–discharge values, and by considering the principle of superposition, the selected ANN is trained for estimating monthly water table variations that have been obtained for each agricultural zone. The suitable variables, which significantly affect the groundwater table variations, have been selected

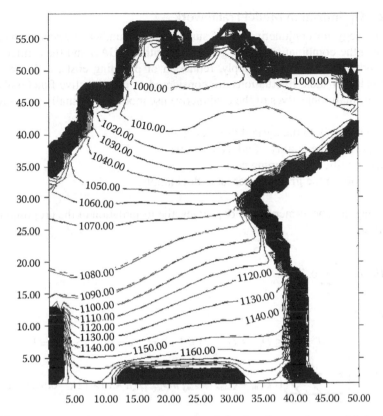

FIGURE 8.16 Comparison between observed and simulated groundwater table levels.

using a trial and error process. Finally, the total monthly groundwater discharge and the average groundwater table level in aquifer at the beginning of the month have been considered as variables; and the other factors that have a negligible effect on the groundwater table variations are assumed to be constant for each month. The general form of the response function of the average groundwater table variation in month t in each zone is estimated as follows:

$$\Delta H_t = purelin(w'_{2t} \times tansig((w_{1t} \times h_t) + b_{1t}) + b_{2t}) \tag{8.11}$$

where

ΔH_t is the vector of the groundwater table variations in month t (m) in agriculture zones (negative values refer to water table drawdown)

w'_{2t} is the weight parameter in the second layer of the ANN developed for month t

w_{1t} is the weight parameter in the first layer of the ANN developed for month t

h_t is the input matrix, which consists of the average groundwater table level at the beginning of each month and discharge average from agriculture zones

b_{1t} is the bias parameter in the first layer of the ANN developed for month t

b_{2t} is the bias parameter in the second layer of the ANN developed for month t

8.3.4.2 Optimization Model Framework

A methodology for conjunctive use of surface and groundwater resources is developed using the combination of the genetic algorithms (GAs) and the artificial neural networks (ANNs). Water supply, reduction of pumping costs, and controlling the groundwater table fluctuations are considered as the objective functions of the model. The main objectives of the conjunctive use models are usually as follows:

- Water supply to the agricultural demands
- Improving the quality of allocated water
- Minimizing the pumping costs
- Controlling the groundwater table fluctuations

Considering the above-mentioned objectives, the formulation of the proposed model is as follows:

$$Minimize \sum_{i=1}^{4} \alpha_i Z_i$$

Subject to:

$$Z_1 = \begin{cases} 0 & if \ \ D_{imy} \le Q_{imy} + G_{imy} \quad i=1,\dots,n, \quad y=1,\dots,15, \quad m=1,\dots,12 \\ \displaystyle\sum_{i=1}^{n}\sum_{y=1}^{15}\sum_{m=1}^{12} \left((D_{imy} - Q_{imy} - G_{imy})/D_{imy} \right)/(n \times 15 \times 12), & \text{otherwise} \end{cases}$$

$$Z_2 = \sum_{i=1}^{n}\sum_{y=1}^{15}\sum_{m=1}^{12} \left(G_{imy} H_{imy} / (G_{im} H_{im})_{max} \right)/(n \times 15 \times 12)$$

$$Z_3 = \left(\sum_{i=1}^{n} \left(\sum_{y=1}^{15}\sum_{m=1}^{12} \Delta L_{imy} \right)/\Delta L_{max} \right)/n$$

$$Z_4 = \left(\sum_{i=1}^{n} \left(\sum_{y=1}^{15}\sum_{m=1}^{12} C_{imy} \right)/c\,max \right)/n \tag{8.12}$$

$$\Delta L_{imy} \le \Delta L_{max}, \quad y=1,\dots,15, \quad m=1,\dots,12 \tag{8.13}$$

$$RD_{imy} = R_{imy} - Q_{imy} \tag{8.14}$$

$$\sum_{i=1}^{2} RD_{imy} \ge E_{my,min} \tag{8.15}$$

$$\sum_{i=1}^{4} \alpha_i = 1 \tag{8.16}$$

$$Q_{imy} \le R_{imy}, \quad i = 1,\ldots,n, \quad y = 1,\ldots,15, \quad m = 1,\ldots,12 \tag{8.17}$$

$$\Delta L_{imy} = f(\tilde{G}_{my}, \tilde{O}_{my}, \tilde{M}_{my}) \tag{8.18}$$

$$\Delta C_{imy} = f'(\tilde{G}_{my}, \tilde{O}_{my}, \tilde{M}_{my}, c_{my}) \tag{8.19}$$

where

G_{imy} is the volume of groundwater extracted in the agriculture zone i in month m of year y (m³/day)

Q_{imy} is the volume of surface water allocated to agriculture zone i in month m of year y (m³/day)

C_{imy} is the average concentration of the water quality indicator in the allocated water to agriculture zone i in month m of year y (mg/L)

c_{imy} is the average concentration of the water quality indicator in the return flow to agriculture zone i in month m of year y (m³/day)

D_{imy} is the agricultural water demand in zone i in month m of year y (m³/day)

ΔL_{imy} is the variation of the groundwater table level in month m of year y in the agriculture zone i (m) (draw down is considered to be negative)

ΔL_{max} is the maximum allowable cumulative groundwater table fluctuation (m)

C_{max} is the maximum allowable concentration of water quality variable in allocated water (mg/L)

C_{im} is the volume of groundwater extracted in the agriculture zone i in month m (m³/day)

H_{imy} is the groundwater table level in agriculture zone i in month m in year y

$(C_{im} \times G_{im})_{max}$ is the maximum value of $(C_{im} \times G_{im})$ in different years

R_{imy} is the surface water flow rate in zone i in month m of year y

$E_{im,max}$ is the environment water demand in month m of year y

RD_{imy} is the unused surface water in zone i in month m of year y

Z_1 is the loss function of water allocation

Z_2 is the loss function of pumping cost

Z_3 is the loss function of groundwater table fluctuation

Z_4 is the loss function of the quality of allocated water

α_i is the relative weight of the objective function i

m is the number of months in the planning horizon

n is the number of agricultural zones in the study area

y is the number of years in the planning horizon

The objective functions Z_1 to Z_4 should be normalized between 0 and 1. For this purpose the relative weights α_i can be assigned much easier based on the relative importance of the objectives. Based on the Equation 8.13, the monthly variation of the water table level is limited to a maximum level. Equation 8.18, which is called the response function, presents the monthly groundwater table variations in each zone, which is a function of the set of the volumes of the groundwater extracted in month m of year y in the agricultural zones ($\tilde{G}_{my}$), the outflow at the boundaries, and the groundwater discharge through springs and Qantas in month m of year $y(\tilde{Q}_{my})$, recharge by direct precipitation, allocated surface water, and also recharge by absorption wells ($\tilde{M}_{my}$). This equation should be formulated using a comprehensive groundwater simulation model (Karamouz et al., 2007).

8.3.4.3 Results and Discussion

The cumulative groundwater table variations in each zone, which has experienced a total fluctuation of more than ±20 m in the last 20 years, is limited to ±5 m over the planning horizon (15 years). The relative weight of Z_1 to Z_4 are considered to be 30, 10, 35, and 25, respectively. Table 8.8 presents the reliability of the water supply to the agricultural water demands, and the share of ground and surface water resources allocation. Based on the results presented in Table 8.8, only 79% of the total water demand can be allocated and more than 64% of the allocated demand is supplied by the groundwater resources. Figure 8.17 shows the cumulative variation of groundwater table in the agricultural zones based on the optimal operating polices. In Zone 2, less groundwater is extracted because of a lower concentration of water quality variable in the surface water of this zone (see Figures 8.18 through 8.20). It can be seen in Figure 8.19, most of the allocated water to Zone 2 is supplied from surface water resources due to the availability of the surface water and best water quality in this zone as well as to reduce the pumping cost. Figure 8.21 presents the total allocated water to Zone 1. The peak value shown in Figure 8.21 is due to the overallocating of the groundwater to Zone 1 to control the groundwater table in this zone. Based on the conjunctive use polices, the groundwater is allocated to Zone 3 to control the increasing level of the groundwater table in this agricultural zone. Results of this study show that the proposed model can be effectively used for conjunctive use of surface and groundwater resources considering the water quality issues.

TABLE 8.8

The Share of Surface and Groundwater Resources in Water Allocation to Different Zones

Zone	Percent of Demand Supplied by Groundwater	Percent of Demand Supplied by Surface Water	Reliability of Water Supply (without Quality)	Reliability of Water Supply (with Quality)
1	79	21	99	95
2	14	86	85	77
3	100	0	75	65

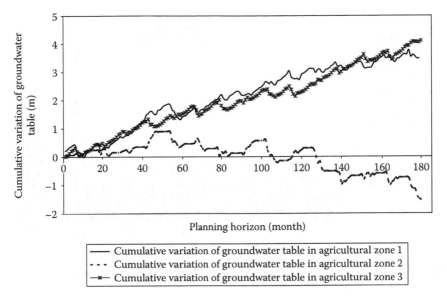

FIGURE 8.17 Cumulative variation of the groundwater water table level in the agricultural zones.

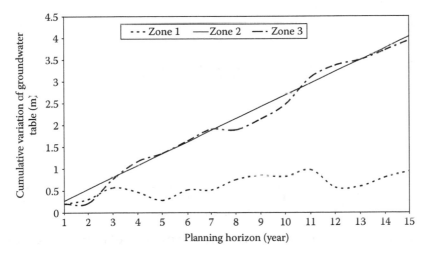

FIGURE 8.18 Cumulative variation of the groundwater level in the agricultural zones.

8.3.5 Case Study 2: Water Allocation from the Aquifer-River System Using a Conflict Resolution Model

The proposed optimization model is applied to the Lenjanat aquifer and Zayandeh-Rud River in central Iran, as shown in Figure 8.22. About 750 million cubic meters of water per year is provided for agricultural consumption and 220 million cubic meters of water per year is provided for industrial consumption. A water demand of more than 800 million cubic meters is supplied from Zayandeh-Rud River.

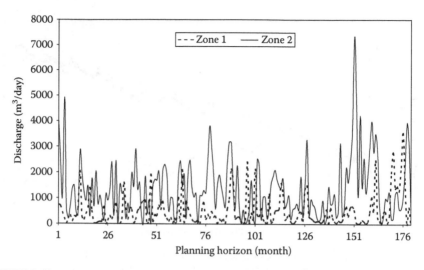

FIGURE 8.19   Monthly allocated surface water to agricultural zones 1 and 2.

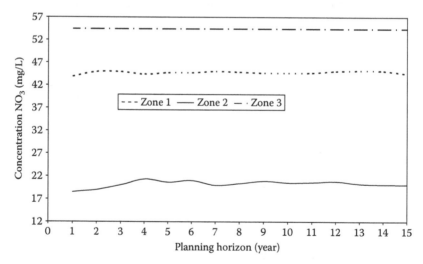

FIGURE 8.20   Average annual NO_3 concentration of the allocated water to the agricultural zones.

The area is located in Gavkhouni Basin in central Iran. The return flows from agricultural and industrial sectors are 15% and 10% of the allocated water, respectively. The transmissivity in Lenjanat unconfined aquifer varies from 300 to 2400 m^2/day. The average coefficient of hydraulic conductivity in this aquifer is about 7 m/day. There are about 40 piezometric wells in this aquifer.

Lenjanat aquifer is mainly recharged by inflow at the boundaries, precipitation, infiltration from the bed of river, and return flows from industrial and agricultural

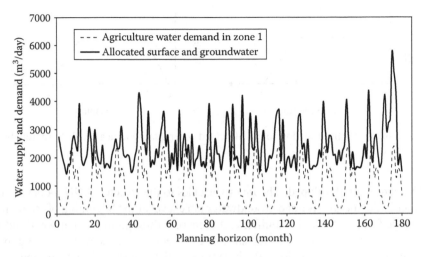

FIGURE 8.21 Monthly allocated water to agricultural zone 1.

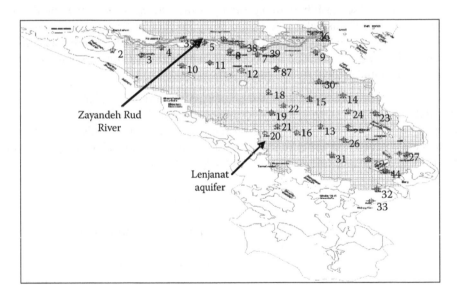

FIGURE 8.22 Studied area and the location of river and piezometric wells in the region.

sectors. The discharge from the aquifer is through water extraction from wells, springs, and Qanats as well as groundwater outflow and evapotranspiration.

Zayandeh-Rud River has an average of 1300 million cubic meters of discharge per year. The industrial and agricultural return flows are the most important sources of water pollution in the region. The recharged water from the aquifer to the river is also the cause of increasing water salinity. The TDS variation could be a good index for assessing these pollutants and therefore it was used in this study. PMWIN model is used to simulate the aquifer and then based on the simulation results, two seasonal

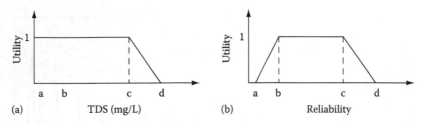

FIGURE 8.23 Utility functions of (a) allocated water quality indicator, TDS; (b) reliability of water supply.

ANN models are developed. The input variables to the selected ANN model are the number of the month, average groundwater table level at the beginning of the month, discharge from wells in the aquifer, and recharges through to the aquifer including agricultural drainage as well as recharge due to precipitation, head water in the river, concentration of water quality at the beginning of the month, and TDS concentration of the river and recharge (Karamouz et al., 2006).

In this study, the Nash bargaining theory is used to incorporate the conflicting interests of the decision makers. A typical utility function in Nash bargaining theory is illustrated in Figure 8.23. The utility function for different sectors are determined based on the available data and the results of some brain storming sessions with the main decision makers and stakeholders of the system.

Environmental sector: The quality and quantity of water in the Lenjanat Aquifer and Zayandeh-Rud River is the main concern for this sector. The utility function related to water quality is determined based on the river ecosystems needs and instream flow requirements. The utility function of water quality ($f_{r,c,m}$) and quantity ($f_{r,q,m}$) in the river in month m are described, respectively, as follows:

$$f_{r,c,m}(C_{d,m}) = \begin{cases} 0 & \text{if } C_{d,m} \geq 2500 \text{ mg/L} \\ 1 - \dfrac{(C_{d,m} - 1200)}{1300} & \text{if } 1200 \leq C_{d,m} < 2500 \text{ mg/L} \\ 1 & \text{if } C_{d,m} \leq 1200 \text{ mg/L} \end{cases} \quad (8.20)$$

$$f_{r,q,m}(Q_{d,m}) = \begin{cases} 1 & \text{if } Q_{g,m} \geq 30 \text{ MCM} \\ 1 - \dfrac{(30 - Q_{d,m})}{20} & \text{if } 10 \leq Q_{g,m} < 30 \text{ MCM} \\ 0 & \text{if } Q_{g,m} < 10 \text{ MCM} \end{cases} \quad (8.21)$$

where $C_{d,m}$ and $Q_{d,m}$ the concentration of water quality indicator and discharge in the river in month m.

The utility function of water quality in the aquifer and water table fluctuations in month m are described, respectively, as follows:

$$f_{G,c,m}(C_{G,m}) = \begin{cases} 1 & \text{if } C_{G,c,m} < 1200 \text{ mg/L} \\ 1 - \dfrac{(C_{G,m} - 1200)}{1300} & \text{if } 1200 \leq C_{G,c,m} < 2500 \text{ mg/L} \\ 0 & \text{if } C_{G,c,m} \geq 2500 \text{ mg/L} \end{cases} \quad (8.22)$$

$$f_{G,l,m}(l_m) = \begin{cases} 1 & \text{if } 0 \leq l_m < 3 \text{ cm} \\ 1 - \dfrac{(l_m - 3)}{2} & \text{if } 3 \leq l_m < 5 \text{ cm} \\ 0 & \text{if } l_m \geq 5 \text{ cm} \end{cases} \quad (8.23)$$

where $C_{G,m}$ and l_m are the concentration of water quality indicator in the aquifer and the water table fluctuation in month m.

Industrial sector: The industry sector is one of the most important water consumers in the study area and the quantity and quality of allocated water to this sector is one important issue of the water policy in the region. The utility function of industrial sector is considered a function of reliability of water supply (I_m) as follows:

$$f_{I,m}(I_m) = \begin{cases} 1 & \text{if } I_m > 85 \\ 1 - \dfrac{(85 - I_m)}{55} & \text{if } 30 < I_m \leq 85 \\ 0 & \text{if } 0 < I_m \leq 30 \end{cases} \quad (8.24)$$

The utility function of water quality for this sector is considered as follows:

$$f_{I,c,m}(C_{I,m}) = \begin{cases} 1 & \text{if } C_{I,m} \leq 1500 \\ 1 - \dfrac{(C_{I,m} - 1500)}{1500} & \text{if } 1500 < C_{I,m} \leq 3000 \\ 0 & \text{if } C_{I,m} > 3000 \end{cases} \quad (8.25)$$

where $C_{G,m}$ and I_m are the concentration of water quality indicator and reliability of demand supply for industrial sector.

Agricultural sector: The main objectives of this sector are water supply, water with acceptable quality to agricultural demands, and reduction of waste removal cost. The utility of this sector related to the water supply is based on the water supply reliability. Also, the utility functions related to water quality and quantity of agricultural sector are expressed as follows:

$$f_{A,m}(A_m) = \begin{cases} 1 & \text{if } A_m > 80 \\ 1 - \dfrac{(80 - A_m)}{60} & \text{if } 20 < A_m \leq 80 \\ 0 & \text{if } 0 < A_m \leq 20 \end{cases} \quad (8.26)$$

$$f_{A,c,m}(C_{A,m}) = \begin{cases} 1 & \text{if } C_{A,m} \leq 1500 \\ 1 - \dfrac{(C_{A,m} - 1500)}{1500} & \text{if } 1500 < C_{A,m} < 3000 \\ 0 & \text{if } C_{A,m} \geq 3000 \end{cases} \qquad (8.27)$$

where $C_{A,m}$ and A_m the concentration of water quality indicator and reliability of demand supply for agricultural sector.

8.3.5.1 Optimization Model Framework

Based on the Nash bargaining theory, the utilities and the objectives of the decision makers of the system as well as their relative authorities should be considered in water allocation to different sectors. The Nash bargaining based objective function in the optimization model is considered as:

$$Maximize: Z = \prod_{m=1}^{12 \times year} (f_{A,m}(A_m) - d_{A,m})^{w_A} (f_{A,c,m}(C_{A,m}) - d_{A,c,m})^{w_{Ac}} (f_{r,c,m}(C_{d,m}) - d_{r,c,m})^{w_{rc}}$$

$$\times (f_{r,q,m}(Q_{d,m}) - d_{r,q,m})^{w_{rq}} (f_{G,l,m}(l_m) - d_{G,l,m})^{w_{Gl}} (f_{G,c,m}(C_{G,m}) - d_{G,c,m})^{w_{Glc}}$$

$$\times (f_{I,m}(I_m) - d_{I,m})^{w_I} (f_{I,c,m}(C_{I,m}) - d_{I,c,m})^{w_{Ic}} \qquad (8.28)$$

Subject to:

$$Q_{I,m} = Q_{I,r,m} + Q_{I,G,m} \qquad m = 1, 2, \ldots, 10 \times year$$

$$Q_{A,m} = Q_{A,r,m} + Q_{A,G,m} \qquad (8.29)$$

$$I_m = \frac{Q_{I,m}}{D_{I,m}}$$

$$A_m = \frac{Q_{A,m}}{D_{A,m}} \qquad (8.30)$$

$$C_{I,m} = \frac{(Q_{I,r,m} \times C_{r,m} + Q_{I,G,m} \times C_{G,m})}{Q_{I,m}}$$

$$C_{A,m} = \frac{(Q_{A,r,m} \times C_{r,m} + Q_{A,G,m} \times C_{G,m})}{Q_{A,m}} \qquad (8.31)$$

$$C_{d,m} = \frac{C_{r,m}(Q_{u,m} - Q_{r,m} - IN_{G,m}) + C_{I,w,m}Q_{I,w,r,m} + C_{A,w,m}Q_{A,w,r,m} + IN_{r,m}C_{G,m}}{Q_{u,m} - Q_{r,m} + Q_{I,w,r,m} + Q_{A,w,r,m} + R_{r,m}} \qquad (8.32)$$

$$Q_{d,m} = Q_{u,m} + Q_{I,w,r,m} + Q_{A,w,r,m} - Q_{r,m} - IN_{G,m} + IN_{r,m} \tag{8.33}$$

$$IN_{R,m} = f(Q_{G,m}, h_{r,m}, R_{G,m})$$
$$IN_{G,m} = f(Q_{G,m}, h_{r,m}, R_{G,m}) \tag{8.34}$$

$$C_{G,m} = f(Q_{G,m}, R_m, C_{w,G,m}, h_{r,m}, C_{r,m})$$
$$l_m = f(Q_{G,m}, h_{r,m}, R_{G,m}) \tag{8.35}$$

where

$Q_{I,r,m}$, $Q_{I,G,m}$ is the allocated water to industrial sector from surface and groundwater resources in month m

$Q_{A,r,m}$, $Q_{A,G,m}$ is the allocated water to agricultural sector from surface and groundwater resources in month m

$D_{I,m}$, $D_{A,m}$ is the water demands of industrial and agricultural sectors in month m, respectively

$C_{r,m}$, $C_{G,m}$ is the water qualities of river and aquifers in month m

$C_{w,G,m}$ is the concentration of return flow to the aquifer in month m

l_m is the water table fluctuation in month m

$R_{G,m}$ is the volume of recharge to the aquifer in month m

$Q_{d,m}$, $Q_{u,m}$ is the discharges in the downstream and upstream of the river in month m, respectively

$C_{I,w,m}$, $C_{A,w,m}$ is the concentrations of industrial and agricultural return flow in month m

$Q_{I,w,G,m}$, $Q_{A,w,G,m}$ is the discharge volumes of industrial and agricultural return flow to the aquifer in month m

$h_{R,m}$ is the water depth in the river in month m

$Q_{e,G,m}$ is the discharge to the aquifer due to precipitation in month m

$IN_{R,m}$ is the discharge to the aquifer due to river interaction in month m

$IN_{G,m}$ is the discharge to the river due to aquifer interaction in month m

In the formulation, the objective function is the multiplication of each of the utility functions subtracted from their point of disagreement for every sector (agricultural, industrial, and domestic), river and aquifer interaction and groundwater variation in the aquifer for specific target. The model constrains are including the continuity equation, mass balance, upper and lower limits, water quality, and water table simulation models, which are linked to the optimization model.

The water quality variations in an aquifer and river can be implicitly obtained using a groundwater quality simulation model and a river water quality. Variations of groundwater table and water quality in each month are functions of the time series of recharges to the groundwater, water withdrawal from wells and interaction between river and aquifer that can be implicitly obtained using a comprehensive groundwater simulation model. River water quality at control points in each month is a function of the return flows quality and quantity and time series of upstream water quantity and quality that can be implicitly obtained using a river water quality simulation model.

8.3.5.2 Results and Discussion

In this study, 10 years of monthly streamflow quality and quantity data (1990–2000) are used to allocate water from the Zayandeh-Rud River and Lenjanat aquifer, considering quantity and quality issues. In this study, GA is used to solve the optimization model. In this model, the gene values are the monthly allocated surface water and groundwater to the agricultural and industrial sectors.

Figure 8.24 shows the time series of allocated water to the agricultural sector and industrial sectors, respectively. The results show that 86% and 90% of water demand in the agricultural sector and the industrial sector have been provided during 10 years operating period.

Figure 8.25 shows the TDS concentration in allocated water to the different sectors. As shown in Figure 8.25, the utility function related to this restriction has been

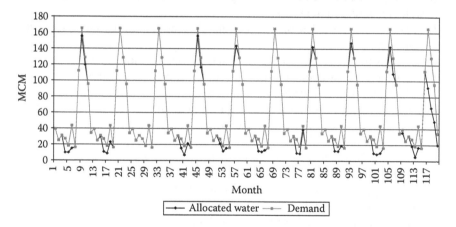

FIGURE 8.24 Allocated water to agricultural sector during the operating period.

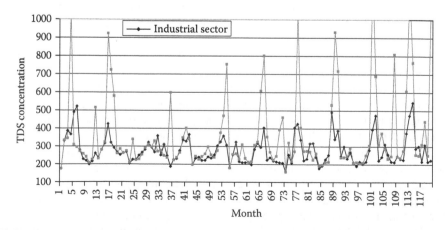

FIGURE 8.25 TDS concentration in allocated water to the agricultural.

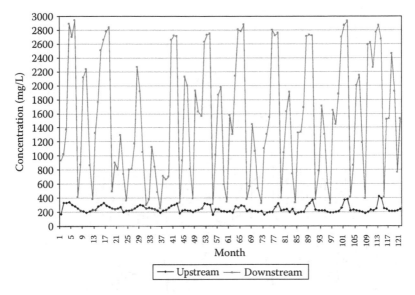

FIGURE 8.26 TDS concentration in the upstream and downstream of studied area and industrial sectors.

satisfied. This is the result of the TDS concentration in water resources of the studied area which is above.

Figure 8.26 shows the TDS concentration in river in both upstream and downstream of the studied area aquifer. Agricultural and industrial wastewater and water seepage from the aquifer to the river is the cause of the increase in TDS concentration downstream. The cumulative groundwater table variation in the studied aquifer, which has experienced a total fluctuation of more than 2.5 m in the last 10 years could be kept under 1.1 m over the planning horizon using this model.

8.3.6 Case Study 3: Conjunctive Use and Crop Pattern Management

The proposed model is applied to the plains, surrounding Tehran, the capital city of Iran. Tehran plains are located between 34°-52' and 36°-21' of northern latitude and between 50°-10' and 53°-9' of eastern longitude. The area is bounded by four major watersheds namely Kavir, Mazandaran, Sefidrood, and Hamedan-Markazy, respectively, as shown in Figure 8.27.

In the study area the weather is assumed to be affected by Mediterranean systems in winter and warm Arabian systems in summer. The average amount of precipitation in Varamin plain is about 131.4 mm/year and the temperature varies from −14.4°C to 45.5°C. The minimum and maximum relative humidity rates are 15% and 75% in summer and winter, respectively.

The surface water resources of Varamin are limited to Jajroud and Shour Rivers in the West and North of the plain, respectively (see Figure 8.28). Effluents of Tehran wastewater treatment plant No. 6 is also transferred to this plain by a canal to be used

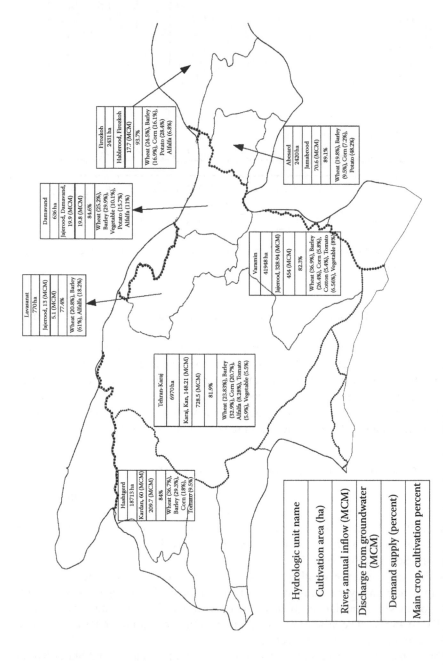

FIGURE 8.27   Characteristics of hydrologic units of Tehran province.

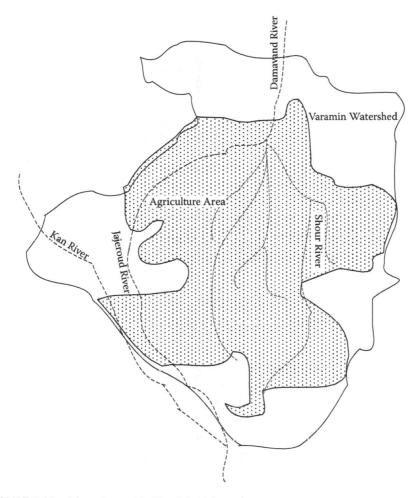

FIGURE 8.28   Rivers located in Varamin plain.

for irrigation. Table 8.9 shows the average monthly discharge of Jajroud and Shour
Rivers upstream of the Varamin plain.

8.3.6.1   Optimization Model Structure

In this study, a genetic algorithm model is developed to optimize crop pattern
of agricultural lands considering the water allocation priorities and surface and
ground water availability. The objective function of this model is to maximize the
net benefits of the agricultural products considering allowable water table fluctuations, the pumping costs, as well as the impacts of water supply deficit in crop
production. The model is formulated in a way that both crop pattern and water
allocation from surface or groundwater resources are simultaneously optimized
(Karamouz et al., 2008). The main objective of the proposed model is to maximize

TABLE 8.9

Average Monthly Discharge of Jajroud and Shour Rivers in Upstream of the Varamin Plain

River Name		Average Monthly Discharge (MCM)										
	October	November	December	January	February	March	April	May	June	July	August	September
Jajroud	6.1	11.45	12.91	11.25	14.89	29.58	58.79	56.47	21.65	4.78	4.55	2.16
Shour	8.27	20.68	28.56	22.26	24.53	23.92	22.99	21.22	5.93	3.69	0.74	1.42

the difference between gross benefit of crops production and the pumping costs. The main constraints of this model are

- Maximum incremental and cumulative allowable water table fluctuations
- Total area of agricultural lands
- Availability of surface water
- Instream flow requirements in the rivers

Considering the above objectives and constraints, the formulation of the problem is as follows:

$$\text{Maximize } Z = B - C \tag{8.36}$$

Subject to:

$$B = \sum_{f=1}^{y} \sum_{j=1}^{n} \sum_{p=1}^{c} (Ya_{pif} \times T_p \times \alpha_{pj} \times A_j) \tag{8.37}$$

$$C = \sum_{j=1}^{n} \sum_{f=1}^{y} \sum_{k=1}^{m} \left(\frac{G_{jfk} \times H_{jfk} \times hr_{jfk}}{\eta \times 0.102} \times \text{Pr} \right) + \left(\frac{S_{jfk} \times H'_{jfk} \times hr'_{jfk}}{\eta \times 0.102} \times \text{Pr}' \right) \tag{8.38}$$

$$D_{jfk} = \sum_{p=1}^{c} \left(\alpha_{pj} \times A_j \times d_{pjfk} \right), \quad j = 1, \dots, n, \quad k = 1, \dots, m, \quad f = 1, \dots, y \tag{8.39}$$

$$\sum_{p=1}^{c} \alpha_{pj} = 1, \quad j = 1, 2, \dots, n \tag{8.40}$$

$$S_{jfk} \le 0.7 \times I_{jfk}, \quad j = 1, \dots, n, \quad k = 1, \dots, m, \quad f = 1, \dots, y \tag{8.41}$$

$$\Delta L_{jfk} = \frac{\left(Ch_{jfk} + \alpha \times \left(\sum_{j=1}^{n} \left(S_{jfk} + G_{jfk} \right) \right) - \sum_{j=1}^{n} G_{jfk} - GD_{jfk} \right)}{(Ss_j \times A_j)}, \tag{8.42}$$

$$j = 1, \dots, n, \quad k = 1, \dots, m, \quad f = 1, \dots, y$$

$$Ya_{pif} = Ym_p \times \prod_{k=1}^{m} \left(1 - Ky_{pk} \left(1 - \left(\frac{ETa_{pjfk}}{ETm_{pjfk}} \right) \right) \right), \quad p = 1, \dots, c, \ j = 1, \dots, n, \ f = 1, \dots, y \tag{8.43}$$

$$RF_{jfk} = 0.3 \times (S_{jfk} + G_{jfk}), \quad j = 1,\ldots,n, \quad k = 1,\ldots,m, \quad f = 1,\ldots,y \quad (8.44)$$

$$\left| \Delta L_{jfk} \right| \leq \Delta Lmax_j, \quad j = 1,\ldots,n, \quad k = 1,\ldots,m, \quad f = 1,\ldots,y \quad (8.45)$$

$$S_{jfk} + G_{jfk} \geq D_{jfk}, \quad j = 1,\ldots,n, \quad k = 1,\ldots,m, \quad f = 1,\ldots,y \quad (8.46)$$

$$\sum_{f=1}^{y} \sum_{k=1}^{m} \Delta L_{jfk} < 5, \quad j = 1,2,\ldots,n \quad (8.47)$$

where

Z is the net benefit of the crop cultivation

B is the gross benefit of crop cultivation in the planning horizon

C is the pumping cost of surface and groundwater in the planning horizon

G_{jfk} is the volume of groundwater withdrawal from wells in zone j of month k and year f(MCM)

S_{jfk} is the volume of surface water withdrawal from rivers in zone j of month k and year f(MCM)

D_{jfk} is the monthly crops water demand in zone j of month k and year f (MCM)

ΔL_{jfk} is the water table fluctuations in zone j of month k and year f (m)

$\Delta Lmax_j$ is the maximum monthly allowable fluctuation of water table in zone j (m)

ETa_{pjfk} is the water allocation to crop p of the zone j of month k and year f

ETm_{pjfk} is the water demand of crop p of the zone j of month k and year f

Ss_j is the storage coefficient of aquifer in zone j

Ch_{jfk} is the aquifer recharge in zone j of month k and year f (MCM)

GD_{jfk} is the aquifer discharge through drainage in zone j of month k and year f (MCM)

A_j is the agricultural area in zone j (ha)

α_{pj} is the area allocated percent for crop p in zone j (%)

d_{pjfk} is the monthly crop water demand for crop p in zone j of month k and year f (m)

I_{jfk} is the discharge of river in zone j of month k and year f (MCM)

Ya_{pjf} is the actual yield for crop p in zone j and year f (kg/ha)

Ym_p is the maximum yield for crop p (kg/ha)

Ky_{pk} is the sensitive Coefficient for crop p in the month k

T_p is the price of crop p (Rial/kg)

RF_{jfk} is the Volume of recharged water to the groundwater from total water used for irrigation in zone j of month k and year f (MCM)

$(G_{jfk} \times H_{jfk} \times hr_{jfk})/(\eta \times 0.102)$ is the power needed for groundwater pumping (kWh)

$(S_{jfk} \times H'_{jfk} \times hr'_{jfk})/(\eta \times 0.102)$ is the power needed for surface water pumping (kWh)

H_{jfk} is the depth of water table in zone j of month k and year f(m)

H'_{jfk} is the head of water derived from river in zone j of month k and year f (m)

Pr is the price of electricity needed for water pumping from groundwater

Pr' is the price of electricity needed for water pumping from surface water

η is the pumping efficiency (%)

hr_{jfk} is the Total number of hours of pumping from groundwater in zone j of month k and year f (h)

hr'_{jfk} is the total number of hours of pumping from surface water in zone j of month k and year f (h)

c is the number of crops in zone j

n is the number of zones

m is the number of months

y is the number of years in planning horizon

Constraint (8.37) is used for estimating the total income based on actual crop production and selling price of different crops. Constraint (8.38) is formulated for estimating the cost of water pumping. Constraint (8.39) calculates the monthly water demand according to the area allocated to each crop and the monthly net demands estimated based on the regional conditions. Constraints (8.40) and (8.41) are used for considering the total area and surface water availability in the region. Constraint (8.42) shows the relation of water table fluctuations with different characteristics of the aquifer. To calculate the amount of crop production, a production function developed by Dorenbos and Kassam (1979), which is a function of actual and potential evapotranspiration extended by Ghahraman and Sepaskhah (2004) as well as allocated water and crop water demand supply is used in this study. The amount of water returns to groundwater and maximum allowable water table fluctuation are considered in constraints (8.44) and (8.45), respectively. Constraints (8.46) and (8.47) show allocated water resources for irrigation demands and the maximum cumulative fluctuation of allowable groundwater level.

Due to the large number of decision variables and nonlinearity of the problem, GA is used to find the optimal solution. The schematic of one chromosome used in the GA model in the proposed optimization model is shown in Figure 8.29. The number of genes in a chromosome for a y year planning horizon with P dominant crops is equal to $P + 2 \times m \times y$. The first set of P genes presents the percentage of each crop in the plain. The remaining genes are filled with the amount of allocated water from surface and ground water resources for m months of y years. One of the main advantages of the proposed model is to consider the time series of monthly crop water demand, which allows incorporation of demand uncertainty in the development of

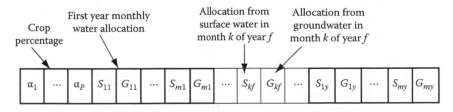

FIGURE 8.29 Schematic of a chromosome used in the proposed GA model.

optimal operation policies. A production function has also been used to estimate the production rate with respect to the amount of water allocation and water supply deficits (Karamouz et al., 2009).

8.3.6.2 Results and Discussion

Obtained optimal crop pattern is presented in Table 8.10. In this scenario, the coverage area of barely and cucumber, which have high yield with low water demand, has increased to improve the region income. Components of the objective function including the benefit and cost corresponding to the optimal crop pattern are presented in Table 8.11. Maximum fraction of cost is obtained thought wheat cultivation out of optimal crop pattern. As shown in Table 8.11, cucumber is the most beneficial crop, although its cultivation has considerable cost. The total benefit of optimal option is about (million U.S.$) 440, while the benefit of the existing condition is about (million U.S.$) 79.

Investigation of the optimal water allocation scheme in comparison with water demand variation during 11 years of the study period (Figure 8.30) shows that in certain months when the surface water is more than the water demand, the preference option in water allocation is surface water. However, in some months with low demand for avoiding the water table increasing beyond the desired level, the total allocated

TABLE 8.10
Components of Objective Function in Scenario A

Crops	Percent (%)	Benefit (Million Dollars)	Cost (Million Dollars)
Wheat	12	16.6	7.99
Barley	47	554	93.5
Cucumber	39	45.3	16.6
Tomato	2	2.53	1.36
Maize	0	0	0
Total	100	618.1	119.45

TABLE 8.11
Crop Pattern, Net Benefit, and Allocated Surface and Groundwater in Observed Condition and Optimal Results

	Actual Observed	Optimal Allocated Area
Crop pattern (%)		
Wheat	38	12
Barley	29	47
Cucumber	4.5	39
Tomato	6.5	2
Maize	22	0
Net benefit (million dollar)	79	440
Allocated surface water (MCM)	399	228
Allocated groundwater (MCM)	454	217

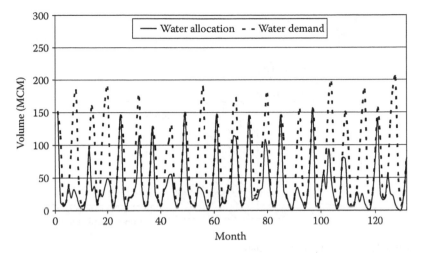

FIGURE 8.30 Comparison of water demand and allocation.

water exceeds demand. The reliability, resiliency, and vulnerability of supplying water demand are calculated as 50.0%, 36.4%, 24.1 MCM/month, respectively.

8.4 OPERATION OF GROUNDWATER RESOURCES IN SEMIARID REGION

In arid and semiarid regions, water demand is supplied by storing water in holes as the only water source for domestic use and small-scale irrigation. Even today, people in some regions use water holes in sand rivers. Coarse sand and gravel in sand rivers can store the water in the voids between the solids of sand and then it can be extracted from the holes.

However, flooding water is usually stored for a short period. In order to prevent pulling water downstream in a sand river, the underground dykes can be created. As it is shown in Figure 8.31, creating sand dams increase the volume of sand and water by forcing flash floods to deposit coarse sand.

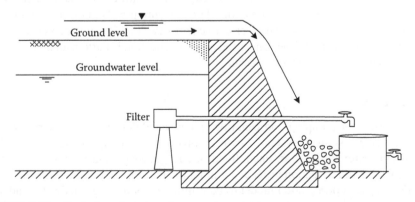

FIGURE 8.31 Typical sand-storage dam.

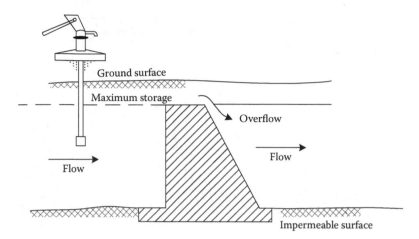

FIGURE 8.32   Typical subsurface dam.

Sand dams are one type of groundwater dam. It is a weir of suitable size constructed across the stream bed that leads to deposit the sand carried by heavy flows during the rain. Thus, a reservoir is made that is filled up with sand (Onder and Yilmaz, 2005; Wipplinger, 1958).

Another type of groundwater dam is subsurface dam, which is built underground to store underground water and to interrupt the natural flow of groundwater as shown in Figure 8.32. The underground reservoirs have certain advantages compared to surface reservoir as follows: (1) water is stored underground and hence the land above the underground dam can be utilized, (2) the evaporation losses are much less for water stored underground, (3) risk of contamination of the stored water from the surface is reduced because the ground layer can filter the contamination, and (4) disaster caused by the cutoff wall failure is prevented.

However, groundwater dam has some disadvantages: (1) construction of cut-off wall needs carefully controls and (2) the operation and maintenance of the intake facilities are expensive.

Groundwater dams are effective in parts of the world, which have a large variance of groundwater during the rainy and dry seasons with sufficient groundwater inflow to the underground area and the negligible contribution of landuse in groundwater contamination. It should have high effective porosity, sufficient thickness needs, and an impermeable bedrock layer under the aquifer. The dam is built in an underground valley to reduce the construction costs.

PROBLEMS

8.1 Groundwater pumping and canal diversion schedules are to be determined for the water resources system shown in Figure 8.33 (Longenbaugh, 1970). The groundwater basin is an unconfined aquifer; seepage and percolation of precipitation and irrigation water account for the majority of groundwater recharge. Develop an optimization model to determine the optimal groundwater and surface water allocation. Assume that the system objective is to minimize the total

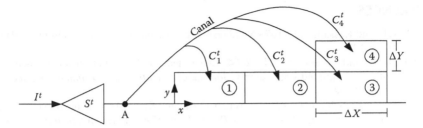

FIGURE 8.33 Water resources system in Problem 8.1.

TABLE 8.12
Recorded Discharge at a Stream Gage Station

Time	q (m³/sec)	Time	q (m³/sec)	Time	q (m³/sec)
1	580	6	2670	11	1930
2	531	7	3900	12	1240
3	705	8	4300	13	920
4	1012	9	3800	14	815
5	1700	10	2940	15	765

cost of meeting the forecasted agricultural water demand and that the stream can be represented by a constant head boundary condition.

8.2 The data presented in Table 8.12 are for a storm at a stream gage station. The area of the watershed is about 9000 km². Compute the direct runoff and base flow distribution and the volume of direct runoff for each of the following: (a) The constant-discharge method assuming direct runoff (b) the constant-slope method.

8.3 The daily discharge rates from recession curves for six storms are presented in Table 8.13. Estimate the parameters of fitting the depletion curve in this region.

TABLE 8.13
Recorded Discharge for Six Storms

Storm 1	Storm 2	Storm 3	Storm 4	Storm 5	Storm 6
2100	440	1310	250	1040	900
1890	410	1210	220	920	780
1710	375	1120	196	810	700
1570	355	1040	181	760	630
1460	330	975	170	680	580
1460	312	925	159	640	550
1400	298	880	150	610	525
1330	286	840	144	580	505
1210	273	800	138	560	485
1160		770	130	535	460
		740	126		440
					425

REFERENCES

Dorenbos, H. and Kassam, A.H. 1979. Yield response to water. *Irrigation and Drainage Paper, FAO*, 33, 193.

Ghahraman, B. and Sepaskhah, A.R. 2004. Linear and non-linear optimization models for allocation of a limited water supply. *Journal of Irrigation and Drainage Engineering*, 124(5), 138–149.

Karamouz, M., Kerachian, R., Zahraie, Ba., and Zahraie, Be. 2002. Conjunctive use of surface and groundwater resources, in *Proceedings of ASCE Environmental and Water Resources Institute Conference*, Roanoke, VA, May 19–22, 2002.

Karamouz, M., Hashemi Olya, R., Moridi, A., and Ahmadi, A. 2006. Water allocation from river system considering the interaction between River and aquifer. *ASCE Environmental and Water Resources Institute Conference*, Delhi, India, December, 2006.

Karamouz, M., Tabari, M.M., and Kerachian, R. 2007. Application of artificial neural networks and generic algorithms in conjunctive use of surface and groundwater resources. *Water International, IWRA*, 32(1), 163–167.

Karamouz, M., Zahraie, B., Kerachian, R., and Eslami, A. 2008. Cropping pattern and conjunctive use management: A case study. *Irrigation and Drainage*, 57, 1–13.

Karamouz, M., Ahmadi, A., and Nazif, S. 2009. Development of management schemes in irrigation planning: Economic parameters and crop pattern consideration. *International Journal of Science & Technology, Scientia Iranica*, 16(6), 457–466.

Longenbaugh, R.A. 1970. Determining optimum operational policies for conjunctive use of ground and surface water using linear programming. *ASCE, 18th Annual Specialty Conference*, Minneapolis, MN.

Onder, H. and Yilmaz, M. 2005. Underground dams, A tool of sustainable development and management of groundwater resources. *European Water*, 11/12, 35–45.

U.S. Army Corps of Engineers. 1999. *Engineering and Design, Groundwater Hydrology*. EM 1110-2-1421, Dept. of the Army, Washington, DC.

Willis, R. and Yeh, W.W.-G. 1987. *Groundwater Systems Planning and Management*. Prentice-Hall Inc., Englewood Cliffs, NJ, pp. 165–342.

Winter, T.C., Harvey, J.W., Franke, O.L., and Alley, W.M. 1998. *Ground Water and Surface Water: A Single Resource*, USGS (U.S. Geological Survey), 1139, Denver, Colorado, 79 p.

Wipplinger, O. 1958. *The Storage of Water in Sand, South West Africa Administration*. Windhoek, South West Africa, 107 p.

Yang Y.S., Kalin, R.M., Zhang, Y., Lin, X., and Zou, L. 2001. Multi-objective optimization for sustainable groundwater resource management in a semiarid catchment. *Hydrological Sciences Journal*, 46(1), 55–72.

9 Aquifer Restoration and Monitoring

9.1 INTRODUCTION

Pollution of groundwater can result from many activities, including leaching from municipal and chemical landfills, accidental leaking of chemicals or waste materials, and improper underground injection of liquid wastes from septic tank systems. On the other hand, population increase and failure in economically developing surface water supplies have increased the demand for groundwater usage. Until recently, the primary reason to deal with polluted aquifers was that once an aquifer becomes polluted, its water usage must be limited. However, due to the increase in water demand, this approach is not reasonable anymore. Therefore, the development of appropriate methodologies for aquifer cleanup and restoration has been considered as a new approach to meet the supplying needs.

When a contaminant is applied to a field, depending on the type of the contaminant, a portion of it can be adsorbed by layers of soil, a portion might be volatilized, a portion might be degraded by microorganisms, a portion might be taken by plants uptake, and the rest of that may reach the underlying aquifers. Therefore, due to the type of the contaminant applied to a field, the amount that reaches the underlying aquifer might be significantly less than the initial amount. But this diluted amount of contaminant can still cause quality problems for the underlying aquifer. Furthermore, some toxics and chemicals as well as some dissolved contaminants are not degraded or adsorbed and almost all of what is applied to the field will reach the aquifer.

Aquifers suspected to be polluted or the ones likely to become contaminated must be monitored. As soon as some concentrations of a pollutant are observed in the aquifer or unsaturated zone, restoration programs or reformatory actions may be set to protect the aquifer against contamination.

In this chapter, the partitioning process of the contaminants is explained to determine the amount of the applied pollutions to a field to reach the underlying aquifer followed by natural ways that may reduce the concentration of some contaminants are described. The potential ways for groundwater pollution control as well as groundwater treatment methods are also explained as the most important elements of groundwater restoration. The methods and processes of monitoring the quality status of water in unsaturated zone and groundwater are explained in the last part of this chapter.

9.2 PARTITIONING PROCESS

Since contaminants move through soil layers to reach groundwater, it is important to know how much of the pollutant is adsorbed to the soil particles and how much is in solution or can potentially be dissolved into pore water. In the dissolved phase, the pollutants can be transported to aquifers and surface waters.

The term partitioning is used to divide the total pollutant mass between the particulate (adsorbed on particulate matter) and dissolved fractions.

Adsorption of the contaminant into the particulate form immobilizes contaminants and makes them biologically unavailable in most cases. Dissolved or dissociated (ionized) contaminants are mobile with soil pore water. The master variables that affect the interactions between the soluble and particulate fractions of the contaminant are the pH, the anion–cation composition of the soil-pore-water solution, and the particulate and dissolved organic carbon (Salomons and Stol, 1995).

Hydrophilic or water-soluble compounds have higher bioavailability and leach more easily into groundwater. On the other hand, hydrophobic or water-insoluble compounds are immobile and they accumulate in soils.

The mobility of an organic chemical in soils is related to the octanol–water partitioning coefficient, K_{OW}. For nonionic chemicals and some other chemicals, the partitioning can be correlated to the octane partitioning coefficient. Rao and Davidson (1982) presented the values of K_{OW} for different pesticides.

For describing the proportion between the dissolved (ionized) and adsorbed (precipitated) particulate fractions, among several proposed mathematical formulations of adsorption equilibrium (isotherms), the Langmuir, Freundlich, and linear isotherms are most widely accepted.

The Languor adsorption model is expressed as

$$r = \frac{Q^0 b C_d}{1 + b C_d} \qquad (9.1)$$

where

r is the adsorbed concentration of the contaminant ($\mu g/g$)

C_d is the dissolved concentration of the contaminant in water or pore water of soil ($\mu g/L$)

Q^0 is the adsorption maximum at the fixed temperature ($\mu g/g$)

b is a constant related to the energy of net enthalpy of adsorption ($L/\mu g$)

The general form of Freundlich isotherm is

$$r = K C_d^{1/n} \qquad (9.2)$$

where K and n are constants.

If $b C_d$ in the Langmuir isotherm is much less than 1 or the exponent in the Freundlich equation is close to 1, the isotherm becomes linear. This isotherm is expressed as

$$r = \Pi C_d \qquad (9.3)$$

where $\Pi = b Q^0$ is the partition coefficient and has units of L/g if r is in $\mu g/g$ and C_d is in $\mu g/L$.

The total concentration of a pollutant in the soil is then a sum of the dissolved and adsorbed particulate concentrations:

$$C_T = \theta C_d + r\rho_{bd} \tag{9.4}$$

where
C_T is the total concentration of the pollutant in the soil (μg/L)
θ is the water content of the soil
ρ_{bd} is the bulk density of solids (g/L) and is calculated as $\rho_{bd} = 2.65\,(1-n)$

In this relation, 2.65 (g/cm³) is the density of soil particles and n is the porosity.

The relation of the dissolved (pore water) concentration of a chemical and the total concentration in the soil is obtained from the following equation:

$$C_d = \frac{1}{\theta + \Pi\rho_{bd}} C_T \tag{9.5}$$

Example 9.1

Phosphorus 9 is applied and distributed uniformly to a field through the first 25 cm of soil's depth at a rate of 105 kg/h. The soil's maximum adsorption of phosphorus is 310 μg/L. The porosity of the soil and the specific density of the soil sample are 45% and 1460 kg/m³ (=g/L), respectively. Estimate the approximate groundwater contamination by phosphorus during a long period of rain (the soil is saturated). Assume that constant b is equal to 0.0018 L/μg.

Solution

The total inorganic P content of the soil is

$$C_T = \frac{105\ \text{kg/ha} \times 10^9\ \mu\text{g/kg}}{0.25\ \text{m} \times 10^4\ \text{m}^2/\text{ha} \times 1{,}000\ \text{L/m}^3}\ 42{,}000\ \mu\text{g/L}$$

This concentration is distributed between adsorbed and dissolved fractions as

$$C_T = r\rho_{bd} + C_d\theta = \frac{Q^0 b C_d}{1 + b C_d}\rho_{bd} + C_d\theta$$

Since the soil is saturated, θ = porosity. Therefore:

$$42{,}000\ \mu\text{g/L} = \frac{310\ \mu\text{g/L} \times 0.0018\ \text{L/}\mu\text{g} \times C_d}{1 + 0.0018\ \text{L/}\mu\text{g} \times C_d} \times 1{,}460\ \text{g/L} + C_d \times 0.45$$

Solving this equation, the groundwater contamination by phosphorus from the application of this fertilizer is 56.8 μg/L.

9.3   RETARDATION

In subsurface zones, organic chemicals move more slowly than dissolved contaminants since they are retarded by adsorption. To incorporate adsorption to groundwater contaminant transport calculations, retardation factor is introduced. Retardation factor is an empirical term derived from partitioning coefficient and is the ratio of total chemicals to dissolved chemicals:

$$R = \frac{\text{Total chemicals}}{\text{Dissolved chemicals}} = \frac{\theta C_d + r\rho_{bd}}{\theta C_d} \tag{9.6}$$

where R is the retardation factor. In the above equation, r can be substituted by its corresponding value (ΠC_d) and we have

$$R = 1 + \frac{\rho_{bd}\Pi}{\theta} \tag{9.7}$$

The chemical solute moves only in dissolved phase and the adsorbed phase in not moving. Therefore, the ratio of the velocity of water to the velocity of the solute can also be determined by R:

$$V_{Solute} = \frac{V_{Water}}{R} \tag{9.8}$$

The downward velocity of water/or dissolved phase is equal to infiltration rate divided by the soil water content.

So far, we determined how fast a solute moves downward in the unsaturated zone. It is also important to determine how far the solute goes downward. Imagine a depth of water, d, is added to the soil at a constant water content, θ. The depth to which the water moves down, D, is computed as

$$D = \frac{d}{\theta} \tag{9.9}$$

If this water contains a chemical with the concentration C_d, and assuming this concentration remains constant, the chemical moves downward to the depth:

$$D' = \frac{D}{R} = \frac{d}{\theta R} \tag{9.10}$$

Example 9.2

A pesticide is applied to an agricultural field. The soil water content and bulk density are 0.30 and 1600 kg/m³, respectively. Assuming the partitioning coefficient is 1.25 L/kg, what fraction of the total chemical may reach the groundwater? How far can the dissolved portion go downward in a day? Imagine the water table depth is 1.2 m, how long does it take for the chemical to reach the water table? Assume the soil water content remains constant.

Solution

$$\rho_{bd} = 1600\,\text{kg/m}^3 = 1.6\,\text{kg/L}$$

$$R = 1 + \frac{\rho_{bd}\Pi}{\theta} = 1 + \frac{1.6\,\text{kg/L} \times 1.25\,\text{L/kg}}{0.30} = 7.67$$

$$\text{Dissolved fraction} = \frac{1}{R} = \frac{1}{7.67} = 0.13 = 13\%$$

The depth to which the chemical reaches:

$$D' = \frac{d}{\theta R} = \frac{1.2}{0.3 \times 7.67} = 0.52\,\text{cm}$$

$$t = \frac{180}{0.52} = 346\,\text{days}$$

Therefore, the chemical reaches the water table after 346 days.

9.4 NATURAL LOSSES OF CONTAMINANTS

Generally, not all of the contaminants applied to a field reach the underlying aquifer. There are different paths that the pollutants may be removed from soil before reaching aquifers. These paths include volatilization, biological degradation, and plant uptake, which are explained in the following section:

9.4.1 VOLATILIZATION

Volatilization is the loss of chemicals in vapor form from soil or water surfaces to the atmosphere. The potential of volatilization is mainly related to the saturated vapor pressure of the chemical in the air above the interface. It should be noted that volatilization does not occur from submerged aquatic sediments.

The relation between the vapor density and corresponding concentration of the chemical in water solution can be computed by Henry's law:

$$C_d = K_H C_g \tag{9.11}$$

where
K_H is Henry's constant for the chemical and is considered dimensionless
C_g is the vapor density of the chemical (μg/L)

To include the vapor phase of a chemical in the partitioning equation, Equation 9.11 is expanded as

$$C_d = \frac{1}{\theta + \Pi\rho_{bd} + (n-\theta)/K_H} C_T \tag{9.12}$$

9.4.2 Biological Degradation

The breakdown of a chemical by microorganisms to more simple compounds is interpreted as biological degradation. In this process, the chemical is ultimately transformed to carbon dioxide, water, methane, ammonium, and other simple by-products. Decomposition of the compound into mineral products is called mineralization. On the other hand, transformation is a change to another form, which may or may not be less environmentally hazardous. Chemicals that are not biodegradable are called persistent in the environment. The removal of persistent chemicals is only by transferring to another medium (from soil to water or air) or by burial. Some microorganisms or fungi can also cause biotransformation of chemicals. These organisms require energy and organic carbon source as well as nutrients for their growth.

In general, the degradation rate of chemicals can be calculated using he first-order reaction as the most accepted model:

$$C_t = C_0 e^{-k_d t} \qquad (9.13)$$

where
C_t and C_0 are the chemical masses at time t and 0, respectively
k_d is the overall degradation coefficient (day^{-1})
t is the time (day)

Given the values of C_t and C_0 from experiment, the value of k_d can be calculated using the above relation. Then the half-life of organic chemicals, which is the mean time required for 50% degradation of the original chemical in the soil, or their persistence in the soil is computed as

$$t_{1/2} = -\frac{\ln(0.5)}{k_d} = \frac{0.69}{k_d} \qquad (9.14)$$

9.4.3 Plant Uptake

Plants uptake nutrients and some chemicals and immobilize them. The nutrient and chemical uptake process is a part of the overall transpiration process of plants. Nutrients and chemicals are transported into plant tissues from the dissolved or ionic pool of chemicals in pore water. Therefore, chemical and nutrient adsorbed on soil particles or precipitated in soils are not removed from the system by plant uptake. These adsorbed pollutants must first be transferred into pore water by disruption or dissolution.

Example 9.3

A pesticide, atrazine, is applied to an agricultural field at a rate of 1.5 kg/ha. This field is irrigated at a rate of 2 cm/day and the crop water use is 0.45 cm/day. Assuming soil water content remains constant at 0.35, the partitioning coefficient is 0.97, soil bulk density is 1.5 g/cm³, and the water table depth is 2.5 m. Determine the load of dissolved atrazine that reaches to the water table. The decay factor for atrazine is 0.042 (day⁻¹).

Solution

First, the fraction of adsorbed and dissolved atrazine is calculated:

$$R = 1 + \frac{\rho_{bd}\Pi}{\theta} = 1 + \frac{1.5 \times 0.97}{0.35} = 5.16$$

$$\text{Dissolved fraction} = \frac{1}{R} = \frac{1}{5.16} = 0.194 = 19.4\%$$

$$\text{Adsorbed fraction} = 100 - 19.4 = 80.6\%$$

Out of 2 cm/day water that is applied to the field, 0.45 cm/day is used by the crop. Therefore, the depth to which water reaches in a day is

$$D = \frac{d}{\theta} = \frac{2 - 0.45}{0.35} = 4.4\,\text{cm}$$

The depth to which atrazine reaches in a day is

$$D' = \frac{D}{R} = \frac{4.4}{5.16} = 0.85\,\text{cm}$$

The time needed for atrazine to reach the water table at the depth of 2.5 m is

$$t = \frac{250\,\text{cm}}{0.85\,\text{cm}} = 294 \text{ days}$$

Using the first-order reaction equation and given the value of decay factor, the concentration of dissolved atrazine that will remain in water when it reaches the water table is computed:

$$C_t = C_0 e^{-kt} = 1.5\,\text{kg/ha} \times e^{-(0.042 \times 294)} = 6.5 \times 10^{-6}\,\text{kg/ha}$$

9.5 GROUNDWATER POLLUTION CONTROL

To be able to perform any aquifer restoration activities, it is important to have information about the best available technologies. In other words, what technology can be applied to prevent, abate, or clean up polluted groundwater? In this section, the information on institutional measures and source control strategies for preventing or minimizing the groundwater pollution is provided. The information involves physical control measures such as well and interceptor systems, surface capping and liners, and physical containment barriers such as sheet piling, grouting, and slurry walls.

9.5.1 INSTITUTIONAL TOOLS

The main purpose of using institutional tools is to protect groundwater resources. Institutional tools include regional groundwater management, land zoning, effluent charges/credits, guidelines, aquifer standards, and criteria.

9.5.1.1 Regional Ground Water Management

Regional management of ground water can be an effecting approach in prevention of long-term quality and quantity polluting activities. Public education and some regulatory enforcement are some possible functions of this approach.

Public education could include proper use of agricultural fertilizers, pesticides, herbicides, chemicals, waste disposal guidelines, and lists of reputable ground water professionals such as well drillers. Minimal enforcement capabilities could help prevent such polluting activities as overuse or abuse of chemicals, excessive groundwater withdrawal, etc.

9.5.1.2 Land Zoning

Land zoning is related to optimum site selection and disallows industrial or waste disposal activities in groundwater recharge areas.

9.5.1.3 Effluent Charges/Credits

Economic incentives are important tools to protect groundwater resources. On the other hand, by applying charges to polluting effluents applied to or disposed on the soil, the quality of effluent from certain facilities may be improved and in consequence, less pollution load is transferred to groundwater resources. Concurrently, credits can be applied to those facilities that move to improve the groundwater resources such as those employing artificial recharge measures.

9.5.1.4 Guidelines

Establishing guidelines or standards for construction and operation of groundwater polluting activities is one of the most effective tools to protect aquifers. Many states are now adopting standards for construction of waste disposal facilities, and this can be expanded to include oil well drilling, mining and minerals extraction, on-site sewage disposal, and other activities.

9.5.1.5 Aquifer Standards and Criteria

Other effective institutional tools to protect aquifers from pollution is establishment and application of groundwater quality standards and criteria for aquifers, and modification of current surface water beneficial use standards to be able to apply them for groundwater resources.

9.5.2 Source Control Strategies

The best solution to pollution is prevention (Novotny, 2003). Source control strategies represent attempts to minimize or prevent groundwater pollution before a potential polluting activity is initiated. These strategies can be considered as another institutional tool. By applying source control strategies, the volume of wastewater or the pollution load from an effluent can be reduced.

Source control strategies are applied to the design of new facilities as a component of one or more groundwater pollution prevention steps. However, some of the strategies can be used as pollution reduction tools for existing facilities.

Knox et al. (1986) summarized some common types of source control strategies that can be applied to prevent groundwater pollution at waste disposal facilities. These strategies are as follows:

1. Volume reduction strategies
   A. Recycling
   B. Resource recovery
      • Materials recovery
      • Waste-to-energy conversion
   C. Centrifugation
   D. Filtration
   E. Sand drying beds
2. Physical/chemical alteration strategies
   A. Chemical fixation
      • Neutralization
      • Precipitation
      • Chelation
      • Cementation
      • Oxidation–reduction
      • Biodegradation
   B. Detoxification
      • Thermal
      • Chemical—ion-exchange, pyrolysis, etc.
      • Biological—activated sludge, aerated lagoons, etc.
   C. Degradation
      • Hydrolysis
      • Dechlorination
      • Photolysis
      • Oxidation
   D. Encapsulation
   E. Waste segregation
   F. Co-disposal
   G. Leachate recirculation

The only criterion applicable to these strategies is that they all require extensive monitoring and recordation. The advantages of source control strategies include the following:

1. Enhancing groundwater quality by reducing the pollution load applied to the aquifer
2. Reducing the stabilization time of waste disposal facilities
3. Decreasing the cost of aquifer restoration and groundwater treatment

The disadvantages of source control strategies are the following:

1. It is hard to convince industries to apply these strategies because of the lack of public knowledge about the environmental hazards
2. Installation and maintenance costs
3. Monitoring and skilled operator requirements

9.5.3 STABILIZATION/SOLIDIFICATION STRATEGIES

By structurally isolating the waste material in a solid matrix prior to landfilling, the threat to groundwater from land disposal of waste materials can be reduced. This process, known as stabilization/solidification, is becoming significantly popular for hazardous and radioactive waste disposal. In this process, the waste in a solid matrix is chemically fixed. This reduces the exposed surface area and minimizes leaching of toxic constituents. In immobilization process, toxic components chemically react to form compounds immobile in the environment and/or the toxic material in an inert stable solid entrap. Thus, stabilization and solidification have different meanings, although the terms are often used interchangeably. From a definitional perspective, stabilization refers to immobilization by chemical reaction or entrapping while solidification means the production of a solid, monolithic mass with enough integrity to be easily transported. These processes may overlap or take place within one operation. An example is cementation where the process both stabilizes by producing insoluble heavy metal compounds, and solidifies into a formed mass while entrapping the pollutants.

When a material is chemically stabilized, a substance which is more resistant to leaching and also more amenable to the solidification process is provided. By chemically fixing the hazardous waste constituents, their release will be minimized in the event of a breakdown of the solid matrix. pH adjustment is the simplest stabilization process. In most industrial sludge, toxic metals are precipitated as amorphous hydroxides that are insoluble at an elevated pH. By carefully selecting stabilization system of suitable pH, the solubility of any metal hydroxide can be minimized. Certain metals can also be stabilized by forming insoluble carbonates or sulfides. It should still be taken into account that there is a probability for these metals to be remobilized because of changes in pH or redox conditions after they have been introduced into the environment.

By microencapsulation or macro-encapsulation, stabilized wastes can be solidified into a solid mass. Microencapsulation refers to the dispersion and chemical reaction of the toxic materials within a solid matrix. Therefore, any breakdown of the solid material only exposes the material located at the surface to potential release to the environment. Macro-encapsulation is the sealing of the waste in a thick, relatively impermeable coating layer. Plastic and asphalt coatings, or secured landfilling, are considered to be macro-encapsulation methods. Breakdown of the protective layer with macro-encapsulation could result in a significant release of toxic material to the environment.

By stabilization/solidification processes, a material is produced which does not pose a threat to the land on which it is disposed. The resultant stabilized/solidified material is impervious, having a good dimensional stability and load-bearing characteristics.

It should have satisfactory wet-dry and freeze-thaw weathering resistance. These properties plus optimum size and shape will make the material easily transportable.

Because of their chemical and toxic components, most of stabilized industrial wastes cannot be used on agricultural land; however, they might be suitable for building sub-foundations or as subgrade materials under roads or runways. There is no optimum stabilization/solidification process applicable to every type of hazardous waste. Each individual waste must be characterized and bench tests and pilot studies performed to determine the suitability of a given process.

Some common solidification/stabilization systems, suggested by Knox et al. (1986) are as follows:

1. Solidification through cement addition
2. Solidification through the addition of lime and other pozzolanic materials
3. Techniques involving embedding wastes in thermoplastic materials such as bitumen, paraffin, or polyethylene
4. Solidification by addition of an organic polymer
5. Encapsulation of wastes in an inactive coating
6. Treatment of the wastes to produce a cementitious product with major additions of other constituents
7. Formation of a glass by fusion of wastes with silica

9.5.4 Well Systems

For groundwater pollution control purposes, the subsurface hydraulic gradient is manipulated through injection or withdrawal of water. In well systems, the movement of water phase is controlled directly and concentration of pollutants in the subsurface water is controlled indirectly. This approach is often referred to as plume management. There are three main classes of well systems. These three classes are well point systems, deep well systems, and pressure ridge systems. There is injection of water in the latter one and water withdrawal in the other ones. All of these three systems require the installation of several wells at selected sites. These wells are then pumped (or injected) at specified rates so as to make the movement of the water phase and associated pollutants.

The use of well systems is the most common method of groundwater pollution control. Well systems represent the most assured means of controlling subsurface flows contaminants. These systems can be installed readily, can sometimes include recharge of aquifer, and have high design flexibility. In addition, construction costs for these systems can be lower than artificial barriers. However, operation and maintenance costs are high and well systems require continued monitoring after installation. Besides, clean water may also be removed along with polluted water in withdrawal systems and these systems may require surface treatment prior to discharge.

9.5.5 Interceptor Systems

In interceptor systems, a trench is excavated below the water table and a pipe is placed in the trench. The function of these drains or trenches is similar to an infinite line

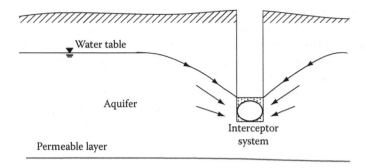

FIGURE 9.1   Hydraulic gradient toward interceptor system.

of extraction wells; that is, they affect a continuous zone of depression, which runs the length of the drainage trench (Kufs et al., 1983). Figure 9.1 shows an interceptor trench. Usually the trench receives perforated pipe and coarse backfill material for efficiency of collecting incoming water and transporting it. Interceptor systems are classified as preventive measures (leachate collection systems) or abatement measures (interceptor drains).

Interceptor systems provide a means of collecting leachate without the use of impervious liners and since flow to under-drains is by gravity, operation costs are relatively cheap. Design of under-drains is flexible and construction methods are simple. However, these systems are not recommended for poorly permeable soils and it is not feasible to situate under-drains beneath an existing site. In addition, interceptor systems require continuous and careful monitoring to assure adequate leachate collection. The following disadvantages can also be considered for interceptor systems:

1. When dissolved constituents are involved, it may be necessary to monitor groundwater down-gradient of the recovery line (Quince and Gardner, 1982).
2. Open systems require safety precautions to prevent fires or explosions.
3. Interceptor trenches are less efficient than well-point systems.
4. Not useful for deep disposal sites (U.S. Environmental Protection Agency, 1982).

9.5.6   SURFACE WATER CONTROL, CAPPING, AND LINERS

Surface water control, capping, and liners are three different technologies that are usually used in conjunction with each other. The three technologies serve different purposes in preventing groundwater pollution. Using surface water control tools, the amount of surface water flowing onto a site is minimized, which results in reducing the amount of potential infiltration. By capping of a site, the infiltration of any surface water or direct precipitation coming onto a site is reduced. Impermeable liners provide groundwater protection by inhibiting downward flow of leachate and other pollutants by adsorption processes. These technologies are preventive measures, although surface water control and capping have also been used successfully to abate groundwater pollution problems (Emrich et al., 1980). One of the approaches for placement of liners after a polluting activity has been initiated is the use of grouts. Grouting will be discussed in a subsequent section.

Surface water control measures are relatively inexpensive means of minimizing future groundwater pollution problems. Surface water flow in streams and channels is easily minimized by using structural techniques such as diversion dams and drainage ditches (Smith, 1983). The design of a site utilizing surface caps or impermeable liners, or both, is based on a water balance for the site.

There are four combinations of capping and liners that could be used (in conjunction with surface water control measures) at a waste site. Figure 9.2 depicts these combinations. In this figure, Q_C is the rate of water collected by the system and Q_R is the rate of water released from the system. Type (a) is called a natural attenuation site and neither of capping nor lining is utilized in this type. It relies entirely on the pollutant-attenuating capacity of the native soil. Type (b) utilizes an engineered liner (impermeable liner) to maximize the amount of leachate collected and minimize the amount escaping. The engineered cover (impermeable cap) sites are designed to minimize infiltration into the waste source by maximizing surface runoff and evapotranspiration (Type (c)). The sites that have the combination of engineered cover and liner (total containment) are designed to minimize infiltration and maximize the collected leachate (Type (d)).

Natural attenuation systems do not have leachate collection, transport, and treatment costs; but they require unique hydrogeologic setting and it is difficult to obtain regulatory acceptance in these systems. Engineered liner systems decrease hydrogeologic impact and stabilize waste quickly. However, the construction costs are higher and surface discharge is likely in these systems. Engineered cover systems decrease hydrogeologic impact after closure. The construction costs relative to liners are reduced in these systems but closure costs are increased. There is no leachate control during site operations and these systems need long-term monitoring. Engineered cover systems has lesser environmental impacts and they are more

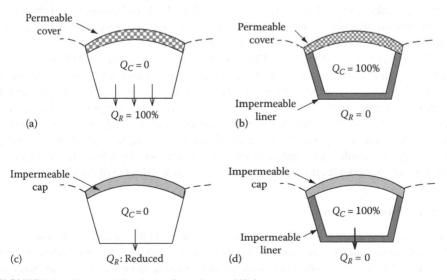

FIGURE 9.2 Four combinations of capping and lining systems.

acceptable socially and politically. They also minimize post-closure leachate collection, transport, and treatment costs. However, they have higher engineering and construction costs and need high-quality clay or synthetic material. In addition, these systems need more time for waste stabilization.

9.5.7 SHEET PILING

In sheet piling, a thin impermeable permanent barrier for flow is created by driving lengths of steel that connect together into the ground. Timber and concrete can also be used if the groundwater is not polluted. Pile driving requires a relatively uniform, loose, boulder-free soil for ease of construction. Steel sheet piles are also effective and economical for groundwater control. As the size of a project increases, application of sheet piling becomes uneconomical.

Construction of sheet piles can be economical and is not difficult, since excavation is not necessary. There is no maintenance required after construction. However, driving piles through ground containing boulders is difficult. In addition, for steel sheet piles, certain chemicals may attack the steel and the steel sheet piling is not initially watertight.

9.5.8 GROUTING

The grouting technology is used to increase soil-bearing capacity or to decrease the permeability of soils. This technology has been used in the construction industry for years. In the grouting process, a liquid, slurry, or emulsion is injected under pressure into the soil. This fluid occupies the available pore spaces. It then solidifies, resulting in a decrease in the original soil permeability and an increase in the soil-bearing capacity. Grouts are usually applied in two types: particulate and chemical. Particulate grouts consist of water plus particulate material that solidifies within the soil matrix. Chemical grouts consist of two or more liquids that forms a gel when they are combined together. Particulate grouts have higher viscosities than chemical grouts and are therefore better suited for larger pore spaces, whereas chemical grouts are better suited for smaller pore spaces (USEPA, 1998).

Grouts have been used for over 100 years in construction and soil stabilization projects. They can be used successfully for groundwater pollution control. There are many kinds of grouts available that suit a wide range of soil types. However, grouting is limited to granular types of soils that have a pore size large enough to accept grout fluids under a pressure which is small enough to prevent significant pollutant migration before implementation of grout program. Highly layered soil profiles are not suitable for grouting since they may result in incomplete formation of a grout envelope. In addition, the presence of high water table and rapidly flowing ground water limits groutability.

9.5.9 SLURRY WALLS

Slurry walls encapsulate an area to prevent groundwater pollution and restrict the movement of contaminated groundwater. In this method, a trench is first excavated around an area. This trench is then filled with an impermeable material. This technology is regarded as an effective method of isolating hazardous waste and preventing migration

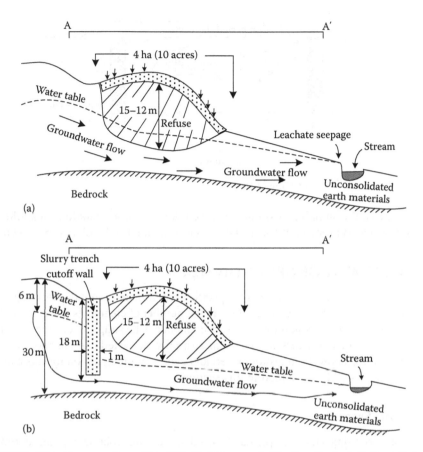

FIGURE 9.3 Cross section of landfill (a) before and (b) after the installation of slurry trench cutoff wall. (From Tolman, et al., *Guidance Manual for Minimizing Pollution from Waste Disposal Sites*, EPA-600/2-78-142, August, U.S. Environmental Protection Agency, Cincinnati, OH, 1978.)

of pollutants (Pearlman, 1999). Slurry walls can either be placed up-gradient from a waste site (Figure 9.3) to prevent flow of groundwater into the site, or placed around a site (Figure 9.4) to prevent movement of polluted groundwater away from a site.

Slurry walls have been applied to the Rocky Mountain Arsenal in Colorado and the Gilson Road Hazardous Waste Dump in New Hampshire (Knox et al., 1986).

The construction methods for slurry walls are simple and adjacent areas are not usually affected by ground water drawdown. Slurry wall have low maintenance requirements. They eliminate risks due to pump breakdowns or power failure. If bentonite is used in the slurry wall, it will not deteriorate with age. The most widely used technique for containment is the soil-bentonite slurry wall. It is typically the most economical, utilizes low-permeability backfill, and usually allows reuse of all or most of the material excavated during trenching (Kresic, 2008).

However, bentonite deteriorates when exposed to high ionic strength leachates. Since bentonite is available mostly in the west, in applying bentonite slurry walls

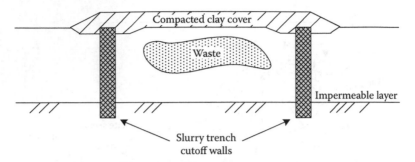

FIGURE 9.4   Isolation of existing buried waste.

in other areas, the shipping cost must also be taken into consideration. In addition, some construction procedures for slurry walls are patented and will require a license.

9.6   TREATMENT OF GROUNDWATER

The type of treatment required for site specific contaminated groundwaters depends on the contaminants being removed. Due to the number of chemicals that potentially exist in groundwater as contaminants; the treatment can be simple or extremely complex. Some treatment processes such as air stripping, activated carbon, and biological treatment for organics in groundwater, and chemical precipitation for inorganics in groundwater will be explained in this section.

9.6.1   Air Sparging

In situ air sparging (IAS) is primarily used to remove volatile organic carbons (VOCs) from the saturated subsurface. The removal process is done through stripping. The basic IAS system strips VOCs by injecting air into the saturated zone to promote contaminant partitioning from the liquid to the vapor phase. Figure 9.5

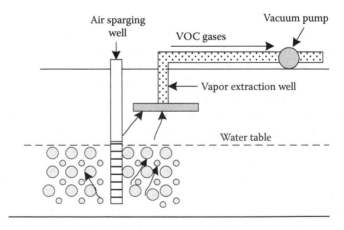

FIGURE 9.5   Typical in situ air sparging system.

depicts a typical IAS system. Since injected air, oxygen, can induce the activity of indigenous microbes, IAS can be effective in increasing the rate of natural aerobic biodegradation. In addition, by injecting a non-oxygenated gaseous carbon source, anaerobic environments can be created to provide denitrification conditions for nitrate removal from the groundwater. Enhanced degradation of organic compounds, such as chlorinated VOCs, to daughter products would result in increased volatility, which could improve the effectiveness of stripping and phase transfer during IAS (USACE, 2008).

IAS is relatively easy to implement and the equipment necessary for IAS is inexpensive and easily obtained. It is a well-known process to regulatory agencies. Therefore, IAS is one of the most practiced engineered technologies for in situ groundwater remediation (Kresic, 2008). However, the presence and distribution of preferential airflow pathways, the degree of groundwater mixing, and potential precipitation and clogging of the soil formation by inorganic compounds are important aspects to be taken into consideration in designing an IAS system.

In air stripping, a substance is transferred from the solution in water to solution in a gas, and then it is transferred to atmosphere. The rate of mass transfer can be calculated, using the following equation:

$$M = K_L A (C_L - C_g) \tag{9.15}$$

where
    M is the mass of substance transferred per unit time and volume (g/h/m³)
    K_L is the coefficient of mass transfer (m/h)
    A is the effective area (m²/m³)
    C_L is the concentration in liquid phase (g/m³)
    C_g is the concentration in gas phase (g/m³)
    $(C_L - C_g)$ is the driving force

The driving force is the difference between actual conditions in the air stripping unit and conditions associated with equilibrium between the gas and liquid phases. If equilibrium exists at the air–liquid interface, the liquid phase concentration is related to the gas phase concentration by Henry's law:

$$C_{g.eq} = K_H C_{l.eq} \tag{9.16}$$

where
    $C_{g.eq}$ is the equilibrium concentration in gas phase (g/m³)
    $C_{l.eq}$ is the equilibrium concentration in liquid phase (g/m³)
    K_H is the Henry's law constant

If a compound has a high Henry's law constant, it is stripped from water easier than the one with a lower Henry's law constant.

9.6.2 Carbon Adsorption

Organic molecule can be held on the surface of activated carbon by physical and/or chemical forces. Due to the balance between the forces that keep the compound in solution and the forces that attract the compound to the carbon surface, the quantity of a compound that can be adsorbed by activated carbon is determined. The following factors can affect this balance:

1. There is an inverse relation between adsorptivity and solubility.
2. The pH has a significant impact on the adsorptive capacity. For example, organic acids adsorb better under acidic conditions, while amino compounds prefer alkaline conditions.
3. Aromatic and halogenated compounds adsorb better than aliphatic compounds.
4. Adsorption capacity (but not the rate of adsorption) and temperature have an inverse relationship.

When activated carbon begins adsorbing organic chemicals, their concentration will decrease from an initial concentration, C_0, to an equilibrium concentration, C_e.

Using Freundlich and Langmuir isotherms, the approximate capacity of the carbon for the application can be estimated and a rough estimate of the carbon dosage required can be provided. Activated carbon adsorption is a successful way for removing organics from contaminated groundwater.

9.6.3 Adding Chemicals and In-Situ Chemical Oxidation

Adding chemicals to contaminated aquifers is one the effective ways for the removal of inorganic compounds. The carbonate system, the hydroxide system, and the sulfide system are three common basic types of chemical addition systems. Among these chemicals, the sulfide system removes the most inorganics, with the exception of arsenic (because of the low solubility of sulfide compounds). But the side effect is that sulfide sludge might be oxidized to sulfate when exposed to air, resulting in resolubilization of the metals. In the carbonate system, soda ash is used to adjust the pH to 8.2 and 8.5. This system is difficult to control in actual condition. The hydroxide system has almost the best performance for the removal of inorganics and metals. The system uses either lime (CaOH) or sodium hydroxide (NaOH) as the chemical to adjust the pH upwards.

In in-situ chemical oxidation (ISCO) method, groundwater or soil contaminants are transformed into less harmful chemical species by the introduction of a chemical oxidant into the subsurface. In the remediation of contaminated groundwater using ISCO, oxidants and potential amendments are directly injected into the source zone and down-gradient plume. The oxidant chemicals react with the contaminants and produce harmless substances such as carbon dioxide, water, etc. ISCO is a suitable method for removing BTEX, MTBE, TPH, chlorinated solvents (ethenes and ethanes), PAHs, PCBs, chlorinated benzenes, phenols, and organic pesticides. The most commonly used oxidants for remediation purposes are as follows:

- Permanganate (MnO_4^-)
- Hydrogen peroxide (H_2O_2) and iron (Fe) (Fenton-driven or H_2O_2-derived oxidation)
- Persulfate ($S_2O_8^{2-}$)
- Ozone (O_3)

The persistence of the oxidant in the subsurface is important, since this affects the contact time for advective and diffusive transport and ultimately the delivery of oxidant to targeted zones in the subsurface (Huling and Pivetz, 2006). For example, permanganate persists for long periods of time, but H_2O_2 can persist in soil and aquifer material for minutes to hours. Therefore, the diffusive and advective transport distances may be relatively limited for H_2O_2. The rates of oxidation reactions are dependent on many variables that must be considered simultaneously, including temperature, pH, concentration of the reactants, catalysts, reaction by-products, and system impurities such as natural organic matter (NOM) and oxidant scavengers (ITRC, 2005).

9.6.4 THERMAL TECHNOLOGIES

In situ thermal heating methods are generally used for enhanced oil recovery. These methods include hot water injection, steam injection, hot air injection, and electric resistive heating (ERH). Soil and groundwater zones contaminated by chlorinated solvents can be remediated by these methods. Volatile and semi-volatile organic contaminants (SVOCs), such as PCBs, PAHs, pesticides, and various fuels, oils, and lubricants can also be treated by in situ thermal technologies.

A thermal heating system typically consists of a series of injection and extraction wells. The injection wells are usually placed in clean areas around the source zone, if possible, to minimize the risk of contaminant spreading. Figure 9.6 expresses a schematic of thermal heating system. In this figure, hot water, steam, or hot air is injected into the subsurface to dissolve, vaporize, and mobilize contaminants that are then recovered. Then, mobilized contaminants are extracted from the subsurface

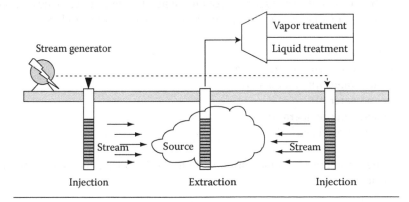

FIGURE 9.6 Schematic of thermal heating systems. (From USEPA, *In Situ Thermal Treatment of Chlorinated Solvent; Fundamentals and Field Applications*, EPA 542/R-04/010, Office of Solid Waste and Mergency Response, U.S. Environmental Protection Agency, Washington, DC, 2004.)

using vapor and liquid extraction equipment. Extracted vapors and liquids are then treated using conventional aboveground treatment technologies.

In all thermal technologies, the viscosity of nonaqueous phase liquids (NAPLs) are lowered and the vapor pressure and solubility of VOCs or SVOCs are increased. In consequence, the removal of NAPL is enhanced. For the removal and treatment of mobilized contaminants, a vapor extraction or liquid extraction system, in case of steam injection, must be integrated with other thermal technologies.

9.6.5 BIOREMEDIATION TECHNIQUES

In Section 9.4.2, biological treatment of contaminants was explained as a natural way for contaminant removal. Bioremediation can also be enhanced by adding some materials or microorganisms artificially to remove organic chemicals. In application of biological means for in situ treatment of organic chemicals in the saturated zone, growth and reproduction of microorganisms, which are capable of degrading the chemicals, are enhanced. Volatile organic carbons (VOCs) are the most frequently occurring type of contaminant in soil and groundwater at hazardous waste sites in the United States. In situ enhanced bioremediation of halogenated VOCs, and in particular of chlorinated aliphatic hydrocarbons (CAHs), is the most rapidly growing technology for groundwater remediation (Kresic, 2008). Bioremediation technologies have been developed and implemented to reduce the cost and time required to clean up those sites. These technologies are usually less expensive than the other available alternatives, they do not require waste extraction or excavation, and are more publicly acceptable since they rely on natural processes to treat contaminants (USEPA, 2000).

In bioremediation technology, the concentration of electron acceptors, electron donors, and nutrients in groundwater are adjusted to stimulate biodegradation of contaminants by indigenous microorganisms. An electron acceptor is a compound capable of accepting electrons during oxidation–reduction reactions. Microorganisms obtain energy by transferring electrons from electron donors to an electron acceptor. Bioremediation also includes the addition of microbes or exogenous microorganisms into the subsurface to degrade specific contaminants. In-site characteristics that control aquifer bioremediation are (Testa and Winegardner, 2000) as follows:

- Hydraulic conductivity
- Soil structure and stratification
- Groundwater mineral content
- Groundwater pH
- Groundwater temperature
- Microbial presence

In designing an aquifer bioremediation system, the following issues must be taken into account:

- Electron acceptor and nutrient requirements must be estimated.
- Location of injection, extraction, and monitoring wells must be selected carefully.

- The rate of electron acceptor and nutrient addition must be specified.
- The site must be equipped with failure control and alarm systems for safety and operational concerns.
- After setting up the system, it must be monitored at least weekly. In the monitoring process, some information such as groundwater electron acceptor concentration, nutrient concentration, pH, conductivity, and groundwater levels must be recorded. By monitoring groundwater and aquifer media samples, the remedial progress is controlled.

9.6.6 Dissolved Phase (Plume) Remediation

Many contaminated sites have multiple point sources of groundwater contamination. These sources may form individual plumes of individual contaminants, individual plumes of mixed contaminants from identifiable sources, or merged plumes of various contaminants from multiple sources, which might be identifiable. Developing an effective prescription to remediate these complex contaminated sites is sometimes impossible. Therefore, some strategies such as pump and treat or application of permeable reactive barriers might be selected to deal with these contaminated sites.

9.6.6.1 Pump and Treat

Pump and treat is the most widely used groundwater remediation technology today and is an important component of groundwater remediation efforts (USEPA, 2007). This technology is only criticized because of its inability to reduce contaminants to levels required by health-based standards in 5–10 years. However, since the complete aquifer restoration is an unrealistic goal for many contaminated sites, pump and treat systems can be considered as effective practices for aquifer remediation. It should still be taken into account that pump and treat systems, especially when combined with in situ restoration technologies, can also result in the full restoration at some sites with relatively simple characteristics.

Pump and treat systems are primarily used to (1) control the movement of contaminated groundwater and prevent the continued expansion of the contaminated zone (hydraulic containment) and (2) reduce the dissolved contaminant concentrations in groundwater in a degree that complies with cleanup standards (treatment) (USEPA, 1996).

Figure 9.7 shows the zone of hydraulic containment, capture zone, which is the three-dimensional region that contributes the groundwater extracted by a well. This zone is different from radius of well influence. The capture zone embraces the volume of porous media from which groundwater flows toward the well(s). This zone is eventually extracted by the well(s). The radius of well influence is the zone in which the hydraulic head is lowered in result of the well(s) operation. However, the groundwater flow direction in this zone is not reversed from the pre-pumping direction. In the well capture zone, enough monitoring wells must be installed to show the actual gradient reversal in all three dimensions.

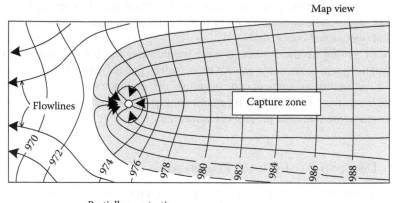

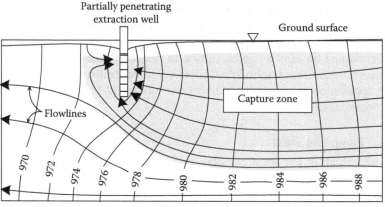

FIGURE 9.7 Illustration of capture zone in map view (top) and cross-sectional view (bottom) for a pump-and-treat system. (From USEPA, *A Systematic Approach for Evaluation of Captire Zones at Pump and Treat Systems: Final Project Report*, EPA/600/R-08/003, U.S. Environmental Protection Agency, Office of Research and Development, Washington, DC, p. 38.)

9.6.6.2   Permeable Reactive Barriers

Permeable reactive barriers are not barriers to the water; they are barriers to the contaminant. In this method, reactive materials are placed in the subsurface across the path of a plume of contaminated groundwater to create a passive treatment system. When the polluted groundwater passes through these materials, treated water comes out the other side. Figure 9.8 sketches a plume being treated by a permeable reactive barrier.

Permeable reactive barriers may be installed using trenches or by injecting reactive materials into the subsurface via boreholes. These barriers may contain metal-based catalysts for degrading volatile organics, nutrients, and oxygen for microorganisms to enhance biodegradation or other agents.

The barrier can be a precipitation wall or a sorption wall. The former reacts with the contaminants to form insoluble products, which may remain in the wall and the latter adsorbs contaminants to the wall surface.

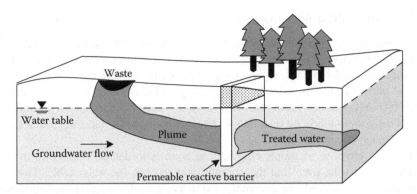

FIGURE 9.8 Plume being treated by a permeable reactive barrier. (From USEPA, *Evaluation of Subsurface Engineered Barriers at Waste Sites*, EPA 542-R-98005, Office of Solid Waste and Emergency Response, U.S. Environmental Protection Agency, Washington, DC, 1998.)

Granular iron has been the most widely used reactive media in full-scale barriers. This material abiotically degrades a variety of chlorinated solvent compounds such as perchloroethylene (PCE) and trichloroethylene (TCE) (Wilkin and Puls, 2003). The process induces substitution of chlorine atoms by hydrogen in the structure of chlorinated hydrocarbons.

9.7 MONITORING SUBSURFACE WATER

The purpose of water quality monitoring is to obtain quantitative information about the qualitative characteristics of water using statistical sampling. In other words, the purpose of water quality monitoring is to define the physical, chemical, and biological characteristics of water.

The growing concern for the quality of surface and groundwater has put significant demands on optimal design of monitoring networks and extraction of suitable information from collected data. Monitoring is one of the most important steps of water quality management since the process of decision making is based on available information related to water quality characteristics. Houlihan and Lucia (1999) categorized the purposes of water quality monitoring as follows:

- Detection monitoring programs that are used to detect an impact to surface and groundwater quality.
- Assessment monitoring programs that are used to assess the nature and extent of detected contaminants and to collect data that may be needed for remediation of contaminants.
- Corrective action monitoring programs that are used to assess the impact of remediation or pollution control program of contaminant concentrations.
- Performance monitoring programs that are used to evaluate the effectiveness of an element on remediation systems in meeting its design criteria.

9.7.1 VADOSE-ZONE MONITORING

So far it was understood that restoration of contaminated groundwater is so complex and costly. Therefore, it is always desirable to prevent groundwater against becoming contaminated. In this regard, it must be determined if groundwater contamination is likely to occur in the future so that reformatory action could be taken early enough to prevent contamination from reaching the water table. For example, a groundwater monitoring system might be installed to detect leaks from the facilities when a landfill or a lagoon is constructed.

The main purpose of monitoring vadose zone is to detect leaks from overlain facilities prior to the time that the contaminant reaches the water table. This way, reformatory actions could be accomplished to avoid contamination of groundwater. Examples of these reformatory actions are draining a lagoon and repairing the liner.

Contaminants in the vadose/unsaturated zone move in both the liquid and vapor phases. Depending on their relative density and air-pressure gradients, vapors may move in any direction. But liquids generally move only downward. Vadose-zone monitoring systems for liquids are usually installed beneath the facility to be monitored due to the direction the liquid would be seeping. It is difficult to install monitoring devices after constructing facilities. Hence, these devices are usually installed prior to the time that the facility is constructed. Since vapors can migrate laterally, gas-monitoring wells can be placed to the side of active facilities. Therefore, it is easier to install gas-monitoring devices than liquid-monitoring devices in the vamoose zone. Gas-monitoring wells can also be located beneath a landfill or lagoon if they are installed before construction. But after construction, they can be placed to the side of an active facility.

Gas monitoring wells terminate in the unsaturated zone. Figure 9.9 shows a typical gas monitoring well. These wells are usually used to detect the movement of methane from municipal landfills. Gas-monitoring wells can also detect the vapors of volatile organic chemicals in the vamoose zone. Most of the volatile organic chemicals are suspected carcinogens and they move rapidly in groundwater.

There are also some indirect means to monitor vadose zone. For example, by monitoring the changes in soil moisture in a lagoon that has aqueous liquid, it can be determined if there is a potential leak. If the lagoon holds a brine, changes in the electrical conductivity in the soil beneath the landfill could mean a leak. Temperature changes in the vadose zone could also indicate a leak.

The vadose zone can also be sampled by direct sampling of the soil. The moisture content as well as chemical composition of the moisture extracted from the sample is tested. Soil samples can also be leached with a standard leaching solution and the leachate is tested. If the soil sample is placed in a vial, a good indication of the organic in the soil can be obtained by testing the volatile organic in the headspace in the vial.

Generally the devices that are placed in the vadose zone to collect a sample of the liquid in the pores fall into two categories:

1. *Suction lysimeters*: these devices apply tension to a porous ceramic cup to induce the pore water that is under negative pressure to enter the porous cup and then it will be transported to the surface. The suction lysimeter

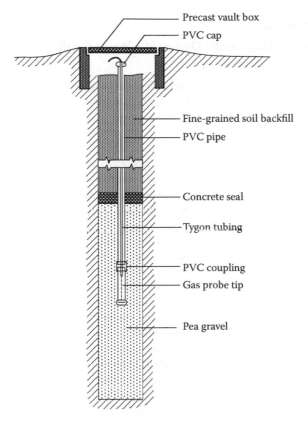

FIGURE 9.9 Gas monitoring well in vadose zone. (From Wilson, L.S., *Ground Water Monit. Rev.*, 3(1), 155, 1983.)

has been used for many years in agricultural applications and landfills to monitor for leachate. The vacuum that can be applied is limited to 0.5–0.8 bar. This limitation in the vacuum causes failure of collecting liquids in some applications because the lysimeter becomes clogged. Furthermore, if the suction lysimeter is not located under a leak, that leak may not be detected.

2. *Collection lysimeters*: These devices have a membrane liner buried in the vadose zone and the liquid is collected on the liner until the soil above it becomes saturated, forcing the liquid to flow to a wet well. The liquid is then sampled in this well. Collection lysimeters fail less than the other type. They can be made any size and can be placed under spots where failure of a system component is more likely.

9.7.2 GROUNDWATER MONITORING

Sanders et al. (1987) proposed general guidelines for the establishment of a monitoring system (Table 9.1). As shown in this table, network design is one of the five steps. Identifying the information to be obtained and statistical methods for converting the

TABLE 9.1

Major Steps in the Design of a Water Quality Monitoring System

Step 1	Evaluate information expectations
	Water quality concerns
	Information goals
	Monitoring objectives
Step 2	Establish statistical design criteria
	Development of hypotheses
	Selection of statistical methods
Step 3	Design monitoring network
	Where to sample
	What to measure
	How frequently to sample
Step 4	Develop operation plans and procedures
	Sampling procedures
	Laboratory analysis procedures
	Quality control procedures
	Data storage and retrieval
Step 5	Develop information reporting procedures
	Types and timing of reports
	Reporting formats
	Distribution of information
	Monitoring program evaluation
	(Are information expectations being met?)

Source: Sanders, T.G. et al., *Design of Networks for Monitoring Water Quality,* Water Resources Publications, Littleton, CO, 1987.

data to the useful information are considered in the first two steps and the third step deals with monitoring network design. Monitoring operation plans are specified in the forth step. In the fifth step, the information related to original monitoring objectives is determined.

9.7.3 WATER QUALITY VARIABLE SELECTION

Selection of the water quality variables is one of the most important steps in design and operation of water quality monitoring networks. The quality of water body is usually characterized by interrelated sets of physical, chemical, and biological variables. Therefore, the relationships between water quality and quantity variables are often complex. Many variables are needed to completely describe the quality of water body and it is economically unfeasible to measure all of them. Therefore, the variables should be ranked and a minimum number of water quality variables should be selected for sampling. The variables can be scored and ranked based on the multi-criterion decision-making methods. The variables should be selected in both time

TABLE 9.2

Hierarchical Ranking of Water Quantity, Water Quality Random Variables for Purposes of Information Procurement on Water Quality Processes

Primary, Basic Variables	Water Quantity Carrier Variables	First Level Variables
Quantity discharge , water level, small volume in a body of water		
Secondary Associated	Quality variables of aggregated effects	Second level variables
Examples	Temperature, pH, turbidity, BOD, DO, cations anions, conductivity, chlorides, radioactivity	
Aggregate-Producing Quality Variables	Quality variables that produce aggregated effects	Third level variables
Example1	Radioactivity-producing variables: strontium, cesium, tritium, etc.	
Example 2	Turbidity-producing variables: suspended matter, colloids, biota groups, dissolved minerals, etc.	
Or Species Producing Effects in Aggregation	Specific compounds most detailed classification of quality variables	Fourth level variables
Examples	Minerals affecting turbidity: iron oxides, manganese compounds, alumina, etc.	

Source: Sanders, T.G. et al., *Design of Networks for Monitoring Water Quality,* Water Resources Publications, Littleton, CO, 1987.

and space. One problem in economical selection of water variables is that the water quality variables are time–space stochastic processes that cannot be recorded continuously by an instrument at a site.

Establishing the relationship between water quality and quantity variables and analyzing the dependencies between water quality variables are necessary to select representative variables. These correlations can be used to produce information for variables that are not regularly monitored.

Sanders et al. (1987) proposed a hierarchical ranking of water quantity and quality variables that can be efficiently used in variable selection for monitoring network (Table 9.2). In the hierarchical classification of variables, the optimal cutoff point to select suitable variables is determined considering the objectives of monitoring, sample collection laboratory analysis costs, and correlation structure between variables.

9.7.4 Geostatistical Tools for Water Quality Monitoring Networks

Geostatistics is referred to the studies of a phenomenon that fluctuate in space and/or time. Geostatistics offers a collection of deterministic and statistical tools for modeling spatial and/or temporal variability. The basic concept of geostatistics is to determine any unknown value Z as a random variable Z, with a probability distribution, which models the uncertainty of Z. The random variable is a variable that can take a variety of values according to a probability distribution. One of the most famous methods of geostatistics is kriging that is used widely in the field of environment, mining, surveying, and water resources engineering studies. The ability of regionalizing spatiotemporal process has made these methods efficient tools for designing a water quality monitoring network. It suggests the most suitable points for locating monitoring stations, considering the Kriging estimation errors (estimation variances) for the regionalized components of the concentration of an indicator water quality variable.

9.7.4.1 Kriging

Kriging is a collection of generalized linear regression techniques for minimizing the estimation variance, defined from a prior model for covariance. It is of importance to have an area-wide distribution of drought information before applying a drought-control measure to the entire basin. Kriging could be used in its spatial form to determine the drought-affected areas in a region or for future installation of new gage stations in an ungaged area. Kriging methods may be used to study the temporal variation of a process. In this case a temporal domain, time variogram and covariance function could be used to predict variables of unobserved data. Regionalization data resulting from dynamic processes in space/time uses variogram and covariance functions, allowing prediction through space and time.

Kriging in the simplest form applies first-order analysis based on the following intrinsic hypotheses (Bogardi et al., 1985):

The expectation of $Z(x)$, a single realization of the parameter z, exists and is independent of the point x:

$$E[Z(x)] = m \tag{9.17}$$

Or if there is a known function $m(x)$ in which,

$$E[Z(x)] = m(x) \tag{9.18}$$

A variable such as $Y(x)$ could be used as

$$Y(x) = Z(x) - m(x) \tag{9.19}$$

and

$$E[Y(x)] = 0, \quad \text{for all } x \tag{9.20}$$

For all vectors h, the increment $[Z(x + h) - Z(x)]$ has a finite variance independent of x:

$$\text{Var}[Z(x+h) - Z(h)] = 2\gamma(h) \tag{9.21}$$

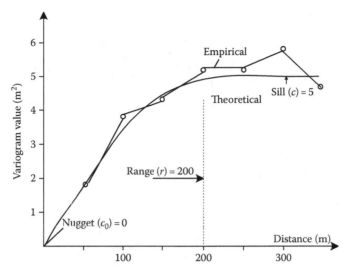

FIGURE 9.10   A pair of empirical and theoretical variograms.

The γ function is called the variogram. Variograms are obtained from the following three parameters:

1. "Nugget" effect, which can partly be explained by the measurement error (C_0)
2. "Sill," which is the total variance of the data beyond a certain distance or "range" (C)
3. "Range" or zone of influence (R) (see Figure 9.10)

The first step in the kriging analysis is to estimate a semi-variogram.

In all applications, the purpose is to analyze the correlation between neighboring measurements of a property and thereby derive the spatial variability of such a property. The functional description of this spatial structure, if it exists is called the semi-variogram in geostatistics. The computational form of semi-variogram is

$$\gamma(h) = \frac{1}{2n} \sum [(z(x_i) - z(x_i + h))^2] \tag{9.22}$$

where
  $\gamma(h)$ is the semi-variogram value for the distance h
  $z(x_i)$ is the measurement of location x_i
  $z(x_i + h)$ is the measurement at location $x_i + h$
  n is the number of measurement pairs separated by a vectorial distance, h

The empirical variogram could be replaced by a theoretical variogram model to be used in a kriging problem. A theoretical semi-variogram model consists of an isotropic nugget effect and any positive linear combination of the following standard semi-variogram models:

Spherical model:

$$\gamma(h) = c.Sph\left(\frac{h}{a}\right) = \begin{cases} c.\left[1.5\frac{h}{a} - 0.5\left(\frac{h}{a}\right)^3\right], & \text{if } (h \leq a) \\ c, & \text{if } (h \leq a) \end{cases} \qquad (9.23)$$

where
    a is an actual range
    c is the positive variance contribution or sill

Exponential model:

$$\gamma(h) = c.Exp\left(\frac{h}{a}\right) = c.\left[1 - \exp\left(-\frac{3h}{a}\right)\right] \qquad (9.24)$$

Gaussian model

$$\gamma(h) = c\left[1 - \exp\left(-\frac{(3h)^2}{a^2}\right)\right] \qquad (9.25)$$

Power model

$$\gamma(h) = ch^\omega \qquad (9.26)$$

where ω is a constant between 0–2. Hole effect model:

$$\gamma(h) = c\left[1.0 - \cos\left(\frac{h}{a}\pi\right)\right] \qquad (9.27)$$

Consider the estimate of an unknown (unsampled) value $z(u)$ from neighboring data values $z(u_a)$, $a = 1, ..., n$. The random function model $z(u)$ is stationary with mean m and covariance $C(h)$. In its simplest form, also known as simple kriging, the algorithm considers the following linear estimator:

$$Z_{SK}^*(u) = \sum_{\alpha=1}^{n} \lambda_\alpha(u)Z(u_\alpha) + \left(1 - \sum \lambda_\alpha(u)\right)m \qquad (9.28)$$

where
    Z_{SK}^* is the simple kriging estimator
    $\lambda_\alpha(u)$ are the weights to minimize the error variance, also called "estimation variance"

The minimization of the error variance in a set of normal equations results:

$$\sum_{\beta=1}^{n} \lambda_{\beta}(u)C(u_{\beta} - u_{\alpha}) = C(u - u_{\alpha}), \quad \forall \alpha = 1, ..., n \tag{9.29}$$

where
C is the stationary covariance
$C(0)$ is the stationary variance of $Z(u)$

The corresponding minimum estimated variance, or kriging variance, is

$$\sigma_{SK}^{2}(u) = C(0) - \sum_{\alpha=1}^{n} \lambda_{\alpha}(u)C(u - u_{\alpha}) \geq 0 \tag{9.30}$$

Ordinary kriging is the most commonly used method of kriging, whereby the sum of weights $\sum_{\alpha=1}^{n} \lambda_{\alpha}(u)$ is limited to 1. This allows building an estimator $Z_{OK}^{*}(u)$ that docs not require prior knowledge of the stationary mean m, and remains unbiased in the sense that

$$E\{Z_{OK}^{*}(u)\} = E\{Z(u)\} \tag{9.31}$$

There is a relationship between the weights λ and semi-variograms introduced earlier. This relationship could be represented in a matrix form as follows:

$$\begin{pmatrix} 0 & \gamma_{12} & \gamma_{13} & \cdots & \gamma_{1n} & 1 \\ \gamma_{21} & 0 & \gamma_{23} & \cdots & \gamma_{2n} & 1 \\ \vdots & \vdots & \vdots & \ddots & \vdots & \vdots \\ \gamma_{n1} & \gamma_{n2} & \gamma_{n3} & \cdots & 0 & 1 \\ 1 & 1 & 1 & \cdots & 1 & 0 \end{pmatrix} \begin{pmatrix} \lambda_{0}^{1} \\ \lambda_{0}^{2} \\ \vdots \\ \gamma_{0}^{n} \\ \mu \end{pmatrix} = \begin{pmatrix} \gamma_{10} \\ \gamma_{20} \\ \vdots \\ \gamma_{n0} \\ 1 \end{pmatrix} \tag{9.32}$$

where
λ_{0}^{i} represents the weight of station i
0 is for the origin station

In the following section, a water quality-monitoring network in a case study is presented.

9.7.5 Location of Water Quality Monitoring Stations

The location of a permanent monitoring station is very important because when the collected samples are not representative of water body, the other activities in the water quality monitoring system can be inconsequential. The factors that can affect the location of water quality monitoring station are different in ground and surface water resources because length is the dominant space coordinate in rivers, while depth and width have the same role in groundwater.

Groundwater monitoring objectives should be defined before monitoring because it can affect the needed procedures, techniques, and analyses. The main objectives of groundwater monitoring are as follows:

- Ambient trend monitoring
- Source monitoring

Ambient trend monitoring is used to discover the spatial and temporal trends of groundwater quality variables. The cost of ambient trend monitoring is highly related to the number of sampling stations and the sampling frequency.

Source monitoring determines the qualitative characteristics and the rate of pollutant movement. The flow rate can be determined using flow net. The pollutant's mass flow rate can be determined as

$$R_{total} = \sum_{i=1}^{n} q_i c_i \tag{9.33}$$

where
  R_{total} is the total mass flow rate
  n is the total number of sections
  q_i is the flow rate in section i
  c_i is the water quality variable concentration in section i

A suitable groundwater-monitoring network should provide the spatial and temporal characteristics of groundwater resources pollution from sources of pollution such as plumes.

Hudak and Loaiciga (1993) proposed a quantitative, analytical based hydrologic approach for monitoring network design (sampling location) in a multilayered groundwater flow system. In this method, which is described below, susceptibility to contamination of points in the model domain is quantified using weight values. This network design methodology consists of three steps as follows:

1. Domain definition and discretization
2. Determination of relative weights for candidate monitoring sites
3. Selection of an optimal configuration for monitoring network

9.7.5.1 Definition and Discretization of the Model Domain

The model domain includes the contaminant source and a surrounding area, which could be affected by pollutant sources and within which monitoring wells are selected. Zones that encompass the pathways of contaminant migration are defined by advection envelopes. Advection envelopes are extended from hydrogeologic outlets of pollutant sources based upon the known or inferred variation of hydraulic head (Figure 9.11). The two flow lines that originate from separate ends of hydrogeological

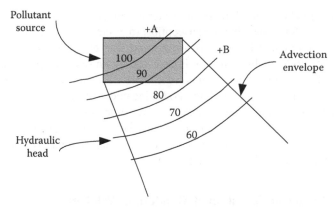

FIGURE 9.11 Advection envelope extended from hypothetical contaminant. (From Hudak, P.F. and Loaiciga, H.A., *Water Resour. Res.*, 29(8), 2835, 1993.)

outlet can be considered as boundaries of advection envelopes. Monitoring well sites can be classified to "background" and "detection" monitoring sites. Detection monitoring wells are constructed for early detection of contamination of pollutant source, but background monitoring sites determine the background water quality and are located in the up gradient zone of the model domain.

The background monitoring wells should not be located adjacent to contaminant source and advection envelope. They also should not be intersected by an outward normal extended from an advection envelope (Hudak and Loaiciga, 1993).

9.7.5.2 Calculation of Weights for Candidate Monitoring Sites

The relative weight of each monitoring site should be calculated based on the location of the site relative to the pollutant source and advection envelope. The proposed model can be used for designing a monitoring network for an uncontaminated aquifer (before pollutant is released from the pollutant source). Therefore, the existing contamination concentrations are not used in the monitoring wells configuration.

The relative weight of each candidate monitoring well in a multilayer aquifer is determined as (Hudak and Loaiciga, 1993)

$$W_{jk} = \frac{1}{D(s)_{jk}\, D(e)_{jk}} \tag{9.34}$$

where j is a real index of potential well site, k: HSI (hydrostratigraphic interval) index; a HSI is defined as a layer, which has relatively uniform hydraulic conductivity, $D(s)_{jk}$ is the horizontal distance from contaminant source boundary to the node j in $HSI\ k$, $D(e)_{jk}$ is the perpendicular distance from node j in $HSI\ k$ to closest boundary of advection envelope boundary (equal to $D(e)(\max)$ if node j cannot be intersected by a perpendicular from an advection envelope), $D(e)_{jk}(\max)$ is the maximum value of $D(e)_{jk}$ among nodes for which the quantity can be measured.

Based on the above equation, nodes located close to an area with higher probability of contamination have higher weights. To reduce the computational efforts, the boundaries of pollutant source and advection envelopes are approximated by nodes within a grid of candidate groundwater monitoring wells. The value of $D(e)_{jk}$ is calculated only for nodes that are susceptible to contamination from plum spreading, which are interested by an outward normal line extended from an advection envelope. Other nodes, which are not intersected by an outward perpendicular line from an advection envelope, are assigned the maximum value of $D(e)_{jk}$ calculated for all other nodes. On the other hand, the nodes located within the advection envelope are assigned the minimum value of $D(e)_{jk}$. For example in Figure 9.11, however nodes A and B have equal distance from advection envelope, node B has greater probability to be contaminated because node A is located in an upgradient area and should be assigned the maximum of $D(e)_{jk}$ $(D(e)(max))$.

9.7.5.3 Optimal Configuration of Monitoring Well Network

Optimal configuration of monitoring sites can be obtained using a mathematical model. The objective function of this model is preferential location of monitoring wells at points with high probability of contamination. The model formulation can be as follows (Hudak and Loaiciga, 1993):

$$Max\ Z = \sum_{k\varepsilon K}\sum_{j\varepsilon J_{k-uk}} W_{jk}x_{jk} - \sum_{k\varepsilon K}\sum_{j\varepsilon J_{uk}} W_{jk}x_{jk} \qquad (9.35)$$

subject to

$$\sum_{j\in J_k} x_{jk} \geq P_k\,(min), \quad \text{for each } k\varepsilon K \qquad (9.36)$$

$$\sum_{j\in J_{uk}} x_{jk} = P_{uk}, \quad \text{for each } k\varepsilon K \qquad (9.37)$$

$$\sum_{k\varepsilon K}\sum_{j\varepsilon J_k} x_{jk} = P \qquad (9.38)$$

$$x_{jk} = \begin{cases} 0, & \text{otherwise} \\ 1, & \text{if a well is installed at node } j \text{ in HSI } k \end{cases}, \quad \text{for each } j\varepsilon J_k, k\varepsilon K \qquad (9.39)$$

where
  j is the area l index of potential well site
  J_k is the set of potential well sites in HSI k
  J_{k-uk} is the set of potential well sites, excluding sites in upgradient zone, in HSI k
  J_{uk} is the set of potential well sites in upgradient zone in HSI k
  k is the HSI index
  K is the set of HSIs
  W_{jk} is the weight for node j in HSI k

$P_k(min)$ is the minimum number of wells to be located in HSI k

P_{uk} is the number of wells allocated to upgradient zone in HSI k

P is the total number of wells to be located

The second term of Equation 9.35 ensures that the nodes with lowest weight will be selected for background monitoring in upgradient zone. Equations 9.36 and 9.37 constraint the minimum and total number of wells in upgradient zone in each HSI. Constraint 9.38 ensures that all the wells are located throughout the model domain and constraint 9.39 shows that variable x_{jk} is a binary integer.

The total number of monitoring wells in each layer could be determined by budgetary constraints or regulatory requirements. Budget constraint can be written as

$$\sum_{k \varepsilon K} \sum_{j \varepsilon J_k} C_{jk} x_{jk} \leq R \qquad (9.40)$$

where

C_{jk} is the construction cost of a well at monitoring site j in HSI k

R is the total available budget

The proposed formulation can be solved using integer programming techniques. This model can define the location of the P monitoring wells including P_{uk} upgradient wells. For more details, the reader is referred to Hudak and Loaiciga (1993).

9.7.5.4 Kriging Method in Designing Optimal Monitoring Well Network

The following steps can be performed to find the optimal locations of a monitoring network (Karamouz, 2009):

First the water quality indicators are selected considering the characteristics of pollution loads and water uses as well as the existing spatial and temporal variations of the water quality variables of the system. Then, the spatial distribution of the existing water quality monitoring stations is investigated. After that, the data of the selected water quality variables are standardized and the time series are plotted in order to determine a pattern (monthly, seasonal, etc.) for their temporal variations.

Among the mathematical or statistical time series models, a model which best fits the time series of the water quality data is selected to represent the temporal trend of the concentration of the water quality variables. Then, the parameters of this model will be regionalized using the Kriging method and the experimental variograms of the regional parameters are plotted and the best theoretical variograms are fitted to them. A sample of these variograms is shown in Figure 9.12. For example, if the trend of the variables has sine and cosine cycles, the Fourier series is fitted to the series. Therefore, the trend component is determined as

$$Z(t) = a \cos\left(\frac{2\pi t}{T}\right) + b \sin\left(\frac{2\pi t}{T}\right) \qquad (9.41)$$

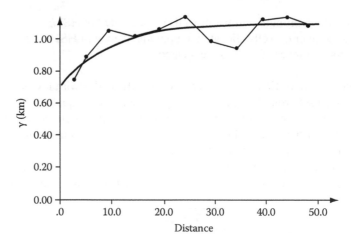

FIGURE 9.12 The experimental and theoretical variograms for the stochastic (residual) component of the COD time series at a water quality monitoring station in Karoon River. (From Karamouz, M. et al., *J. Environ. Model. Assess.*, 14(6), 705–714, 2009.)

In this equation, a and b are the coefficients of the Fourier series and T is the number of periods (e.g., 12 for monthly data). The regional parameters, which have to be regionalized using the Kriging analysis, will then be as follows:

- The cosine coefficient (a) of the trend component
- The sine coefficient (b) of the trend component
- The parameters of the theoretical variograms of the residuals (range and nugget effect)

After regionalizing the parameters, the Kriging analysis is done for the regional parameters using the theoretical variograms in order to calculate the values of estimations and estimation variances. Figure 9.13 depicts the typical results of the Kriging estimation and estimation variance for regionalized parameters as two dimensional color pixel plots.

Ultimately, a set of potential points for the monitoring stations is proposed considering the spatial variations of the Kriging estimation variances for the regionalized parameters, accessibility of stations, locations of the main point loads, locations of the main water withdrawals, locations of the main tributaries, locations of the development projects, and locations of existing stations and the length of their water quality and quantity data.

9.7.5.5 Genetic Algorithms in Designing Optimal Monitoring Well Network

To design the monitoring network using genetic algorithms (GA) optimization model, a water quality simulation model is usually linked to the optimization. The simulation model acts as a tool to check the performance and results obtained from the GA optimization model. As discussed in Chapter 7, GA is a method to identify the optimal solutions. This method can be used to evaluate if the existing monitoring wells is

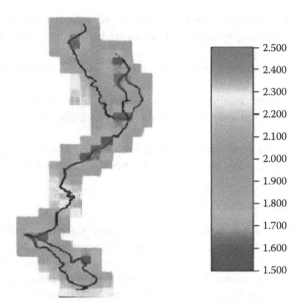

2.500
2.400
2.300
2.200
2.100
2.000
1.900
1.800
1.700
1.600
1.500

FIGURE 9.13 Two-dimensional kriging estimation variance plots for the cosine coefficient (a) in the deterministic (trend) component of the COD time series. (From Karamouz, M. et al., *J. Environ. Model. Assess.*, 14(6), 705–714, 2009.)

performing reasonably well. In this regard, the historical data from the existing monitoring wells are considered. Besides the existing monitoring wells, some data may also be collected from piezometric and operation wells, which are likely to be useful for the purpose of monitoring. These wells are selected based on the engineering judgment.

The simulation water quality model is calibrated based on the observed data. Then, the stations are determined using GA. The simulated water quality variables should be best fitted to the historical spatial and temporal variations of the selected water quality variables. Therefore, the main objective of the model is to minimize the absolute difference between the observed and the simulated concentration of the selected water quality variables. Karamouz et al. (2009) proposed the following objective function for an optimization model to design a river water quality monitoring network:

$$\text{Minimize: } Z = \sum_{p=1}^{n_p} \sum_{t=1}^{T} \sum_{s=1}^{n_s} w_p \frac{\left| C_{t,s,p} - C'_{t,s,p} \right|}{C_{t,s,p}} \tag{9.42}$$

where
p is the index of water quality variables selected for water quality monitoring
n_p is the total number of selected water quality variables
t is the time step (month)
T is the total number of time steps in the planning horizon
s is the river section
n_s is the total number of potential sampling stations

$C_{t,s,p}$ is the observed or simulated concentration of water quality variable p in river section s in time step t; simulated values are based on the real values of point load quality and quantity

$C'_{t,s,p}$ is the concentration of pth water quality variable in station s at time step t

It is obtained from water quality simulation model using the monthly average values of point loads quality and quantity, w_p is the relative weight of the pth water quality variables.

This objective function minimizes the difference between the observed and simulated values of the considered water quality variable concentration. It can be solved using GA.

9.7.6 SAMPLING FREQUENCY

Sampling frequency shows how often water quality samples should be gathered. Detection of stream standard violation and determination of temporal water quality variation are main criteria for sampling frequency analysis. The selection of sampling frequency requires some background information on behavior of the random variables, which are important in water quality. The analyst may face three situations:

- There is water quality data from the stream, which is under study.
- There is only water quality data from a stream in a region, which has similar characteristics to the study area.
- There is no existing data from the stream under consideration and similar streams.

When there is no data available, a two stage monitoring program should be utilized. In this case, design data, collected in the first stage, can be used for suitable design of a monitoring network and to estimate the sampling frequency. When the water quality data is available only for the similar rivers in the region, the water quality data should be generated for the stream, which is under study. In this case, the similarity between their geological and hydrologic characteristics, distribution of pollutant sources and pollutant characteristics are usually considered for water quality data generation.

9.7.6.1 Methods of Selecting Sampling Frequencies

Sampling frequency design can vary from single station, single water quality variable to multiple station, multiple water quality variable. Sanders et al. (1987) proposed statistical methods for each case that are presented in the next sections.

9.7.6.2 Single Station and Single Water Quality Variable

In this case, selected sampling frequency for a specified water quality variable at a station should provide the desired confidence interval around the annual mean. If the collected samples can be assumed independent, the variance of sample mean is given as

$$Var(\bar{x}) = \frac{\sigma^2}{n} \tag{9.43}$$

where
  σ^2 is the population variance
  n is the number of samples

The confidence interval for the population mean, μ, is as follows:

$$\bar{x} - Z_{\alpha/2}\left(\frac{\sigma^2}{n}\right)^{1/2} \le \mu \le \bar{x} + Z_{\alpha/2}\left(\frac{\sigma^2}{n}\right)^{1/2} \tag{9.44}$$

where $Z_{\alpha/2}$ is the standard normal deviate corresponding to the probability of $\alpha/2$. With rearranging terms, the number of samples required, n, to estimate the mean with a known confidence level, α, is as follows:

$$n \ge \left[\frac{Z_{\alpha/2}\sigma}{\mu - \bar{x}}\right]^2 \tag{9.45}$$

When sample standard deviation is used instead of σ, the $Z_{\alpha/2}$ is replaced with Student's t statistic, $t_{\alpha/2}$:

$$n \ge \left[\frac{t_{\alpha/2}s}{\mu - \bar{x}}\right]^2 \tag{9.46}$$

In this case, the computation of n becomes an iterative problem because when t distribution is used, the number of degrees of freedom, which is $n - 1$, must be known, but it is not known. Therefore, for solving the problem, an initial value for n is estimated and degrees of freedom are determined and a new estimation of n is obtained using Equation 9.46. By repeating this procedure, the values of n converge to a desired value. When the number of samples is large (more than 30), σ can be estimated with s, without any significant error in determining the sampling frequency.

Example 9.4

Select the sampling frequency for a monitoring well, which monitors BOD concentration. The annual mean and variance of BOD concentration based on historical data are 6.01 mg/L and 8.1 (mg/L)^2. The desired confidence interval width is assumed to be 3 mg/L with 95% confidence level.

Solution
The confidence interval width $= 2 \times 1.96 \times \dfrac{\sigma}{\sqrt{n}}$

$$2 \times 1.96 \times \frac{\sigma}{\sqrt{n}} = 3$$

$$n = \left[\frac{2\times1.96}{3}\right]^2 \sigma^2 = \left[\frac{2\times1.96}{3}\right]^2 \times 8.1$$

$$n = 13.8 \approx 14 \text{ samples/year}$$

9.7.6.3 Single Station and Multiple Water Quality Variables

If the problem of sampling frequency selection includes more than one water quality variable, a compromise programming should be used to obtain the most acceptable confidence interval considering all water quality variables.

One of the simple ways of doing this is to compute a weighted average of confidence interval widths for several water quality variables. The relative weights of water quality variables can be selected by engineering judgment so that the contributions of all water quality variables are comparable.

Example 9.5

Select sampling frequency for a monitoring well, which has four water quality variables. Assume that the average 95% confidence interval width about the mean will be equal to one fourth of average of means. The historical population statistics of variables are presented in Table 9.3.

Solution

The width of 95% confidence interval should be one fourth of average of the means. Thus

$$\frac{R_1 + R_2 + R_3 + R_4}{4} = \frac{1}{4}\left[\frac{\mu_1 + \mu_2 + \mu_3 + \mu_4}{4}\right]$$

where R_i is the width of confidence interval in the station i that can be computed using following equation:

$$R_i = 2(1.96)\frac{\sigma_i}{\sqrt{n}}, \quad i = 1, 2, 3, 4$$

$$\frac{1}{4}(\mu_1 + \mu_2 + \mu_3 + \mu_4) = \frac{2\times1.96}{\sqrt{n}}(\sigma_1 + \sigma_2 + \sigma_3 + \sigma_4)$$

$$n = \left[4\times2\times1.96\frac{(16.91 + 4.25 + 6.16 + 7.75)}{(41 + 10 + 40 + 34)}\right]^2 = 19.35 \approx 20 \text{ samples/year}$$

TABLE 9.3

Mean and Variance of Four Water Quality Variables

Variable	Mean (mg/L)	Variance (mg/L)^2
BOD	41	286
TN	10	18.1
Ca^{+2}	40	38
Mg^{+2}	34	60

9.7.6.4 Multiple Stations and Single Water Quality Variable

As the designer usually wants to obtain equal information from each sampling station of monitoring network, it can be a basis for selection of sampling frequency. If based on historical data, the means of concentrations in the stations are approximately equal, the sampling frequencies should be selected so that they provide equal confidence interval about means in different stations. In a simple procedure, which was proposed by Ward et al. (1979), total numbers of samples (N) are allocated to stations based upon their relative weights, using the following equation:

$$n_i = w_i N \qquad (9.47)$$

where
   n_i is the allocated samples to station i
   w_i is the relative weight of station i
   N is the total number of samples

The relative weights show the relative importance (priority) of stations and can be estimated based on historical data. For example, relative weights based on historical mean or historical variance can be computed using following equations:

$$w_i = \frac{\mu_i}{\sum_{i=1}^{N_s} \mu_i} \qquad (9.48)$$

$$w_i = \frac{\sigma_i^2}{\sum_{i=1}^{N_s} \sigma_i^2} \qquad (9.49)$$

where
   μ_i and σ_i^2 are historical mean and variance of water quality variable at station i
   N_s is the total number of stations in water quality monitoring network

9.7.6.5 Multiple Stations and Multiple Water Quality Variables

The design problem encountered most frequently in practice involves more than one water quality variable as well as more than one sampling station. In this case, a separate sampling frequency should be selected for each water quality variable at each station and a weighted average of them can be considered as station's sampling frequency. Sanders et al. (1987) expressed that a weighted average variance over the water quality variables can be computed for each station and used in Equation 9.49. In the case of multiple variables, it is not possible to obtain equal confidence interval widths for water quality variables in all station if all variables at a given station are sampled by the same frequency.

Example 9.6

The historical means and variances for three water quality variables in four stations have been calculated and shown in Table 9.4. Select the sampling frequency for each station using the weighting factors that are based on historical variances. The total number of samples per year is considered to be equal to 40.

Solution

The total number of samples should be allocated to each station for each water quality variable using weighting factors. The relative weight of each station for each variable is as follows:

$$w_i = \frac{\sigma_i^2}{\sum_{i=1}^{4} \sigma_i^2}, \quad i = 1, 2, 3, 4 \tag{9.50}$$

The weighting factors for different water quality variables based on the variances, have been presented in Table 9.5. Table 9.6 shows the average number of samples estimated for each station.

9.7.7 ENTROPY THEORY IN WATER QUALITY MONITORING NETWORK

Water quality monitoring systems are designed to obtain quantitative information about temporal and spatial distribution of water quality variables and thus on the physical, chemical, and biological characteristics of water resources. Due to the important role of rivers for supplying water demands and also the considerable

TABLE 9.4
Means and Variance for Three Water Quality Variables

	TN		BOD		TP	
Station	Mean (mg/L)	Variance	Mean (mg/L)	Variance	Mean (mg/L)	Variance
1	2.4	0.50	4.51	3.41	0.21	0.01
2	3.3	0.90	5.01	8.04	0.22	0.01
3	5.4	1.31	3.50	1.52	1.28	0.01
4	3.8	1.20	4.41	3.66	0.23	0.02

TABLE 9.5

Relative Weights of Stations in Example 9.6

Station	TN	BOD	TP
1	0.13	0.21	0.2
2	0.23	0.48	0.2
3	0.33	0.09	0.2
4	0.31	0.22	0.4

TABLE 9.6

Allocation of Samples via Proportional Sampling Based on Variances

Station Number	Number of Samples Allocated			Average (TN,BOD, and TP)
	TN	BOD	TP	
1	5	8	8	7
2	9	19	8	12
3	13	4	8	8
4	13	9	16	13

capital and operational costs of sampling and monitoring systems, the design of an optimal monitoring network should be considered optimal as well as economic constraints. The entropy theory could be used to eliminate the stations with redundant information in designing the water quality monitoring network.

9.7.7.1 Entropy Theory

The entropy (or information) theory, developed by Shannon (1948), recently has been applied in many different fields. The Entropy Theory has also been applied in hydrology and water resources for measuring the information content of random variables and models, evaluating information transfer between hydrological processes, evaluating data acquisition systems, and designing water quality monitoring networks.

There are four basic information measures based on entropy: marginal, joint, and conditional entropies and transinformation. Shannon and Weaver (1949) were the first to define the marginal entropy, $H(X)$, of a discrete random variable X as

$$H(x) = \sum_{i=1}^{n} p(x_i) \log p(x_i) \tag{9.51}$$

Here n represents the number of elementary events with probabilities $p(x_i)$ ($i = 1,\ldots, N$). The total entropy of two independent random variables X and Y is equal to the sum of their marginal entropies.

$$H(x,y) = H(x) + H(y) \tag{9.52}$$

where X and Y are stochastically dependent, their joint entropy is less than the total entropy of (2). Conditional entropy of X given Y represents the uncertainty remaining in X when Y is known, and vice versa.

$$H(x|y) = H(x,y) - H(y) \tag{9.53}$$

Transinformation is another entropy measure that measures the redundant or mutual information between X and Y. It is described as the difference between the total entropy and the joint entropy of dependent X and Y.

$$T(x,y) = H(x) + H(y) - H(x,y) \tag{9.54}$$

or

$$T(x,y) = H(x) - H(x|y) = H(y) - H(y|x) \tag{9.55}$$

The above expression can be extended to the multivariate case with M variables (Harmancioglu and Alpaslan, 1992). The total entropy of independent variables X_m ($m = 1, \ldots, M$) equals

$$H(X_1, X_2, \ldots, X_m) = \sum_{m=1}^{M} H(X_m) \tag{9.56}$$

If the variables are dependent, their joint entropy can be expressed as

$$H(X_1, X_2, \ldots, X_m) = H(X_1) + \sum_{m=2}^{M} H(X_m | X_1, X_2, \ldots, X_{m-1}) \tag{9.57}$$

It is sufficient to compute the joint entropy of the variables to estimate the conditional entropies of Equation 9.57 as the latter can be obtained as the difference between two joint entropies; for example

$$H(X_m | X_1, X_2, \ldots, X_{m-1}) = H(X_1, X_2, \ldots, X_m) - H(X_1, X_2, \ldots, X_{m-1}) \tag{9.58}$$

Finally, when the multivariate normal distribution is assumed for $f(X_1, X_2, \ldots, X_M)$, the joint entropy of X, with X being the vector of M variables, can be expressed as

$$H(x_1, x_2, \ldots, x_M) = \left(\frac{M}{2}\right) \ln 2\Pi + \left(\frac{1}{2}\right) \ln|C| + \frac{M}{2} - M \ln(\Delta x) \tag{9.59}$$

where

    $|C|$ is the determinant of the covariance matrix C

    Δx is the class interval size assumed to be the same for all M variables

Design of water quality monitoring networks is still a controversial issue, for there are difficulties in the selection of temporal and spatial sampling frequencies, the variables to be monitored, the sampling duration and the objectives of sampling (Harmancioglu et al., 1999). Entropy theory can be used in the optimal design of water quality monitoring systems. By adding new stations and gathering new information, the uncertainty (entropy) in the quality of the water is reduced. In other words, transinformation can show the redundant information obtained from a monitoring system which is due to spatial and temporal correlation among the values of water quality variables. Therefore, this index can be effectively used for optimal location of the monitoring stations as well as determining the sampling frequencies.

9.8 SUMMARY

Important concepts of aquifer restoration such as institutional tools, groundwater pollution prevention methods as well as groundwater treatment methods were reviewed in this chapter. Besides these methods, the chapter first explains how to calculate the portion of the total applied contaminant to a field that reaches the groundwater and the retardation rate, caused by adsorption. In addition to adsorption, there are some natural processes that may dilute contaminants before they reach an aquifer. The main natural processes causing dilution of contaminants, such as volatilization, biological degradation, and plant uptake were discussed in this chapter. On the other hand, since the process of decision making is based on available information related to water quality characteristics, unsaturated zone and groundwater monitoring processes were discussed as one of the most important steps of aquifer quality management.

PROBLEMS

9.1 What fraction of the total contaminant may reach the underlying unconfined aquifer if water table is at 1.8 m below the soil surface? Assume that the soil water content and bulk density are 0.35 and 1540 kg/m^3, respectively. The partitioning coefficient is 0.98.

9.2 In the previous problem, determine how far the dissolved portion can go downward in a day. In addition, compute how long it takes for the contaminant to reach groundwater? Assume that the irrigation rate is 2 cm/day and crop water use is 0.45 cm/day.

9.3 A chemical is applied to an agricultural field. Determine the load of the dissolved portion of the chemical that reaches groundwater.
- Application rate of the chemical: 3.2 kg/ha.
- Irrigation rate: 1.5 cm/day
- Crop water use: 0.38 cm/day
- Soil water content: 0.40
- Partitioning coefficient is 0.89

- Soil bulk density: $1600\,kg/m^3$
- Depth of water table 2.0 m.
- Decay factor of the chemical: 0.056 (day^{-1}).

9.4 The annual mean and variance of historical data of TDS concentrations in a water quality monitoring well are 750 mg/L and 238 (mg/L)^2, respectively. If the desired confidence interval width is 25 mg/L with 95% confidence level, what sampling frequency do you suggest for the well?

9.5 A fertilizer is applied and uniformly distributed to a field. The characteristics of the underlying soil and the fertilizer are presented below. What is the approximate groundwater contamination by the fertilizer if the soil is saturated?

- Application rate of the fertilizer: 140 kg/h
- Depth of application: 0.20 m
- Soil's maximum adsorption: 270 μg/L
- Soil porosity: 32%
- Specific density of soil: $1540\,kg/m^3$
- Constant b: 0.0022 L/μg

9.6 The concentrations of calcium and magnesium are measured at a monitoring well. Select the sampling frequency for this well based on the historical population statistics presented in Table 9.7.

9.7 Concentrations of total nitrogen and BOD are being measured in three monitoring wells. Using the weighting factors, based on historical variances, determine the sampling frequency for each well. Assume that in total 30 samples are taken per year. The historical means and variances for these water quality variables in the wells are presented in Table 9.8.

TABLE 9.7

The Concentrations of Calcium and Magnesium in Problem 9.6

Variable	Mean (mg/L)	Variance (mg/L)^2	95% Confidence Interval Width
Ca^{+2}	47	62	23
Mg^{+2}	41	35	22

TABLE 9.8

Means and Variances for Water Quality Variables in Problem 9.7

	TN		BOD	
Station	Mean (mg/L)	Variance	Mean (mg/L)	Variance
1	4.2	1.10	7.8	2.90
2	2.8	0.75	6.9	5.45
4	3.2	0.95	8.3	4.60

9.8 An herbicide is applied to a golf course at a rate of 1.5 kg/ha. This field is irrigated with the rate of 2.5 cm/day. If the partitioning coefficient and soil bulk density are 1.2 and 1.670 kg/m³, respectively, determine the fraction of the herbicide that is adsorbed to soil and the fraction that may reach the groundwater. Assume that the soil is at saturation.

9.9 In the previous problem, determine the amount of herbicide left behind when it reaches the water table at 2.5 m below the soil surface. Assume that the decay constant is 0.018. How long does it take for the herbicide to reach the water table? The crop water use is 0.42 cm/day.

REFERENCES

Bogardi, I., Bardossy, A., and Duckstein, L. 1985. Multicriterion network design using geostatistics. *Water Resources Research*, 21, 199–208.

Emrich, G.H., Beck, W.W., and Tolman, A.L. 1980. *Top-Sealing to Minimize Leachate Generation: Case Study of the Windham*, Connecticut Landfill, SMC-Martin, King of Prussia, PA.

Harmancioglu, N.B. and Alpaslan, N. 1992. Water quality monitoring network design: A problem of multi-objective decision making, *Water Resource Bulletin*, 28(1), 179–192.

Harmancioglu, N.B., Fistikoglu, O., Ozkul, S.D., Singh, V.P., and Alpaslan, M.N. 1999. *Water Quality Monitoring Network Design*. Kluwer Academic Publishers, Boston, MA.

Houlihan, M.F. and Lucia, P.C. 1999. *Groundwater Monitoring: The Handbook of Groundwater Engineering*, J.W. Delleur ed. CRC Press, Boca Raton, FL.

Hudak, P.F. and Loaiciga, H.A. 1993. An optimization method for monitoring network design in multilayered groundwater flow systems. *Water Resources Research*, 29(8), 2835–2845.

Huling, S.G. and Pivetz, B.E. 2006. *In-situ Chemical Oxidation*. Engineering Issue, EPA/600/R-06/072, U.S. Environmental Protection Agency, Office of Research and Development, National Risk Management Research Laboratory, Cincinnati, OH.

ITRC. 2005. *Technical and Regulatory Guidance for In Situ Chemical Oxidation of Contaminated Soil and Groundwater*. 2nd edn. In Situ Chemical Oxidation Team, Interstate Technology & Regulatory Council, Washington, DC.

Karamouz, M. 2004. *Assessment of Water Quality Management in the Province of Khuzestan in Iran, the World Bank*.

Karamouz, M., Kerachian, R., Akhbari, M., and Haafez, B. 2009. Optimal design of river water quality monitoring networks: A case study. *Journal of Environmental Modeling and Assessment*, 14(6), 705–714.

Knox, R.C., Counter, L.W., Kinconnon, D.F., Stover, E.L., and Ward, C.H. 1986. *Aquifer Restoration: State of the art*, Noyes Publication, Park Ridge, New Jersey.

Kresic, N. 2008. *Groundwater Resources: Sustainability, Management, and Restoration*, McGraw-Hill, New York.

Kufs, C., Wagner, K., Rogoshewski, P., Kaplan, M., and Repa, E. 1983. Controlling the migration of leachate plumes, in *Ninth Annual Research Symposium*, EPA-600/9-83-018, pp. 87–113.

Novotny, V. 2003. *Water Quality: Diffuse Pollution and Watershed Management*, John Wiley and Sons Publishing, NY.

Pearlman, L. 1999. *Subsurface Containment and Monitoring Systems: Barriers and Beyond (Overview Report)*. U.S. Environmental Protection Agency, Office of Solid Waste and Emergency Response, Technology Innovation Office, Washington, DC.

Quince, J.R. and Gardner, G.L. 1982. Recovery and treatment of contaminated ground water, in *The Second National Symposium on Aquifer Restoration and Ground Water Monitoring*, pp. 58–68.

Rao, P.S.C. and Davidson, J.M. 1982. Estimation of pesticide retention and transformation parameters required in nonpoint source pollution models, in *Environmental Impact of Nonpoint Source Pollution*. M.R. Overcash and J.M. Davidson eds., Ann Arbor Science, Ann Arbor, MI.

Salomons, W. and Stol, B. 1995. Soil pollution and its mitigation: Impact of land use changes on soil storage of pollutants, in *Nonpoint Pollution and Urban Stormwater Management*, ed. V. Novotny, Technomic Publishing, Lancaster, PA.

Sanders, T.G., Ward, R.C., Loftis, J.C., Steele, T.D., Ardin, D.D., and Yevjevich, V. 1987. *Design of Networks for Monitoring Water Quality*. Water Resources Publications, Littleton, CO.

Shannon, C.E. 1948. A mathematical theory of communication. *Bell System Technical Journal*, 27, 379–423.

Shannon, C.E. and Weaver, W. 1949. *The Mathematical Theory of Communication*. University of Illinois Press, Urbana, IL.

Smith, M.E. 1983. *Sanitary Landfills, Conference on Reclamation of Difficult Sites*. University of Wisconsin Extension, Madison, WI.

Testa, S.M. and Wingardner, D.L. 2000. *Restoration of Contaminated Aquifers*, 2nd edn., CRC Press/Lewis Publishers, Boca Raton, FL.

Tolman, A.L., Ballestero, A.P., Jr., Beck, W.W. Jr., and Emrich, G.H. 1978. *Guidance Manual for Minimizing Pollution from Waste Disposal Sites*. EPA-600/2-78-142, August, U.S. Environmental Protection Agency, Cincinnati, OH.

USACE. 2008. *Engineering and Design: In-situ Air Sparging*. Manual No. 1110-1-4005, Washington, DC. http://140.194.76.129/publications/eng-manuals/em 1110-1-4005/toc.htm

U.S. Environmental Protection Agency. 1979. *Survey of Solidification/Stabilization Technology for Hazardous Industrial Wastes*. EPA-600/2-79-056, US. EPA, Cincinnati, OH.

USEPA. 1989. *Risk Assessment Guidance for Superfund, Vol. I; Human Health Evaluation Manual (Part A): Interim Final*. EPA/540/1-89/002, Office of Emergency and Remedial Response, U.S. Environmental Protection Agency, Washington, DC.

USEPA. 1996. *Pump-and-Treat Ground-Water Remediation: A Guide for Decision Makers and Practitioners*. EPA/625/R-95/005, Office of Research and Development, U.S. Environmental Protection Agency, Washington, DC.

USEPA. 1998. *Evaluation of Subsurface Engineered Barriers at Waste Sites*. EPA 542-R-98005, Office of Solid Waste and Emergency Response, U.S. Environmental Protection Agency, Washington, DC.

USEPA. 2000. *Presenter's Manual for: Superfund Risk Assessment and How You Can Help: A 40-Minute Videotape*. EPA/540/R-99/013, Office of Solid Waste and Emergency Response, U.S. Environmental Protection Agency, Washington, DC.

USEPA. 2004. *In Situ Thermal Treatment of Chlorinated Solvent; Fundamentals and Field Applications*, EPA 542/R-04/010. Office of Solid Waste and Mergency Response, U.S. Environmental Protection Agency, Washington, DC, various pages.

USEPA. 2007. *Treatment technologies for Site Cleanup: Annual Status Report*, 12th edn. EPA-542-R-07-012, U.S. Environmental Protection Agency, Office of Solid Waste and Emergency Response, Washington, DC.

Ward, R.C., J.C. Loftis, K.S. Nielsen, and Anderson, R.D. 1979. Statistical evaluation of sampling frequencies in monitoring networks. *Journal Water Pollution Control Federation*, 51, 2292–2300.

Wilkin, R.T. and Puls, R.W. 2003. Capstone Report on the Application, Monitoring, and Performance of Permeable Reactive Barriers for Groundwater Remediation: Volume 1-Performance Evaluation at Two Sites. EPA/600/R-03/045a. U.S. Environmental Protection Agency, Washington, DC.

Wilson, L.S. 1983. Monitoring in the Vadose Zone: Part III, *Groundwater Monitoring Review*, 3(1), 155–166.

10 Groundwater Risk and Disaster Management

10.1 INTRODUCTION

Water is a renewable natural resource that cleanses and redistributes itself through natural cycles. However, its global quantity is finite. Increasing demands for water, the fact that there is no substitute for this essential resource, and decreasing water availability due to overuse, pollution, and inefficient water management lead to worldwide water disaster.

Besides the increasing demand for water, water shortages, water pollution, flood disasters, and aging water infrastructures are becoming more serious. Therefore, the need for disaster planning and management is growing fast. These problems are especially more significant in developing countries. The developed societies are also becoming more vulnerable since they are highly dependent on water. The rapid expansion of water sectors and inadequate institutional and infrastructural setups are worsening these problems.

Water-related problems grew significantly in the 1980s due to rapid urban and industrial growth. More concerns about water-related problems were created in the 1990s that grabbed the attention of researchers to global issues. Efforts by governments and international organizations have resulted in action plans being put into practice to remedy the problems. In this maze of complexity, groundwater is a target for both threats and hazards aimed at more widespread vulnerabilities, but it is more stable when the region is facing contained and noninvasive threats. Surface water resources and storage facilities have more vulnerability to direct structural threats.

In the previous chapters, water-related problems due to human activities and the strategies to decrease the negative influences of these activities on water resources were discussed. However, it should be noted that natural causes are the source of many disasters. The problem is more devastating in developing countries, where the social infrastructure is not sufficient. Devastating floods, droughts, and hurricanes have always caused regional disasters. Nonetheless, the complexity of societies and urban settings as well as means of transportation and communication has caused greater vulnerability for water disaster. Climate change also intensifies flood and drought impacts all around the world with adverse effect on water resources, supplies, and public water services and utilities.

In this chapter, risk as an overlay of hazard and vulnerability is discussed. Disasters are described as the realization of risk. Both natural and man-made hazards and vulnerability are changing as well as our adaptive capacity due to climate change and our more interconnected complex world. These are presented in this context in the following sections and in Chapter 11.

10.2 GROUNDWATER RISK ASSESSMENT

Risk assessment includes a wide variety of disciplines and the perspectives of engineers, hydrogeologists, environmental scientists, chemists, microbiologists, geologists, geographers, and public administrators. Furthermore, it includes the impact of groundwater pollution on human health. Some elements of risk analysis and assessment are briefly discussed in this section.

10.2.1 ELEMENTS OF RISK ASSESSMENT

There are two basic elements associated with risk assessment. These elements are risk quantification or estimation and risk evaluation (Lee and Nair, 1979). In risk quantification, the hazard is defined, and the initiating events that would cause the hazard are identified. Furthermore, the response and consequences to the receptor system is evaluated, and probabilities of occurrence are determined.

Figure 10.1 shows hypothetically the extent of probable hazards and vulnerabilities. Risk is the overlay and intersection of the two sets. Other components such as social and climate sensitivities can form another layer that should be considered for an assessment of risk. The adaptive capacity of a system such as a groundwater supply system or the entire aquifer to face different threats and hazards is directly related to the measures that can be taken to reduce future vulnerabilities. Figure 10.2 shows the need to consider the impact of social and climate change in assessing future vulnerabilities.

Risk assessment can be used as a tool in evaluating the potential impacts on public health of hazardous materials in remote or disposal facilities. See Masters and Ela (2008) for more details. As an example, McTernan and Kaplan (1990) assessed the potential impacts of a hypothetical release from a landfill/disposal site to groundwater and its subsequent impact on public health. In evaluating the potential exposure to contaminations present in groundwater, it is important to define the characteristics of the aquifer and the extent of physical space and chemical content to be evaluated. The U.S. Environmental Protection Agency's (EPA) groundwater strategy (USEPA, 1986) sets the guideline for assessing the risks associated with potential water supplies.

The National Research Council (1983), when dealing with health hazards, divided risk assessment into the following four steps: hazard identification, dose–response assessment, exposure assessment, and risk characterization. Risk assessment process steps are shown in Figure 10.3.

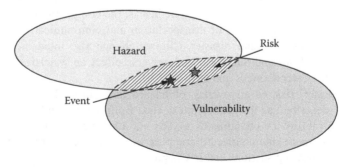

FIGURE 10.1   Risk as an overlay of hazard and vulnerability.

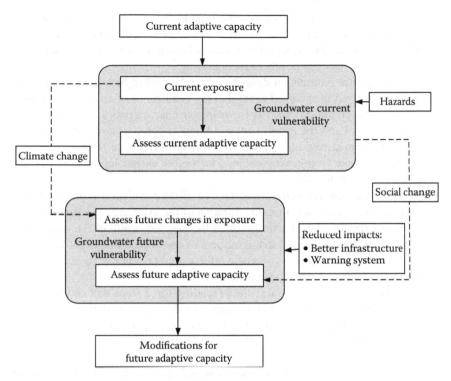

FIGURE 10.2 Main steps in groundwater vulnerability and the adaptive capacity and action approach.

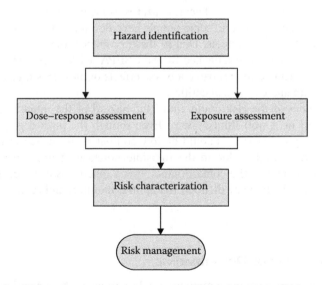

FIGURE 10.3 Risk assessment process steps. (From Masters, G.M. and Ela, W.P., *Introduction to Environmental Engineering and Science*, Prentice Hall Inc., Simon and Schuster, Englewood Cliffs, NJ, 576 p., 2008.)

- *Hazard identification* determines the linkage between a particular chemical to particular health effects, such as cancer or birth defects. As the availability of human data is so often scarce, this step focuses on the toxicity of a chemical in animals or other test organisms.
- *Dose–response assessment* describes the relationship between the dose of an administered or received agent and the occurrence of an adverse health effect. Some conditions such as the response characteristics of being carcinogenic or not and the experiment condition of being a one-time acute test or a long-term chronic test lead to a diversity of response relationships for any given agent. In this assessment, a method of extrapolating animal data to humans is included.
- *Exposure assessment* determines the size of the population exposed to the toxicant under consideration, the duration of the exposure, and the concentration of the toxicant. Factors such as the age and health of the subjected population, the likelihood that members of the population might be pregnant, and whether or not synergistic effects might occur due to exposure to multiple toxicants should be considered.
- *Risk characterization* contains the preceding three steps that estimate the magnitude of the public-health problem. In this step, the data collected in the previous steps are combined to generate a statement of risk. The final statement of risk can be qualitative or qualitative.

One of the most controversial and sensitive issues dealing with groundwater contamination is the hazard of gradual exposure to the fumes and vapors released through the unsaturated zone. Volatile hydrocarbons have a tendency to evaporate from the surface of a water table. In many industrial zones or landfills converted to residential areas, occupants have experienced health hazards, some with carcinogenic potential. As a result of gradual exposure, some people have developed reproductive damage, neurological, and/or birth defects. Due to these problems, the development of risk assessment criteria has been launched by the USEPA. Risk assessment is a process to measure and estimate the probable adverse effects of past, existing, and/or future hazards to human and ecological entities.

Another aspect of the risk could entail the hazard of the intentional contamination of a well or a well fluid system. Even though the hazard of groundwater contamination by terrorist acts could be low compared to surface water, the vulnerability could be higher due to the invisible notation of groundwater and the widespread movement of the contaminants. It is a much slower process, but, once contaminated, it will be very difficult to capture, monitor, and remediate the long lasting effects.

10.2.2 Assessment of Dose–Response

In dose–response assessment, a mathematical relationship between the toxic materials that humans are subject to and the risk as an unhealthy response to that dose is obtained. In dose–response curves, the severe toxicity amount is measured in

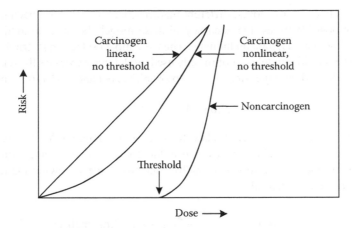

FIGURE 10.4 Hypothetical dose–response curves.

milligrams per kilogram of body weight. These curves are obtained due to the chronic toxicity that explains the organism's prolonged exposure over a substantial fraction of its life. The horizontal axis in these curves is a dose with the unit of the average milligrams of content per kilogram of body weight per day (mg/kg-day). The average exposure of a dose is over a whole lifetime. For instance, the human exposure time could be 70 years, the life expectancy period. The vertical axis is the response (risk) that is the probability of the occurrence of adverse health effects.

For a carcinogenic response, there is always a routine practice to assume that exposure to any amount of the carcinogen will cause the probable occurrence of cancer. For noncarcinogenic responses, a threshold dose is assumed below which there will be no response. From these two assumptions and as is shown in Figure 10.4, it can be concluded that the application of the dose–response curves and the methods for carcinogenic and noncarcinogenic effects are different. Dose–response curves for carcinogens are assumed to have no threshold; that is, any exposure produces some chance of causing cancer. It should be noted that the same chemical may cause both kinds of response.

A scaling factor is presented through which the dose–response data obtained from animal bioassays can be extrapolated for humans as well. Sometimes, if the dose amounts for human and animal per unit of body weight are the same, the scaling factor will be assumed equal. The resulting human dose–response curve is described by the standard mg/kg-day units for dose. Also, for differences in the rates of chemical absorption, animal and human responses have to be adjusted. In most cases, due to insufficient data, the absorption rate is assumed to be same.

10.2.3 RISK ASSESSMENT METHODS AND TOOLS

The method of extrapolating from the high doses to the low doses is the most debatable aspect of dose–response curves for carcinogens. The high doses are implemented to test animals, and, for human exposure, the low dose is considered.

To extrapolate the low doses, different mathematical models have been proposed, but it is not possible to test the accuracy of these models because no model can be validated without actual data on humans. Then, the model selection is strictly a policy decision. The one-hit model is a commonly used model that expresses the relationship between the dose (d) and the lifetime risk (probability) of cancer, $P(d)$ (Crump, 1984).

$$P(d) = 1 - e^{-(q_0 + q_1 d)} \tag{10.1}$$

where q_0 and q_1 are parameters picked to fit the data. This model is based on the assumption that a single chemical exposure (hit) could produce a cancer tumor.

By substituting $d = 0$ into Equation 10.1 and using the following mathematical expansion for an exponential,

$$e^x = 1 + x + \frac{x^2}{2!} + \cdots + \frac{x^n}{n!} \cong 1 + x \quad \text{(for small } x) \tag{10.2}$$

P can be obtained assuming that the background cancer causing rate is small:

$$P(0) = 1 - e^{q_0} \cong 1 - [1 + (-q_0)] = q_0 \tag{10.3}$$

Using the exponential expansion again, the lifetime probability of cancer for small dose rates can be expressed as

$$P(d) \cong 1 - [1 - (q_0 + q_{1d})] = q_0 + q_1 d = P(0) + q_1 d \tag{10.4}$$

For low doses, the added risk of cancer is

$$\text{Added risk} = A(d) = P(d) - P(0) = q_1 d \tag{10.5}$$

In order to protect public health, the EPA chooses a modified multistage model. It is a linearized model at low doses with the constant of proportionality picked to result in 95% probability of overestimating the risk. See Masters and Ela (2008) for more details.

10.2.3.1 Potency Factor for Cancer-Causing Substances

To study chronic toxicity, a low dose is implemented over a primary part of one's lifetime. At low doses, where the dose–response curve is linear, the curve slope is referred to as the potency factor (PF):

$$\text{PF} = \frac{\text{Risk}}{\text{CDI}} \tag{10.6}$$

where
    Risk is the incremental lifetime risk
    CDI is the average milligrams of toxicant absorbed per kilogram of body weight
        per day in mg/kg-day

Risk is expressed as probability and has no units; therefore, the potency factor unit is the inverse of the CDI unit.

By knowing the chronic daily intake CDI (based on exposure data) PF (from Table 10.1), the lifetime, incremental cancer risk can be found as follows:

$$Risk = CDI \times PF \qquad (10.7)$$

According to the USEPA, this value should be less than 10^{-6} or one person in a million. Potency factors in Equation 10.7 are obtained from an EPA database on toxic substances referred to as the Integrated Risk Information System (IRIS). A short list of the PF for some of these chemicals (for both oral and inhalation exposure) is shown in Table 10.1.

TABLE 10.1

Toxicity Data for Selected Potential Carcinogens

Chemical	Category	Potency Factor Oral Route (mg/kg-day)$^{-1}$	Potency Factor Inhalation Route (mg/kg-day)$^{-1}$
Arsenic	A	1.75	50
Benzene	A	2.9×10^{-2}	2.9×10^{-2}
Benzol(a)pyrene	B2	11.5	6.11
Cadmium	B1	—	6.1
Carbon tetrachloride	B2	0.13	—
Chloroform	B2	6.1×10^{-3}	8.1×10^{-2}
Chromium VI	A	—	41
DDT	B2	0.34	—
1,1-Dichlorocthylene	C	0.58	1.16
Dieldrin	B2	30	—
Heptachlor	B2	3.4	—
Hexachloroethane	C	1.4×10^{-2}	—
Methylene chloride	B2	7.5×10^{-3}	1.4×10^{-2}
Nickel and compounds	A	—	1.19
Polychlorinated biphenyls (PCBs)	B2	7.7	—
2,3,7,8-TCDD (dioxin)	B2	1.56×10^{5}	—
Tetrachloroethylene	B2	5.1×10^{-2}	$1.0–3.3 \times 10^{-3}$
1,1,1-Trichloroethane (1,1,1-TCA)	D	—	—
Trichlorocthylene (TCE)	B2	1.1×10^{-2}	1.3×10^{-2}
Vinyl chloride	A	2.3	0.295

To calculate CDI, the following expression is used:

$$\text{CDI (mg/kg-day)} = \frac{\text{ADD}}{W} \qquad (10.8)$$

where
ADD is the average daily dose in (mg/day) over an assumed 70 year lifetime
W is the body weight in kg

Example 10.1

An underground storage tank has been leaking benzene to the groundwater of a city that only makes basic treatment. A 70 kg person is exposed in different ways including drinking 1.8 L of water with a benzene concentration of 0.007 mg/L every day for 7 years. Assume the standard 70 year lifetime. Find the upper-bound cancer risk for this individual.

Solution

From Table 10.1, we see that benzene is a probable human carcinogen with a potency factor of 2.9×10^{-2} (mg/kg-day)$^{-1}$ for both oral and inhalation exposure. Using Equation 10.8, the chronic daily intake is

$$\text{CDI (mg/kg-day)} = \frac{\text{ADD}}{W}$$

$$= \frac{0.007 \text{ mg/L} \times 1.8 \text{ L/day} \times 7 \text{ year}}{70 \text{ kg} \times 70 \text{ year}} = 0.000018 \text{ mg/kg-day}$$

Using Equation 10.7, the incremental lifetime cancer risk is

$$\text{Risk} = \text{CDI} \times \text{PF}$$

$$= 0.000018 \text{ mg/kg-day} \times 2.9 \times 10^{-2} \text{ (mg/kg-day)}^{-1} = 0.52 \times 10^{-6}$$

So, over a 70 year period, the upper-bound estimate of the probability that a person will be affected by adverse health from this drinking water is about 0.52 in 1 million, which is less than the EPA acceptable limit of 10^{-6}.

An event tree is another tool that provides a structure for risk quantification. Event trees provide a structured approach for risk quantification, as shown in Figure 10.5. In this figure, assume that the initiating event is the placement of hazardous materials at a site with two liners. Systems 1 and 2 represent the primary (top) liner and the

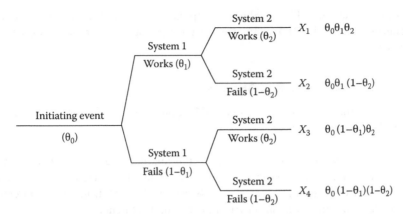

FIGURE 10.5 Risk quantification using an event tree. (From Lee, W.W. and Nair, K., Risk quantification and risk evaluation, in *Proceedings of National Conference on Hazardous Material Risk Assessment, Disposal and Management*, Information Transfer, Inc., Silver Spring, MD, pp. 44–48, 1979.)

secondary (bottom) liner, respectively. If System 1 fails and System 2 works, there is no damage. But, if both systems fail, there is the possible release of the hazardous materials to the subsurface environment. Figure 10.5 shows how the probabilities of each branch of the event tree can be calculated.

Figure 10.6 shows an actual event tree provided by Lee and Nair (1979) for a risk assessment of liquefied natural gas (LNG) spills in a coastal harbor. Decision trees can also be developed as a part of a risk quantification process.

In risk evaluation, socially acceptable risks are determined. One approach for risk evaluation is to compare the quantified risk with various possible existing

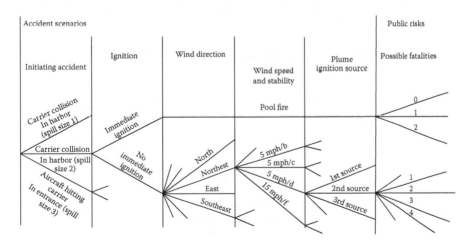

FIGURE 10.6 Event tree analysis of LNG spills. (From Lee, W.W. and Nair, K., Risk quantification and risk evaluation, in *Proceedings of National Conference on Hazardous Material Risk Assessment, Disposal and Management*, Information Transfer, Inc., Silver Spring, MD, pp. 44–48, 1979.)

risks in everyday life. This evaluation approach is called the "revealed societal preferences" approach. Risk evaluation is a difficult and controversial task in risk assessment.

10.2.4 ENVIRONMENTAL RISK ANALYSIS

There are four major approaches suggested by Conway (1982) for environmental risk analysis. These approaches are

1. The stochastic/statistical approach: In this approach, large amounts of data are obtained under a variety of conditions. Then, the correlations between the input of a certain material and its observed concentrations and effects are determined in various environmental compartments.
2. The model ecosystem approach: In this approach, which is also called the "microcosm" approach, a physical model of a given environmental situation is constructed. A chemical is then applied to the model. The fate and effects of the chemical are observed through the model.
3. The "deterministic" approach: A simple mathematical model is used in this approach to describe the rates of individual transformations and the transport of the chemical in the environment. Figure 10.7 depicts a sample of this approach.
4. The "baseline chemical" approach: In this approach, transformations, transports, and effects are measured as in the deterministic approach. Then, the results are compared with data on chemicals of known degrees of risk.

Cornaby et al. (1982) suggested an approach for risk analysis of environmental pathways from a waste site, like the one depicted in Figure 10.8. This approach integrates human and ecological aspects of environmental concerns. Figure 10.9 shows the overall steps in this approach.

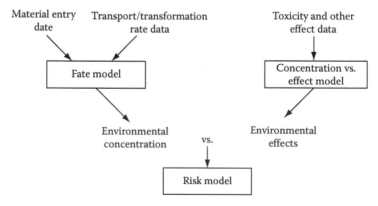

FIGURE 10.7 Sample of the deterministic approach to environmental risk analysis. (From Conway, R.A., Introduction to environmental risk analysis, Chapter 1, in *Environmental Risk Analysis for Chemicals*, R.A. Conway, ed., Van Nostrand Reinhold Company, New York, pp. 1–30, 1982.)

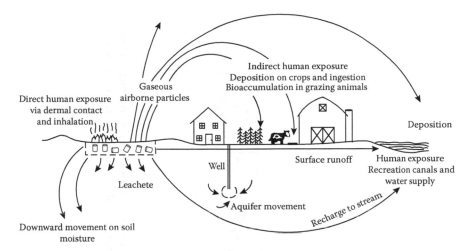

FIGURE 10.8 Environmental pathways from a waste site. (From Schweitzer, G.E., Risk assessment near uncontrolled hazardous waste sites: Role of monitoring data, in *Proceedings of the National Conference on Management of Uncontrolled Hazardous Waste Sites*, Hazardous Materials Control Research Institute, Silver Spring, MD, pp. 238–247, 1981.)

10.3 GROUNDWATER RISK MANAGEMENT

A set of policies, procedures, and practices applied to minimize disaster risks at all levels and locations is called risk management. The aim of risk management is to increase awareness, assessment, analysis, evaluation, and management measures. The framework of risk management includes legal provisions defining the responsibilities for disaster damage and long-term social impacts and losses (Sine, 2004a). In risk management, administrative decisions, technical measures, and the participation of the public should protect water-supply and sanitation facilities from natural disasters. The following actions are required to be performed for risk assessment and disaster formulation (Sine, 2004b):

- Risk awareness and assessment including hazard analysis and vulnerability/capacity analysis
- Knowledge development (education, training, research, and information)
- Institutional frameworks (organizational, policy, legislation, and community action)
- Application of measures such as environmental management, land use and urban planning, protection of critical facilities, application of science and technology, partnership and networking, and financial instruments
- Using early-warning systems including forecasting, public notification of warnings, and preparation measures

In other words, risk management includes preventive, organizational, and technical activities with the aim of preparing the society for disasters and decreasing their consequences. Jeggle (2005) stated "The most important aspects of effective disaster

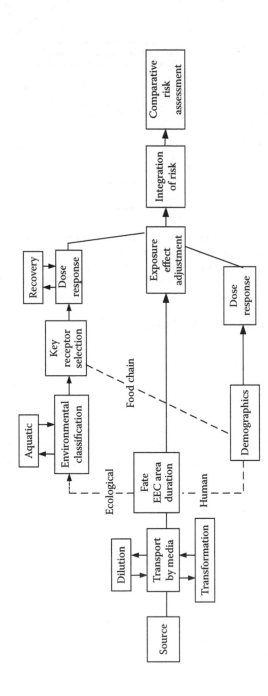

FIGURE 10.9 Overall environmental risk assessment scheme. (From Cornaby, B.W. et al., Application of environmental risk techniques to uncontrolled hazardous waste sites, in *Proceedings of the National Conference on Management of Uncontrolled Hazardous Waste Sites, Hazardous Materials Control Research Institute, Silver Spring, MD, pp. 380–384, 1982.)*

management are precisely those activities that are undertaken and pursued at the times when there is no disaster."

When planning the location and construction of wells, some hydrological data are needed. These data include water level, stream discharge, recurrence of floods, extent of the flooded area, and the morphology of the flood plain. This is especially important when deciding about urban planning in flood plains. The water management authorities should know in advance the measures that are required to be applied for protecting the wells and springs in case of floods or other disasters. In addition, it is important to inform the administration authorities, civil defense, fire brigades, and volunteer organizations (Red Cross and Red Crescent) in time about an expected flood. Then, they can coordinate their activities and inform and protect the population.

In case of a disaster, newer wells in low vulnerable aquifers should be proposed, drilled, and tested to serve as emergency water resources, and the public should be aware of the location of the emergency drinking water resources. If emergency water resources are not available, a system of water supply by mobile cisterns can be established, and drinking water can be distributed by vehicles.

In general, in groundwater risk management, the preventive control and protection of drinking water supplies and the relevant technical devices (e.g., pumping stations) are taken into account with the aim of reducing the impact of disasters on water supply and distribution systems as much as possible.

10.3.1 GROUNDWATER RISK MITIGATION

Risk mitigation can be structural and nonstructural measures used to limit the adverse impacts of natural and technological hazards as well as environmental degradation (Sine, 2004b). These measures include

- Switch off polluted wells to avoid infectious diseases
- Activate emergency water resources, such as deep wells, springs, etc.
- Provide water in mobile tanks in case of no local emergency water resources
- Make people aware of the location of emergency water resources
- Ration water in case of water scarcity threats
- Take immediate actions for the restoration of the affected water supply systems after the hazard

10.4 GROUNDWATER DISASTER

Disaster is the realization of the risk. Groundwater is less vulnerable to short-term natural disasters such as floods, but it is more vulnerable toward long-term man-induced disasters such as overexploitation, gradual drawdown, and widespread contamination. Many regional problems are the direct result of regional climates, geography, and hydrology. Others are as a result of human activities and mismanagement. Traditionally, natural water disaster referred to droughts and floods. Drought is the deficit of water that is required to meet basic demands. Drought also causes inadequate recharge of groundwater. Water supplies in urban areas are frailer during drought events because of limited storage facilities and high water demand variability. Floods are the result of excess runoff.

Besides the quantity of water, its quality could also be the cause of a water disaster. Contamination of water mains and water supply reservoirs by accidental or intentional acts in both developed and developing countries has been taken into consideration by many investigators, which has created a new awareness about water security among public and private agencies dealing with water.

10.4.1 REQUIREMENTS FOR INSTITUTIONAL AND TECHNICAL CAPACITIES

For developing and implementing disaster mitigation and management plans, all the dimensions of a country's institutional and technical capacity building, and whether such capacities are applied in a consistent manner need to be considered. To effectively address disaster prevention, preparedness, emergency response, recovery and mitigation, and national and local institutional and technical capacity building are required.

Institutional capacity building includes governmental authorities, the legal framework, control mechanisms, the availability of human resources and public participation, awareness, and education. Generally, institutional capacity building has the following aspects:

- Governmental authorities are responsible for the coordination of water governance activities and the implementation of water risk management and disaster mitigation plans. The coordination is carried out among policy and decision makers, civil society institutions, water managers and stakeholders, the scientific community, and the general public. Multisectorial national disaster risk mitigation mechanisms have been effective approaches implemented by many countries for disaster reduction and the management of post-disaster rescue activities (Vrba, 2006).
- In disaster management, a legal framework and regulatory status is required to be established to support disaster reduction. This framework is considered the implementation of effective environmental and water protection management and policy. Preventive protection measures of water resources like water supply protection zones and the operation of early warning monitoring systems as well as the public right to information can also be included within this framework.
- Control mechanisms are also required to be established by environmental and water governmental authorities for the protection of water resources against natural and man-made hazards. These mechanisms such as monitoring and warning systems, provide data on natural and anthropogenic hazards.
- Some other important disaster mitigation policies involve public participation, information, and education, which are very effective in the prevention and mitigation of natural and man-induced impacts on water resources.

- Groundwater systems analysis, the identification of potential and existing pollution sources and natural hazards, the establishment of early warning monitoring systems, and in general technical and scientific capacity building, which needs interdisciplinary research and the transfer of knowledge and expertise.

10.4.2 Prevention and Mitigation of Natural and Man-Induced Disasters

There are five phases in a disaster event. These phases include anticipatory, warning, impact, relief, and rehabilitation (Dooge, 2004). Within each phase of disaster prevention and mitigation, the following activities related to public and domestic drinking water supplies can be implemented:

Anticipatory phase: Identification and assessment of the potential risk and vulnerability of existing public and domestic water supply systems as well as the identification, investigation, delineation, and evaluation of groundwater resources resistant to natural hazards.

The warning phase: The activities of this phase are strongly related to the activities in the previous phase. The main activities of the warning phase are the establishment and operation of early warning monitoring systems, geological monitoring systems, and integrated hydro-climatological monitoring systems. The important elements of the warning phase are the formulation of suitable indicators of disaster risk and vulnerability as well as relevant groundwater indicators.

The impact and relief phases: These phases focus on rescue efforts during and after disasters as well as on immediate external help. These efforts include the distribution of drinking water, which requires identification, development, and setting aside rescue activities for the immediate emergency. In many regions affected by natural disasters, water supplies depend on the transport of water by tankers from somewhere outside of the area or on the import of bottled water.

The rehabilitation phase: This is a long-term phase and involves the reconstruction of water supply systems and the water distribution infrastructure, the remediation of polluted water, and the intensive pumping of existing deep wells or developing deeper aquifers of low vulnerability in areas with known properties. Other effective activities are well cleaning, rehabilitation of both well and pumping mechanisms and well dewatering and disinfection, and well deepening (in case of drought). Post-evaluation of all phases of the rescue process, the preparation of plans for rehabilitation, including water management plans, and an assessment of emergency costs can help the situation.

10.4.3 Disaster Management Phases

Disaster management includes all activities, programs, and measures that can be attended to before, during, and after a disaster with main purposes of avoiding

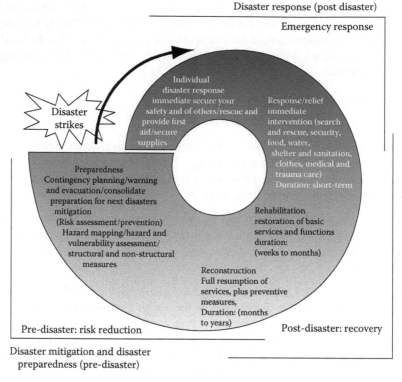

FIGURE 10.10 Disaster mitigation, preparedness, response, and recovery. (From Toshimaru, N., *Are You Prepared? Learning from the Great Hanshin-Awaji Earthquake Disaster, Handbook for Disaster Reduction and Volunteer Activities*, Hyogo University, Kokogawa, Hyogo, Japan 1998.)

a disaster, reducing its impacts, and recovering from its losses. There are three phases in disaster management as shown in Figure 10.10:

1. *Before a disaster (pre-disaster)*: Initiatives are taken to reduce human and property losses caused by a potential disaster. For example, conducting awareness campaigns, strengthening the existing weak structures, providing disaster management plans at the household and community level, etc. These risk reduction measures are called "mitigation and preparedness activities."

2. *During a disaster (disaster occurrence)*: Activities are undertaken to ensure that the needs of victims are met and impacts are minimized. Activities in this stage are called "emergency response activities."

3. *After a disaster (post-disaster)*: Activities are immediately undertaken right after a disaster for the recovery and rehabilitation of affected communities. These activities are termed as "response and recovery activities."

10.5 DISASTER INDICES

Indicators of water supply security can be used to measure the relative level of water disaster. The most common indicators are reliability, resiliency, and vulnerability. A weighted combination of the reliability of supply and resources, the resiliency of the system to return to a satisfactory state of operation, and the vulnerability of the system to a high impact on the available safe water supply determines the state of the systems' readiness for a disaster (Karamouz et al., 2010).

Therefore, relative levels of disaster can be quantified by disaster indices. These criteria can be economic, environmental, ecological, and social. There are different ways to define these indicators. For this purpose, the overall set of criteria is first identified. Then the range of satisfactory values is specified for each indicator. These decisions for identifying the set of criteria and the ranges are subjective and based on human judgments or social goals, not scientific theories. Most criteria are not predefined. However, for many criteria, the duration and magnitude of individual or cumulative failures are important.

10.5.1 RELIABILITY

System performance indices identify optimal situations. If the performance of a system is undesired, the condition is called failure. For example, if a system cannot deliver water with the desired pressure and the demanded quantity and quality, the condition is considered as a failure. To assess the performance of a system, the mean and standard deviation of the outputs of the system are evaluated. However, it is not sufficient. Figure 10.11 shows two systems with totally different performances that have the same mean and standard deviation. There are two failure events in Figure 10.11b but there is no failure situation in Figure 10.11a. Therefore, the probability, severity, and frequency of a system failure cannot be defined by only mean and standard deviation indices.

To evaluate the probability of failure and system performance, reliability is one of the reasonable indices that can be used. In determining the reliability of a system, the random variable of X_t is the system's output. The output can be a failure output or a success output. The probability that X_t belongs to the success output group is the reliability of the system (Hashimoto et al., 1982). The reliability of a system is calculated as

$$\alpha = \text{Prob}[X_t \in S] \quad \forall t \tag{10.9}$$

where S is the set of all satisfactory outputs. Therefore, reliability is the opposite of risk, which is the probability of a system failure. However, reliability is conceptually related to the probability of a system failure. There are several different ways to measure reliability. These ways depend on the needs and relevance of a particular situation. For instance, the expected length of time between consecutive failures may be characterized, such as a 100 year flood event, an event that is expected to

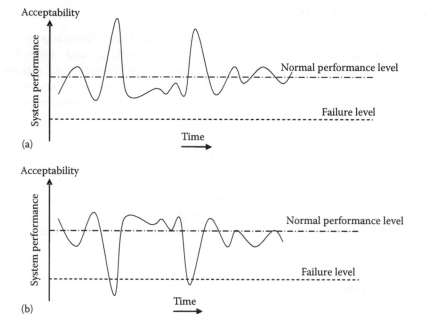

FIGURE 10.11   Comparison of two different functions (a and b) with the same mean and standard deviation. (After Hashimoto, T. et al., *J. Water Resour. Res.*, 18, 14, 1982.)

occur on average once in 100 years. Reliability can also be considered in qualitative rather than quantitative terms. Therefore, there is no one definition and measurement approach applicable in all aspects of influence and applications.

Reliability depends on the complexity of interactions between water-related rights, regulations, laws, and rules. Hence, in order to define water supply reliability, these interactions need to be specified.

For reliability estimation, two different methods can be used.

1. Statistical analysis of recorded data of failures of similar systems
2. Indirect analysis by combination of significant parameter impacts on reliability

In the reliability theory, lifetime (the time between two successive system failures) is usually modeled by the Weibull distribution. The cumulative probability function of the system failures is

$$F(t) = 1 - e^{-\alpha t^{\beta}}, \quad (t > 0) \tag{10.10}$$

where α, $\beta > 0$ are parameters of the distribution that must be estimated. Another definition of reliability is the probability that no failure will occur within the planning horizon (Karamouz et al., 2003). This can be formulated as

$$R(t) = 1 - F(t) = e^{-\alpha t^{\beta}} \tag{10.11}$$

The density function of the time of failure occurrences is obtained by a simple differentiation of Equation 10.10:

$$f(t) = \alpha\beta t^{\beta-1} e^{-\alpha t^{\beta}}, \quad (t > 0) \tag{10.12}$$

It should be noted that in the special case of $\beta = 1$, $f(t) = \alpha e^{-\alpha t}$, which is the density function of the exponential distribution. Because of the forgetfulness property of the exponential distribution, this distribution is rarely used in reliability studies. If X is a exponential variable indicating the time when the system fails, then for all $t, \tau > 0$

$$P(X > t + \tau | X > \tau) = P(X > t) \tag{10.13}$$

This equation illustrates that the probability of the working condition in any time length t is independent of the time length that the system has been working. Equation 10.13 can be rewritten as

$$P(X > t + \tau | X > \tau) = \frac{P(X > t + \tau)}{P(X > \tau)} = \frac{R(t + \tau)}{R(\tau)} = \frac{e^{-\alpha(t+\tau)}}{e^{-\alpha\tau}} = e^{-\alpha t} = R(t) = P(X > t) \tag{10.14}$$

In most practical cases, the breakdown probability increases by the aging of the system. Hence, the exponential variable is inappropriate in such cases.

In case of a failure, the hazard rate can be computed as the ratio of the density function of time between failure occurrences and the reliability function (Karamouz et al., 2003):

$$\rho(t) = \lim_{\Delta t \to 0} \frac{P(t \leq X \leq t + \Delta t | t \leq X)}{\Delta t} \tag{10.15}$$

Therefore,

$$\rho(t) = \lim_{\Delta t \to 0} \frac{P(t \leq X \leq t + \Delta t)}{P(t \leq X)\Delta t} = \lim_{\Delta t \to 0} \frac{F(t + \Delta t) - F(t)}{\Delta t} \frac{1}{R(t)} = \frac{f(t)}{R(t)} \tag{10.16}$$

Example 10.2

Assume the time between drought events in an aquifer system follows the Weibull distribution. Formulate the reliability and hazard rate for this system.

Solution
The reliability is

$$R(t) = 1 - F(t) = e^{-\alpha t^{\beta}}$$

The hazard rate is calculated using Equation 10.16:

$$\rho(t) = \frac{\alpha\beta t^{\beta-1}e^{-\alpha t^{\beta}}}{e^{-\alpha t^{\beta}}} = \alpha\beta t^{\beta-1}$$

It shows that the hazard rate follows an increasing polynomial of t, which means failure will occur more frequently as time goes by (i.e., for larger values of t). In the special case of $\beta = 1$, $\rho(t) = \alpha$, which means that the hazard rate remains constant.

If the hazard rate is given, $F(t)$ or the reliability function can be recalculated (Karamouz et al., 2003). For this purpose, first,

$$\rho(t) = \frac{f(t)}{1-F(t)} = \frac{(F(t))'}{1-F(t)} = -\frac{(1-F(t))'}{1-F(t)} \tag{10.17}$$

Integrating both sides in the interval $[0, t]$:

$$\int_0^t \rho(\tau)d\tau = [-\ln(1-F(\tau))]_{\tau=0}^t = -\ln(1-F(t)) + \ln(1-F(0)) \tag{10.18}$$

Since $F(0) = 0$, the second term in the right hand side of the equation is equal to zero, so

$$\ln(1-F(t)) = -\int_0^t \rho(\tau)d\tau \tag{10.19}$$

Therefore,

$$1 - F(t) = \exp\left(-\int_0^t \rho(\tau)d\tau\right) \tag{10.20}$$

Finally,

$$F(t) = 1 - \exp\left(-\int_0^t \rho(\tau)d\tau\right) \tag{10.21}$$

Example 10.3

Construct the CDF of failure events for the following situations:

1. The hazard rate is constant, $\rho(t) = \alpha$
2. $\rho(t) = \alpha\beta t^{\beta-1}$

Solution

Using Equation 10.21 for $\rho(t) = \alpha$:

$$F(t) = 1 - \exp\left(-\int_0^t \rho(\tau)d\tau\right) = 1 - \exp\left(-\int_0^t \alpha d\tau\right) = 1 - e^{-\alpha t}$$

If $\rho(t) = \alpha\beta t^{\beta-1}$,

$$F(t) = 1 - \exp\left(-\int_0^t \alpha\beta\tau^{\beta-1}d\tau\right) = 1 - \exp\left(-\left[\alpha\tau^\beta\right]_0^t\right) = 1 - e^{-\alpha t^\beta}$$

which demonstrates that the distribution is Weibull.

In contrast with $F(t)$ which is increasing, $F(0) = 0$ and $\lim_{t\to\infty} F(t) = 1$, $R(t)$ is decreasing, $R(0) = 1$ and $\lim_{t\to\infty} R(t) = 0$. In the Weibull distribution,

- If $\beta > 2$, then $\rho(0) = 0$, $\lim_{t\to\infty}\rho(t) = \infty$ and $\rho(t)$ is a strictly increasing and strictly convex function.
- If $\beta = 2$, then $\rho(t)$ is linear.
- If $1 < \beta < 2$, then $\rho(t)$ is strictly increasing and strictly concave.
- If $\beta = 1$, then $\rho(t)$ is constant.
- If $\beta < 1$, the hazard rate is decreasing in t, in which case defective systems tend to fail early.

$$\rho'(t) = \frac{f'(t)(1 - F(t)) + f(t)^2}{(1 - F(t))^2} \tag{10.22}$$

which is positive if the numerator is positive.

$$f'(t) > -\frac{f(t)^2}{1 - F(t)} = -f(t)\rho(t) \tag{10.23}$$

If the inequity of Equation 10.23 is satisfied, $\rho(t)$ locally increases; otherwise $\rho(t)$ decreases.

The above discussions about reliability are applicable to a simple system with one component. The actual natural systems have several components and the reliability in these systems depends on the configuration of the components in the entire system. The combination of components of systems can be "series," "parallel," or "combined." In a series system, the failure of any component results in a system failure, but, in a parallel system, failure occurs only if all components fail simultaneously.

FIGURE 10.12 Series combination of system components.

Figure 10.12 depicts a typical series combination. In this system, $R_i(t)$ denotes the reliability function of component i ($i = 1, 2, ..., n$). The reliability function of the entire system could be estimated as

$$R(t) = P(X > t) = P((X_1 > t) \cap (X_2 > t) \cap ... \cap (X_n > t)) \qquad (10.24)$$

where
X is the time of the system failure
$X_1, ..., X_n$ are the times of failure for the components

If it is assumed that the failures of the different components occur independently of each other,

$$R(t) = P(X_1 > t)P(X_2 > t)...P(X_n > t) = \prod_{i=1}^{n} R_i(t) \qquad (10.25)$$

Inclusion of a new component into the system may result in a smaller reliability function because it is multiplied by the new factor $R_{n+1}(t)$, which is below one.

Figure 10.13 shows a parallel system. As mentioned before, the system fails if all components fail. The system failure probability can then be calculated as

$$P(X \le t) = P((X_1 \le t) \cap (X_2 \le t) \cap ... \cap (X_n \le t)) \qquad (10.26)$$

If the components fail independently of each other,

$$R(t) = 1 - F(t) = 1 - P(X \le t) = 1 - \prod_{i=1}^{n} P(X_i \le t) = 1 - \prod_{i=1}^{n} F_i(t) = 1 - \prod_{i=1}^{n} (1 - R_i(t))$$

$$(10.27)$$

where F and F_i are the cumulative distributions up to the first failure of the system and component i, respectively. In parallel systems, the reliability of the system increases by inclusion of a new component in the system.

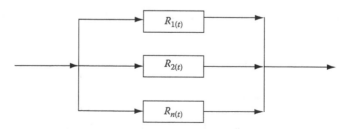

FIGURE 10.13 Parallel combination of system components.

In practical cases, usually a combined series and parallel combination between systems components is considered. In the combined case, relations 10.26 and 10.27 need to be combined appropriately.

10.5.2 RESILIENCY

Once failure occurs in a system, the time taken for the system to recover from the failure is described by resiliency. We may measure the duration of an unsatisfactory condition. In general form, the resiliency of a system can be calculated as (Karamouz et al., 2003):

$$\beta = \text{Prob}\left[X_{t+1} \in S \mid X_t \in F \right] \tag{10.28}$$

where S and F are the sets of all satisfactory and unsatisfactory outputs, respectively.

Resiliency originates from ecology. Holling (1973) defined resiliency as "a measure of the persistence of systems and of their ability to absorb changes and disturbances and still maintain the same relationships between populations and state variables." Pimm (1991) defined resiliency as "the speed with which a system disturbed from equilibrium recovers some proportion of its equilibrium." In general, resiliency can be considered as a tendency to stability and a resistance to perturbation. However, there is no general and overall accepted definition for resiliency.

Depending on the field of application, the definition and formulation of resiliency may vary and be specialized. For example, resilient flood risk management is a management approach that aims at giving room to the floods while minimizing the impacts (De Bruijn and Klijn, 2001). In a hazard event, if the less valuable parts are damaged prior to the more valuable parts, the system as a whole is resilient.

10.5.3 VULNERABILITY

The distance of the failed state from the desired condition in a historical time series describes the vulnerability of the corresponding system. Vulnerability can be considered as the maximum distance from the desired situation, the expected value of system failure cases, or the probability of failure occurrence. However, the approach that considers vulnerability as the expected value of a system failure is more common. Hence, the vulnerability of a system can be defined as

$$\text{Vulnerability} = \frac{\text{No. of situations with positive } (X_T - X_t)}{\text{No. of failure situations}} \tag{10.29}$$

where
X_T is the desired situation
X_t is the system situation

Using the probability of failure occurrence, the vulnerability of a system can be calculated by

$$v = \sum_{j \in F} s_j e_j \tag{10.30}$$

where e_j is the probability that x_j corresponds to s_j.

10.6 GROUNDWATER VULNERABILITY

The vulnerability of groundwater in general and natural terms depends on the geological setting of the groundwater system. The deeper the hydrogeological structure and the older the groundwater, the lower its vulnerability. Therefore, the deeply confined aquifers have a lower vulnerability than unconfined aquifers (Figure 10.14).

Hydrogeological investigation, and knowledge of the geological setting of the region are the requirements for the development of groundwater resources in deep aquifers. Wells must be located beyond alluvial plains and the mouth of the well shaft. The casing should be above the level of the expected flood. The main types of aquifers vulnerable to natural hazards and human impacts are (Vrba, 2006)

- Shallow uppermost unconfined aquifers located in unconsolidated fluvial and glacial deposits that are overlain by a permeable unsaturated zone 'of low thickness (less than 10m). These aquifers can be characterized by young groundwater with an interface with surface water.

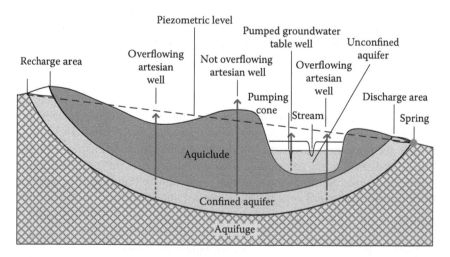

FIGURE 10.14 Example of an unconfined vulnerable aquifer over a confined aquifer in a floodplain. (From Keshari, A.K. et al., Impact of the 26-12-2004 tsunami on Indian coastal groundwater and emergency remediation strategy, in *Groundwater for Emergency Situations*, IHP-VI, Series on Groundwater (12), J. Vrba and B.Th. Verhagen, eds., UNESCO, Paris, pp. 80–85, 2006.)

- Deeper unconfined aquifers located in consolidated rocks, such as sandstones of regional extent, overlain by a permeable unsaturated zone consist of a number of laterally and vertically interconnected groundwater flow systems.
- Karstic aquifers located in carbonate rocks with groundwater flow in conduits; typically with high groundwater flow velocities; springs are important phenomena of the groundwater karstic regime.
- Coastal aquifers—tidal fluctuations, stream flow changes, the gradient and volume of groundwater flow toward the seashore and the geological environment control seawater intrusion into coastal aquifers. In these aquifers, groundwater pumping has a significant impact on the groundwater–seawater interface.
- Recharge areas of all types of aquifers.

The aquifers resistant to natural hazards and human impacts are

- Deep confined aquifers, which are aquifer systems mostly in sedimentary rocks overlain by a thick low permeable or impermeable unsaturated zone.
- Deep mostly confined aquifers, which are aquifer systems overlain by a thick low permeable or impermeable unsaturated zone. Low intensity of groundwater circulation, very limited aquifer replenishment and very large aquifer storage are typical characteristics of these systems. The average annual aquifer renewal in these systems is less than 0.1% of the aquifer storage. Hence, the average renewal period would be at least 1000 years (Margat et al., 2006).
- Water entrapped in the sediments of low permeability at the time of their deposition.

10.6.1 DROUGHT

In the case of water disasters, cyclones, hurricanes, storms, and floods always come to mind first, with less attention to droughts, which could be a disaster with the longest effects and intense social and environmental impacts. Drought is referred to as a sustained and extensive occurrence of low flow thresholds on natural water availability.

Meteorological drought, agricultural drought, hydrologic drought, and socioeconomic drought are different types of droughts. These droughts are defined as follows:

- Meteorological or climatic drought is a period of well below average or normal rainfall that lasts from a few months to several years.
- Agricultural drought is a period when soil moisture is inadequate to meet the demands for crops to initiate and sustain plant growth.
- Hydrologic drought is a period of below-average stream flow that sometimes rivers dry up completely and remain dry for a very long period and there is significant depletion of water in aquifers.
- Socioeconomic drought refers to the situation that physical water shortage affects people and their welfare.

Groundwater is normally the last water resource within the hydrological cycle that reacts to a drought situation, unless surface water is mainly fed by groundwater. In deep aquifers, only major meteorological droughts will cause groundwater droughts. The lag between a meteorological and a groundwater drought can be about months or even years, while the lag between a meteorological and a stream flow drought varies from days to months (Tallaksen and van Lanen, 2004). The following factors have influence on the impact of a meteorological drought on groundwater:

- *Severity and duration of the drought*: In cases that groundwater responds slowly to rainfall deficit, it generally also recovers more slowly after drought. It results in complex and unrelated linkages between rainfall and the impact on groundwater resources.
- *Design and location of the groundwater well or borehole*: Shallow wells may be expected to be more sensitive to recharge variations than deeper boreholes.
- *Hydraulic characteristics of the aquifer*: This refers to the connectivity and linkage of the aquifer to recharge sources and the storage characteristics of the aquifer itself.
- *Excessive demand and source failure*: Decrease of the water table level and depletion of the aquifer during drought events may cause failure of the pump and in consequence, increase of demand on a neighboring source, which results in increasing stress and probability of failure.
- *Long-term increases in demand*: Long-term increase in demand due to population growth, economic change, and the introduction of irrigation or other water-intensive activities can eventually mean that demand will exceed supply.
- *Long-term changes in climate*: Climate change affects recharge processes that puts both high- and low-yielding sources at risk. Increase in the erosion of the soil cover as well as vegetation removal because of climate change may increase runoff and reduce recharge that may result in an inadequate recharge for groundwater recovery.

In many parts of the world, groundwater is increasingly exploited as a resource to supply domestic, industrial, and agricultural demands. Hence, the response of groundwater systems to drought events and their performance under drought conditions is important. When drought affects groundwater systems, first groundwater recharge and later groundwater levels, and eventually groundwater discharge decreases. These droughts are called groundwater droughts and their time scale is generally months to years (Van Lanen and Peters, 2000). The characteristics of the droughts, such as duration, severity, and frequency, are changed by the groundwater system. Therefore, droughts have different impacts on the groundwater levels and discharge rather than groundwater recharge.

The effect of drought on public water supplies necessitates cooperation among all stakeholders, such as water users and local, regional, and national public officials.

For drought management, strategies focus on reducing the impact of droughts by regional analysis and determine the most vulnerable parts of a region. Figure 10.15

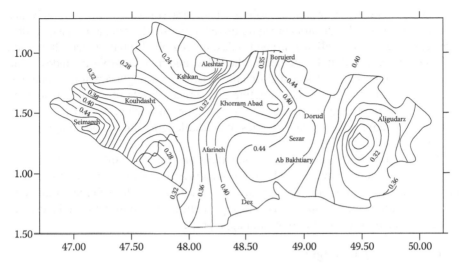

FIGURE 10.15   Vulnerable areas of the Lorestan Province in Iran for a drought with a 10 year return period.

shows the vulnerable areas of the Lorestan Province in Iran for a drought with a 10 year return period. The hatched parts show vulnerable areas that require more attention in the case of a 10 year return period drought. In drought planning and management, these issues must be taken into account—the risk of potential economic and social (psychological) damages resulting from drought, the ecological and economical costs, and the benefits of exercising alternative options such as aquifer recharge and capacity (physical) building (filling the reservoir) schemes. Besides, it must be determined how these options can reduce potential damages and provide opportunity to face future droughts (Karamouz, 2002).

Developing a national or regional drought policy and an emergency response program is essential for reducing societal vulnerability and disaster, thereby also reducing water disaster impacts. Resiliency in the context of drought management has recently received attention (Karamouz et al., 2004).

Drought contingency plans provide a structured response that minimizes the damaging effects caused by water shortage conditions. A common feature of drought contingency plans is a structure that allows rigorous drought response measures to be implemented in successive stages as the water supply diminishes or increases. The termination of each implementation stage should be defined by specific "triggering" criteria. Triggering criteria ensure that (1) timely action is taken in response to a developing situation and (2) the response is appropriate for the level of severity of the situation.

10.6.1.1   Drought Indicators

Effective drought management depends on drought indicators and triggers. Drought indicators can be defined as variables for detecting and characterizing drought conditions. They describe the magnitude, duration, severity, and spatial extent of drought. Typical indicators are based on meteorological and hydrological variables, such as precipitation, stream flows, soil moisture, reservoir storage, and groundwater levels. When several indicators are synthesized into a single indicator on a quantitative scale, the result is called a drought index. In creating drought indices, how each indicator is combined and weighted in the index, and how an arbitrary index value relates to geophysical and statistical characteristics of drought are some challenging issues. Indicator thresholds to define and activate levels of drought responses are called drought triggers. Determining indicators and triggers, and using them in a drought management plan becomes more challenging due to the complexity of drought.

Atmospheric condition, such as days without measurable rainfall, is a basic factor in the regions that have groundwater as their primary or only source of supply to identify drought conditions. Palmer Drought Severity, Standard Precipitation, and Keetch–Byram some commonly used drought indices. These indices can be used to determine local drought conditions and appropriate responses.

10.6.1.2 Drought Triggers

Threshold values of an indicator that identify a drought level are called drought triggers. They determine when management actions should begin and end. Triggers specify the indicator value, the time period, the spatial scale, and the drought level. Drought levels (phases, stages) are categorized as mild, moderate, and severe, extreme drought or stage 1, stage 2, and stage 3 droughts corresponding respectively to the local conditions of a drought "watch," "warning," or "emergency."

Drought indicators and triggers can be used for the following purposes (Steinemann et al., 2005):

To detect and monitor drought conditions
To determine the timing and level of drought responses
To characterize and compare drought events

Drought response triggers are based on an assessment of the water user's vulnerability and they are specific to each water supplier. Groundwater drought triggers can be based on the levels of demand, the capabilities of a water treatment plant or delivery system, and the water levels in monitoring wells that have a record of historical measurements. Trigger criteria are defined on well-established relationships between the benchmark and historical experience. If historical observations are not available, then common sense must prevail until such time that more specific data can be presented.

Groundwater triggers are not identified as easily as factors related to surface-water systems. The reason is the rapid response of stream discharge and reservoir storage to (even short-term) changes in climatic conditions and the slower response of groundwater systems to recharge processes. Wet climatic conditions could have a significant impact on the availability of surface water even over a period of 1 year. But, the underlying aquifers might not show such response as surface waters for much longer periods of time. The response of these aquifers depends on infiltration rates, size, and location of the recharge areas, the distribution of precipitation, and the extent to which aquifers are developed and exploited.

Aquifers that do not receive sufficient recharge to maintain a balance between the recharge and natural discharge or pumping, and so groundwater may be depleted over time. In such aquifers, the level of the water table may decrease over a long period of time, and eventually reaches a point at which the cost of pumping water becomes uneconomical. Therefore, in these situations, the water levels alone may not be a satisfactory drought trigger. Thus, the average annual rate of the water level of the aquifer, and increased well pumping costs due to the water level's drop may be considered as drought triggers.

If historical water level measurements are available, water levels in observation wells in municipal well fields may be considered as drought triggers for municipalities. Water levels below specified elevations for a given period of time might be viewed as rational groundwater indicators of drought conditions.

When a measuring device is installed, the staged pumping level triggers are set for the well. These staged triggers must be above the critical pumping level, which is defined as the lowest pumping level that can be tolerated in a well without deterioration or damage to the well or the pump.

After determining the critical pumping level, two or three staged triggers are established above the critical pumping level to trigger local drought watch, drought warning and drought emergency conditions, and the appropriate emergency response measures. These triggers must be set in a way that sufficient warning for an adequate emergency response is given by them. However, it must be noted that these levels should not trigger response actions so frequently or prematurely that the public loses trust and does not respond. Since drawdowns to the critical pumping level may cause damage to the well or pump, all trigger levels must be above the critical pumping level. The annual fluctuation of groundwater pumping levels is depicted schematically in Figure 10.16. In this figure, the pumping level triggers are set at 40%, 30%, and 20% of the distance between the normal pumping levels and the critical pumping level for

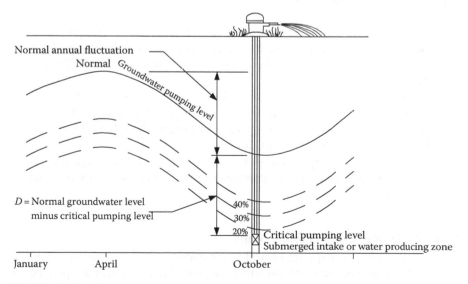

FIGURE 10.16 Groundwater pumping levels and drought triggers in a well.

each month of the year. At 40% of the distance between a normal pumping level and the critical level, a drought watch would trigger; at 30% a warning would trigger; and at 20% an emergency would trigger. These percentage triggers can be considered as initial settings until a more precise set of triggers are developed.

10.6.1.3 Drought Management Strategy Plan

Drought contingency plans must be developed and submitted by local water utilities to the water district to grant well permits and groundwater transfer permits. Actions that must be taken by local utilities to conserve water under given demand scenarios are specified by these plans. These actions include customer awareness, voluntary water conservation, mandatory water use restrictions, and critical water use restrictions.

Groundwater availability is influenced by drought in two ways: (1) recharge from precipitation on the aquifer outcrop and infiltration of surface water are reduced and (2) groundwater pumping is increased to make up for surface water supplies that are more susceptible to drought conditions. To manage district groundwater resources in droughts, the district's vulnerability to drought is evaluated and triggering criteria for specific management zones are defined. Then, drought response goals and measures are determined.

After installation of water level measures in the water district, data of rainfall and water levels in monitoring wells are measured for several years. These data are then evaluated to establish a basis for revising the initial drought management strategy plan. The drought vulnerability assessment is specific to groundwater management zones. It reflects the differences in aquifer depths, local and regional use patterns, recharge area in the district, and other factors.

Due to the water demands, the following are recommended as the district's response goals or trigger levels:

- *Response level 1*: Public awareness and notification of drought conditions. This level could also be triggered by occurrence of when a moderate drought occurs. A moderate drought could be defined as a Palmer Drought Severity Index of −2.0.
- *Response level 2*: Water rate incentives or penalties, i.e., surcharges for excess use. When an extreme drought over a planning period in the water district occurs. This level is triggered. An extreme drought is defined as a Palmer Drought Severity Index of −4.0.
- *Response level 3*: Imposition of water rationing. Permits for high capacity new wells should incorporate provisions for water rate incentives and water rationing.

10.6.1.4 Drought Vulnerability Maps

Drought vulnerability maps are central to the optimum use of groundwater during drought. Regions that are more vulnerable to groundwater drought are pointed out by these maps. They include a variety of decisive factors. Some of these factors are aquifer type, depth of the saturated zone, well and borehole yields, and the amount and variability of rainfall. It should be noted that some regions are more vulnerable to groundwater droughts than others. Some examples of these regions are where there is low water supply coverage and high population density, and dependencies on traditional sources.

In vulnerability maps, different layers are incorporated and superimposed. Two important layers in vulnerability maps are

- A sociologeological dataset that determines the distribution of demands
- A physical dataset that identifies the availability and accessibility of resources

Using these maps eases the identification of vulnerable regions and helps to target them to plan some preparations such as the design of wells and boreholes in pre-drought periods.

10.6.1.5 Drought Early Warning Systems

In comparison with drought vulnerability mapping, early warning systems are based on longer-term meteorological forecasting. Groundwater drought early warning is usually accomplished at two levels: global and regional levels. The global level is in international meteorological scales and considers climate and climate variation, including long-term changes. At the regional level, which is in government, donor, and NGO scales, watershed/aquifer scale signals of groundwater drought and possible consequences are considered (Verhagen, 2006).

Reliable data and long-term observed and antecedent meteorological and hydrological time series, such as water levels and yields, are required for developing a meaningful groundwater drought early warning system. These data are then compiled and analyzed and user needs are determined. Finally, an early warning system is developed to identify drought management areas.

10.6.1.6 Case Study

The Province of Yazd with an area of 131,000 km^2 is located in the central plateau of Iran possessing an arid climate. It is located on the desert belt of the northern hemisphere, with relatively low elevation, far away from surface water bodies, dry anticyclones, topographical conditions coupled with intensive sunlight making this region dry and arid. The province of Yazd is 1568 m above sea level on average, the highest spot in the region (Shirkooh Mountain) is 4075 m and the lowest point (the northeastern desert) is about 670 m above sea level. The average annual rainfall amounts to 109.4 mm in the province and 59.1 mm in Yazd City, the capital of the province. In this region, annual average evaporation is 3193 mm, annual average temperature is 17°C, and annual relative humidity does not exceed 43%. These conditions account for the absence of any permanent surface stream in this area, in fact there is no permanent river but some seasonal rivers originating from Shirkooh Mountain, located in the central part of the province, for a few months in case of a wet year. Therefore, only groundwater supplies limited discharge for the agricultural, industrial, and domestic sectors.

In 2001, the province of Yazd faced a problematic decline in the amount of precipitation that aggravated the ongoing drought and resulted in a considerable depletion of the aquifer water table. During spring, the total rainfall received was less than 5 mm. Therefore, a considerable part of vegetation was dried out. So the villagers, who had to buy forage for the cattle and flock instead of grazing, could no longer afford the installments of the existing loans. The farmers witnessed no yield on their crops due to water scarcity in the region. Furthermore, the orchards dried up despite long time investment and hard work.

Considering the hydrological conditions of the province as well as the fact that 91% of the available groundwater was being consumed by the agricultural sector, the following groundwater protection activities were performed by governmental agencies to alleviate the depletion of groundwater reserves (Yazdi and Khaneiki, 2006):

1. Rehabilitation and repair of Qanats
2. Construction of artificial recharge dams
3. Implementation of cloud seeding projects
4. Intensifying legal penalties in case of over pumping without an official permit
5. Reducing withdrawal from the aquifer by means of modifying and limiting the existing permits
6. Replacing diesel pumps with electric ones to make it possible to control the amount of water by measuring the electricity consumption
7. Installing bulk meters on the wells

10.6.1.7 Qanats

Qanats were first developed in Iran but their use spread to India, Arabia, Egypt, North Africa, Spain, and even to the New world. They are referred to by different names in different areas. In Afghanistan and Pakistan, they are as known as karezes, in North Africa, as foggers, and in the United Arab Emirates, as falaj. In Iran, traditional irrigation has an inherent mechanism to adapt the existing water to environmental conditions. One of the critical concerns of Iran is the evaporation and acute transpiration of surface water under solar radiation. In different regions of Iran, groundwater plays a key role in all forms of development. To make use of the limited amounts of water in arid regions, the Iranians developed man-made underground water channels called Qanats.

A Qanat is a traditional delivery system with over 3000 years of history used to provide a reliable supply of water for irrigation in hot, arid, and semiarid climates. The Qanat system consists of underground channels that convey water from groundwater aquifers in high lands to the surfaces at lower level by gravity. Qanats are constructed as a series of wells connected by gently sloping tunnels. In Qanat systems, a large quantity of water is delivered to the surface without need for pumping. The water drains rely on gravity, with the destination lower than the source, which is typically an upland aquifer. Qanats allow water to be transported over long distances in hot dry climates without losing a large proportion of the water due to seepage or evaporation. Figure 10.17 shows a schematic cross section of a Qanat.

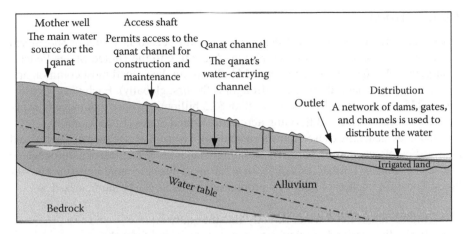

FIGURE 10.17 Schematic cross section of a Qanat.

This mechanism works through technical factors as well as water management. In case of drought, the users of Qanat have two choices to cope with this situation technically. These choices are to extend the gallery of the Qanat by digging it up through the aquifer and adjusting the area of farm lands to the available water. The farm lands are divided into two parts for trees and vegetables. A special ratio should be considered between the areas of orchards and farms so that the maximum amount of the available water would meet the agricultural need for the total area. The area of the cultivated lands and the cropping pattern is adjusted to match the existing discharge and to avoid placing pressure on groundwater sources and in consequence facing drought.

Traditional water management is very flexible to environmental conditions. It is necessary to mention that a Qanat may belong to hundreds of farmers who have an assigned share of water from the Qanats to irrigate their lands. This water-sharing system can match all likely changes in the volume of water during a year, while satisfying agricultural needs. For example, if the volume of water decreases due to a drought, each farmer receives two times more water than his normal share but once every two periods. By doing so, this amount of water can reach every corner of his farm.

Qanats are renewable water supply systems that have sustained agricultural settlement on the Iranian plateau for millennia. The great advantages of Qanats are, no evaporation during transit, little seepage, no raising of the water table, and no pollution in the area surrounding the conduits. The sustainable management of underground reservoirs should be based on a reasonable and economical exploitation of such natural resources. The agricultural sector is the greatest consumer of groundwater. Therefore, existing agricultural methods need to be modified so that more valuable crops are gained with minimum consumption of water. In the study area, if the methods of irrigation are optimized to reach a 5% decrease in water consumption, the water demand for the present industries, mines, towns, and villages can be supplied.

10.6.2 FLOOD

Floods continue to be the most destructive natural hazard in terms of short-term damages and economic losses to a region. According to the United States Federal Emergency Management Agency (FEMA), floods are the second most common and widespread of all natural disasters (drought is the first globally). Each year, an average of 225 people are killed and more than $3.5 billion in property damage is experienced by heavy rains and flooding across the United States (FEMA, 2008).

Floods are generally caused most frequently by a coincidence of meteorological and hydrological circumstances. However, flooding of lands has geological or anthropogenic (caused by humans) reasons, especially in urban areas that have more impermeable surfaces. The main impacts of floods on groundwater systems can be summarized as

- Infiltration of flood water directly into the aquifer
- Infiltration of overland flow into sewer systems and from the sewer system into the groundwater
- Increasing inflow from the recharge areas
- Transferring contaminants from the ground surface to the underlying aquifers

However, increasing urbanization causes decreasing infiltration capacities, and as a consequence, the rate of recharge to aquifers is reduced. On the other hand, flood runoff is accumulated in rivers that cause greater interaction with and recharge of the aquifers.

Flash floods occur even in arid regions where they can cause damage to the communities and where groundwater resources are located in dry stream beds. These floods may cause water supply and polluted wells to collapse groundwater. However, it should be noted that the floods of a longer duration caused by regional rains are different from the flash floods. In some climatic zones, floods are a common phenomenon linked with regular precipitation. The most vulnerable groundwater resources are those in direct contact with the earth's surface where groundwater recharge occurs, such as in sediment deposits of the flood plain, in outcrops of permeable rocks (e.g., sandstone) or in karstified areas.

Flood plains provide a large volume of groundwater recharge, but, due to a high suspended load, their infiltration capacity is low. Nonetheless, since the aquifers recharged by floods are vulnerable to pollution, they are not recommended as substitutions for drinking water sources. Groundwater in the coastal aquifers could be polluted by marine water from the surface due to coastal floods/tsunamis. This groundwater should be carefully utilized without disturbing the underlying brackish water as shown in Figure 10.18.

10.6.2.1 Maps of Groundwater Flood Hazard

Groundwater flooding can cause many problems. Structural damage, sewer system backups, and damaged appliances are three of the most inconvenient consequences. The aim of groundwater flood hazard maps is to indicate areas with need of protection of underground structures against groundwater flood. After regionalizing the

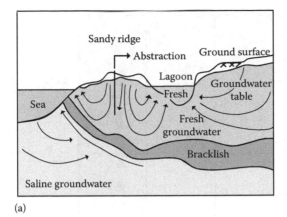

(a)

(b)

FIGURE 10.18   Schematic presentation of a mechanism of tsunami events affecting a coastal groundwater system (a) before and (b) after tsunami. (From Keshari, A.K. et al., Impact of the 26-12-2004 tsunami on Indian coastal groundwater and emergency remediation strategy, in *Groundwater for Emergency Situations*, IHP-VI, Series on Groundwater (12), J. Vrba and B.Th. Verhagen, eds., UNESCO, Paris, pp. 80–85, 2006.)

subsurface hazard potential, it is possible to create maps of groundwater flood hazard (Sommer, 2006). Figure 10.19 depicts an example of hazard potential maps (Sommer and Ullrich, 2005). Flood control and flood prevention strategies can be configured as a pyramid of measures. Figure 10.20 illustrates this pyramid.

Temporary flood control means measures during flood against damages caused by rising groundwater, e.g., groundwater drawdown as well as loading or basement flooding. General flood control refers to prevention measures on building with basements or buildings without basement. Temporary flood control measures are recommended in areas with a long duration of groundwater flood. General flood control measures are recommended in areas with high intensity of groundwater flood. While the building owner is responsible for temporary and general flood control measures, the communal authority should decree restrictions for buildings or prohibition of underground structures by urban land-use planning for areas with high hazard potential.

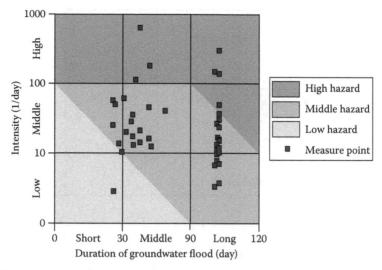

FIGURE 10.19   Example of hazard potential maps.

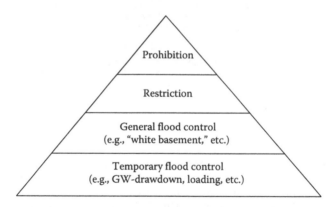

FIGURE 10.20   Control measures against groundwater flood. (From Sommer, T.H. and Ullrich, K., Influence of flood event 2002 on groundwater, Research Report (in German). Environmental Office, City of Dresden, 68 p., 2005.)

10.6.3   LAND SUBSIDENCE

Gradual settling or sudden sinking of the Earth's surface due to the movement of earth materials is called land subsidence. There are three main processes causing land subsidence:

1. Compression (compaction or consolidation) of the interbedded layers of clay and silt within the aquifer system
2. Drainage and oxidation of organic soils
3. Dissolution and collapse of susceptible rocks

The sudden collapse of the ground to form a depression, and also the slow subsidence or compaction of the near-surface sediments and soils are considered subsidence hazards. In comparison with earthquake, volcanic, or landslide events, sudden collapse events are rarely major regional disasters. However, slow subsidence of areas can cause significant economic damage during a longer period of time, even if the types of hazards associated with subsidence are different from that caused by sudden and catastrophic natural events (e.g., floods and earthquakes), because surface sinking is a slow event, extensive damage can occur.

Over extraction of groundwater from aquifers is the major cause of subsidence (Bell, 1997). When large amounts of water are pumped, the aquifer may be compacted due to the reduction of the size and number of open pore spaces that are used to hold water. In consequence, the total storage capacity of the aquifer may be reduced permanently.

For example, if an aquifer has beds of clay or silt, the reduced water pressure results in a loss of support for the clay and silt beds. Since these beds are compressible, they are compacted and become thinner. Therefore, the land surface is lowered. After land subsidence, recharging the aquifer until groundwater returns to the original levels would not result in a recovery of the land surface elevation. The main problems caused by land subsidence are

- Changes in elevation and slope of streams, canals, and drains
- Damage to bridges, roads, railroads, electric power lines, storm drains, sanitary sewers, canals, and levees
- Damage to private and public buildings
- Failure of well casings from forces generated by compaction of fine-grained materials in aquifer systems
- In some coastal areas, subsidence may result in tides moving into low-lying areas that used to be above high-tide levels
- Land subsidence includes loss of levee freeboard, and therefore, reduction in flood protection

To prevent these types of structural damage, it is important to detect land subsidence at very early stages. Because of the rapidly increasing use of groundwater, oil, and gas, major subsidence areas around the world have occurred in the past half century. Especially, most areas of known subsidence are along coasts where the phenomenon becomes more significant when ocean or lake waters start coming further up on the shore.

The accurate monitoring of land subsidence helps to consider recommendations for the sustainable use of the underground resources. In case of a land subsidence occurrence, some recommendations include reducing groundwater use and switching to surface waters where possible, a careful determination of locations for pumping, and artificial recharging. The main objective of artificial recharge is to preserve or enhance groundwater resources. However, there are many other beneficial purposes, the prevention of land subsidence being one of them.

10.6.3.1 Subsidence Calculation

For a unit area of a horizontal plane at depth Z below the ground surface, and utilizing the pressure equilibrium, the upward fluid pressure P_h, and the granular pressure P_i exerted between the grains of the material is resisted by the total downward pressure P_t on the plane due to the weight of the overburden: $P_t = P_h + P_i$. Decreasing the hydrostatic pressure leads to an increase of the granular pressure and a lowering of the water table/piezometric surface. For an aquifer with thickness of Z and P_{i1}, P_{i2} as the granular pressures before and after decreasing the water table or the piezometric surface, respectively, the vertical subsidence can be estimated as

$$S_u = Z \frac{P_{i2} - P_{i1}}{E} \tag{10.31}$$

where E is the modulus of elasticity of the soil layer that is presented in Table 2.5 (Delleur, 1999).

Example 10.4

Consider an unconfined and sandy aquifer with a thickness of 50 and a water table 7 m below the ground surface. Assume that the porosity of the soil is $n = 0.35$, the water content in the unsaturated zone is $\theta = 0.06$, the modulus of elasticity of the sand is $E = 980\,N/cm^2$, and the specific weight of the solids and water are $\gamma_s = 25.1\,kN/m^3$ and $\gamma_w = 9.81\,kN/m^3$ respectively. If the water table drops 15 m, calculate the subsidence.

Solution

The total pressure (granular pressure) at the water table is

$$P_t = 7[(1 - 0.35) \times 25.1 + 0.06 \times 9.81] = 118.3\,kPa$$

The total pressure at the bottom of the layer is

$$P_t = 118.3 + 43 \times [(1 - 0.35) \times 25.1 + 0.35 \times 9.81] = 967.5\,kPa$$

The hydrostatic pressure at the bottom of the layer before decreasing the water table is

$$P_h = 43 \times 9.81 = 421.8\,kPa$$

So, the granular pressure is $967.5 - 421.8 = 545.7\,kPa$.

After decreasing the water table to 15 m, the depth to the water table would be 28 m. The total pressure at the bottom of the layer is

$$P_t = 22[(1 - 0.35) \times 25.1 + 0.06 \times 9.81] + 28[(1 - 0.35) \times 25.1 + 0.35 \times 9.81]$$

$$= 924.8\,kPa$$

The hydrostatic pressure at the bottom of the layer before decreasing the water table is

$$P_h = 28 \times 9.81 = 274.7\,kPa$$

So, the granular pressure is $924.8 - 274.7 = 650.2\,kPa$.

The granular pressure varies linearly from 0.0 at the 7 m depth to 650.2 kPa at the 22 m depth. The average increment in intergranular pressure is $(0 + 650.2)/2 = 325.1\,kPa$, and the settlement in the layer using Equation 10.31 from 7 to 22 m is

$$S_{u1} = Z \frac{P_{i1} - P_{i2}}{E} = 15 \frac{325.1}{9800} = 0.497\,m$$

The subsidence in the layer from 22 to 50 m is

$$S_{u2} = Z \frac{P_{i1} - P_{i2}}{E} = 28 \frac{650.2}{9800} = 1.858 \, \text{m}$$

The total subsidence is thus 0.497 + 1.858 = 2.355 m.

10.6.3.2 Monitoring Land Subsidence

For monitoring land subsidence, the most basic approaches use repeated surveys with conventional or GPS leveling. Permanent compaction recorders or vertical extensometers can also be used for monitoring land subsidence.

In these devices, a pipe or a cable is installed inside a well casing. The pipe extends from the ground surface to some depth through compressible sediments. A table at the ground surface holds instruments that monitor changes in distance between the top of the pipe and the table. Actual land subsidence can be measured if the inner pipe and casing go through the entire thickness of compressible sediments. Groundwater levels must also be measured in addition to measurement of sediments compaction to be able to analyze the data to determine properties that can be used to predict future subsidence.

10.6.3.3 Reducing Future Subsidence

Switching from groundwater to surface water supplies can help to stop land subsidence in some areas where groundwater pumping has caused subsidence. If surface water is not available, possible measures can be reducing water use and determining locations for pumping and artificial recharge. Optimization models coupled with groundwater flow models can be used to develop such strategies.

10.6.3.4 Examples of Subsidence

There is some evidence in Iran showing subsidence has been caused by withdrawal of water from aquifers for industrial and other purposes.

1. In the southeast of Iran, near the city of Iranshahr, two residential areas were established 12 years ago. A few years after that, some cracks were observed around the buildings in these residential areas. The buildings, made up of bricks and cement blocks, had cracks appear all around them in all directions with different sizes. Those made with reinforced concrete, had cracks right at the joints. All crack-causing factors such as faulting, earth movement, folding, subduction, and subsidence were studied and evaluated. In addition, some instruments such as seismographs were installed and geological studies performed. The studies showed that there were no geological and seismic reasons for these cracks. For a long time there had been a few small quakes around the area, but they were not of high magnitude. In addition, the study of faults in the area showed no considerable movement during the past few years. Eventually, technical assessments showed that the water level fluctuations due to water withdrawal from the aquifers and the discharge of waste water into groundwater, as well as the sensibility of clay, which forms the biggest containment of the soil profile, caused the expansions and shrinking, and was the main cause of the cracks and fractures.

2. Mahyar plain is located approximately in the central part of Iran. There are almost 200 km² of agricultural planes and several residential zones in this area. The plain is overlain with almost 150–200 m of alluvium consisting of sand, silt, and clay. The bed rock contains 50 m of clay. Over the past 40 years, almost 250 water wells have been pumping out groundwater.

A high rate of groundwater withdrawal from the aquifer without enough recharge has caused a rapid drawdown of the water table, and in consequence, subsidence of the land. Because of the presence of a few meters of clay on the surface of the plain, the subsidence has caused long and wide cracks on the land surface. These types of land surface cracks can also be observed in Huston, Texas, Goos Creek in North Dakota, Santa Clara in California, and in the Kerman Province in Iran. Reducing the drawdown by an injection of water in the aquifers has ceased or stopped subsidence and cracking in the area.

10.6.4 WIDESPREAD CONTAMINATION

There are two categories of biological and inorganic groundwater contaminants. Biological forms of microorganisms include viruses and bacteria that can cause various diseases, such as cholera. Harmful microorganisms often enter groundwater from leaking sewers and latrines. Therefore, latrines should be situated at a down gradient from supply wells. Figure 10.21 shows different elements of

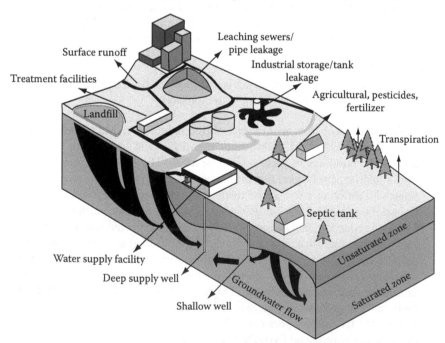

FIGURE 10.21 Threatened hazards of groundwater quality. (From Davies, J. et al., Development of a curriculum and training of supervision teams in borehole construction in Malawi, British Geological Survey Internal Report CR/02/219N, 2002.)

urban life, agricultural, and industrial activities as they pose potential hazards that threaten groundwater quality. Groundwater contamination can occur when a source releases contaminants into the environment. The main sources of groundwater pollution with some of their main characteristics are presented in Table 10.2. The sources include

- The municipal system (urban runoff, sewer leakage, wastewater effluent, septic tanks, sewage sludge, urban runoff, landfill, and latrines)
- Industries (processes, water treatment, industrial effluent, leaking tanks, and pipelines)
- Agriculture (return flows, fertilizers, pesticides, herbalists, and animal wastes)
- Mining (solid wastes and liquid wastes)

Groundwater natural dissolved minerals are primarily expressed as total dissolved solids (TDS) and are derived from soils and rocks. TDS can be estimated by measuring the electrical conductivity of groundwater. TDS in mg/L is equivalent to 0.6 times electrical conductivity (EC) in microSiemens per centimeter (μS/cm). As the TDS increases, the water will look more brackish. The natural TDS of groundwater is normally harmless, although some traces of fluoride or arsenic may be present in groundwater that could be harmful.

However, the major risk to groundwater quality is by pollution coming from various anthropogenic activities. Mining or industrial effluents could infilterate, and pesticides and fertilizers that are spread on the agricultural lands could sink into the ground. The movement of groundwater is usually very slow, and has great filtration capacity for pollutants. But this depends on the properties of the aquifer, fractured or intergranular, and the depth of the water table. Pollutant travels more quickly in fractured aquifers. An aquifer is vulnerable when there is a high risk of contamination. The risk will increase as the number of people and other aquifer water users increase. The risk of contamination is especially serious when many people or industries depend on the aquifer for their water.

The vulnerability of an aquifer to different pollutants is related to the hydraulic characteristics of the aquifer and the characteristic of contaminant attenuation. The groundwater and soil/porous medium and the pollutants determine the vulnerability of a groundwater system to pollution. Land-use planning has a direct impact on the pollution of groundwater resources with negative impacts. It represents the location of high-risk activities.

Any new development in a general land use and zoning structure should follow environmental impact assessment (EIA) legislation to determine the impact of developments on groundwater resources. A wide range of pollutants including bacteria and other microorganisms, major inorganic ions (NO_3, Cl, SO_2, etc.), heavy metals, and a wide range of organic chemicals could contaminate groundwater. The challenges facing engineers to protect groundwater include mechanisms by which pollutants enter groundwater, predictions of the transport of pollutants, and protective measures and legal enforcement issues. In countries such as Sweden (Scharp, 1999), some acts have been exercised to address groundwater protection.

TABLE 10.2

Main Sources of Groundwater Pollution with Some of Their Main Characteristics

Pollution Category	Pollution Source	Main Pollutant	Potential Impact
Municipal	Sewer leakage	Nitrates	Health risk to users, eutrophications of water bodies, odor and taste
	Septic tanks, cesspools, privies	Viruses and bacteria	
	Sewage effluent and sludge	Nitrate, minerals, organic compounds, viruses, and bacteria	
	Storm water runoff	Bacteria and viruses	Health risk to users
	Landfills	Inorganic minerals, organic compounds, heavy metals, bacteria and viruses	Health risk to users, eutrophications of water bodies, odor, and taste
	Cemeteries	Nitrate, viruses and bacteria	Health risk to users
	Feedlot wastes	Nitrate-nitrogen-ammonia, viruses and bacteria	Health risk to users (e.g., metahemoglobinemia)
Agricultural	Pesticides and herbicides	Organic compounds	Toxic/carcinogenic
	Fertilizers	Nitrogen, phosphorous	Eutrophications of water bodies
	Leaked salts	Dissolved salts	Increased TDS in groundwater
Industrial	Process water and plant effluent	Organic compounds heavy metals	Carcinogenic and toxic elements (As, Cn)
	Industrial landfills	Inorganic mineral, organic compounds, heavy metals, bacteria and viruses	Health risk to users, eutrophications of water bodies, odor and taste
	Leaking storage tank (e.g., petrol station)	Hydrocarbons, heavy metals	Odor and taste
	Chemical transport	Hydrocarbons, chemical	Carcinogens and toxic compounds
	Pipelines leaks		
Atmospheric deposition	Coal-fired power stations	Acidic precipitation	Acidification of groundwater and toxic leached heavy metals
	Vehicles emissions		
Mining	Mine tailings and stockpiles	Acidic drainage	

TABLE 10.2 (continued)
Main Sources of Groundwater Pollution with Some of Their Main Characteristics

Pollution Category	Pollution Source	Main Pollutant	Potential Impact
	Dewatering of mine shafts	Salinity, inorganic compounds, metals	May increase concentrations of some compounds to toxic levels
Groundwater development	Saltwater intrusion	Inorganic minerals, dissolved salts	Steady water quality deterioration

Source: Sililo, O.T.N. and Saayman, I.C., *Groundwater Vulnerability to Pollution in Urban Catchments*, The Water Program Division of Water, Environment and Technology CSIR, University of Cape Town, Cape Town, 2001.

In recognition of the importance of protecting groundwater resources from contamination, scientists and resource managers have sought to develop techniques for predicting which areas are more likely than others to become contaminated as a result of human activities at the land surface. The idea being that once identified, areas prone to contamination could be subjected to certain use restrictions or targeted for greater attention.

10.6.4.1 Vulnerability Assessment in Groundwater Pollution

"Groundwater vulnerability is an amorphous concept not a measurable property" (NRC, 1993). It is a tendency or likelihood that contamination may occur. The vulnerability of groundwater to pollution depends on

1. The travel time of infiltrating water carrying contaminants
2. The proportion of contaminants that reaches groundwater
3. The contaminant attenuation capacity of the geological materials through which the contaminated water travels

Barber et al. (1993) subdivided the different approaches for predicting groundwater vulnerability into empirical, deterministic, probabilistic, and stochastic methods (Table 10.3). In empirical methods, maps of various physiographic attributes (e.g., geology, soils, depth to water table) of the region are combined and a numerical index or score is assigned to each attribute. This method is based on experiences and professional judgment. The simplest state of this method is that all attributes are assigned equal weights. Empirical methods that attempt to be more quantitative assign different numerical scores and weights to the attributes. The DRASTIC method (Aller et al., 1987) is the best known example of the experimental method. This method will be explained later in this chapter.

Deterministic approaches use simplified analytical algorithms to arrive at a leaching potential index (LPI). The LPI is expressed in terms of pollutant velocity

TABLE 10.3

Example of Different Approaches to Vulnerability Assessment

Type of Assessment	Scales of Application	Pollution Hazard	Example Identifier	Reference
Empirical	Local	UST-	MATRIX	Orgon DEQ (1991)
	Local	Petroleum	LeGrand	LeGrand (1983)
	Regional and local	Landfill	DRASTIC	Aller et al. (1987)
	Regional	Leakage	GOD	Foster (1987)
	Regional/national	Universal		NRA (1991)
		Universal		Lorber, et al. (1989)
		Universal		
		Aldicarb		
Deterministic	Local/regional	Specific		Bachmat and Collin (1987)
	Regional	Pollutants Pesticides	LPI	Meeks and Dean (1990)
Combined Empirical	Regional	Pesticides	DRASTIC-CLMS	Ehteshami et al. (1991) Banton and Villeneuve (1989)
Deterministic	Regional	Pesticides	DRASTIC-PRZM	
Probabilistic	Regional	Pesticides	VULPEST	Villeneuve et al. (1990)
	Regional	Pesticides	Discriminate	Teso et al. (1989)
Stochastic	Regional possible	Universal pesticides	Analysis weight of evidence models	New LWRRC project
	National			

Source: Barber, C. et al., Assessment of the relative vulnerability of groundwater to pollution: A review and background paper for the *Conference Workshop on Vulnerability Assessment, AGSO*, 14(2/3), pp. 147–154, 1993.

(V/R, where V is the average vadose zone percolation rate, and R is the retardation factor), depth to groundwater (z), and decay constant (x) such that

$$\text{LPI} = 1000 \frac{V}{Rxz} \qquad (10.32)$$

Probabilistic, numerical/stochastic modeling requires numerical solutions to mathematical equations. In stochastic methods, statistics is used to identify combinations of factors that determine vulnerability. These methods usually incorporate data on given areal contaminant distributions and determine contaminant potential for a certain area from which data were drawn.

10.6.4.2 DRASTIC Method

All of the existing aquifer vulnerability assessment methods can be grouped into three major categories—overlay and index methods, process-based methods, and statistical methods (Anthony et al., 1998). These developed GIS maps are useful as preliminary screening tools in setting groundwater management strategies and for policy and decision making_on a regional scale.

GIS maps were developed on a regional scale by the National Well Water Association in conjunction with the EPA in the 1980s for evaluating the pollution potential of groundwater systems on a regional scale. The DRASTIC model follows the category of overlay and index methods. The aquifer vulnerability DRASTIC model is a standardized system that evaluates groundwater pollution potential of hydrogeologic settings. DRASTIC is an acronym for the thematic maps required for the model.

The factors considered in DRASTIC are

- Depth to water table
- Recharge (aquifer)
- Aquifer media
- Soil media
- Topography
- Impact of vadose zone media
- Conductivity (hydraulic) of the aquifer

The DRASTIC model defines ranges and ratings for the classes associated with each of the above thematic maps as well as weights for each thematic map. The significant media types/classes that have bearing on the vulnerability in each map are the ranges, which have associated ratings on a 1–10 scale. The relative importance of each of the above thematic maps on vulnerability is considered as its weight. These ratings and weights are used in obtaining the DRASTIC index, which is calculated as

$$\text{DRASTIC Index} = D_r D_w + R_r R_w + A_r A_w + S_r S_w + T_r T_w + I_r I_w + C_r C_w \quad (10.33)$$

where

D_r is the ratings for the depth to water table
D_w is the weight assigned to the depth to water table
R_r is the ratings for ranges of aquifer recharge
R_w is the weight for the aquifer recharge
A_r is the ratings assigned to the aquifer media
A_w is the weight assigned to the aquifer media
S_r is the ratings for the soil media
S_w is the weight for the soil media
T_r is the ratings for topography (slope)
T_w is the weight assigned to topography
I_r is the ratings assigned to the vadose zone
I_w is the weight assigned to the vadose zone
C_r is the ratings for rates of hydraulic conductivity
C_w is the weight given to hydraulic conductivity

TABLE 10.4

Assigned Importance Weights for Factors in Two DRASTIC Models

Depth to Water Table (D)		Net Recharge (R)		Aquifer Media (A)		Soil Media (S)		Topography (T)		Impact of the Vadose-Zone Media (I)		Hydraulic Conductivity (C)	
Generic Model	Pesticide Model	Generic Model	Pesticide Model	Generic Model	Pesticide Model	Generic Model	Pesticide Model	Generic Model	Pesticide Model	Generic Model	Pesticide Model	Generic Model	Pesticide Model
5	5	4	4	3	3	2	5	1	3	5	4	3	2

Source: Aller, L.T. et al., DRASTIC: A standardized system for evaluating groundwater pollution potential using hydrogeologic settings, EPA/6002/2-85-018, USEPA, 641 p., 1987.

The greater the relative pollution potential, the higher the index that can be calculated. Table 10.4 shows the typical weights for the DRASTIC factors for generic and pesticide models. The DRASTIC index can be converted into qualitative risk categories of low, moderate, high, and very high. GIS maps of the DRASTIC model showing the areas vulnerable to groundwater contamination for a region would have many potential uses, such as the implementation of groundwater management strategies to prevent the degradation of groundwater quality and to locate groundwater monitoring systems (Delleur, 1999).

Example 10.5

This example gives an overview of applying the DRASTIC index to determine groundwater vulnerability in the Rafsanjan subbasin. It is a part of the Daranjir basin and is located in an arid region in the central part of Iran (see Figure 10.22).

The annual average maximum and minimum temperatures are 37.6°C and 0.30°C, respectively. The average rainfall is about 87 (mm/year). About 99% of the total demands (agricultural and domestic demands) in the plain are supplied by groundwater reservoirs (see Table 10.5). Each year, about 200 MCM of water is discharged, in excess of the aquifers safe yield. An annual average of 0.83 cm water drawdown has been reported over the last 10 years.

| Position of Daranjir basin in Iran | Rafsanjan plain |

FIGURE 10.22 Position of the Daranjir basin and the Rafsanjan plain.

TABLE 10.5
Water Supply and Demand Overview of Rafsanjan

Supplied Water to Sectors		Water Withdrawal from Resources			
Agriculture	Domestic	Groundwater	Surface Water	Total	Groundwater Safe Yield
711 MCM	14 MCM	720 MCM	4 MCM	725 MCM	504 MCM
98%	2%	99%	1%		

TABLE 10.6
Rating and Evaluation of DRASTIC Factors

Depth to Water Table (m)	Rating	Net Recharge (mm)	Rating	Topography (Slope)	Rating	Hydraulic Conductivity (m/day)	Rating	Aquifer Media	Rating	Impact of the Vadose-Zone Media	Rating	Soil Media	Rating
0–1.5	10	0–50.8	1	0–2	10	0.04–4.1	1	Massive shale	2	Confining layer	1	Thin or absent	10
1.5–4.6	9	50.8–101.6	3	2–6	9	4.1–12.3	2	Metamorphic igneous	3	Silt/clay	3	Gravel	10
4.6–9.1	7	101.6–177.8	6	6–12	5	12.3–27.8	4	Weathered-meta morphic igneous	4	Shale	3	Sand	9
9.1–15.2	5	177.8–254	8	12–18	3	28.7–41	6	Glacial till	5	Limestone	3	Peat	8
15.2–22.8	3	>254	9	>18	1	41–82	8	Bedded sandstone, limestone	6	Sandstone	6	Shrinking clay	7
22.8–30.4	2					>82	10			Bedded limestone, sandstone	6	Sandy loam	6
>30.4	1							Massive sandstone	6			Loam	5
								Massive limestone	8	Sand and gravel silt	6	Silty loam	4
								Sand and gravel	8	Sand and gravel	8	Clay loam	3
								Basalt	9	Basalt	9	Muck	2
								Karsts limestone	10	Karsts limestone	10	Nonshrinking Clay	1

With the following data layers on the ArcView software, determine the aquifer sensitivity using the DRASTIC model: depth to aquifer, net recharge, aquifer media, soil media, topography, impact of the vadose zone, hydraulic conductivity.

Solution

The DRASTIC model for assessing groundwater vulnerability requires seven criteria—depth to the water table, the recharge rates, aquifer permeability, the soil type, the topography, the impact of the vadose zone, and the conductivity of the vadose zone. DRASTIC evaluates the pollution potential based on a weighted combination of these hydrogeologic settings. Each factor is assigned a weight based on its relative significance in affecting the pollution potential. Each factor is further assigned a rating for different ranges of the values. The assigned importance weights for factors are derived from Table 10.4.

The rating ranges are from 1 to 10 and the weights are from 1 to 5. The DRASTIC index is computed by a summation of the products of rating and weights for each factor (see Equation 10.33). The results for rating and evaluation of DRASTIC factors are shown in Table 10.6 and Figure 10.23. Each of these parameters is part of a weighted model, resulting in a map of relative sensitivity within the region with a rated scale of 1–10. Areas with a rating of 8, 9, or 10 are highly sensitive to contamination.

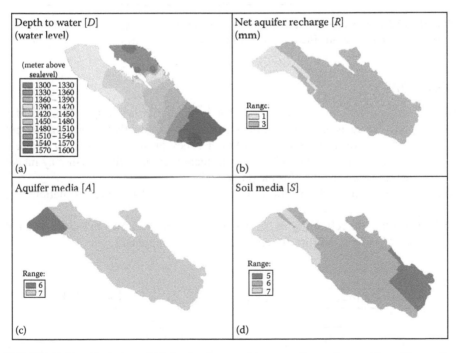

FIGURE 10.23 Variation of (a) depth of water, (b) net aquifer recharge, (c) aquifer media, (d) soil media,

(continued)

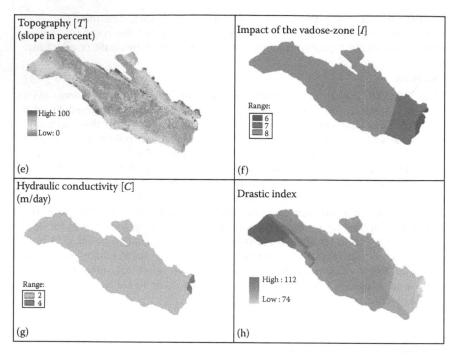

FIGURE 10.23 (continued)   (e) topography, (f) impact of the vadose zone, (g) hydraulic conductivity, and (h) drastic index in the Rafsanjan plain.

PROBLEMS

10.1   The concentration of a toxic chemical in an aquifer is 9.8 mg/L. This aquifer is being used as the source of drinking water. Assume that the average weight of adult women in the area is 60 kg and they drink 1.5 L of water per day, determine the total amount of the chemical that an adult woman may intake after 25 years and the chronic daily intake?

10.2   Construct the CDF of failure occurrences in two hazard rate situations. (1) $\rho(t) = t\alpha$ (2) $\rho(t) = e^{t-1}$.

10.3   Calculate the reliability of the water supply system shown in Figure 10.24 at $t = 0.5$ using reliability functions of system components.

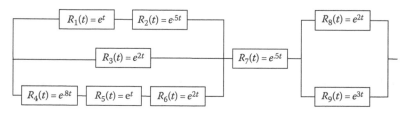

FIGURE 10.24   Water supply system in Problem 10.3

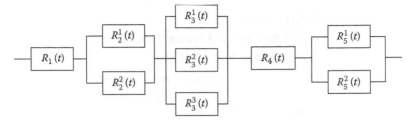

FIGURE 10.25 Water supply system in Problem 10.4

10.4 Calculate the reliability of the water supply system shown in Figure 10.25 at $t = 0.1$ assuming that $R_1(t) = R_4(t) = e^{-2t}$ and $R_2^i(t) = R_3^j(t) = R_5^i(t) = e^{-t}$, $(1 \le i \le 2; 1 \le j \le 3)$.

10.5 TDS concentration data of removed water from a well is presented in Table 10.7. The standard value of TDS concentration in this region is considered as 1400 (mg/L). Calculate the reliability, resiliency and vulnerability of water quality requirement.

10.6 The monthly water withdrawals from an aquifer and monthly water demand during three years are presented in Table 10.8. Calculate the reliability, resiliency and vulnerability of this aquifer of water supply.

10.7 Consider an aquifer with 30 m thickness. The water table in the aquifer is 2 m below the ground surface. After pumping, the water level dropped 5 m in the aquifer. Compute the approximate the land subsidence using the following data: are $\gamma_s = 25.3\,kN/m^3$ and $\gamma_w = 9.81\,kN/m^3$, $n = 0.25$, $E = 980\,N/cm^2$, $\theta = 0.04$.

10.8 Using GIS data layers of Depth to Aquifer, Net Recharge, Aquifer Media, Soil Media, Topography, Impact of Vadose Zone, Hydraulic Conductivity in your region, Determine the aquifer vulnerability using DRASTIC model.

TABLE 10.7

TDS Concentration of Removed Water from a Well (mg/L)

	Year 1	Year 2	Year 3	Year 4
January	1505	1422	1230	1556
February	1215	1498	1409	1545
March	1559	1410	1371	1316
April	1244	1526	1449	1600
May	1532	1210	1574	1486
Jun	1260	1468	1387	1549
July	1480	1300	1206	1317
August	1228	1549	1472	1275
September	1312	1360	1487	1300
October	1574	1319	1406	1539
November	1309	1200	1568	1491
December	1468	1438	1271	1506

TABLE 10.8

Water Release and Water Demand During Three Years (MCM)

	Water Release		Water Demand	
January	8	8	8	10
February	7	6	5	11
March	7	7	6	12
April	8	9	5	12
May	10	10	7	13
Jun	11	11	9	12
July	10	9	10	11
August	12	10	12	9
September	13	12	11	7
October	10	8	9	8
November	12	10	11	9
December	11	11	10	9

REFERENCES

Aller, L.T., Bennet, T., Lehr, J.H., and Petty, R.J. 1987. DRASTIC: A standardized system for evaluating groundwater pollution potential using hydrogeologic settings. EPA/6002/2-85-018, USEPA, Washington DC, 641 p.

Anthony, J.T., Emily, L.I., and Frank, D.V. 1998. Assessing Groundwater Vulnerability using Logistic Regression. In Proc. for the Source Water Assessment and Protection 98 Conf., Dallas, TX, USA 1998, 157–165.

Bachmat, Y. and Colin M. 1987. Mapping to assess groundwater vulnerability to pollution. In *Proceeding of the Conference on Vulnerability of Soil and Groundwater to Pollutants*. W. van Duijvenbooden & H.C.T. van Walgeningh eds. National Institute for Public Health and Environmental Hygiene, The Hague, 1987.

Banton, O. and Villeneuve, J-.P. 1989. Evaluation of groundwater vulnerability to pesticides: a comparison between the pesticide DRASTIC index and the PRJZM leaching quantities. *Journal of Contaminant hydrology*, 4, 285–296.

Barber, C., Bates, L.E., Barren, R., and Allison, H. 1993. Assessment of the relative vulnerability of groundwater to pollution: A review and background paper for the *Conference Workshop on Vulnerability Assessment, AGSO,* 14(2/3), pp. 147–154.

Bell, J.W. 1997. Las Vegas valley: Land subsidence and fissuring due to groundwater withdrawal. Nevada Bureau of Mines and Geology, Reno, NV. Impact of climate change and land use in the southwestern United States.

Conway, R.A. 1982. Introduction to environmental risk analysis, Chapter 1, in *Environmental Risk Analysis for Chemicals*, R.A. Conway, ed. Van Nostrand Reinhold Company, New York, pp. 1–30.

Cornaby, B.W. et al. 1982. Application of environmental risk techniques to uncontrolled hazardous waste sites, in *Proceedings of the National Conference on Management of Uncontrolled Hazardous Waste Sites*, Hazardous Materials Control Research Institute, Silver Spring, MD, pp. 380–384.

Crump, K.S. 1984. A new method for determining allowable daily intakes. *Fundamental and Applied Toxicology.* 4, 854–871.

Davies, J., Coffing, J., Robins, N.S., Kulubanga, G.H., Mandowa, W., and Hankin, P. 2002. Development of a curriculum and training of supervision teams in borehole construction in Malawi. British Geological Survey Internal Report CR/02/219N.

De Bruijn, K.M. and Klijn, F. 2001. Resilient flood risk management strategies, in *Proceedings of the IAHR Congress*, Beijing, China, September 16–21, 2001, ISBN 7-302-04676-X/ TV, L. Guifen and L. Wenxue, eds., Tsinghua University Press, Beijing, China, pp. 450–457.

Delleur, J.W. 1999. *The Handbook of Groundwater Engineering*. CRC Press/Springer-Verlag, Boca Raton, FL.

Dooge, J. 2004. *Ethics of Water Related Disasters*. Series on Water and Ethics, Essay 9. UNESCO, France, Paris.

Ehteshami, M., Peralta, R.C., Eisele, H., Dear, H., and Tindalt, T. 1991. Assessing groundwater pesticide contamination to groundwater: a rapid approach. *Ground Water*, 29, 862–868.

Federal Emergency Management Agency (FEMA). 2008. http://www.fema.gov/library

Foster, S.S.D. 1987. Fundamental concepts in aquifer vulnerability, pollution risk and protection strategy. In *Proceeding of Conference on Vulnerability of Soil and Groundwater to Pollutants*. W. van Duijvenbooden & H.G. van Walgeningh eds., National Institute of Public Health and Environmental Hygiene. The Hague, 1987.

Hashimoto, T., Stedinger, J.R., and Loucks, D.P. 1982. Reliability, resiliency and vulnerability criteria for water resources performance evaluation. *Journal of Water Resources Research*, 18(1), 14–20.

Holling, C.S. 1973. Resilience and stability of ecological systems. *Annual Review of Ecology and Systematic*, 4, 1–24.

Jeggle, T. 2005. Introduction to risk assessment process, in *Disaster Prevention and Reduction*, eds. Vrba, J. and Salamat, A.R. Czech University of Agriculture, Prague, Czech Republic, CD ROM. 9 p.

Karamouz, M. 2002. Decision support system and drought management in Lorestan Province. Final Report to Lorestan Water Board, Iran.

Karamouz, M., Szidarovszky, F., and Zahraie, B. 2003. *Water Resources Systems Analysis*. Lewis Publishers, CRC Publishing, Boca Raton, FL.

Karamouz, M., Torabi, S., and Araghi Nejad, Sh. 2004. Analysis of hydrologic and agricultural droughts in central part of Iran. *ASCE, Journal of Hydrologic Engineering*, 9(5), 402–414.

Karamouz, M., Moridi, A., and Nazif, S. 2010. *Urban Water Engineering and Management*, CRC Press, Boca Raton, FL, 628 p.

Keshari, A.K., Ramanathan, A., and Neupane, B. 2006. Impact of the 26-12-2004 tsunami on Indian coastal groundwater and emergency remediation strategy, in *Groundwater for Emergency Situations*. IHP-VI, Series on Groundwater (12). J. Vrba and B.Th. Verhagen, eds., UNESCO, Paris, pp. 80–85.

Le Grande, H.E. 1983. A standardised system for evaluating waste disposal sites. National Water Well Association, Worthington, Ohio, (USA).

Lee, W.W. and Nair, K. 1979. Risk quantification and risk evaluation, in *Proceedings of National Conference on Hazardous Material Risk Assessment, Disposal and Management*. Information Transfer, Inc., Silver Spring, MD, pp. 44–48.

Lorber, M.N., Cohen, S.Z., Noren, S.E., and De Buchannanne, G.D. 1989. A national evaluation of leaching potential of Aldicarb, Pt 1. An integrated assessment methodology. *Groundwater Monitoring Review*, 9(4), 109–125.

Margat, J., Foster, S., and Loucks, P. 2006. *Non-Renewable Groundwater Resources: A Guidebook on Socially-Sustainable Management for Water Policy-Makers*. IHP-VI, Series on Groundwater. UNESCO, Paris, p. 10.

Masters, G.M. and Ela, W.P. 2008. *Introduction to Environmental Engineering and Science*. Prentice Hall Inc., Simon and Schuster, Englewood Cliffs, NJ, 576 p.

McTernan, W.F. and Kaplan, E. 1990. *Risk Assessment for Groundwater Pollution Control*. Task Committee, Committee on Water Pollution Management, American Society of Civil Engineers, New York, 368 p.

Meeks, Y.J. and Dean, J.D. 1990. Evaluating groundwater vulnerability to pesticides. *Journal of Water Resources Planning and Management*, 116, 693–707.

National Research Council (NRC). 1983. *Risk Assessment in the Federal Government. Managing the Process*. National Academy Press, Washington, DC.

National Research Council (NRC). 1993. *Groundwater Vulnerability Assessment. Predicting Relative Contaminant Potential under Conditions of Uncertainty*. National Academy Press, Washington, DC, 193 p.

National Rivers Authority (NRA). 1992. Policy and practice for the protection of groundwater. HMSO, London, UK, 52pp.

Oregon, D.E.Q. 1991. UST Clean-up Manual: Clean-up rules for leaking petroleum UST systems and numeric soil clean-up levels for motor fuel and heating oil. July 1991. Oregon Department of Environmental Quality, Oregon, USA.

Pimm, S.L. 1991. *The Balance of Nature? Ecological Issues in the Conservation of Species and Communities*. University of Chicago Press, Chicago, IL.

Scharp, C. 1999. Strategic groundwater protection—A systematic approach. PhD dissertation, Royal Institute of Technology, Stockholm, Sweden, p. 48.

Schweitzer, G.E. 1981. Risk assessment near uncontrolled hazardous waste sites: Role of monitoring data, in *Proceedings of the National Conference on Management of Uncontrolled Hazardous Waste Sites*. Hazardous Materials Control Research Institute, Silver Spring, MD, pp. 238–247.

Sililo, O.T.N. and Saayman, I.C. 2001. *Groundwater Vulnerability to Pollution in Urban Catchments*. The Water Program Division of Water, Environment and Technology CSIR, University of Cape Town, Cape Town.

Sine. 2004a. *Water for People Water for Life—The United Nations World Water Development Report, Part IV: Management Challenges: Stewardship and Governance, 11 Mitigating Risk and Coping with Uncertainty*. UNESCO, Paris, p. 290.

Sine. 2004b. *Living with Risk. A Global Review of Disaster Reduction Initiatives*. ISDR, C.D., 2004 version, United Nations, 429 p.

Sommer, T.H. 2006. Groundwater management-a part of flood risk management, in *Proceedings of the International Workshop Groundwater for Emergency Situations*, Tehran, Iran, October 29–31, 2006.

Sommer, T.H. and Ullrich, K. 2005. Influence of flood event 2002 on groundwater. Research Report (in German). Environmental Office, City of Dresden, 68 p.

Steinemann, A., Hayes, M.J., and Cavalcanti, L. 2005. *Drought Indicators and Triggers. Drought and Water Crises: Science, Technology, and Management Issues*. Marcel Dekker, New York.

Tallaksen, L.M. and van Lanen, H.A.J. 2004. Hydrological drought, in *Processes and Estimation Methods for Streamflow and Groundwater*, Development in Water Science 48, Elsevier, the Netherlands, 579 pp.

Teso, R.R., Younglove, T., Peterson, M.R., Sheeks, D.L., and Gallavan, R.E. 1989. Soil taxonomy and surveys: classification of aerial sensitivity to pesticide contamination of groundwater. *Journal of Soil and Water Conservation*, 43(4), 1348-352.

Toshimaru, N. 1998. *Are You Prepared? Learning from the Great Hanshin-Awaji Earthquake Disaster, Handbook for Disaster Reduction and Volunteer Activities*. Hyogo University, Kakogawa, Hyogo, Japan.

USEPA. 1986. Quality criteria for water 1986. EPA 440/5-86-001, Washington, DC.

Van Lanen, H.A.J. and Peters, E. 2000. Definition, effects and assessment of groundwater droughts, in *Drought and Drought Mitigation in Europe*. J.V. Vogt and F. Somma eds., Kluwer Academic Publishers, Dordrecht, the Netherlands, pp. 49–61.

Verhagen, B.T. 2006. Drought: Identification, investigation, planning and risk management, in *Proceedings of the International Workshop Groundwater for Emergency Situations*, Tehran, Iran, October 29–31, 2006.

Villeneuve, J-.P., Banton, O., and LaFrance, P. 1990. A probabilistic approach for assessing groundwater vulnerability to contamination by pesticides: the VULPEST model. *Ecological Modelling*, 51, 47–58.

Vrba, J. 2006. Groundwater for emergency situations—UNESCO-IHP VI Project, in *Proceedings of the International Workshop Groundwater for Emergency Situations*, Tehran, Iran, October 29–31, 2006.

Yazdi, A.A.S. and Khaneiki, M.L. 2006. The drought of 2001 and the measures taken by yazd regional water authority, in *Proceedings of the International Workshop Groundwater for Emergency Situations,* Tehran, Iran, October 29–31, 2006.

11 Climate Change Impacts on Groundwater

11.1 INTRODUCTION

Groundwater has use in different areas such as domestic demand, agriculture, irrigation, and industry. Some developed and successful methods have brought water from under the ground to the surface and ever since, the groundwater use has gradually increased. It is, however, common for the dominant role of groundwater in the freshwater part of the hydrological cycle to be overlooked. Groundwater comprises two-thirds of the freshwater resources on Earth and, without considering the polar ice caps and glaciers, it accounts for all usable freshwater. In other words, the increasing water demand for different activities such as agriculture, industry, habitants use, and recreational water leads to a scarcity of clean and safe drinking water in many regions of the world. In addition, human activities play an important role in climate change, affecting global water resources. Increase in the emission of greenhouse gases (GHGs), as well as changes in land cover due to urbanization, desertification, and deforestation, has already affected the Earth's climate and, therefore, will have an impact on the hydrologic cycle in the future (IPCC, 2001).

The management of groundwater as an important maintainable development has not been considered properly and it has resulted in the depletion and degradation of the resource. Without proactive governance, the damaging effects of defective management will surpass the social gains made so far (Mukharji and Shah, 2005). In addition, long-term replenishment of groundwater resources is controlled by long-term climate conditions. Therefore, proper mitigation and adaptation to the negative impacts of climate change on groundwater is proposed by groundwater management. It is necessary to maintain the particular resilience of larger aquifers as a buffer for climate change and contingency supply capabilities as well as adequate water quality for human consumption. Studies predicting the climate for different intervals point toward changes in future climate, which in turn will influence regional hydrological cycles and consequently the quantity and quality of regional water resources (Gleick, 1989). These changes should be considered in terms of their impact both directly and indirectly. In fact, variations in the major long-term climate variables such as air temperature, precipitation, and evapotranspiration will have a direct impact on changes in groundwater levels, reserves, and quality. For instance, some climate change impacts such as variability of precipitation may add to existing pressure on groundwater resources to fill eventual gaps in surface water availability. In vulnerable areas, impacts on groundwater resources may render the only available reserved freshwater unavailable or unsuitable for use in the near future (IPCC, 2007). Direct interaction of groundwater with surface water resources, such as lakes and rivers, expresses another aspect of the direct relationship between groundwater resources and climate change. Climate change will influence groundwater indirectly through the recharge process with water extracted from aquifers, and by impeding recharge capacities in some areas. All these impacts would alter hydrological systems, regional land use, and agricultural practices. Therefore, two important factors, credible prediction of changes in the major climatic variable and precise estimation of groundwater recharge, should be done to cope with potential impacts of climate change on groundwater.

Groundwater is vulnerable to a number of threats, including climate change and pollution. Protecting it from these threats is vital. A physical approach is required to estimate the impact of changes in climatic parameters on groundwater recharge. Temporal variation and its impact on the hydrologic cycle as well as spatial variation of surface and subsurface properties should be considered.

11.2 CLIMATE CHANGE

Climate is the accumulation of daily and seasonal weather conditions over a long period of time at a location or in a region. It includes average weather conditions, as well as the variability of elements and includes the occurrence of extreme events (Lutgens and Tarbuck, 1995). In other words, humidity, air temperature and pressure, wind speed and direction, cloud cover and type, and the amount and form of precipitation are all basic elements which extract the atmospheric characteristics of climate and weather. These elements establish the variables by which weather patterns and climatic types are described. The timescale is the main difference between weather and climate. Climate is the trends in weather patterns over an extended period of time, not days, or weeks, or months but years. Indeed, weather changes create almost diverse weather conditions at any given time and place. A broad definition of climate system is an interactive system consisting of five major components including the atmosphere, land surface, hydrosphere, biosphere, and cryosphere, or snow and ice. Figure 11.1 shows the components of a global climate system which includes their process and interaction.

These five major components are influenced by changes in various external factors that affect climate such as volcanic eruptions and solar variations. Solar radiation powers the climate system. It passes through the clear atmosphere. Some part of the sunlight is absorbed by the Earth's surface and atmosphere. The other part is reflected back to space where most of it is absorbed by GHGs (see Figure 11.2). The most important GHGs are water vapor and carbon dioxide. Without the existence of these gases, the Earth would be cold. But if the greenhouse effect becomes stronger, it will make the Earth warmer than usual. Consequently, in both conditions living would be impossible.

The greenhouse effect is caused by slow changes in the Earth's orbit around the sun or the change in sunlight absorption by the atmosphere due to the increased concentration of GHGs (Mitchell, 1989; Mitchell et al., 1995). The impact of the greenhouse effect depends upon the composition of certain gases in the atmosphere such as water vapor, carbon dioxide, nitrous oxide, and methane. These gases increase global temperature, and related climate changes such as evaporation and precipitation

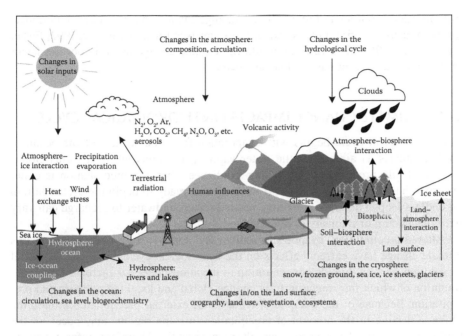

FIGURE 11.1 Components of the global climate system. (From Intergovernmental Panel on Climate Change (IPCC), Climate change 2001: The scientific basis, http://www.ipcc.ch, 2001.)

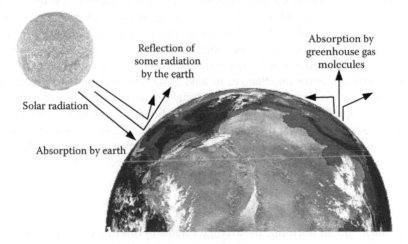

FIGURE 11.2 The greenhouse effect.

(EPA, 2007a). Generally, some other factors which may lead to the greenhouse effect include natural processes within the climate system such as changes in ocean circulation, and human activities such as burning fossil fuels, urbanization, and deforestation, which change the atmosphere's composition and the land surface. Generally, all these factors affect the hydrological cycle and, consequently, influence water-storage patterns that exchange among aquifers, streams, rivers, and lakes. Mall et al. (2007) evaluated climate change and the different aspects of its impact on surface water and groundwater resources in India. Climate change can also have an important impact on the organisms living in these aquatic systems (see Firth and Fisher, 1992; Fisher and Grime, 1996). As a result, considering climate change as an important factor influencing water resources will be necessary.

11.3   CLIMATE CHANGE IMPACTS ON HYDROLOGICAL CYCLE

The hydrological cycle begins with evaporation from the surface of the ocean or land, continues as air carries the water vapor to locations where it forms clouds and eventually precipitates out. It then continues when the precipitation is either absorbed into the ground or runs off to the ocean, ready to begin the cycle again in an endless loop. The amount of time needed for groundwater to recharge can vary with the amount or intensity of precipitation.

Altered precipitation, evapotranspiration, and soil moisture patterns defined as climate change have extreme effects on the hydrological cycle. Since the Industrial Revolution, carbon dioxide concentration in the atmosphere has increased, the continuation of which may meaningfully change global and local temperature and precipitation. Because of the unequal distribution of external precipitation around the globe, a decline in the amount of precipitation in some regions may occur, or the timing of wet and dry seasons may alter significantly. Therefore, information on the local or regional impacts of climate change on hydrological processes and coastal water resources is becoming more important. The key changes to the hydrological cycle associated with greenhouse effects were identified by Prof. Andrew Goudie of Oxford University, which include changes in seasonal pattern and intensity of precipitation under different conditions; changes in snow and rainfall share of precipitation; increase in evapotranspiration and decrease in soil moisture; increase in the extent of saltwater intrusion; changes in land cover due to changes in precipitation and temperature; sea-level rise due to melting glacial ice, decrease in water density and, consequently, increase in coastal inundation and wetland loss; increase the probability of occurrence of fire; and finally, decrease in transpiration and increase in water-use efficiency due to impacts of CO_2 on plant physiology. Figure 11.3 presents different aspects of climate change impacts on water resources.

11.3.1   FLOODS AND DROUGHTS

In response to global warming, the average volume of runoff will intensify considerably. Indeed, a rising in temperature due to the greenhouse effect leads to increase in evaporation, in the intensity of water cycling, and consequently causes a high amount of moisture in the air. As a result, the increase in the magnitude and frequency of weather events such as precipitation in tropical storms increases the severity of hydrological extremes such as the variability of stream flows, risk of floods and droughts occurrence, and erosion due to high velocity of flow. It should be mentioned that increase in flood frequencies may increase groundwater recharge in some floodplains, and also could increase the release of urban and agricultural waste into groundwater systems, and affect the groundwater quality of aquifers.

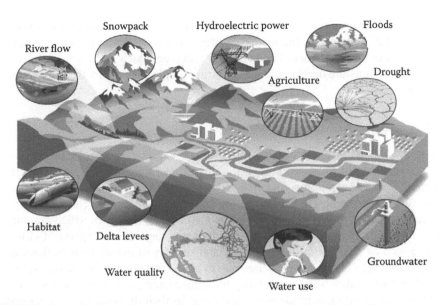

FIGURE 11.3 Different impacts of climate change on hydrological cycle components. (California Department of Water Resource, 2007.)

On the other hand, in some regions the precipitation intensity will decrease with the expected climatic changes during the twenty-first century and, therefore, droughts in these regions will grow longer in duration, which could imperil species and also could put harvests at risk and decrease food production in the world. Wigley and Jones (1985) used a simple water-balance model to conclude that runoff is most sensitive to climate changes, and changes in precipitation have more impact on runoff changes than the changes in evaporation. Also, the relative change in runoff is always greater than that in precipitation.

In addition, as a consequence of drought, the level of groundwater development will change unexpectedly and constantly. In fact, use of groundwater resources to soften the severity of drought impacts is the most effective advance planning for that purpose. On the other hand, droughts can also have an impact on water levels in aquifers because of decrease in rainfall intensity, increase in evaporation, and a decline in percolation.

11.3.2 AGRICULTURAL DROUGHTS

One aspect of drought impacts is crop damage which depends upon the climatic changes, and agriculture is the first economic subdivision affected by drought. Considering the relationship between the harvest and water shortage is the assessment of agricultural droughts. The lack of moisture to support average crop yield on a farm results in agricultural drought. It can also be a result of a long period of low precipitation and/or the type of water usage. Other climatic factors such as high temperature, high wind, and low relative humidity are related to droughts in many regions of the world, and as mentioned before, climate change globally causes these variations and can influence local agriculture. Moreover, agriculture uses a large amount of groundwater in the world. The development of tube wells due to agricultural drought shows the increasing dependency of agriculture on groundwater. Agricultural development has increased recharge which has led to salinity. In addition, soil moisture decline and agricultural drought will impact the groundwater recharge.

11.3.3 WATER USE

Potential changes in climate factors such as temperature and precipitation could affect water use and water resource availability. Under these impacts, demand management, resource expansion, and changes in resource management play significant roles to satisfy demand. The use of water is expressed by industrial, commercial, and agricultural generation, and is managed by a number of measures which are applicable under conditions with and without climate change. Some aspects of climate change such as changes in the frequency and severity of events such as flood and drought, and also higher temperatures due to global warming, increase evaporation and, therefore, reduce water availability for humans and ecosystems. Globally, one-third of the population depends on groundwater for their drinking water. Therefore, groundwater and particularly coastal aquifers are used to complement the available surface water.

11.3.4 WATER QUALITY

Climate change is expected to have significant impacts on water quality. As mentioned in Section 11.3.1, severe floods due to climatic change transport pollutants into water bodies, overload storm and sewage systems, and damnify wastewater treatment devices and, therefore, increase the risks of water pollution. In addition, water quality is influenced by temperature and water amount. Warmer air temperature will result in warm water. Higher temperatures decrease oxygen levels dispersed, which can endanger aquatic ecosystems. Also, the quality of groundwater is being degraded by a variety of factors including chemical contamination. For instance, more frequent intensive precipitation events will release more contaminants and sediments into lakes and rivers and result in declining water-resource quality including surface water and groundwater. Also, global warming increases the melting of snow and contributes to sea-level rise which affects groundwater quality. This will cause saltwater intrusion, particularly in coastal aquifers, with the amount of intrusion depending upon local groundwater gradients.

11.3.5 HABITAT

Global warming threatens people and habitats today, and it is difficult for plants and animals to keep up with the rapid change of habitats. Some species such as polar bears and even trout and salmon are being adversely affected by global warming. It is likely that through direct energy transfer from the atmosphere, climate change will affect lake, reservoir, and stream temperatures. In temperatures higher than their habitat threshold, some species such as salmon and trout will be endangered.

Indeed, high temperatures decrease the dissolved oxygen, increase metabolic rates, and result in the decline of the salmon's resistance to pests. In addition, warmer temperatures can increase the infection rate or vulnerability of fish to illness. Changes in temperature regimes also affect ice cover, and result in sea-level rise which leads to the flooding of low-lying areas. This is likely to imperil the riparian habitats, and to inundate some agricultural lands of residents located on the banks of the river. Moreover, groundwater pumping affects the riparian habitat when it causes the water table to drop beyond the reach of riparian plants.

11.3.6 HYDROELECTRIC POWER

The impacts of climate change on river flows are uncertain, but any considerable changes would influence hydroelectric generation. The amount of electricity produced by hydroelectric generation and the timing of power production depends upon climate change. Hydroelectric generation is more sensitive to changes in river flows than other types of water systems. As a consequence, the reduction in hydroelectric generation may be significantly more than the reduction in river flow. Changes in the magnitude and timing of river runoff, together with high evaporation from reservoirs will have adverse effects on producing hydroelectric power.

In addition, climate change will affect both electricity demand and supply. In fact, in high temperatures the winter heating demands will decrease, and the summer cooling demand will increase. In addition, hydroelectric generation is one form of water use for energy production, and changes in the producing of hydroelectric power will contribute to the water use.

11.3.7 SNOWPACK

Hydroclimatic change results in general change in snowpack and snow cover. Temperature affects the winter snowfall and, consequently, accumulated snowpack in complex ways. Increasing temperature contributes to higher moisture availability and total precipitation, and then greater snow accumulation, but if temperatures increase intensely, snowmelt will increase and, consequently, snowpack will reduce.

As a result, reduced snowpack and earlier spring snowmelt influence other hydrologic variables such as the amount and timing of runoff, evapotranspiration, and soil moisture (Hamlet et al., 2007; Moore et al., 2007). Also, surface water supplies depending upon snowmelt are influenced by changes in the intensity of the precipitation occurring due to climate change. Indeed, warmer temperature leads to a shift from falling snow to rain, changes in snow accumulation, and earlier melting. All these events affect the amount and timing of water supply. Moreover, groundwater recharge from snowmelt is influenced by snowmelt mechanisms and local recharge conditions. During snowmelt, soils are frozen, upland infiltration is reduced, and sinking of snowmelt water increases. The characteristics of the groundwater basin and the soil moisture condition affect the interaction between infiltrating melt water and streamflow.

11.3.8 RIVER FLOW

Climate change affects the streamflow timing through temperature increase and change in the amount of precipitation. Changes in the onset or duration of rainy seasons can affect the river flow. A lot of hydrological studies have been made to assess the impacts of climate change on streamflow and runoff. Global warming increases the amount of evaporation and precipitation. Changes in these hydrological factors will alter the severity of some hydrological events such as increase in the variation of streamflow. As a result of the intensifying of streamflow and change in the soil moisture, the infiltration rate is affected. This change can influence groundwater recharge.

11.4 GENERAL CLIMATE CHANGE IMPACTS ON GROUNDWATER

The potential impacts of climate change on water resources have been recognized for some decades but there has been little research relating to groundwater (IPCC, 2001). Groundwater plays a pivotal role in the economy of developing countries, especially in arid and semiarid regions, and restricting access to this source leads to impaired development. In these regions agriculture and industry also depend on groundwater, so guidelines of coping with their development must incorporate the impacts of climate change on groundwater resources. Stuyt (1995) simulated the effect of sea-level rise and climate change on groundwater salinity and agro-hydrology in a low coastal region of the Netherlands. As a result, simulated crop growth was increased significantly by temperature and CO_2 increase, without increase in the demand for irrigation. Jacob et al. (2001) studied the impacts of climate change on groundwater quality and quantity in the United States. Different aspects including agricultural chemicals, wastewater treatment, industrial pollution, and saltwater intrusion in coastal aquifers are explained. As a result, effects of climate change on groundwater influence riparian systems that support a high percentage of biological diversity. Scanlan (2005) evaluated the impact of land-cover and land-use changes on groundwater recharge and quality in the southern United States. Based on the strong correlation between these factors and groundwater in this study, it is suggested that managed changes in these factors can be used to control groundwater recharge and groundwater quality. IGES (2007) assessed potential impacts of climate change on groundwater in Asia and some structural and institutional adaptations were discussed to minimize risk. Ranjan (2007) assessed climatic variability such as temperature, precipitation, and land-use change and their impacts on saltwater intrusion in coastal aquifers. The results showed that the groundwater recharge increases in arid areas, because deforestation leads to reduced evapotranspiration even though it favors runoff. Hetzel (2008) discussed some challenges and possibilities such as protection of groundwater, securing soil infiltration, and water saving to use and develop groundwater under climate change in Africa. Generally, impacts of climate variability on groundwater resources can be divided into three aspects including changes in the timing and magnitude of recharge, the interaction between groundwater and surface waters, and water retreats. Using general circulation models to predict the future climate and to quantify these impacts leads to some uncertainties. For example, if

CO_2 concentration level rises, it is predicted that the land surface and soil profile will change due to climate change and water passing through the soil and entering aquifers; therefore, some regions will encounter a large increase in the rate of groundwater recharge. But, the spatial variability of recharge and other physical properties is significant to assess under climatic changes. Change in the timing of recharge is important to manage both water use and utilization, and land-use allocation.

Shrestha and Kataoka (2008) have mentioned some direct and indirect impacts of climate change on groundwater resources. Some direct impacts include variation in the amount and intensity of precipitation and evapotranspiration, which will change recharge rates; sea level rising, leading to saltwater intrusion into the low-lying river deltas; and variation in atmospheric CO_2 level that may change the rate of dissolved carbonate and karst formation. Some indirect impacts include changes in land cover which may change recharge rate; increase of some hydrological events such as floods and droughts, which will decline the surface water reliability and quantity and, consequently, increase the groundwater exploitation. On the other hand, the increase in these events may deteriorate groundwater quality of alluvial aquifers, and change soil properties which may impact the percolation rate above aquifers.

11.4.1 HYDROLOGICAL COMPONENTS AFFECTING THE GROUNDWATER

Variation in temperature, precipitation, and consequently, evaporation, and soil moisture, as hydrological components, affect groundwater recharge, demand, and other factors related to groundwater. These impacts and their possible effects on groundwater could be explained as follows.

11.4.1.1 Temperature and Evaporation

In recent times, several studies around the globe show that climate change is likely to impact significantly upon freshwater resource availability.

Climate change refers to a significant change in the intensity of climatic factors such as temperature or precipitation lasting for an expanded period. Most climate-change scenarios present increase in GHG concentrations through 2100, with a continued increase in average global temperatures (IPCC, 2007, as found in EPA, 2007b). It is expected that global warming has several impacts on water resources such as declining snowpack and increasing evaporation, which affects the water supply and its demand rate (Field et al., 2007, p. 619).

The natural resources of groundwater recharge are rainfall and snowmelt. In fact, the recharged water is only to be abstracted, otherwise we run into the trouble of deficit in groundwater supply. Changes in rainfall trend due to climate change will affect the groundwater recharge, and water scarcity will become a more important problem than today. One reason for this problem is high variation in the rate of evaporation, which depends upon temperature and relative humidity. This variation impacts the amount of water available to replenish groundwater supplies. Groundwater reduction can be due to some pivotal impacts of global warming containing more intense rainfall with shorter duration leading to more runoff and less infiltration, increased evapotranspiration to atmosphere, and increased irrigation. The second reason is that as the temperature increases the demand for water will increase, especially in the countries like India, Africa, Bangladesh, South America, etc. The increase in demand leads to more pressure on the groundwater use, and therefore leads to its depletion. Furthermore, the other climate factor that is sensitive to temperature is the spatial and temporal distribution of snow accumulation (Deng et al., 1994). Snowmelt is principally based on the energy balance at the air–snow interface, and on the physical characteristics of the snowcaps (Harms and Chanasyk, 1998). However, the amount and distribution of the infiltrating snowmelt are influenced by the size of the frost layer. Snowmelt or rain may be redistributed by either surface runoff or interflow on top of the frozen soil layer depending on tile topography (Johnson and Linden, 1991).

11.4.1.2 Precipitation

Climatic factors such as wind and temperature make precipitation patterns very dynamic and unbalanced in space and time, leading to huge temporal variability in water resources, especially in groundwater recharge. Therefore, determining the precipitation rate is difficult from region to region due to its spatial and temporal variation. For instance, early snowmelt and diminished late-summer streamflows due to global warming, influence aquifer recharge for underground water supplies. Also, higher amount of precipitation in winter could increase the intensity of storm water runoff instead of slow recharge. In summer, higher temperature with less rainfall could increase the evaporation and transpiration in surface and plants, drying the soil. As a consequence, groundwater levels descend, and therefore, some wells will go dry. On the other hand, accumulated snowfalls in winter are made available for infiltration and runoff when a threshold temperature above zero is reached. Thus, estimated recharge is altered due to variations in groundwater level despite the use of specific yield.

Climate change models predict an increase in the global annual precipitation in the coming century, although changes in its pattern will vary from region to region (IPCC, 2007, as found in EPA, 2007b). However, these projections from climate models are uncertain and not completely reliable.

There are some other impacts of changing precipitation patterns on groundwater. For example, increase in rainfall frequency with extending impervious surfaces could increase urban flood risks producing more pollution which will affect the groundwater quality degradation. Furthermore, the diversity of precipitation amounts in different regions impacts on the water-supplying resources such as groundwater to apply for freshwater.

After a rainfall, infiltration to the lower strata of soil reaches the groundwater recharge. Then, water level in an observation well in shallow unconfined aquifers rises right away. There are three reasons leading to this. First, the infiltrating water around the casing may leak into the well through its perforations unless the well is properly sealed. Second, the rainfall may directly enter the well unless it is capped. And third, the entrapped air pressure within the zone of aeration may inure (Todd, 1980) because the pores near the grimed surface are sealed by the rainwater and the underlying air is compressed by the infiltrating water.

Observation wells usually are capped and sealed near the top, and the perforations are located near their lower ends. Thus, the entrapped air would be the most probable cause for water-level rise due to rainfall. As a result, the water rise can be computed by the use of the following equation (Todd, 1980):

$$\Delta h = \frac{d_m}{d_v - d_m} \frac{p_a}{\gamma_w} \qquad (11.1)$$

where

Δh is the water rise above the water table due to air entrapment

d_m is the depth of the saturated zone of rain infiltration

p_a is the atmosphere pressure $p_a \cong 1000\gamma_w$ ($\gamma_w = 1\,g/cm^3$)

d_v is the vertical distance between the ground surface and the water table, which is the depth of the vadose zone

After a short period of time, Δh scatters as a result of the escape of air.

Example 11.1

On a global scale, climate warming results in increased precipitation and some part of the precipitation absorbs to the ground. The infiltration leads the water level in an observation well in an unconfined aquifer to rise to 3.33 m after a rainfall. If the depth of the vadose zone is 7.7 m, find the depth of the zone of compressed air above the water table.

Solution

Based on the Equation 11.1, average zone of compressed air is equal to the difference between d_v and d_m. $\Delta h = 3.33$ m and $d_v = 7.7$ m.

Assume $d_v - d_m = X$, Then $d_m = X - d_v$

$$p_a \cong 1000\,\gamma_w$$

$$\Delta h = \frac{d_m}{d_v - d_m} \frac{p_a}{\gamma_w}$$

$$3.33 = \frac{X - d_v}{X} \frac{1000\gamma_w}{\gamma_w}$$

$$d_v - d_m = 7.73\,m$$

Then, the depth of the zone of compressed air will be 7.73 m.

11.4.1.3 Soil Moisture

The water stored in the soil is basically important to agriculture and is an influence on the rate of actual evaporation, groundwater recharge, and runoff generation. Climate change affects soil characteristics through changes in waterlogging or cracking, which in turn may affect soil moisture storage properties. The frequency and intensity of freezing, impacts the infiltration soil capacity.

A warmer climate will intensify the hydrologic cycle, changing rainfall and runoff magnitude and timing. In addition, higher temperature due to warm air holds more

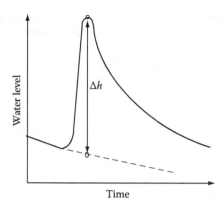

FIGURE 11.4 Principle of water-table fluctuation method for estimating recharge. (From Kresic, N., *Groundwater Resources: Sustainability, Management, and Restoration*, McGraw-Hill, New York, 2008.)

moisture and accelerates evaporation of surface moisture. Moreover, the increase in temperature and evaporation alters the existent moisture in the soil, which could contribute to longer and more severe droughts. Therefore, change in climate will affect the soil moisture and groundwater recharge. Especially in unconfined aquifers with water tables near ground surface, the increase in temperature leads to fluctuations. Both processes of evaporation and/or transpiration cause a discharge of groundwater into the atmosphere.

After rainfall occurs, water table rises, which is the most accurate indication of actual aquifer recharge. The recharge rate can be assumed to be equal to the product of water-table rise and specific yield (Figure 11.4):

$$R = S_y \Delta h \tag{11.2}$$

where
  R is recharge (L)
  S_y is specific yield (dimensionless)
  Δh is water-table rise (L)

This method estimates the groundwater recharge based on water-level rise. The value of specific yield in many cases should be assumed, which increases the uncertainty in applying this approach. When water-table fluctuation method is applied, the frequency of water-level measurement is considered. Manglik and Rai (2000) applied the finite Fourier sine transform solution and predicted the water-table fluctuation in an unconfined aquifer. Zaidi et al. (2007) calculated water budget for a semiarid crystalline rock aquifer located in India. They used the double water-table fluctuation method. In this method, the hydrological year is divided into dry and wet seasons influenced by the monsoon characteristics of rainfall in the area. Then, these two divisions are applied at the following water budget equation twice a year:

$$R + RF + Q_{in} = E + PG + Q_{out} + S_y \Delta h \tag{11.3}$$

where two parameters, natural recharge, R, and specific yield, S_y, are unknown. Water-table fluctuations, Δh, are measured, and other components of the water budget are independently estimated: RF is the irrigation return flow, Q_{in} and Q_{out} are horizontal inflows and outflows in the basin, respectively, E is evaporation, and PG is groundwater extraction by pumping. As the equation is applied for wet and dry seasons to estimate recharge and specific yield, the method is referred as "double water table fluctuation." This decreases the uncertainties due to specific yield estimation at a large scale.

11.4.2 Direct Impacts of Climate Change on Groundwater

Global warming influences hydrological components as mentioned in the previous section. Changes in pattern, intensity, and rate of these components can affect some items of groundwater as recharge and hydraulic conductivity which are explained as follows.

11.4.2.1 Groundwater Recharge

Groundwater recharge is a hydrologic process where soil water infiltrates naturally through the water cycle or anthropologically through the artificial groundwater recharge into the zone of saturation water. This process takes place in vadose zone. Figure 11.5

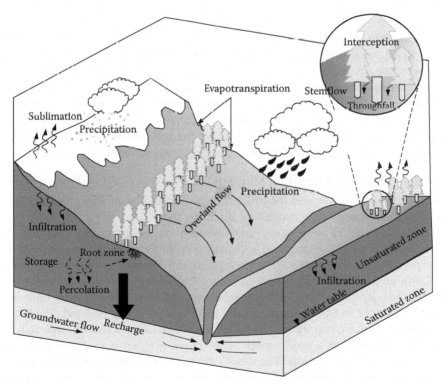

FIGURE 11.5 The hydrologic cycle. (From Delleur, J.W., *The Handbook of Groundwater Engineering*, CRC Press LLC, Boca Raton, FL, 988 p., 2007.)

illustrates the main processes. In deeper confined aquifers, recharge can occur both laterally and vertically through aquitard leakage in the saturated zone (Gerber and Howard, 2000).

To clarify the context of groundwater recharge, the difference between infiltration and percolation is given. Infiltration occurs through a unit area of soil surface, whereas percolation occurs through the slow passage of water through the soil profile in the unsaturated zone. Soil scientists often know the deep-soil water percolation to be groundwater recharge.

The resource of natural recharge is precipitation. According to Lerner et al. (1990), precipitation can recharge groundwater in three ways: direct recharge is the percolation of precipitation vertically through the unsaturated zone; localized recharge, such as a mediatory recharge, is the depression of concentrated surface water under the ground; and indirect recharge is the water which flows through the river or lake beds to the water table.

Direct recharge contributes more than other mentioned recharges to the groundwater system (Knutssen, 1988). In humid climates, surface water usually works as discharge areas rather than recharge areas and therefore less is used as general production of groundwater recharge. Climate change can influence the water levels on the surface, recharge and discharge.

As shown in the hydrologic cycle in Figure 11.5, precipitation is formed either as rain or snow. The rainfall intensity and land cover lead water to be channeled to the ground via streamflow or infiltrate through soils. Then, the infiltrated water penetrates to the lower strata of soil through the vegetative root zone or evaporative zone roots. Some parts of the infiltrated water will be taken by roots and the remaining will percolate deeper to become groundwater recharge in the saturated zone. Also, the percolated water evaporates from the vadose zone based on the soil moisture content and its amount in different parts of the unsaturated zone. Generally, variations in the effective rainfall amount will change the recharge duration. Also, the effective rainfall, rivers, and lakes reload aquifers, and the replenishment rate may be high through the fissures or low by infiltrating through soils or rocks.

11.4.2.2   Recharge Estimation Methods

The precipitation recharge is the link between climate change and groundwater at the land surface and tile recharge and discharge that occurs at surface water bodies. Different studies were done, and diverse recharge estimation methods are used. Zhang et al. (1997) prepared a technical report for recharge estimation in the Liverpool Plains. They used APSIM (Agricultural Production System Simulator, McCown et al., 1996). This model combines the plant growth and soil water balance models. Rushton et al. (2005) estimated groundwater recharge using daily soil moisture balance in two contrasting case studies, Nigeria and England. These two case studies indicate that the soil moisture balance model is a reliable approach for recharge estimation in a wide variety of situations. Singhal et al. (2010) estimated groundwater recharge in India. The water-table fluctuation (WTF) method is used to estimate rainfall recharge in the study area during monsoon season. Also, specific yield has been calculated using groundwater equation for dry season. The difference between the results of the RF infiltration and WTF methods is calculated as 40%.

Moreover, Lerner et al. (1990) explained five main methods estimating direct recharge from precipitation including direct measurement, empirical methods, water budget methods, oracular approaches, and tracers. By comparing methods, and considering the widespread application of them in different areas, and also, based on the data requirements, water budget method can be employed. This method is on the basis of the soil moisture balance. All the components of the water balance equation are estimated by groundwater recharge from the residual terms in the general daily water balance:

$$R = P - ET \pm O \pm \Delta S \qquad (11.4)$$

where
   R is the recharge
   P is precipitation
   ET is actual evapotranspiration
   O is the lateral surface runoff in and out of the area
   ΔS represents the change in water storage in the unsaturated zone

One of the advantages of this method is its application where precipitation data are available. Climate, soil, and vegetation data are used to explain and direct water inflow, outflow, and changes in the daily storage. The main disadvantage of this method is that recharge is estimated as the residual term in a balance equation, but the other budget terms usually are estimated with substantial error. Risser (2008) applied this method to estimate mean annual recharge for landscapes with various physical characteristics.

11.4.2.3   Aquifers Recharging

Climate change influences various types of aquifers to be recharged differently. The main two types of aquifers are unconfined and confined. In the unconfined aquifer, the recharge sources are local rainfall, rivers, and lakes, and the recharge rate will be influenced by the infiltration of overlying rocks and soils. Some samples of the climate change impact on recharge rate in unconfined aquifers have been studied in France, Kenya, Tanzania, Texas, New York, and the Caribbean islands. Bouraoui et al. (1999) mentioned considerable decline in groundwater recharge near Grenoble, France, almost entirely due to increase in evaporation rate during the recharge season.

Shallow unconfined aquifers are recharged by seasonal streamflows and can be reduced directly by evaporation along floodplains. Therefore, recharge changes will be evaluated by changes in the duration of flow of these streams, and the penetrability of the overlying beds, but by increasing evaporative demands, the groundwater storage will lower. It is indicated that unconfined aquifers are sensitive to local climate change, abstraction, and seawater intrusion. However, determining the recharge amount using the characteristics of the aquifers themselves as well as overlying rocks and soils is difficult.

On the other hand, a confined aquifer is described by an impermeable overlying bed, and local rainfall is not considered as an effective factor in the aquifer.

This aquifer is recharged from lakes, rivers, and rainfall with diverse rates from a few days to decades.

Efforts have been made to estimate the recharge rate using carbon-14 isotopes and other modeling techniques. Indeed, these techniques are applicable only for aquifers that are recharged from short distances and after short durations. However, calculating the recharge rate from long distances and considering the impacts of climate change is complex. Generally, it is necessary to emphasize research on modeling techniques, aquifer attributes, recharge rates, and seawater intrusion, as well as monitoring of groundwater abstractions. This research will help to better understand and evaluate the impacts of climate change and sea-level rise on recharge and groundwater resources.

11.4.2.4 Hydraulic Conductivity

Research by scientists and climatologists reveal that anthropogenic factors such as changes in land use with overgrazing and burning of forests, and the abnormal spreading of certain gases impact the atmosphere, which leads to global warming. The released gases are referred to as GHGs including carbon dioxide (CO_2) and methane (CH_4). The greenhouse entraps the heat from the sun, and then increases evapotranspiration and precipitation rates in some areas.

Although little research has been done in estimating the climate change impacts on groundwater, more attention can be expected to estimate the increase or decrease of recharge rates and amounts. Also, in some studies, the role of increased pore pressure in the opening of fractures in rocks with consequent increase in hydraulic conductivity is explained.

It has been known that a slight increase in fracture width, increases the flow rate through a fracture by a power of 3; this is known as the cubic law (Witherspoon et al., 1980) and is given as

$$Q = \left(\frac{\rho g b^2}{12\mu}\right)(bw)\left(\frac{\partial h}{\partial l}\right) \tag{11.5}$$

where
  Q is the volumetric flow rate
  ρ is the density of water
  g is acceleration of gravity
  b is aperture width
  μ is dynamic viscosity of water
  w is fracture width perpendicular to flow direction
  h is the hydraulic head
  l is the length of the flow path through the fracture

The fracture is assumed to possess smooth sides, and laminar flow exists. A consistent set of units must be used. This equation can be significant in determining the relationships between the increased precipitation due to climate change and the water-table rise during a period of increased recharge. Reichard (1995) estimated a significant increase of recharge rate due to climate change and using this equation determined what the impacts of an increased recharge-induced water-table rise are on the hydraulic conductivity of an aquifer. His studies showed that the increase of the hydraulic conductivity of fractured aquifers is as much as 15%–30% under such conditions (Delleur, 2007).

11.4.3 INDIRECT IMPACTS OF CLIMATE CHANGE ON GROUNDWATER

Climate change affects different components of the hydrological cycle as mentioned in Section 11.3. These variations have some indirect impact on groundwater. In fact, the changes in temperature, precipitation, and evaporation lead to sea-level rise, saltwater intrusion, and land-cover change which indirectly affect groundwater recharge. To make clear, these impacts are explained as follows.

11.4.3.1 Sea-Level Rise

Sea-level rise is one aspect of global warming. Although future sea-level rise estimation will be unreliable, it will be significant and continuous for centuries. Due to sea-level rise, low-lying coastal areas will be inundated, and flooding risks and wind damage from coastal storms will increase. Also, melting-back of sea ice will lead to an indirect contribution to sea level. Generally, there are three major procedures by which climate change directly impacts sea level. First, water density decreases as temperature rises. Due to climate change, ocean temperatures increase, and consequently, the water will expand, causing sea-level rise as a result of thermal expansion.

Based on the IPCC studies it is indicated that sea-level rising would have broad effects on coastal environments and infrastructure. These impacts include saltwater intrusion into groundwater aquifers, coastal flooding and coastal erosion. Salinization of aquifers affects directly the drinking water supplies in coastal areas. It increases the salinity of both surface water and groundwater. The intrusion rate depends upon local groundwater gradients; shallow coastal aquifers are at greatest risk. In low-lying islands with high flooding risk due to sea-level rise, groundwater is very sensitive to change. In addition, coastal erosion and sedimentation will produce pollution and, therefore, will impact the groundwater quality.

Some other factors may also account for sea-level rise which include groundwater pumping for human use, wetland drainage, and deforestation in response to the global warming. It is obvious that these changes influence the available water supply in the world and groundwater quality along the coastal plains.

11.4.3.2 Saltwater Intrusion

As the surface water has become scarce in the world due to global warming, a high amount of available freshwater lies underground. To meet the increasing water demand in the world, groundwater complements available surface water. In coastal regions aquifers are considered as important water resources. Saltwater intrusion is one of the major problems in groundwater. It this event, freshwater in coastal aquifers is replaced by saltwater as a result of the motion of a saltwater body into the freshwater aquifer. Saltwater intrusion reduces the fresh groundwater availability in coastal aquifers. One factor leading to salinization is climate change. The climate

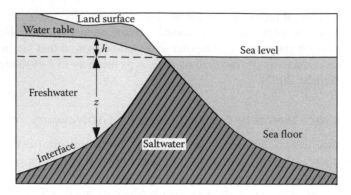

FIGURE 11.6   Saltwater intrusion wedge.

change prediction for the current century is indicated as higher global temperatures and sea-level rise, which emphasize the risks of seawater intrusion. Particularly, in regions where the rate of precipitation becomes less, groundwater salinity may increase by more intensive evaporation and anthropogenic pressures. Considering these influencing factors it is important to study the problem of saltwater intrusion and estimate the saltwater intrusion due to climate changes and change in land-use pattern. In the management aspect of coastal groundwater resources, quantitative understanding of the movement and mixing between freshwater and saltwater are necessary.

In fact, as coastal aquifers are operational reservoirs in water-resource systems, the prediction of the aquifer behavior under different conditions will be significant. Therefore, useful tools should be developed to facilitate this prediction especially when planning salinity management strategies. As a result, the studies on salinization of groundwater are necessary in managing groundwater systems in coastal areas. Figure 11.6 shows the saltwater–freshwater interface, and also, Ghyben–Herzberg equation and the island freshwater zones are presented to explain the relationship between freshwater and saline water.

11.4.3.3   Ghyben–Herzberg Relation between Freshwater and Saline Waters

The interface between saline and freshwater occurs at a depth below sea level about 40 times the height of freshwater head above sea level. It can be defined as the contact zone between the seawater intruding into rocks of a coastal area and the overlaying fresh groundwater in a coastal aquifer flowing seaward. The location of the interface can be approximated according to the Ghyben–Herzberg model (Drabbe and Badon Ghyben, 1889; Herzberg, 1901). This formula is based on the assumption that the potential lines in the freshwater regime are strictly vertical. The hydrostatic balance between freshwater and saline water can be illustrated by the U-tube. Pressures on each side of the tube must be equal; therefore,

$$\rho_s g z = \rho_f g (z + h_f) \tag{11.6}$$

where p, z, and h_f are solving for z and yields

$$z = \frac{\rho_f}{\rho_s - \rho_f} h_f \qquad (11.7)$$

which

z is the depth of the interface below sea level
ρ_s is the density of the saline water
ρ_f is the density of the freshwater
g is the acceleration of gravity
h_f is the elevation of the water table above saline water body elevation (Todd, 1980)

$$z = 40h_f \qquad (11.8)$$

Example 11.2

In an unconfined coastal aquifer, the lower true interface is 95 m below the water table. If the densities of the freshwater and saltwater are 1 and 1.03 g/cm³, respectively, find the water-table elevation above sea level at the apex of the saltwater wedge using the Ghyben–Herzberg principle.

Solution

$$Z + h_f = 95\,\text{m}, \quad \rho_s = 1.03\,\text{g/cm}^3, \quad \text{and} \quad \rho_f = 1.000\,\text{g/cm}^3$$

$$\text{Using Equation 11.7, } 95 - h_f = \frac{\rho_f}{\rho_s - \rho_f} h_f = \frac{1}{0.03} h_f$$

$$h_f = 2.77/92.23\,\text{m} \quad \text{and} \quad z = 92.23\,\text{m}$$

11.4.3.4 Island Freshwater Lens

On small islands, fresh groundwater is a distinct part of water floating on underlying seawater. Freshwater is the result of recharge from precipitation. Climate change impacts the precipitation amount and as a consequence, the recharge rate will change. Therefore, the freshwater amount is influenced by climate change and global warming. Based on theoretical developments, it is predicted that the thickness of the freshwater body near the center of the island is the largest, which diminishes toward the coast where the freshwater is discharged into the sea. As shown in Figure 11.7, the freshwater accumulation is similar to the shape of a lens (Figure 11.7) (McWhorter and Sunada, 1977).

An equation can be derived to express the relation between the depth of the freshwater lens and hydraulic conductivity, the island size, and the recharge as shown in the following. This equation is for a circular island, but the approach can be applied for a long island aquifer with a one-dimensional, normal flow. As shown in Figure 11.7, the discharge at a distance r from the center of the island is estimated as follows:

$$Q(r) = -2\pi K(z+h)r\frac{dh}{dr} = \pi r^2 W \qquad (11.9)$$

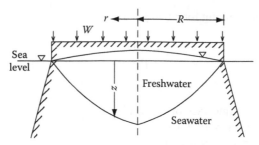

FIGURE 11.7 Freshwater lens on a circular island. (From McWhorter, D.B. and Sunada, D.K., *Ground-Water Hydrology and Hydraulics*, Water Resource Publication, LLC, Highlands Ranch, CO, 290 p., 1977.)

Replacement of h by $(\Delta\rho/\rho)z$, and integration subject to $z = 0$ at $r = R$, yields

$$z = \left\{ \frac{W(R^2 - r^2)}{2K\Delta\rho(1 + \Delta\rho/\rho)} \right\}^{1/2}$$ (11.10)

Apparently, considerable freshwater supplies probably are found on large islands, with large recharge and small K.

Example 11.3

On many islands, the major source of potable water is the freshwater lens. Extraction systems often consist of a large number of closely spaced shallow wells with very small discharges (on the order of 20 m³/day). In some instances, shallow trenches are pumped in order to minimize the drawdown per unit discharge. Assume that sea-level rise occurs due to climate change, consider an approximately circular lens of radius R on which the interface, at the center of the lens, is 20 m below sea level. Also, considering the sea-level rise impacts on the recharge rate, its value is 45 cm/year. A well field is to be developed near the center of the lens. Plans call for the well field to cover 20% of the lens area. Calculate the withdrawal rate for the well field that will eventually result in a depth to the interface of 5 m at the center of the lens (McWhorter and Sunada, 1977).

Solution

The well field is idealized as a circle of radius R_i. Furthermore, withdrawal of groundwater by the well field is simulated by a uniform withdrawal at rate W_i. In the well field

$$Q(r) = \pi r^2 (W - W_i) = -2\pi K(z + h)r \frac{dh}{dr}, \quad r \le R_i$$ (11.11)

Substituting $h = (\Delta\rho/\rho)z$ into Equation 11.11 and integrating subject to $z = z_0$ at $r = 0$ and $z = z_i$ at $r = R_i$ gives

$$z_0^2 = \frac{\rho(W - W_i)R_i^2}{2K\Delta\rho(1 + \Delta\rho/\rho)} + z_i^2$$ (11.12)

The next step is to compute z_i^2. This is accomplished by analyzing the flow for $r \geq R_i$ where the discharge at any r is

$$Q(r) = (W - W_i)\pi R_i^2 + W\pi(r^2 - R_i^2) = -2\pi K(z+h)r\frac{dh}{dr}, \quad r \geq R_i \quad (11.13)$$

Integration, subject to $z = z_i$ at $r = R_i$ and $z = 0$ at $r = R$, gives

$$z_i^2 = \frac{\rho}{2K\Delta\rho(1+\Delta\rho/\rho)}\left\{W(R^2 - R_i^2) - 2W_iR_i^2\ln\left(\frac{R}{R_i}\right)\right\} \quad (11.14)$$

which is combined with Equation 11.12 to yield

$$z_0^2 = \frac{\rho R^2}{2K\Delta\rho(1+\Delta\rho/\rho)}\left[W - W_i\left(\frac{R_i}{R}\right)^2\left\{1+\ln\left(\frac{R}{R_i}\right)^2\right\}\right] \quad (11.15)$$

after some arrangement.
  Since the area, A, of the lens and the area, A_i, of the well field are proportional to their respective radii squared, Equation 11.15 can be written as

$$z_0^2 = \frac{\rho R^2}{2K\Delta\rho(1+\Delta\rho/\rho)}\left[W - W_i\left(\frac{A_i}{A}\right)\left\{1+\ln\left(\frac{A}{A_i}\right)\right\}\right] \quad (11.16)$$

Before development of the well field, $A_i = 0$. In the limit as A_i approaches zero, Equation 11.16 reduces to

$$z_0^2 = \frac{\rho R^2 W}{2K\Delta\rho(1+\Delta\rho/\rho)}$$

But, $z_0^2 = (20)^2$ before development of the well field and $W = 0.3$ m/year. Therefore,

$$\frac{\rho R^2}{2K\Delta\rho(1+\Delta\rho/\rho)} = \frac{(20)^2}{0.45} = 889\,\text{m/year}$$

Thus, Equation 11.16 becomes

$$z_0^2 = 889\left[W - W_i\left(\frac{A_i}{A}\right)\left\{1+\ln\left(\frac{A_i}{A}\right)\right\}\right]$$

which can be solved for W_i:

$$W_i = \frac{W - (z_0^2/889)}{A_i/A\{1+\ln(A/A_i)\}}$$

Substituting $A_i/A = 0.2$, $W = 0.45$ m/year, and $z_0 = 5$ results in

$$W_i = \frac{0.45 - (25/889)}{0.2\{1 + \ln(5)\}} = 0.808 \, \text{m/year}$$

Thus, the withdrawal rate is 8.08×10^5 m³/year/km². The remaining portion is contributed by lateral inflow from land outside the well field. If the well field is expanded to cover essentially all of the lens area, the maximum permissible withdrawal rate is reduced to

$$W_i = 0.45 - \left(\frac{25}{889}\right) = 0.42 \, \text{m/year}$$

11.4.3.5 Land Use and Land Cover

Land-use and land-cover changes can alter the hydrological system. Land-use and land-cover changes impact the hydrologic processes through variation in evapotranspiration regime and/or extreme change in surface runoff. Therefore, land-cover type and land use have an important effect on infiltration and, consequently, the groundwater recharge. Precipitation also varies both temporarily and spatially, and is the primary source of groundwater. Precipitation does not depend upon vegetation type, and evapotranspiration and surface runoff are connected to land-cover characteristics. These impacts depend upon climatic changes. Urban areas also increase the amount of impermeable regions and change the land surface from its natural state. Therefore, it can have a deep impact on groundwater recharge but, urbanization has an adverse impact on the groundwater quality rather than on its quantity (Foster, 2001).

Generally, understanding the impact of land-use and land-cover changes on the atmospheric components of the hydrologic cycle is necessary for an optimal management of natural resources, especially groundwater, which is the largest freshwater resource on Earth. Recharge variations due to land-cover changes can impact groundwater quality negatively, because in semiarid and arid regions, thick unsaturated zones contain accumulated salts that can permeate into underlying aquifers.

Recharge quantification and its controls are important for water-resource evaluation and for analyzing groundwater sensitivity to contamination. Areas with high recharge rates are particularly most prone to pollution. In arid areas, the combination of climatic impacts and land-use changes leads to increased groundwater recharge due to deforestation, which reduces evapotranspiration despite the fact that it favors runoff. Extended research and intelligent decision-makers are required to understand the impact of land-use and land-cover changes on groundwater.

11.5 GROUNDWATER QUANTITY AND QUALITY CONCERNS

The groundwater quality and quantity are depleted and deteriorate by a variety of factors including chemical contamination. Another concerning factor is saltwater intrusion for groundwater quality, particularly in coastal areas where changes in freshwater flows and sea-level rise will both occur. All these factors are influenced by climate change and variability, and because of the importance of groundwater, especially as water supply resource, depletion and deterioration of groundwater due to climate change impacts are explained separately as follows.

11.5.1 DEPLETION OF GROUNDWATER TABLE

The groundwater volume and quality depend upon recharge conditions. The groundwater quantity is controlled by the amount of annual precipitation, land-surface characteristics, vegetation type, and soil attributes. Groundwater supplies are less prone to variations in climate than surface water, but, they are more influenced by long-term trends. More regular or prolonged droughts lead to shortages of surface water supplies, and then, groundwater serves as a buffer pressure as groundwater supplies increase. In addition, declining groundwater levels may affect the seasonal flow intensity and change the temperature regimes that are required to support critical habitat, especially wetlands. The interactions between groundwater and surface water are poorly understood in most areas. Variations in precipitation and temperature regimes may impact the aquifers that are difficult to identify.

11.5.2 DETERIORATION OF GROUNDWATER QUALITY

Groundwater is a vital source of water for public supply, agriculture, and industry in different countries. Climate change impacts indirectly lead to degradation of groundwater quality. Increasing contamination and salinization of groundwater due to sea-level rise are the most important climate change impacts which worsen groundwater quality. These two impacts are explained as follows.

11.5.2.1 Increasing Groundwater Contamination

Groundwater quality contamination is related to public health. Drinking water containing contamination poses a serious challenge to water managers. Agricultural chemicals and wastewater treatment by-products such as nitrates also affect groundwater quality in some areas. Changes in streamflow and consequently sediment regimes, and impacts on human health and ecosystem function are some important water-quality issues. Increase in precipitation intensity can increase risk of contamination. As mentioned before, climate change leads to variations in rainfall regime. This change alters the flow volume and runoff frequency and timing. For instance, low flows will decline contamination dilution capacity, and thus, higher pollutant concentrations. In areas with overall decreased runoff, water quality deterioration will be even worse. On the other hand, extreme streamflows change the sediment and erosion rates on the surface. These sediments may include some pollutant components which can percolate underground. Also, the different extremities of flows rate can affect land cover and infiltration rate, making the infiltration easier. As a result, the groundwater quality will be at risk of increasing contamination due to climatic variations. Furthermore, lower rates of groundwater recharge, flow, and discharge, and higher aquifer temperatures may increase the levels of bacterial, pesticide, nutrient, and metal contamination. Correspondingly, increased flooding could increase the flushing of urban and agricultural waste into groundwater systems, especially into unconfined aquifers, and further deteriorate groundwater quality.

11.5.2.2   Salinization of Groundwater by Sea-Level Rise

As temperature rises globally, the melting of ice sheets and glaciers will increase and will lead to sea-level rise. When sea-level rising happens, saltwater will penetrate farther inland and upstream in low-lying river deltas (IPCC, 1998). Salinization endangers surface water, urban water, and groundwater supplies, damaging ecosystems, and coastal farmland (IPCC, 1998). Furthermore, lower rainfall reduces groundwater head and exacerbates the impacts of sea-level rise. Saltwater intrusion into limestone aquifers is higher than alluvial aquifers. Saltwater intrusion in coastal aquifers is another key groundwater quality concern. As groundwater pumping increases for municipal demand along the coast, freshwater recharge in coastal areas declines, and sea level rises. Thus, groundwater aquifers are affected by seawater infiltration. In areas with low land surface elevations, the relative sea level has been greater. Further increases in sea level may accelerate salinity intrusion into aquifers and affect coastal ecosystems.

11.6   ADAPTATION TO CLIMATE CHANGE

An effort has been made to mitigate GHG emissions to prevent risky climate change. The other stage of the international effort must cope with unavoidable impacts. Adaptations to climate are always made by changing human settlement and agricultural patterns and other aspects of their economies and lifestyles. Climate change due to human activities gives a complex new dimension to this age-old challenge. In the future, the success of human adaptation to climate will depend upon producing greater adaptive capacity with certain patterns of development such as revealing people to higher levels of climate risk.

Climate changes have important impacts on water supplies and water quality. There is already widespread acceleration of ice melting and shifting streamflow from spring to winter peaks in many areas. These shifts could affect the water availability for agriculture and other uses. In coastal aquifers, sea-level rise will result in saltwater intrusion which potentially decreases water resource availability. Changes in precipitation regime due to global warming and related climate change are likely to result in frequent hydrological events such as floods and droughts and increased demand for irrigation water. These events significantly increase the weather-related risks faced by human settlements, and water and power supply failures. First, it is necessary to evaluate the potential impacts of climate change on water resources and their management, based on an agreement that global warming is growing and about its regional dynamics and scale.

However, as the current rainfall predictions are indicative and general rather than definitive, climate change predictions are uncertain and unreliable. This uncertainty limits the utility for indicating the type of strategic challenges. This would be consistent with the underlying analysis of energy flows that underpin predictions of global warming, and is of great importance for water managers who necessarily focus on extreme events. Because of the diverse factors which are combined and influence the available water, predicting impacts of climate change on it will be difficult. For instance, if temperature increases, the evaporation rate from the soil and transpiration rate from plants will increase, therefore, less water will flow as runoff or infiltrates through the underground aquifers; but if the rainfall intensity increases, a larger amount of water

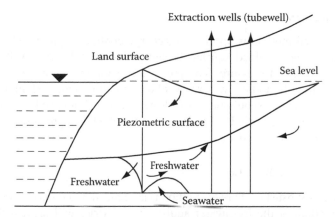

FIGURE 11.8 Schematic diagram of injection well barrier in confined aquifer in a coastal plain.

will flow as floods or seep into the deeper layers of soil underground. Although efforts have been made for predicting the variations of streamflows and groundwater regimes under climate change scenarios, there is a wide range of possible results.

Some adaptation techniques have been developed and applied in the water sector over decades. For instance, in coastal aquifers there is a problem of seawater intrusion due to climate change in groundwater. The introduced technique to control seawater intrusion is to use recharge well barriers through a line of injection tube wells driven parallel to the coast. This mechanism sets up a pressure ridge which pushes the saline front seawards. Figure 11.8 presents a schematic diagram through a confined aquifer system in coastal plain areas and injection well barrier measure for control of seawater intrusion.

Some methods of controlling seawater intrusion into aquifers are modification of groundwater pumping and extraction pattern, artificial recharge, injection barrier, extraction barrier, and subsurface barrier. Each method is described as follows.

11.6.1 Artificial Recharge

There are a number of methods that can be applied in order to raise the level of groundwater table. These methods have been developed to recharge groundwater artificially. For instance, a surface water spreading method should be used in an unconfined groundwater along coastal plain, whereas the well-recharging method is used for confined aquifers.

11.6.2 Modification of Groundwater Extraction Pattern

The landward migration of seawater occurs in the pumping pattern of wells. Therefore, changing the situation of the pumping wells will be a necessary factor. Such wells should be dispersed inland to reestablish the groundwater flow gradient toward the sea. At the same time, the volume of pumped water should decrease from such wells to produce a positive water effect in a freshwater aquifer.

11.6.3 Injection Barrier

The injection well method is applied to recharge confined aquifers where water is injected into a deep confined aquifer through a line of recharging wells along the coast at predetermined pressure. The pressure ridge along the coast is formed by injected water in order to flow water toward the sea. In this method, high quality of water is required, which should be available nearby or it should be imported for well injection recharge. These kinds of wells are required when a pressure ridge to push groundwater seaward is needed.

11.6.4 Subsurface Barrier

The impervious subsurface barrier method is assembled analogous to the coast and through the extent of the freshwater aquifer. This barrier will prevent the flow of seawater from entering into the aquifer. Materials such as clay, asphalt, cement, and bentonite are utilized to build barriers.

11.6.5 Tidal Regulators

This method is utilized to control the discharge of sweet water of a river or stream into the sea. Such structures store fresh water for injection and prevent the inflow of saline water into the river. This increases the water table near the structures and also supplies freshwater along the crest.

Monitoring groundwater around such recharging, namely, salinity entering these structures, is an important factor to manage the availability of freshwater and groundwater.

In this method, dams and tanks should be checked. Also, information such as hydrology, runoff, reservoir level, likely submergence area, command area, geology, geography, soil, and drainage network is collected before proposing a suitable design for tidal regulators.

11.7 CLIMATE CHANGE SIMULATION

Quantifying the atmospheric systems has set up processes that occur on a wide range of time and space scales. Some numerical models of these processes are applied for various purposes, such as predictions of global weather and global climate change. The climate system has dealt with large changes in a geological timescale due to changes in atmospheric composition, location of continents, and the Earth's orbital parameters. The current focuses are climate changes on timescales of seasons to centuries. The current research agenda includes the seasonal climate prediction from an observed initial state and the projection of future climate change. Projecting the climate system in the spatial scales is difficult, both in the atmosphere and on the terrestrial surface.

The high complexities of the Earth's climate result in the uncertainty about changes which will arise from the increase in carbon dioxide concentrations. To increase the reliability of predicted climate and to decrease the errors in predictions, more research and refinement of the global circulation models GCMs are needed. GCMs describe the atmosphere in terms of a rectangular grid covering the Earth with cubes of air above which

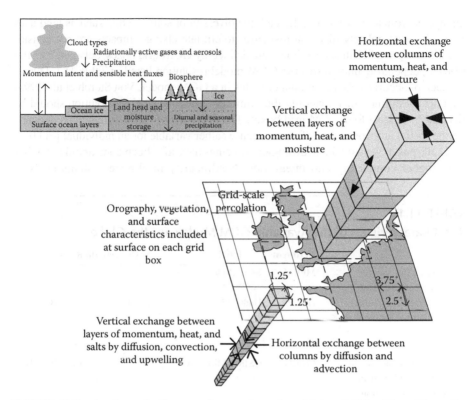

FIGURE 11.9 A schematic discrete of atmosphere for utilizing GCMs. (From Viner, D. and Hulme, M., Climate change scenarios for impact studies in the UK: General circulation models, scenario construction models and application for impact assessment. Report of Department of Environment, Climatic Research Unit, University of East Anglia, Norwich, U.K., 70 p., 1992.)

currently are 2°–5° grids in latitude and longitude, 6–15 vertical levels, which are assumed representative of volume elements in the atmosphere as presented in Figure 11.9.

In the Earth's climate simulation by GCMs, a high number of variables describing the physical and chemical properties of the atmosphere, oceans, and continents are considered. By developing the technology over the past decade, the quality of GCMs has improved vividly. However, some available feebleness should be corrected to improve the GCMs' precision. Much work on these weaknesses needs to be done to simulate the behavior of the Earth's oceans in the GCMs more accurately. Moreover, there are some ambiguities such as the physics of clouds, which are poorly understood, that add another measure of uncertainty to the GCMs. In the following sections, the procedure of developing climate change scenarios, GCM models structure and the downscaling procedure are introduced.

11.7.1 SPATIAL VARIABILITY

A sufficient number of climate variables on the spatial scales to assess the climate changes are estimated by a number of scenarios. Some scenarios result in a uniform change in climate over an area. Indeed, the diversity of some factors such as land uses which have completely different responses to climate change impacts are not considered. The regional changes in climate are defined by GCM grid boxes, between 250 and 600 km, depending upon a specific GCM model resolution listed in Table 11.1. Due to the lack of accuracy about regional climates in a GCM model, Von Storch et al. (1993) support the opinion that generally the minimum effective spatial resolution should be defined by at least four GCM grid boxes. The precision of GCM simulations depends on the spatial autocorrelation of a particular weather variable for an individual grid box.

In the past few years the hydrological schemes have also been considered in GCMs with higher complexity and integration. Nonlinearity in climate systems leads to

TABLE 11.1
The Characteristics of the Available GCM Models for the Globe

	Acronym	Model	SRES Scenario Runs					
Max Planck Institute für Meteorologie Germany	MPIfM	ECHAM4/ OPYC3				A2		B2
Hadley Centre for Climate Prediction and Research, U.K.	HCCPR	HadCM3		A1FI		A2	B1	B2
						A2b		
						A2c		B2b
Australia's Commonwealth Scientific and Industrial Research Organisation Australia Important notice concerning the CSRIO data sets	CSIRO	CSIRO-Mk2	A1			A2	B1	B2
National Center for Atmospheric Research, United States	NCAR	NCAR-CSM				A2		
		NCAR-PCM	A1B				A2	
Geophysical Fluid Dynamics Laboratory, Untied States	GFDL	R30				A2		B2
Canadian Center for Climate Modelling and Analysis, Canada	CCCma	CGCM2				A2		B2
						A2b		B2b
								B2c
						A2c		
Center for Climate System Research National Institute for Environmental Studies, Japan	CCSR/ NIES	CCSR/NIES AGCM + CCSR OGCM	A1	A1FI	A1T	A2	B1	B2

Source: Intergovernmental Panel on Climate Change (IPCC), Fourth Assessment Report, http://www. ipcc.ch, 2007.

difficulty in using a deterministic estimation of precipitation from GCMs for hydro-logical modeling purposes. In this aspect, methods to transform output from GCM to local weather variables are needed. A weakness of numerical models of natural systems is its impossibility to validate the described nonclosed systems. However, the need to improve the quality of these models and the benefits in operational use is pivotal. Although the spatial resolution of GCMs is increasing, they still simulate climate at a very coarse spatial scale on the order of 250×250 km, which is not enough for the evaluation of climate change effects in urban areas. In the following sections, some methods of improving the spatial precision of the GCM outputs are described.

11.7.1.1 Downscaling

The GCMs have unreliability in simulating climate variables and the application of a stochastic model in space and time will overcome this. Even high spatial resolution of GCMs needs to downscale the results from such models. Downscaling methods are still under development and more work needs to be done to increase the accuracy of such models. It creates a connection between the large-scale circulation defined as predictors and local weather variables defined as predictands. But downscaling pertains more to the process of moving from the large-scale predictor to the local predictand. Figure 11.10 shows a schematic of moving from the large scale to the small scale.

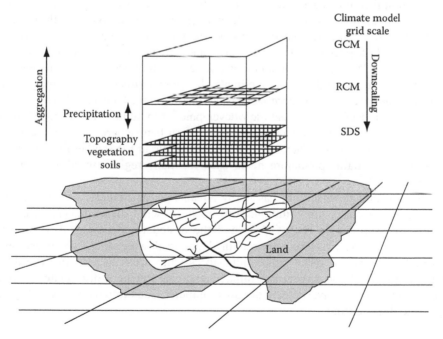

FIGURE 11.10 Schematic drawing of the basic principles of downscaling from GCM output. (From Wilby, R.L. and Dawson, C.W., *Using SDSM Version 3.1—A Decision Support Tool for the Assessment of Regional Climate Change Impacts*, Climate Change Unit, Environment Agency of England and Wales, Nottingham, U.K. and Department of Computer Science, Loughborough University, Leicestershire, U.K., 2004.)

Downscaling techniques calculate sub-grid box scale changes in climate as a function of larger-scale climate or circulation statistics. The two main approaches to downscaling have used either regression relationships between large-area and site-specific climates (e.g., Wigley et al., 1990) or relationships between atmospheric circulation types and local weather (e.g., Von Storch et al., 1993). The disadvantage of the downscaling approach is to calibrate the statistical relationship. In doing so, it requires large amounts of observed data which can be computationally very intensive. Also, as unique relationships need to be derived for each site or region, these methods are very time consuming. A basic assumption in the application of downscaling methods is that the observed statistical relationships will continue to be valid in the future under conditions of climate change. There are a wide range of techniques used for downscaling. The statistical downscaling is usually used for downscaling which is further described in the Section 11.7.2.

11.7.1.2   Regional Models

To involve the use of high-resolution regional climate models, an alternative approach is involved. Regional climate models (RCMs) are typically assembled at a much finer resolution than GCMs (often 50 km) with the domain limitation to continents or subcontinents. In such applications, detailed information at spatial scales down to 10–20 km may be established at temporal scales of hours or less. However, RCMs are still compelled at their boundaries by the coarse-scale output from GCMs. Therefore, in an area, RCM can have the performance similar to that of the driving GCM and also, their performance in relation to downscaling techniques has not yet been assessed completely. Indeed, there are some difficulties in creating the interface between the GCM and the regional model. Moreover, the costs of setting these models for a new region and running a climate change experiment are extremely high; therefore, they are not easily accessible. Nevertheless, in the long term this technique could result in useful results and their development should be encouraged. Most of the regional models are employed on the northern hemisphere, whereas many of the areas that are most vulnerable to global change lie in the southern hemisphere. Therefore, it remains premature to be used for many regions in climate impact assessments.

The dynamic downscaling has many advantages particularly around factors such as temperature, precipitation, soil moisture, wind direction, and strength which are available in both GCMs and the nested finer-scale grid. However, supercomputer systems are required in dynamic downscaling, which are less available to perform these simulations for all the areas around the world to evaluate the local and or regional impacts and consequences of climate change. Then, increasing the availability of supercomputer leads to wide accessibility of dynamic downscaling.

11.7.2   Statistical Downscaling

Developing the statistical downscaling is assuming that an empirical relationship exists between atmospheric processes at different spatial and temporal scales (Wilby et al., 1998). These relationships can depend on predictor variables and broadscale atmospheric variables, such as mean sea-level pressure (MSLP) and geopotential height, through a transfer function. The following are some of the criteria that make a downscaling method successful (Hewitson and Crane, 1996):

- Large-scale predictors are sufficiently reproduced by GCMs.
- For periods outside the calibration period, the relationship between the predictors and the predictands are valid.
- Climate change signals are adequately incorporated in the predictors.

As no predictor meets the three criteria, predictors which fulfill one or two criteria are selected. The disadvantage is that the degree of freedom in the predictor set is increased, and the results become more difficult to analyze. Hewitson and Crane (1996) outlined the major steps of downscaling, presented as follows:

1. Reduction or processing of the predictor
2. Comparison of GCM circulation with observed circulation
3. Spatial analysis and selection of predictor and predictand domain
4. Temporal analysis of the predictor and the predictand
5. Temporal manipulation of the predictor, e.g., time-lagging
6. Calibration of transfer function between predictor and predictand
7. Evaluating the relationship between predictor and predictand
8. Application of the developed model to GCM data

Items 3–7 are repeated until an optimum result is reached for the target objective. During this procedure, predictor variables and weighting functions are selected, concerning the downscaling method used. It is possible that all these steps are not taken in each method, but they are necessary to be considered to reach credible results from downscaling. Three statistical downscaling methods are analog methods, regression methods, and conditional probability approaches. The last one has two main subdivisions containing weather patterns and weather states. These methods are largely used in anticipated hydrological impact studies under climate change scenarios. The main features of statistical downscaling are personal computers which can be used for the calculations, there is no need for detailed knowledge about the physical processes, and long and homogeneous time series are needed for fitting and confirming the statistical relationship.

In the second feature, the exact formulation of the processes is not necessarily considered for statistical downscaling. The main advantages of the statistical downscaling approach are easy implementation and cheap computation. In situations where a low-cost, rapid assessment of highly localized climate change impacts is required, statistical downscaling (currently) represents the more promising option. The integrated feature of downscaling procedure is that it is calibrated to the local level. Furthermore, with very few parameters these methods will be attractive for many hydrological applications (Wilby et al., 2000). Precipitation is a nonlinear process, and precipitation fields are characterized by unusual scaling laws. Also, in assessing the spatial behavior of precipitation, a dependency to the timescale is shown. Downscaling of precipitation using stochastic models helps to better understand the probabilistic structure of precipitation. These models have long been used to model hydrologic events, such as runoff. If they are derived from the atmospheric processes that drive precipitation, they can predict governing precipitation patterns.

Given the wide range of regional and statistical downscaling techniques, there are a number of model comparisons using general data sets and model diagnostics. Until recently, these studies just compared statistical-versus-statistical downscaling models or regional-versus-regional models. However, a growing number of studies have compared statistical-versus-regional models and in Table 11.2 their relative strengths and weaknesses are summarized.

TABLE 11.2
Main Strengths and Weaknesses of Statistical Downscaling and Regional Model

	Statistical Downscaling	Regional Model
Strengths	• Station-scale climate information from GCM-scale output	• 10–50 km resolution climate information from GCM-scale output
	• Cheap, computationally undemanding, and readily transferable	• Respond in physically consistent ways to different external forcings
	• Ensembles of climate scenarios permit risk/uncertainty analyses	• Resolve atmospheric processes such as orographic precipitation
	• Applicable to "exotic" predictands such as air quality and wave heights	• Consistency with GCM
Weakness	• Dependent on the realism of GCM boundary forcing	• Dependent on the realism of GCM boundary forcing
	• Choice of domain size and location affects results	• Choice of domain size and location affects results
	• Requires high-quality data for model calibration	• Requires significant computing resources
	• Predictor–predictand relationships are often nonstationary	• Ensembles of climate scenarios seldom produced
	• Choice of predictor variables affects results	• Initial boundary conditions affect results
	• Choice of empirical transfer scheme affects results	• Choice of cloud/convection scheme affects (precipitation) results
	• Low-frequency climate variability problematic	• Not readily transferred to new regions or domains
	• Always applied off-line, therefore, results do not feedback into the host GCM	• Typically applied off-line, therefore results do not always feedback into the host GCM

Source: Wilby, R.L. and Dawson, C.W., *Using SDSM Version 3.1—A Decision Support Tool for the Assessment of Regional Climate Change Impacts*, Climate Change Unit, Environment Agency of England and Wales, Nottingham, U.K. and Department of Computer Science, Loughborough University, Leicestershire, U.K., 2004.

In the application of statistical downscaling, seasonal dependencies are considered an important point. For the regions defined seasonally based on temperate variations, the driving forces for the precipitation differ over the year, and result in the intra-annual deviations in downscaling. A greater extent of precipitation during summer months is driven than during winter months which should be considered in developing classification schemes. A solution for this problem can be dividing the year into four seasons. A crucial example can be the growth season of plants as there is a nonlinear relationship between the amount of rain and the water-use efficiency of plants.

Furthermore, downscaling methods must be valid in a troubled climate to be useful in future studies of climate change. The methods are assumed to estimate the difference between climate situations and must, therefore, work under a situation different from the one under which it was developed. Also, the climate should be stationary and no major shift in the physical processes governing the climate must occur or else it will result in a downscaling scheme outside the calibrated and validated range. To assess the ability of a model to deal with different climate situations, the model is calibrated on data from the driest seasons and then is validated on the wettest seasons in the calibration period and conversely. This is a simple test of the method's sensitivity to different climate situations. The stability of model performance in extreme years is evaluated. Long time series containing many different situations are used in the downscaling. These situations will be more frequent in a perturbed future climate. If the method can model those situations and if the range of variability of the large-scale variable in a future climate is of the same order as today, the problem will be minimized. There is some difficulty in dealing with stability and non-stationarity in downscaling and one should be careful when applying a model to the output of GCM simulations. If the climate is non-stationary, the underlying assumptions of the statistical link is unaffected by the climate change for model validity. The physical laws governing the climate are not likely to change, but the physical processes are parameterized for a specific climate. A multivariate approach is recommended to overcome the problem with non-stationarity of climate. In this approach, sufficient data should be available to ensure statistical integrity of the results. To consider the non-stationarity, a stochastic model is conditioned on the observed inter-annual variability.

The common method for downscaling is a regression-based method. Wilby and Wigley (1997) defined this method as a model involving the establishing of linear or nonlinear relationships between sub-grid-scale parameters and coarser-resolution predictor variables. Artificial Neural Networks ([ANN], Hewitson and Crane, 1996; Olsson et al., 2004) and Canonical Correlation Analysis ([CCA], Von Storch et al., 1993) are the common regression-based methods in this field. The models and techniques can be different from simpler multiple regression schemes to more sophisticated models, such as a method to link the covariances of atmospheric circulation and of local weather variables in a bilinear way (Burger, 1996). These methods can also be classified according to the used parameters in the downscaling procedure. In one approach, the same variable is used for the large-scale predictors as for the local-scale predictands, e.g., temperature, but in most studies, different parameters are used. Four steps are followed in regression methods:

1. Identification of a large-scale parameter G (predictor) as a controller of the local parameter L (predictand). If L is calculated for climate experiments, G should be simulated well by climate models.
2. Developing a statistical relationship between L and G.
3. Validation of the relationship with independent data.
4. If the relationship is confirmed, G is derived from GCM experiments to estimate L.

The idea that the predictor should be well simulated by GCMs is a key issue in this method. As mentioned before, statistical downscaling of climate variables is used to determine the impacts of climate change. Although the GCM is still the best tool to predict the future climate, but the assumption that the statistical relationships derived will remain fixed in the future is a weak point in this context.

11.8   DOWNSCALING MODELS

In this section, two downscaling models, SDSM and LARS-WG, which are widely employed in climate change studies, are introduced.

11.8.1   STATISTICAL DOWNSCALING MODEL

SDSM is a statistical downscaling model that enables to construct the climate change scenarios for individual sites at daily timescales, using grid resolution GCM output. Large-scale predictor variable information has been prepared for a large North American window. Some employed climate variables in the SDSM are listed, which include mean temperature, mean sea-level pressure, vorticity near surface, specific humidity, etc.

Since the choice of predictors largely determines the character of the downscaled climate scenario, this remains one of the most challenging stages in the development of any statistical downscaling model. The decision process is also complicated by the fact that the explanatory power of individual predictor variables varies both spatially and temporally.

In order to assemble and calibrate the model, the identified large-scale predictor variables are used for determining multiple linear regression relationships between these variables and local station data. It also regulates the average and variance of downscaled daily rainfall considering factors such as bias, and variance to confirm the simulated values to observed one. The first step in the application of SDSM, to determine that rainfall should occur daily, is as follows:

$$\omega_i = \alpha_0 + \sum_{j=1}^{n} \alpha_j \hat{u}_i^{(j)} \qquad (11.17)$$

where
  ω_i is the candidate in a random process generator that indicates a state of rainfall occurring or not occurring on day i
  $\hat{u}_i$ is the normalized predictor on day i
  α_j is the estimated regression coefficient

Precipitation on day i occurs if $\omega_i \leq r_i$, where r_i is a stochastic output from a linear random-number generator.

In the second step, the value of rainfall in each rainy day is estimated using z-score as follows:

$$Z_i = \beta_0 + \sum_{j=1}^{n} \beta_j \hat{u}_i^{(j)} + \varepsilon \tag{11.18}$$

where
  Z_i is the z-score for day i
  β_j are estimated regression coefficients for each month
  ε is a normally distributed stochastic error term

$$y_i = F^{-1}[\phi(Z_i)] \tag{11.19}$$

where
  ϕ is the normal cumulative distribution function
  F is the empirical function of the y_i daily precipitation amounts

The standardized predictors used in this method are provided by subtracting the mean and dividing them with the standard deviations over the calibration period.

In downscaling, a multiple linear regression model is developed between selected large-scale predictors and local-scale predictands such as temperature or precipitation. Based on the correlation analysis, partial correlation, scattered plots, and considering the physical sensitivity between predictors and predictands, the suitable large-scale predictors are chosen. Higher correlation values imply a higher degree of association and lower P-values indicates more chances for association between variables. It is important to know that the P-values lesser than .05 as threshold do not necessarily mean that the result can be statistically significant. On the other hand, the high P-value would indicate that the predictor–predictand correlation could be chance related (Wilbey and Dawson, 2004).

Example 11.4

A rain gage of an area called Station A is located at 60° 34′ longitude and 26° 12′ latitude. The precipitation data for 10 years from 1994 to 1999 are available in this rain gage. The monthly data are given in Table 11.3. Produce the precipitation data of Station A under the HadCM3B2 scenario outputs from 2010 to 2014 using the SDSM.

TABLE 11.3
Monthly Observed Precipitation for Station A for Period of 1992–2001

	January	February	March	April	May	June	July	August	September	October	November	December
1994	1.20	0.00	0.08	0.20	0.00	0.00	0.77	1.50	0.00	0.05	0.00	0.69
1995	0.00	0.24	0.00	0.61	0.00	0.24	2.51	0.20	0.27	0.36	0.05	3.01
1996	2.35	0.63	0.87	0.00	0.45	0.15	0.00	0.00	0.00	0.08	0.53	0.00
1997	0.86	0.00	3.26	0.10	0.00	1.42	0.00	0.00	0.00	3.52	0.64	2.45
1998	1.60	0.13	3.85	0.00	0.00	0.00	0.00	0.10	0.00	0.00	0.00	0.00
1999	0.00	0.75	1.30	0.00	0.00	0.00	1.32	0.00	0.00	0.00	0.00	0.00

Solution

The downscaling procedure at Station A is described step by step as follows:

Step 1: The input file is prepared, the daily precipitation data are sorted, based on the time in two columns, first the date and second the precipitation amount (the monthly data are interpolated to develop the daily data needed as input of SDSM).

Step 2: Then the process of quality control is run to test the format of the prepared input file.

Step 3: In this step the 16 available weather variables are screened to find the variables that could be used as precipitation predictand. For Station A, data start from January 1, 1994 and end at December 31, 1999. The analysis is done in an annual period under unconditional process.

At the end of this analysis, the results of explained variance and correlation matrix between precipitation and selected weather variables are given (Figures 11.11 and 11.12).

Two variables are selected for precipitation simulation as follows:

Relative humidity at 850 hPa height
Near surface specific humidity

The related correlation analysis of these variables is given in Figures 11.11 and 11.12. The P-value is low and partial correlations are high, which show the accuracy of the downscaling process.

Step 4: Model is calibrated using the observed data and the selected signals. At the end of this step, the R-squared and standard error are reported (Figures 11.13 and 11.14).

```
Results                                                                    [X]
File  Help
←  🖶  (i)
Back Print Help

RESULTS: EXPLAINED VARIANCE

Analysis Period: 01/01/1961 - 31/12/1990
Significance level: 0.05

Total missing values: 4648

Predictand: GHASR.DAT
```

Predictors:	JAN	FEB	MAR	APR	MAY	JUN	JUL	AUG	SEP	OCT	NOV	DEC
ncepp500as.dat	0.017	0.048	0.016	0.011						0.009	0.035	0.021
ncepp8_zas.dat		0.016								0.011		0.011
ncepp850as.dat	0.021	0.051	0.015								0.040	0.019
ncepp8thas.dat	0.042		0.025			0.013				0.020	0.011	0.011
ncepp8zhas.dat	0.091	0.038	0.072		0.012	0.010		0.016		0.013	0.012	0.017
ncepr500as.dat	0.025	0.029	0.034		0.033	0.015	0.036	0.066			0.024	0.026
ncepr850as.dat	0.209	0.234	0.186	0.041	0.040	0.034	0.054	0.064	0.020	0.018	0.053	0.111
nceprhumas.dat	0.206	0.267	0.220	0.053	0.055	0.034	0.035	0.053	0.014	0.028	0.061	0.117
ncepshumas.dat	0.160	0.179	0.178	0.038	0.034	0.024	0.031	0.046		0.019	0.040	0.080
nceptempas.dat		0.014	0.017									

FIGURE 11.11 Results of correlation analysis between predictors and daily precipitation.

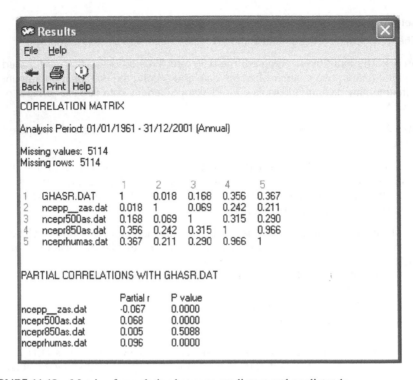

FIGURE 11.12 Matrix of correlation between predictors and predictands.

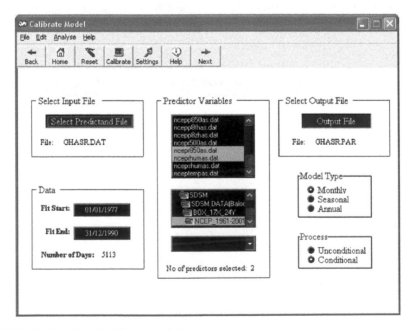

FIGURE 11.13 Model calibration window.

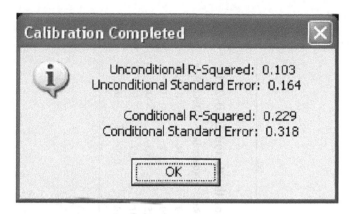

FIGURE 11.14 Model calibration results.

Step 5: Weather generator. In this step, data for considered period and selected climate change scenario (here HadCM3B2) are developed (see Table 11.4). The start date of data generating and amount of data to generate according to considered years will be entered. In this step you can choose production of one or more ensembles wherein one ensemble is selected.

Step 6: In this step, the statistical analysis of the observed and developed data through tables and graphs are given (Figure 11.15).

11.8.2 LARS-WG MODEL

One of the most important downscaling models is Long Ashton Research Station Weather Generator (LARS-WG). The input data of this model are daily time series of climate variables such as precipitation, solar radiation, and maximum and minimum temperatures. This model is a stochastic weather generator that simulates the weather data at a single site under both current and future climate conditions (Racsko et al., 1991; Semenov and Brooks, 1999; Semenov et al., 1998).

Two main purposes of the development of the stochastic weather generator are

1. To prepare a means of simulating synthetic weather time series with statistical characteristics analogous to the observed statistics at a site, and long enough for the assessment of risk in hydrological or agricultural applications
2. To simulate weather time series to unobserved locations, through the interpolation of the weather generator parameters obtained from running the models at neighboring sites

Stochastic weather generator cannot be used in weather forecasting as a predictive tool, but it can generate an identical time series of synthetic weather to the observations.

The semiempirical distributions are used for the lengths of wet and dry day series, daily precipitation, and daily solar radiation. The semiempirical distribution Emp = $\{a_0, a_i; h_i, i = 1, \ldots, 10\}$ is a histogram having ten intervals, $[a_{i-1}, a_i)$, where $a_{i-1} < a_i$, and h_i denotes the number of events from the observed data in the ith interval. Random

TABLE 11.4

Average Monthly Simulated Precipitation (mm) at Station A HadCM3B2 Scenario for the Period 2010–2014

	January	February	March	April	May	June	July	August	September	October	November	December
2010	0.01	0.09	0.77	2.35	0.94	0.14	0.27	0.13	0.05	0.09	0.34	0.12
2011	0.02	0.12	1.19	1.79	0.72	0.41	0.27	0.08	0.04	0.20	0.12	0.01
2012	0.02	0.12	2.24	1.46	0.30	0.60	0.27	0.11	0.03	0.15	0.63	0.01
2013	0.06	0.12	1.42	1.84	0.20	0.17	0.27	0.04	0.03	0.31	0.18	0.00
2014	0.06	0.58	2.33	1.64	0.29	0.11	0.27	0.06	0.04	0.19	0.00	0.02

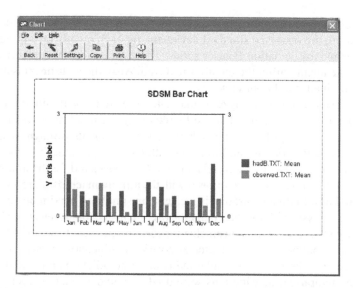

FIGURE 11.15 Comparison of the HadCM3B2 scenario results and observed precipitation.

values are chosen by first selecting of the intervals, and then selecting a value within that interval from the uniform distribution. This kind of distribution is flexible and can approximate a wide variety of shapes by adjusting the intervals $[a_{i-1}, a_i)$. However, this flexibility leads to specify 21 parameters, 11 values indicating the interval bounds and 10 values indicating the number of events within each interval, in comparison with, for instance, 3 parameters for the mixed-exponential distribution used in an earlier version of the model to define the dry and wet series (Racsko et al., 1991).

Based on the expected properties of the weather variables, the intervals $[a_{i-1}, a_i)$ are selected. For solar radiation, the intervals $[a_{i-1}, a_i)$ are divided into the minimum and maximum values of the monthly observed data, whereas for the period of dry and wet series and for precipitation, the interval size regularly increases as i increases. In the last two cases, there are typically many small values and a few very large ones and the use of a very coarse resolution for small values is prevented by this choice of interval structure.

The simulation of precipitation occurrence is modeled as alternate wet and dry series. A day with precipitation >0.0 mm is identified as a wet day. The random length of each series is selected from the wet or dry semiempirical distribution when the series starts in a month. To determine the distribution, observed series are allocated to the month. The precipitation value for a wet day is generated using the semiempirical precipitation distribution for the particular month, which does not depend on the length of the wet series or the amount of precipitation in previous days.

Daily minimum and maximum temperatures are considered as stochastic processes with daily means and daily standard deviations conditioned on the wet or dry status of the day. The technique used to simulate the process is very similar to that presented in Racsko et al. (1991). The seasonal cycles of means and standard deviations are modeled by finite Fourier series of order 3 and the residuals are approximated by a normal distribution. The Fourier series for the mean is fitted to the observed mean values for each month. To give an estimated average daily standard deviation, the observed standard deviations for each month are adjusted before fitting the standard deviation Fourier series. To do this, the estimated effect of the changes in the mean within the month is removed. Also, the adjustment is calculated using the obtained fitted Fourier series.

The observed residuals analyze a time autocorrelation for minimum and maximum temperatures, which are obtained by removing the fitted mean value from the observed data. To simplify this procedure, constant amounts throughout the entire year are assumed for both dry and wet days with the average value from the observed data being used. A preset cross-correlation of 0.6 is considered for the minimum and maximum temperature residuals. In cases with the minimum temperature less than 0.1, the simulated minimum temperature is greater than simulated maximum temperature. Based on the analysis of daily solar radiation over many locations, the normal distribution for daily solar radiation is unsuitable for certain climates (Chia and Hutchinson, 1991). The distribution of solar radiation also varies significantly on wet and dry days.

Therefore, in order to describe solar radiation solar radiation on wet and dry days, separate semiempirical distributions were used. An autocorrelation coefficient was also calculated for solar radiation and was assumed to be constant throughout the year. Solar radiation is modeled independently of temperature. The sunshine hours are accepted by LARS-WG as an alternative to solar radiation data.

The three main parts of the software consist of

1. *Analysis*: In the first step, the observed station data is analyzed to calculate the weather generator parameters. LARS-WG observed inputs are precipitation and maximum and minimum temperature, and sunshine hours. In this analysis, semiempirical distributions of the observed data are used for wet and dry series duration, precipitation amount, and solar radiation. Also, using Fourier series, maximum and minimum temperatures are calculated. The resulting parameter file is used in the generation process.
2. *Generator*: In this step, the synthetic weather data are generated from the combination of a scenario file containing information about changes in precipitation amount, wet and dry series duration, mean temperature, temperature variability, and solar radiation with the parameter files in step (1). The scenario file will contain the appropriate monthly changes if LARS-WG is being used to generate daily data for a particular scenario of climate change.
3. *QTest*: To evaluate the properness of simulating observed data, LARS-WG provides the QTest option. In this step, the statistical characteristics of the observed data are compared with those of synthetic data generated using the parameters derived from the observed station data. Several statistical tests are used to determine whether the distributions, mean values, and standard deviations, of the synthetic data are significantly different from those of the original observed data set.

The main use of LARS-WG is in the generation of daily data from monthly climate change scenario information. The advantage of this software is that a number of different daily time series representing the scenario can be generated by using a different random number to control the stochastic component of the model. Hence, these time series vary on a day-to-day basis, although they have the same statistical characteristics. This permits risk analyses to be undertaken (Canadian Climate Impacts and Scenarios, 2006).

Example 11.5

A rain gage of an area called Station A is located at 47° 4' longitude and 38° 26' latitude. The precipitation data for 10 years from 1986 to 2007 are available in this rain gage. Produce the precipitation data of Station B under the HadCM3A2 scenario outputs from 2011 to 2030, 2046 to 2065, and 2080 to 2099 using the LARS model. Show the result of 2011–2020 in a table.

Solution

Step 1: For applying the LARS model, two input files should be developed as follows.

Station A.st file: This file includes sites name, latitude, longitude, and altitude of the Station A rain gage, the directory path location and name of the file containing the observed weather data for the Station A, and the format of the observed weather data in the file which here is the year (YEAR), Julian day (JDAY, i.e., from 1 to 365), minimum temperature (MIN; °C), maximum temperature (MAX; °C), precipitation (RAIN; mm), and solar radiation (RAD; MJm-2day-1) or sun shine hours (SUN). Note that if you have SUN as an input variable, LARS-WG automatically converts it to global radiation (Figure 11.16).

(Station A.sr) file: Contains the daily data of rainfall, maximum temperature, minimum temperature, and solar radiation at the Station A in the order mentioned at Station A.st file (Figure 11.17).

Step 2: After preparing the mentioned files for calibrating the model, the SITE ANALYSIS option is selected and Station A.st is addressed to model. Two files called Station A.sta and Station A.wg are created. These two files provide the statistical characteristics of the data and parameter information which will be used by LARS-WG to synthesize artificial weather data in the GENERATOR process.

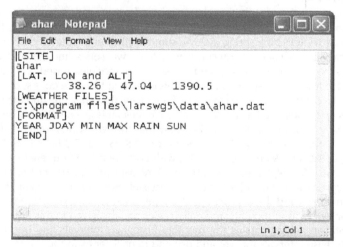

FIGURE 11.16 A layout of basic characteristics of station A (Station A.st file).

```
Untitled - Notepad
File  Edit  Format  View  Help
1987    259     9        26.4     0     10.7
1987    260     10.2     22.6     0     10.2
1987    261     8.2      21.6     0     0
1987    262     7.6      13       0     5.2
1987    263     7.4      17       0     6.8
1987    264     5        27.8     0     10.7
1987    265     5.2      27       0     10.6
1987    266     5.2      26.8     0.1   2.2
1987    267     4        18.8     0     10.4
1987    268     5        27.4     0     8.7
1987    269     10       30.4     0     9.5
1987    270     12       30.2     0     5.3
1987    271     8.4      22       0     5.7
1987    272     7.4      24.5     0     10
1987    273     8        27       0     9.8
1987    274     10.8     28.6     0     9.7
1987    275     13.6     29       0     10
1987    276     11.8     28.8     0     6.7
1987    277     3.4      11.8     1     0
1987    278     0.2      2.6      14    0
1987    279     0.2      1.4      14    0
1987    280     1        6.5      1     0
1987    281     4        6        3.6   0
1987    282     3.6      8        0     1.5
1987    283     2.2      9.7      0     8
1987    284     0.4      10       0     8.2
1987    285     -0.6     9.8      0     10.2
1987    286     1        15.6     0     8.2
1987    287     0.6      12.4     0     9.2
1987    288     5.2      11.8     0     2.4
1987    289     8.4      12.5     0     4.4
1987    290     7.6      13.2     0     5.4

                                            Ln 42, Col 1
```

FIGURE 11.17 The input data of station A for LARS model (Station A.sr).

In Station A.sta file (Figure 11.18), the first few lines give the site name and location information. This is followed by the statistical characteristics of the observed weather data.

First, the empirical distribution characteristics for the length of wet and dry series of days in the observed data are obtained. This information is given in four lines by season (i.e., winter [DJF], spring [MAM], summer [JJA], and autumn [SON]), where the first two lines of each section correspond to the wet series, and the last two lines represent the dry series. The wet and dry series are modeled based on histograms constructed from the observed data. The histograms in this case consist of 10 intervals and the cutoff points for each interval are given in the first line and the number of events in the observed data falling into each interval is given in the second line.

So, using Station A.sta, it can be seen that the wet series intervals (hi) are $0 \leq h1 < 1$, $1 \leq h2 < 2$, $2 \leq h3 < 3$, $3 \leq h4 < 4$, $4 \leq h5 < 5$, $5 \leq h6 < 6$, $6 \leq h7 < 7$, $7 \leq h8 < 8$, $8 \leq h9 < 9$ and $9 \leq h10 < 10$, with corresponding frequencies of occurrence of 236, 161, 44, 16, 8, 3, 0, 2, 0, and 0, respectively (Figure 11.19). The same data are given for precipitation. The summary statistics (min, max, mean, standard deviation) of wet and dry day, precipitation, minimum and maximum temperature, and solar radiation in monthly and daily scales are also given in this file.

File Station A.wg contains the monthly histogram intervals and frequency of events in these intervals for wet and dry and rainfall series in four lines similar to Station A.sta, four Fourier coefficients, $a[i]$ and $b[i]$, $i = 1,...,4$, for the means and

```
 aharWG - Notepad                                                              _ □ X
File  Edit  Format  View  Help
[VERSION]
LARS-WG5.11
[NAME]
aharWG
[LAT, LON and ALT]
     38.26    47.04    1390.50
[YEARS]
     1        50
[SERIES seasonal distributions: WET and DRY]
[DJF]
     554    23    2.18    1.29    2.22    1.32
     0.000  0.048  0.095  0.143  0.190  0.238  0.286  0.333  0.381  0.429  0.476  0.524  0.571
     1.00   1.00   1.00   1.00   1.00   1.00   1.00   1.00   2.00   2.00   2.00   2.00   2.00
     347    23    5.67    5.43    5.78    5.52
     0.000  0.048  0.095  0.143  0.190  0.238  0.286  0.333  0.381  0.429  0.476  0.524  0.571
     1.00   1.00   1.00   1.00   1.00   1.00   2.00   2.00   3.00   3.00   4.00   4.00   5.00
[MAM]
     680    23    2.71    1.27    2.75    1.31
     0.000  0.048  0.095  0.143  0.190  0.238  0.286  0.333  0.381  0.429  0.476  0.524  0.571
     1.00   1.00   2.00   2.00   2.00   2.00   2.00   2.00   2.00   2.00   2.00   2.00   3.00
     677    23    4.25    4.42    4.32    4.99
     0.000  0.048  0.095  0.143  0.190  0.238  0.286  0.333  0.381  0.429  0.476  0.524  0.571
     1.00   1.00   1.00   1.00   2.00   2.00   2.00   2.00   2.00   3.00   3.00   3.00   3.00
[JJA]
     336    23    2.08    1.01    2.05    0.95
     0.000  0.048  0.095  0.143  0.190  0.238  0.286  0.333  0.381  0.429  0.476  0.524  0.571
     1.00   1.00   1.00   1.00   1.00   1.00   2.00   2.00   2.00   2.00   2.00   2.00   2.00
     354    23    12.73   15.70   12.43   14.97
     0.000  0.048  0.095  0.143  0.190  0.238  0.286  0.333  0.381  0.429  0.476  0.524  0.571
     1.00   1.00   1.00   2.00   2.00   2.00   3.00   4.00   4.00   4.00   5.00   6.00   8.00
[SON]
     387    23    2.14    1.21    2.20    1.17
     0.000  0.048  0.095  0.143  0.190  0.238  0.286  0.333  0.381  0.429  0.476  0.524  0.571
     1.00   1.00   1.00   1.00   1.00   1.00   1.00   2.00   2.00   2.00   2.00   2.00   2.00
     378    23    8.42    7.65    8.42    7.72
     0.000  0.048  0.095  0.143  0.190  0.238  0.286  0.333  0.381  0.429  0.476  0.524  0.571
     1.00   1.00   2.00   2.00   2.00   3.00   3.00   4.00   5.00   5.00   5.00   7.00   8.00
[SERIES statistics: max, N of observations, mean and sd for WET and DRY]
     7.00   7.00   8.00   7.00   9.00   7.00   4.00   4.00   4.00   6.00   8.00   7.00
     212    170    213    238    228    175    99     68     93     141    148    172
     2.05   2.49   2.85   2.54   2.77   2.51   1.79   1.41   1.43   2.35   2.30   2.09
     1.39   1.25   1.10   1.15   1.48   0.97   0.89   0.71   0.72   0.94   1.38   1.24
     26.00  22.00  16.00  15.00  50.00  65.00  103.00 62.00  40.00  39.00  38.00  31.00
     212    171    210    232    235    184    103    72     85     139    148    165
     5.41   5.53   4.43   3.72   4.60   7.20   18.61  17.49  10.55  7.87   7.99   6.21
     4.84   4.90   3.29   2.86   6.15   10.07  20.15  14.95  10.01  6.62   7.13   6.52
[RAIN distributions]
     419    23    2.63    2.77    2.87    2.95
     0.000  0.107  0.344  0.390  0.417  0.453  0.490  0.526  0.562  0.599  0.635  0.672  0.708
     0.20   0.50   1.00   1.00   1.10   1.10   1.40   1.60   1.90   2.10   2.30   2.80   3.10
     430    23    2.37    2.86    2.47    2.90
     0.000  0.253  0.463  0.493  0.522  0.552  0.582  0.612  0.642  0.672  0.702  0.731  0.761

                                                                          Ln 1, Col 1
```

FIGURE 11.18 The summary statistics produced through LARS for station A at seasonal scale.

standard deviations of different combinations of minimum and maximum temperature, and solar radiation on wet and dry days (Figure 11.19).

Step 3: For model validation, Test option is selected from the QTest module in LARS model. The QTest option carries out a statistical comparison of synthetic weather data generated using LARS with the parameters derived from observed weather data in calibrated model. Selecting the Test option, first, name of Station A would be selected, then the simulation period (20 years) and random segment related to stochastic predicting model will be selected (Figure 11.20). The performance of the model in simulating the observed precipitation pattern in the calibration period is evaluated in Figure 11.20. The model output matches closely the observed values (scenario A2).

Step 4: The Generator option in LARS model is used to simulate synthetic weather data. This option is used to generate synthetic data which have the same statistical characteristics as the observed weather data, or to generate synthetic weather data corresponding to a scenario of climate change.

First the scenario related to the considered area would be determined. Here, scenarios obtained according to GCM climatic model, namely ECHO-G A1 for grid 17–31, are used. Initial inputs of this stage are name and the address of GCM scenario.

After selecting the site and setting up the appropriate scenario files, there are three further options to be completed. The first is the Scaling factor. This factor can be used if a climate change scenario is implemented for a particular future time period, say 2100, to obtain data for an earlier time period without having to create

```
aharWG - Notepad                                                    _ □ X
File  Edit  Format  View  Help
[[VERSION]
LARS-WG5.11
[NAME]
aharWG
[LAT, LON and ALT]
    38.26    47.04    1390.50
[YEARS]
     1        50
[SERIES WET]
    212      23       2.05     1.39     2.06     1.39
    0.000    0.048    0.095    0.143    0.190    0.238    0.286    0.333    0.381    0.429    0.476    0.524    0.571
    1.00     1.00     1.00     1.00     1.00     1.00     1.00     1.00     1.00     1.00     2.00     2.00     2.00
    170      23       2.49     1.25     2.52     1.28
    0.000    0.048    0.095    0.143    0.190    0.238    0.286    0.333    0.381    0.429    0.476    0.524    0.571
    1.00     1.00     1.00     1.00     2.00     2.00     2.00     2.00     2.00     2.00     2.00     2.00     2.00
    213      23       2.85     1.10     2.80     1.10
    0.000    0.048    0.095    0.143    0.190    0.238    0.286    0.333    0.381    0.429    0.476    0.524    0.571
    1.00     2.00     2.00     2.00     2.00     2.00     2.00     2.00     2.00     2.00     2.00     3.00     3.00
    238      23       2.54     1.15     2.59     1.18
    0.000    0.048    0.095    0.143    0.190    0.238    0.286    0.333    0.381    0.429    0.476    0.524    0.571
    1.00     1.00     1.00     2.00     2.00     2.00     2.00     2.00     2.00     2.00     2.00     2.00     2.00
    228      23       2.77     1.48     2.80     1.46
    0.000    0.048    0.095    0.143    0.190    0.238    0.286    0.333    0.381    0.429    0.476    0.524    0.571
    1.00     1.00     1.00     2.00     2.00     2.00     2.00     2.00     2.00     2.00     2.00     2.00     2.00
    175      23       2.51     0.97     2.50     1.01
    0.000    0.048    0.095    0.143    0.190    0.238    0.286    0.333    0.381    0.429    0.476    0.524    0.571
    1.00     2.00     2.00     2.00     2.00     2.00     2.00     2.00     2.00     2.00     2.00     2.00     2.00
    99       23       1.79     0.89     1.84     0.92
    0.000    0.048    0.095    0.143    0.190    0.238    0.286    0.333    0.381    0.429    0.476    0.524    0.571
    1.00     1.00     1.00     1.00     1.00     1.00     1.00     1.00     1.00     1.00     2.00     2.00     2.00
    68       23       1.41     0.71     1.40     0.69
    0.000    0.048    0.095    0.143    0.190    0.238    0.286    0.333    0.381    0.429    0.476    0.524    0.571
    1.00     1.00     1.00     1.00     1.00     1.00     1.00     1.00     1.00     1.00     1.00     1.00     1.00
    93       23       1.43     0.72     1.44     0.75
    0.000    0.048    0.095    0.143    0.190    0.238    0.286    0.333    0.381    0.429    0.476    0.524    0.571
    1.00     1.00     1.00     1.00     1.00     1.00     1.00     1.00     1.00     1.00     1.00     1.00     1.00
    141      23       2.35     0.94     2.41     0.95
    0.000    0.048    0.095    0.143    0.190    0.238    0.286    0.333    0.381    0.429    0.476    0.524    0.571
    1.00     1.00     1.00     2.00     2.00     2.00     2.00     2.00     2.00     2.00     2.00     2.00     2.00
    148      23       2.30     1.38     2.31     1.38
    0.000    0.048    0.095    0.143    0.190    0.238    0.286    0.333    0.381    0.429    0.476    0.524    0.571
    1.00     1.00     1.00     1.00     1.00     1.00     1.00     2.00     2.00     2.00     2.00     2.00     2.00
    172      23       2.09     1.24     2.06     1.23
    0.000    0.048    0.095    0.143    0.190    0.238    0.286    0.333    0.381    0.429    0.476    0.524    0.571
    1.00     1.00     1.00     1.00     1.00     1.00     1.00     1.00     2.00     2.00     2.00     2.00     2.00
[SERIES DRY]
    212      23       5.41     4.84     5.11     4.78
    0.000    0.048    0.095    0.143    0.190    0.238    0.286    0.333    0.381    0.429    0.476    0.524    0.571
    1.00     1.00     1.00     1.00     1.00     1.00     2.00     2.00     2.00     3.00     3.00     5.00     5.00
    171      23       5.53     4.90     5.53     4.86
    0.000    0.048    0.095    0.143    0.190    0.238    0.286    0.333    0.381    0.429    0.476    0.524    0.571
    1.00     1.00     1.00     1.00     1.00     1.00     2.00     3.00     3.00     3.00     4.00     4.00     4.00
                                                                                              Ln 1, Col 1
```

FIGURE 11.19 The summary statistics produced through LARS for station a at monthly scale.

a scenario file for this earlier time period. The Scaling factor allows doing this by assuming that the changes in climate over time are linear. Here the Scaling factor has been considered equal to one and model is simulated in 20 years with random seed of 877. The simulated monthly data for the period of year 2011 to year 2020 are given in Table 11.5.

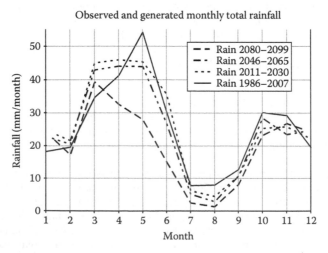

FIGURE 11.20 Generated and observed monthly total rainfall in different intervals.

TABLE 11.5
Generated Precipitation (Scenario A1B for 2011–2020 Period) for 10 Years

Year	January	February	March	April	May	June	July	August	September	October	November	December
									Monthly Total Precipitation (mm)			
1	23.90	27.60	59.80	11.80	46.60	52.80	14.80	4.90	0.50	3.40	16.50	44.50
2	49.90	18.50	9.30	55.90	75.20	35.70	3.60	0.00	3.60	22.90	22.30	44.70
3	22.90	0.90	23.30	60.20	95.00	68.30	5.00	14.60	0.00	17.20	11.60	23.20
4	1.40	23.60	40.50	61.00	26.20	8.70	5.80	0.00	0.00	9.70	0.00	16.00
5	29.10	42.10	27.20	75.50	57.30	44.70	1.70	4.00	8.20	8.30	55.20	21.80
6	19.90	15.10	56.50	50.10	67.20	42.30	0.30	9.20	0.00	46.80	14.60	14.80
7	40.90	32.30	71.20	43.50	60.40	31.70	3.50	3.20	0.00	27.10	24.50	53.10
8	18.10	2.00	55.40	40.00	67.10	0.00	13.90	0.00	7.70	30.90	21.60	10.90
9	6.40	38.30	46.80	32.90	26.50	26.40	1.70	11.10	10.90	54.50	8.20	28.00
10	25.50	3.10	57.80	32.80	35.70	60.80	1.30	0.00	30.30	41.10	25.10	68.40

11.9 CLIMATE CHANGE IMPACTS ON WATER AVAILABILITY IN AN AQUIFER

The impacts of climate change on water suppliers may involve many cause-and-effect relationships which are more remote and uncertain in terms of both the chain of events and timing. For appropriate planning of water systems, a working knowledge about the cause-and-effect relationships that produce impacts is needed. Groundwater is an important part of water-supply resources which is influenced by climate change in different aspects. In the following, an example is presented demonstrating how to combine a groundwater model (MODFLOW) with DSS tool (WEAP). This method could be applied in similar studies to examine the effects of climate change, land-use management, and demand management on groundwater.

Example 11.6

A Main River basin located in a semiarid climate, with precipitation only in the winter months, encompasses 783 km². Most of this precipitation is lost through evapotranspiration. The aquifer consists of fractured crystalline bedrock and alluvial fills in the valleys, and its saturated thickness varies between 35 m in the south to about 53 m in the north. Determine the impacts of climate change on runoff, groundwater recharge, and cell head changes in this watershed.

Solution

The following sections intend to introduce you to the built up model.

Step 1: Installation and running the software

You need a regular PC with Microsoft windows XP or Vista or seven installed as operating system. These steps are required to install and run the model:

1. Run "WEAP 2.2054.exe" and follow the installation wizard. Do not change the installation target (C:Program FilesWEAP21). The installed software is a trial version of WEAP. In order to get the product license, refer to http://www.weap21.org
2. Now you need to create the project into the software directory. To do so, extract "WEAP_MODFLOW.zip" and copy the whole extracted folder named WEAP_MODFLOW to C:Program FilesWEAP21.
3. Run WEAP, go to AreaOpenWEAP_MODFLOW.

Step 2: MODFLOW model:

An existing MODFLOW groundwater model for the Main River basin was taken originally from the GMS Tutorial and modified to the current model. Groundwater flow is toward the spring, the Main River, and its two tributaries. The spring represents the headflow of the Main River, with an average monthly discharge of about 320,000 m³. In the north, outside the model area, the Main River finally drains into a reservoir lake with a constant water level of 304.8 meter above sea level (m.a.s.l.). The topography gradually decreases from south (353 m.a.s.l) to north (335 m.a.s.l) (see Figure 11.21).

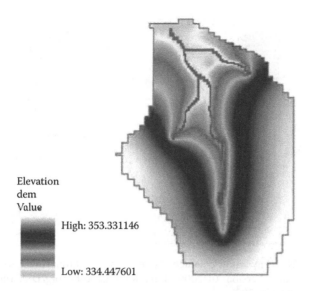

FIGURE 11.21 Topography maps. (From Huber, M. et al., *WEAP-MODFLOW Tutorial*, Version 01/2009, http://www.acsad-bgr.org/DSS-Tutorial/WEAP_MODFLOW_Tutorial.zip, 2009.)

Mathematical model:

Grid: The model consists of one layer (one unconfined aquifer, 60–80 m layer thickness, 35–50 m saturated thickness) with 88 × 70 cells. The cell sizes vary in width and height between 200 and 1200 m (Figure 11.22).

Boundary conditions: Specified-head boundary in the north (reservoir lake level). All other boundaries are no-flow boundaries (Figure 11.22).

Hydraulic conductivity: The four hydraulic conductivity zones for the model are shown in the figure below. The hydraulic conductivity of the aquifer varies between 100 (m/day) along the river valley to 40 (m/day) toward the basin boundary.

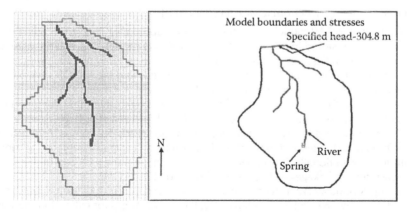

FIGURE 11.22 Model boundaries and stresses. (From Huber, M. et al., *WEAP-MODFLOW Tutorial*, Version 01/2009, http://www.acsad-bgr.org/DSS-Tutorial/WEAP_MODFLOW_Tutorial.zip, 2009.)

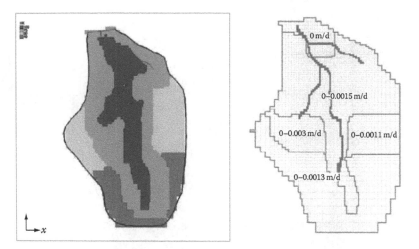

FIGURE 11.23   Model hydraulic conductivity and recharge. (From Huber, M. et al., *WEAP-MODFLOW Tutorial*, Version 01/2009. http://www.acsad-bgr.org/DSS-Tutorial/WEAP_MODFLOW_Tutorial.zip, 2009.)

Recharge: Five different recharge zones have been defined. The values in the zones range from 0 to 0.0015 m/day (Figure 11.23).

Additional package: The spring was modeled as two DRAIN-cells and the rivers as RIVER-cells (blue in Figure 11.23).

Initial head: The initial head ranges from 306 to 327 m.a.s.l., resulting in a depth to water table of 25–36 m (see Figure 11.24).

The model has been built and run on transient state in order to cover 2 years of recharge variation. It consists of 24 monthly stress periods, which contain three equal time steps each. The final balance (see Figure 11.25—units in m³ or m³/day) shows that the model is almost balanced or close to steady state, with recharge as the main inflow and river leakage as the main outflow fraction.

Step 3: Extracting data from a MODFLOW model:

Building up a WEAP-model using the "hard data approach," requires reading out data which are stored in the native MODFLOW2000 model files, like the natural recharge rate, well abstraction, or riverbed conductance. The number and type of information depends upon the model's characteristics and the MODFLOW packages that are defined for the chosen model.

As native MODFLOW files are somewhat confusing in their file structure and complicated to read data from, a tool called Modflow2Weap has been developed to do this job. Modflow2Weap is a slim tool which reads out values from native MODFLOW2000 files and transcribes these values to a spreadsheet-format (*.dbf) output file. Additionally, it is capable to create a shape-file which can be used as the basis for the linkage-file between WEAP and MODFLOW.

Using Modflow2Weap to create the linkage consists of four steps: selecting the MODFLOW name-file, selecting MODFLOW packages for extraction, output options, and running the process. For more information about this tool please refer to Huber et al. (2009).

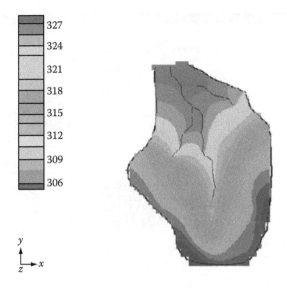

FIGURE 11.24 Initial head. (From Huber, M. et al., *WEAP-MODFLOW Tutorial*, Version 01/2009, http://www.acsad-bgr.org/DSS-Tutorial/WEAP_MODFLOW_Tutorial.zip, 2009.)

```
budget_5f.txt - Editor                                                      _|□|x|
Datei  Bearbeiten  Format  Ansicht  ?
1
  VOLUMETRIC BUDGET FOR ENTIRE MODEL AT END OF TIME STEP  3 IN STRESS PERIOD  24


     CUMULATIVE VOLUMES      L**3        RATES FOR THIS TIME STEP      L**3/T
     ------------------                 ------------------------


        IN:                                IN:
        ---                                ---
            STORAGE =  158684992.0000           STORAGE =      25869.2949
      CONSTANT HEAD =          0.0000     CONSTANT HEAD =          0.0000
             DRAINS =          0.0000            DRAINS =          0.0000
      RIVER LEAKAGE =          0.0000     RIVER LEAKAGE =          0.0000
           RECHARGE =  357105632.0000          RECHARGE =     635431.5625

           TOTAL IN =  515790624.0000          TOTAL IN =     661300.8750

        OUT:                               OUT:
        ----                               ----
            STORAGE =  160023184.0000           STORAGE =     248629.9062
      CONSTANT HEAD =    5145124.0000     CONSTANT HEAD =       6988.5078
             DRAINS =    7527779.5000            DRAINS =      10904.8906
      RIVER LEAKAGE =  343096672.0000     RIVER LEAKAGE =     394787.0938
           RECHARGE =          0.0000          RECHARGE =          0.0000

           TOTAL OUT =  515792768.0000         TOTAL OUT =     661310.3750

           IN - OUT =      -2144.0000          IN - OUT =         -9.5000

  PERCENT DISCREPANCY =        0.00     PERCENT DISCREPANCY =        0.00
```

FIGURE 11.25 Volumetric budget for entire model. (From Huber, M. et al., *WEAP-MODFLOW Tutorial*, Version 01/2009, http://www.acsad-bgr.org/DSS-Tutorial/WEAP_MODFLOW_Tutorial.zip, 2009.)

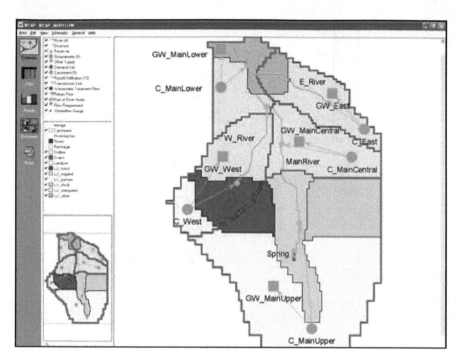

FIGURE 11.26 The schematic of respective data. (From Huber, M. et al., *WEAP-MODFLOW Tutorial*, Version 01/2009, http://www.acsad-bgr.org/DSS-Tutorial/WEAP_MODFLOW_Tutorial.zip, 2009.)

Step 4: Building WEAP model:

A schematic model of the catchment was created in WEAP, as a basis for assigning the respective data in the following section (see Figure 11.26). A new area is created, and the working area boundaries are set. Then, the background layers are loaded and four rivers are digitized. Finally, five catchments and five groundwater nodes are added and named.

WEAP's internal FAO-Rainfall-Runoff model was utilized to calculate the groundwater recharge. The FAO-Rainfall Runoff model is a simple water balance model, which does not consider any soil parameters, based on the following equations. This runoff volume can be fractioned to a surface runoff and a groundwater recharge fraction under Supply and ResourcesRunoff and Infiltrationfrom catchmentInflow and OutflowRunoff Fraction; here it was assigned 40% surface runoff and 60% groundwater recharge. If the evapotranspiration amount is more than the precipitation, the irrigation demand is obtained by subtracting these two values. Also, the actual evapotranspiration is calculated by multiplying the reference evapotranspiration (ET_{ref}) and crop coefficient (K_c).

The calibration target for the Rainfall Runoff model was to get the same recharge values like in the initial stand-alone MODFLOW model. To achieve this, monthly precipitation, evapotranspiration, and crop coefficients have been entered to the model.

The required data for WEAP to run the simulation are Demand Site Area, Crop Coefficients, Monthly Precipitation, Monthly Evapotranspiration, Runoff Fraction, and River Headflows.

Step 5: WEAP scenarios:

Initially the Main River basin was a natural system without any human interference; however, through time various measures can take place to utilize the water resources of the basin or changes of climate and land use may occur. The strength of WEAP is to model scenarios, historic or future ones, so that the user can see the respective impacts and choose the best option for planning in accordance with the decision-makers.

In this section, some changes are introduced as respective scenarios showing the changes and impacts happening from the "natural" CurrentAccount year (2000) to the future (year 2001–2020). As a reference, the scenario "Reference" is kept for comparison (repetition of the CurrentAccountYear) and is at the same time the basis for all other scenarios.

In the "Climate Change" scenario, we will evaluate the impact of climate change on surface runoff, groundwater levels, and spring flow. Declining precipitation accompanied with increasing temperatures is expected in the hypothetical model. Higher temperature will increase the evapotranspiration and decrease infiltration and surface runoff. We compare the respective impacts by creating a Climate Change scenario based on the scenario Reference.

"Climate Change" scenario describes a linear trend in monthly precipitation and evapotranspiration, which will be applied simultaneously, during future 20 years.

Step 6: Results:

Water balance: The result shows the decline of surface runoff (Figure 11.27).
 The groundwater levels for each scenario are shown (Figure 11.28).
 Figure 11.29 shows the groundwater storage volumes.

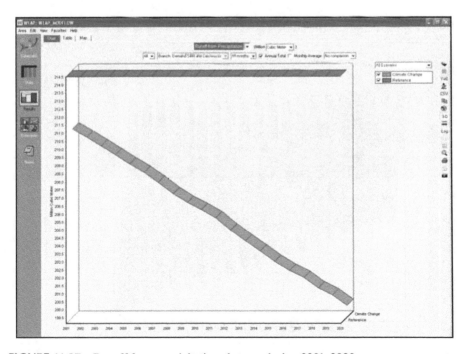

FIGURE 11.27 Runoff from precipitation changes during 2001–2020.

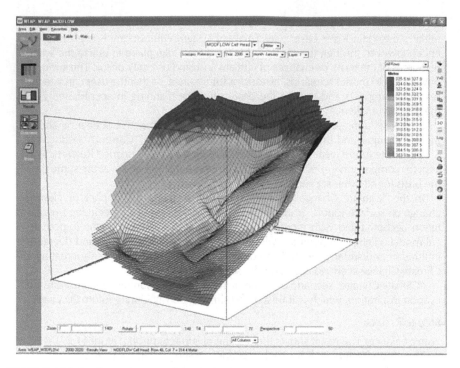

FIGURE 11.28 Groundwater cell head changes.

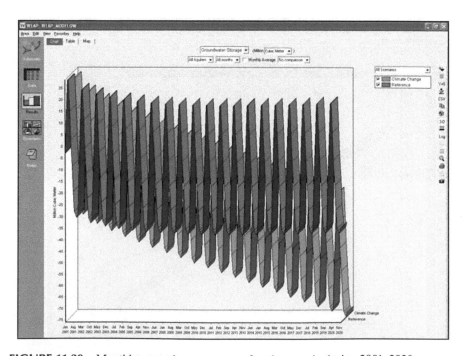

FIGURE 11.29 Monthly groundwater storage of each scenario during 2001–2020.

11.10   CONCLUSION

Climate change can directly affect the water balance of a surface water system; the impact on the groundwater system is indirect and is influenced through the spatially and temporally varying recharge and discharge boundary conditions that link the groundwater system to the surface water system. Groundwater is an irreplaceable resource for livelihoods and agriculture. Adverse impacts of climate change on groundwater resources are expected, including changes in recharge rates, saline intrusion in coastal aquifers, and decreased long-term groundwater storage. Groundwater may increase in importance and help to ameliorate the worst effects of climate change on water resources and sustainable development. However, once seriously damaged, recovering groundwater resources requires vast amounts of funds and time. The role and demand for groundwater will increase as water resources in general become scarcer and less predictable with climate change and other global changes, such as population growth, urbanization, and increasing living standards. Groundwater provides a strategic resource with relatively good and stable properties, in terms of quality and quantity. In arid areas, it may be the primary water source, and in semiarid and less arid areas, it may serve to bridge dry periods and supplement dwindling, poor quality, or mismanaged surface water resources. Groundwater often provides a relatively easily accessible resource, both from a physical/technical point of view, but also from a socioeconomic and equity point of view. Generally, groundwater recharge, discharge, and storage are influenced by climate change. Recharge can occur as a result of precipitation events while both recharge and discharge can occur at surface water bodies. Variation in recharge are including change in diffuse recharge due to change in rainfall, change in diffuse recharge due to land use change, and change in river recharge due to change in river stage. Climate change leads to some changes in the discharge as increase in demand for extraction, and due to the changes in river stage. Moreover, some changes in storage are due to the change in recharge, in extraction, and sea-level rise, which could lead to saltwater intrusion in coastal areas.

It is obvious that the projected climate change resulting in warming, sea-level rise, and melting of glaciers will adversely affect the water balance in different parts of the world, and the quality of groundwater and its recharge rate along the coastal plains. Based on these impacts and also, because of the importance of groundwater as a source for water supply, more studies should be assessed and evaluated to predict and adapt to climate change.

PROBLEMS

11.1 Based on the available GCM data of rainfall compare the results of scenarios B1 and B2 from the HADCM3 model for grid 46×35 in period of 2010–2049. Discuss the possible reasons for the observed differences. Use both 10 and 1 ensemble and compare the differences between the results. Which should be more considered for planning a water resources development project in a city located in this grid?

11.2 Using the precipitation data of a station in your own city, evaluate the changes in mean, maximum and minimum monthly rainfall and wet and dry spells under scenario HadCM3 B2 and A2. Discuss the changes which occurred in each scenario and the intensity of the climate change effects. How should water resources planning changes be planned in response to these changes?

11.3 During a long wet season, the water table raised 10 m (average). The water was increased due to rainfall. Assuming specific yield is 0.2, what is the change in the recharge rate due to the rainfall?

11.4 Describe the difference between the infiltration and percolation and then, explain how climate change leads to change in infiltration rate which influence groundwater.

11.5 Explain the sea level rise impacts on groundwater quality, and mention why these changes would be concerning.

11.6 Mention some adaptation methods to climate change, and explain one related to groundwater.

11.7 Utilize the rainfall, minimum temperature, and maximum temperature data of a station in a city in the LARS model and downscale it. Define different scenarios and evaluate the effect of climate change on the statistics of the considered variables.

11.8 Develop a MLP model to simulate the monthly water demand of an aquifer in the city mentioned in Problem 11.9 as a function of rainfall, minimum, maximum temperature.

11.9 Formulate a simple model based on dynamic hypothesis of climate change impacts on water balance in a region; in this model consider the following causal loops and their interactions. For simplicity, separate primary and secondary impacts considering their interactions and feedbacks.

11.10 Simulate the climate change impacts on groundwater in a catchment located in your country having an area of at least 500 km^2. The input data includes the monthly precipitation 2001–2010. Use the Weap-Modflow model and show the recharge changes during the interval 2010–2030.

11.11 Considering the model simulated in Problem 11.12, generate the storage volume and cell head changes using the climate change scenario in years 2040, 2045, and 2050.

11.12 Use daily temperature data 1980–2000 of station B located at 56° 01′ longitude and 30° 24′ latitude and with elevation 1469 m. Generate the data of station B under HadCM3A2 and B1 scenarios outputs from 2020 to 2025 and using both SDSM and LARS model and compare the difference between the results. Judge which will be more accurate.

11.13 Solve Example 11.4 considering 50 ensembles and use the Briers skill score to choose the best ensemble and then use Levene's test and Wilcoxon rank sum test to determine the best scenario between HADA and HADB. Give the results including P-Value for each month in a table.

REFERENCES

Bouraoui, F., Vachaud, G., Li, L.Z.X., LeTreut, H., and Chen, T. 1999. Evaluation of the impact of climate changes on water storage and groundwater recharge at the watershed scale. *Climate Dynamics*, 15, 153–161.

Burger, G. 1996. Expanded downscaling for generating local weather scenarios. *Climate Research*, 7, 111–128.

California Department of Water Resource. 2007. *Climate Change in California*, State of California, http://www.water.ca.gov/

Canadian Climate Impacts and Scenarios. 2006. *Canadian Institute for Climate Studies.* University of Victoria, Victoria, BC, Canada.

Chia, E. and Hutchinson, M.F. 1991. The beta distribution as a probability model for daily cloud duration. *Agricultural and Forest Meteorology Elsevier*, 56, 195–208.

Deng, Y., Flerchinger, G.N., and Cooley, K.R. 1994. Impacts of spatially and temporally varying snowmelt on subsurface flow in a mountainous watershed: 2. Subsurface processes. *Hydrological Sciences Journal/Journal des. Sciences Hydrologiques*, 39, 521–534.

Delleur, J.W. 2007. *The Handbook of Groundwater Engineering.* CRC Press LLC, Boca Raton, FL, 988 p.

Drabbe, J. and Badon Ghyben, W. 1889. *Nota in verband met de voorgenomen putboring nabji Amssterdam, Tijdschrift van het Koninklijk Instituut van Ingenieurs.* Verhandelingen The Hague, the Netherlands, pp. 8–22.

EPA. 2007a. Climate change website. Climate change–Science: Temperature changes. http://www.epa.gov/climatechange/science/recenttc.html

EPA. 2007b. Climate change website. Climate change–Science: Future temperature changes. http://www.epa.gov/climatechange/science/futuretc.html

Field, C.B., Mortsch, L.D., Brklacich, M., Forbes, D.L., Kovacs, P., Patz, J.A., Running, S.W., and Scott, M.J. 2007. North America, in *IPCC. 2007. Climate Change 2007: Impacts, Adaptation and Vulnerability. Contribution of Working Group II to the Fourth Assessment Report of the Intergovernmental Panel on Climate Change.* M.L. Parry, O.F. Canziani, J.P. Palutikof, P.J. van der Linden, and C.E. Hanson, eds. Cambridge University Press, Cambridge, U.K. and New York. http://www.ipcc.ch/pdf/assessment-report/ar4/wg2/ar4-wg2-chapter14.pdf

Firth, P. and Fisher, S.G. 1992. *Global Climate Change and Freshwater Ecosystems.* Springer-Verlag, New York, 321 p.

Fisher, S.G. and Grime, N.B. 1996. Ecological effects of global climate change on freshwater ecosystems with emphasis on streams and rivers, in *Water Resources Handbook*. L.W. Mays, ed., McGraw-Hill, New York, 1568 p.

Foster, S.S.D. 2001. Interdependence of groundwater and urbanization in rapidly developing cities (Review). *Urban Water*, 3, 185–192.

Gerber, R.A.F. and Howard, K. 2000. Recharge through a regional till aquitard: Three-dimensional flow model water balance approach. *Ground Water*, 38, 411–422.

Gleick, P.H. 1989. Climate change, hydrology, and water resources. *Reviews of Geophysics*, 27, 329–344.

Hamlet, A.F., Mote, P.W., Clark, M.P., and Lettenmaier, D.P. 2007. Twentieth-century trends in runoff, evapotranspiration, and soil moisture in the western United States. *Journal of Climate*, 20, 1468–1486.

Harms, T.E. and Chanasyk, D.S. 1998. Variablity of snowmelt runoff and soil moisture recharge. *Nordic Hydrology*, 29, 179–198.

Herzberg, B. 1901. Die Wasserversorgung einiger Nordseebader, *Journal of Gasbeleutchtung und Wasserversorgung, Munich*, 44, 815–819, 842–844.

Hetzel, F., Vaessen, V., Himmelsbach, T., Struckmeier, W., and Villholth, K.G. 2008. *Groundwater and Changes: Challenges and Possibilities*. BGR and GEUS Publications, Hannover, Germany.

Hewitson, B.C. and Crane, R.G. 1996. Climate downscaling: Techniques and application. *Climate Research*, 7, 85–95.

Huber, M., Wolfer, J., Al-Sibai, M., Al Mahameed, J., Abdallah, A., Al Shihabi, O., and Jnad, I. 2009. *WEAP-MODFLOW Tutorial*, Version 01/2009. http://www.acsad-bgr.org/DSS-Tutorial/WEAP_MODFLOW_Tutorial.zip

IGES. 2007. Groundwater and climate change, no longer the hidden resource. IGES (Institute for Global Environmental Strategies) White Paper, Chapter 7. http://enviroscope.iges.or.jp/modules/envirolib/upload/1565/attach/09_chapter7.pdf

Intergovernmental Panel on Climate Change (IPCC). 1998. *Regional Impacts of Climate Change: An Assessment of Vulnerability*. Cambridge University Press, New York.

Intergovernmental Panel on Climate Change (IPCC). 2001. Climate change 2001: The scientific basis. http://www.ipcc.ch

Intergovernmental Panel on Climate Change (IPCC). 2007. Fourth Assessment Report. http://www.ipcc.ch

Jacobs, K., Adams, D.B., and Gleick, P. 2001. Water resource. Chapter 14, U.S. Global Change Research Program.

Johnson, H. and Lundin, L.1991. Surface runoff and soil water percolation as affected by snow and soil frost. *Journal of Hydrology*, 122, 141–159.

Knutssen, G. 1988. Humid and arid zone groundwater recharge—A comparative analysis, in *Estimation of Natural Groundwater Recharge*. Simmers, I., ed. NATO ASI Series C, Vol. 222 (*Proceedings of the NATO Advanced Research Workshop, Antalya, Turkey, March*), G. Redial Publication Company, Dordrecht, the Netherlands, pp. 493–504.

Kresic, N. 2008. *Groundwater Resources: Sustainability, Management, and Restoration*. McGraw-Hill, New York.

Lerner, D., Issar, A.S., and Simmers, I., eds. 1990. *Groundwater Recharge: A Guide to Understanding and Estimating Natural Recharge*. Heise, Hannover, Germany.

Lutgens, F.K. and Tarbuck, E.J. 1995. *The Atmosphere*. Prentice-Hall, Englewood Cliffs, NJ, 462 p.

Mall, R.K., Bhatla, R., and Pandey, S.N. 2007. Water resources in India and impacts of climate change. *Jalvigyan Sameeksha*, 22, 157–176.

Manglik, A. and Rai, S.N. 2000. Modeling of water table fluctuations in response to time-varying recharge and withdrawal. *Water Resources Management*, 14, 339–347

McCown, R.L., Hammer, G.L., Hargreaves, J.N.G., Holzworth, D.P., and Freebairn, D.M. 1996. APSIM: a novel software system for model development, model testing, and simulation in agricultural systems research. *Agricultural Systems*. 50, 255–271.

McWhorter, D.B. and Sunada, D.K. 1977. *Ground-Water Hydrology and Hydraulics*. Water Resource Publication, LLC, Highlands Ranch, CO, 290 p.

Mitchell, L.F.B. 1989. The greenhouse effect and climate change. *Reviews of Geophysics*, 27, 115–139.

Mitchell, L.F.B., Johns, T.C., Gregory L.M., and Teat, S.F.B. 1995. Climate response to increasing levels of greenhouse gases and sulphate aerosols. *Nature*, 376, 501–504.

Moore, J.N., Harper, J.T., and Greenwood, M.C. 2007. Significance of trends toward earlier snowmelt runoff, Columbia and Missouri Basin headwaters, western United States. *Geophysical Research Letters*, 34(16), L16402, doi:10.1029/2007GL032022.

Mukherji, A. and Shah, T. 2005. Socio-ecology of groundwater irrigation in South Asia: An overview of issues and evidence, in *Selected Papers of the Symposium on Intensive Use of Groundwater*. Valencia, Spain, December 10–14, 2002, IAH Hydrogeology Selected Papers, Balkema Publishers, the Netherlands.

Olsson, J., Uvo, C.B., Jinno, K., Kawamura, A., Nishiyama, K., Koreeda, N., Nakashima, T., and Morita, O. 2004. Neural networks for rainfall forecasting by atmospheric downscaling. *Journal of Hydrologic Engineering*, 9, 1–12.

Racsko, P., Szeidl, L., and Semenov, M. 1991. A serial approach to local stochastic weather models. *Ecological Modeling*, 57, 27–41.

Ranjan, P. 2007. Effect of climate change and land use change on saltwater intrusion. The encyclopedia of eart (EOE) community. http://www.eoearth.org/article/Effect_of_climate_change_and_land_use_change_on_saltwater_intrusion

Reichard, J.S. 1995. Modeling the effects of climatic change on groundwater flow and solute transport. PhD, dissertation, Department of Earth and Atmospheric Sciences, Purdue University, West Lafayette, IN.

Risser, D.W. 2008. Spatial distribution of ground-water recharge estimated with a water-budget method for the Jordan Creek watershed, Lehigh County, Pennsylvania. U.S. Geological Survey Scientific Investigations Report 2008-5041.

Rushton, K.R., Eilers, V.H.M., and Carter, R.C. 2005. Improved soil moisture balance methodology for recharge estimation. *Journal of Hydrology*, 318(1–4), 379–399.

Scanlon, B.R., Reedy, R.C., Stonestrom, D.A., Prudic, D.E., and Dennehy, K.F. 2005. Impact of land use and land cover change on groundwater recharge and quality in the southwestern US. *Global Change Biology*, 11, 1577–1593, doi: 10.1111/j.1365-2486.2005.01026.x.

Semenov, M.A. and Brooks, R.J. 1999. Spatial interpolation of the LARS-WG stochastic weather generator in Great Britain. *Climate Research*, 11, 137–148.

Semenov, M.A., Brooks, R.J., Barrow, E.M., and Richardson, C.W. 1998. Comparison of the WGEN and LARS-WG stochastic weather generators in diverse climates. *Climate Research*, 10, 95–107.

Shrestha, S. and Kataoka, Y. 2008. *Groundwater and Climate Change No Longer the Hidden Resource*. Institute for Global Environmental Strategies (IGES), Hayama, Japan, pp. 159–184.

Singhal, D.C., Israil, M., Sharma, V.K., and Kumar, B. 2010. Evaluation of groundwater resource and estimation of its potential in Pathri Rao watershed, district Haridwar (Uttarakhand), Special Section: Himalayan Bio-Geo Databases–I. *Current Science*, 98(2), p. 25.

Stuyt, L.C.P.M., Kabat, P., Postma, J., and Pomper, A.B. 1995. Effect of sea-level rise and climate change on groundwater salinity and agro-hydrology in a low coastal region of the Netherlands. *Studies in Environmental Science*, 65, 943–946.

Todd, D.K. 1980. *Groundwater Hydrology*. Copyright by John Wiley India Pvt. Ltd, New York, 556 p.

Viner, D. and Hulme, M. 1992. Climate change scenarios for impact studies in the UK: General circulation models, scenario construction models and application for impact assessment. Report of Department of Environment, Climatic Research Unit, University of East Anglia, Norwich, U.K., 70 p.

Von Storch, H., Zorita, E., and Cubasch, U. 1993. Downscaling of global climate change estimates to regional scales: An application to Iberian rainfall in wintertime. *Journal of Climate*, 6, 1161–1171.

Wigley, T.M.L. and Jones, P.D. 1985. Influence of precipitation changes and direct CO_2 effects on streamflow. *Nature*, 314, 149–151.

Wigley, T.M.L., Jones, P.D., Briffa, K.R., and Smith, G. 1990. Obtaining sub-grid-scale information from coarse-resolution general circulation model output. *Journal of Geophysical Research*, 95, 1943–1954.

Wilby, R.L. and Dawson, C.W. 2004. *Using SDSM Version 3.1—A Decision Support Tool for the Assessment of Regional Climate Change Impacts*. Climate Change Unit, Environment Agency of England and Wales, Nottingham, U.K. and Department of Computer Science, Loughborough University, Leicestershire, U.K.

Wilby, R.L. and Wigley, T.M.L. 1997. Downscaling general circulation model output: A review of methods and limitations. *Progress in Physical Geography*, 21, 530–548.

Wilby, R.L., Hassan, H., and Hanaki, K. 1998. Statistical downscaling of hydrometeorological variables using general circulation model output. *Journal of Hydrology*, 205, 1–19.

Wilby, R.L., Hay, L.E., Gutowski, W.J., Arritt, R.W., Takle, E.S., Pan, Z., Leavesley, G.H., and Clark, M.P. 2000. Hydrological responses to dynamically and statistically downscaled climate model output. *Geophysical Research Letters*, 27, 1199–1202.

Witherspoon, P.A., Wang, J.S.Y., Iwai, K., and Gale, J.E. 1980. Validity of the cubic law for fluid flow in a deformable rock fracture. *Water Resource Research*, 16(6), 1016–1024.

Zaidi, F.K., Ahmed, S., Dewandel, B., and Mar´echal, J.-C. 2007. Optimizing a piezometric network in the estimation of the groundwater budget: A case study from a crystalline rock watershed in southern India. *Hydrogeology Journal*, 15, 1131–1145.

Zhang, L., Stauffacher, M., Walker, G.R., and Dyce, P. 1997. Recharge estimation in the liverpool plains (NSW) for input groundwater models, Technical Report. CSIRO Land and Water.

Index